An introduction to the
theory of seismology

An introduction to the theory of seismology

K.E. BULLEN

Formerly Professor of Applied Mathematics, University of Sydney

BRUCE A. BOLT

Professor of Seismology, University of California, Berkeley

FOURTH EDITION

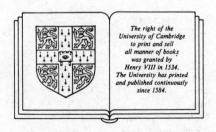

The right of the
University of Cambridge
to print and sell
all manner of books
was granted by
Henry VIII in 1534.
The University has printed
and published continuously
since 1584.

CAMBRIDGE UNIVERSITY PRESS

CAMBRIDGE

NEW YORK NEW ROCHELLE

MELBOURNE SYDNEY

Published by the Press Syndicate of the University of Cambridge
The Pitt Building, Trumpington Street, Cambridge CB2 1RP
32 East 57th Street, New York, NY 10022, USA
10 Stamford Road, Oakleigh, Melbourne 3166, Australia

First published 1947
Second edition 1953
Reprinted 1959
Third edition 1963
Reprinted 1965, 1976
First paperback edition 1979
Fourth edition (revised by Bruce A. Bolt) 1985
Reprinted 1987

Printed in Great Britain at the University Press, Cambridge

Library of Congress catalogue card number: 84–23052

British Library cataloguing in publication data

Bullen, K.E.
An introduction to the theory of seismology –
4th ed.
1. Earthquakes
I. Title   II. Bolt, Bruce A.
551.2′2   QE534.2
ISBN 0 521 23980 X hard covers
ISBN 0 521 28389 2 paper back
[Third edition ISBN 0 521 04367 0 hard covers
              ISBN 0 521 29686 2 paperback]

TM

Contents

To
Sir Harold Jeffreys

Preface

About sixty per cent of this edition is new material. Changes have been introduced throughout the text to shorten and up-date discussions and explanations and to give more emphasis to modern research methods, computer related algorithms, and strong ground-motion problems.

Bullen's death in 1976 prevented him from revising his famous book on the theory of seismology, first published in 1947. The only substantial revision, made in 1963, was quite limited because of strong constraints on the format and length allowed. In the 1970s he was occupied with writing his treatise on the Earth's density, but he had already formed the intention of returning to *Theory of Seismology* at the end of the decade. To this end, he produced extensive subsidiary notes while giving seismology courses in 1971 and 1972 at the International Institute of Seismology and Earthquake Engineering, Tokyo, Japan, and these have been used in making the present revisions.

Any rewriting at this time must take account of the remarkable strengthening and broadening of seismological work in the last two decades. Seismological theory is now so extensive that a reasonably comprehensive introduction in one volume demands a somewhat invidious selection of methods and topics, many of which were hardly even formulated in 1947. Yet the attempt appears worthwhile because Bullen's book has continued as a valuable text for over thirty years.

I have seen my task in undertaking a radical revision as keeping what remains most useful of the 1963 text, but incorporating at the appropriate introductory level the necessary fundamental ideas to fill the most notable gaps. In this edition there are new chapters on the theory of seismic sources, seismic waves through anomalous zones, eigen-vibrations of the Earth, and strong-motion seismology. Other new material has been added, particularly on the theory of the seismograph, damping, density estimation in the Earth, and analysis of travel-times and eigen-vibration data using statistical inverse theory. To make room for this expanded theory, the chapters in the 1963 edition on nuclear explosions and extraterrestrial seismology have been omitted and these topics incorporated more briefly elsewhere. Key references have been reselected and updated and there are new sections giving problems, drawn from many sources, as exercises for the reader.

The principal aim of the book remains as before: to provide an introduction to seismological *theory*. For this reason, the treatment of the extensive observational side of earthquake studies is quite restricted and is introduced only to provide the necessary background. Practical applications of seismology are indicated only briefly, and the more difficult advanced theoretical problems (involving, for example, Cauchy–Riemann theory) are referred to other texts and publications. Several admirable, advanced textbooks on theoretical seismology and elastic wave theory have been published in recent years, among which must be mentioned the works of W. Pilant (1979), K. Aki and P. Richards (1980), J. Hudson (1980), A. Ben-Menahem and S.J. Singh (1981), and B. Kennett (1983). I have endeavoured to select material so that this book meets the requirements of teachers and professionals for a development between elementary physics and advanced mathematics. Based on lecture courses at the University of California, Berkeley, the selection given here should be helpful to upper division undergraduates and beginning graduate students of geophysics and earthquake engineering.

The reader should note several points on format. Equations within each chapter are referred to in that chapter by a single number; equations are referred to from other chapters by two numbers denoting chapter and equation, respectively (e.g. equation (2.4)).

The references have, for overall convenience, been separated at the back of the book into two parts: a short 'Selected bibliography', and 'References'. Entries in both lists are cited in the text by author and year of publication. The reader should therefore consult both lists where necessary.

Because seismology, like a number of large areas of physics, still uses cgs units they have been retained in this edition. A short conversion table to SI(mks) units is printed at the back of the book.

To the best of my ability, the overall approach is meant to remain close to that of my former professor and colleague, K.E. Bullen, and, ultimately, to our common mentor, Sir Harold Jeffreys. I am indebted to many colleagues and research students for help. Indeed, parts of the text are derived directly from joint research articles. W.J. Hannon and R.A. Uhrhammer helped greatly by reading critically through much of the text. Thanks are also due in particular to R.D. Adams, D.R. Brillinger,

V.F. Cormier, P. Dehlinger, J.W. Dewey, L.A. Drake, R. Gutdeutsch, E. Krebs, J.J. Litehiser, T. Mikumo, O.W. Nuttli, P.W. Rodgers, W.D. Smith, W. Stauder, T. Tanimoto, Y.B. Tsai, A. Udias, and J. Verhoogen, all of whom kindly supplied suggestions and comments on parts of the contents.

Berkeley, California 1984   BRUCE A. BOLT

1

The scope of seismology

1.1 Early history

The observations of geologists make it probable that the Earth has suffered earthquakes for at least some hundreds of millions of years.

Early historical records in China contain references to earthquakes as far back as 1800 BC. The ancients attributed earthquakes to super-natural causes; indeed, a writer in the *Philosophic Transactions of the Royal Society of London* as late as AD 1750 apologised to 'those who are apt to be offended at any attempts to give a natural account of earthquakes'.

In Greek science, however, natural causes, such as underground explosions, were considered and Aristotle gave a classification of earthquakes into six types, according to the nature of the earth movement observed; for example, those which caused an upward earth movement, those which shook the ground from side to side, etc. Also, in the year AD 132, the Chinese philosopher Chang Heng devised an artistic earthquake weathercock for indicating the direction of the main impulse due to an earthquake; this instrument is reputed to have detected earthquakes not felt locally, but the internal mechanisms of the device are unknown.

About the middle of the eighteenth century AD, useful observations of earthquake effects began to accumulate. In 1760, John Michell in England published a notable memoir on earthquakes, in which he associated earthquakes with wave motion in the Earth. Most work on earthquakes during the late eighteenth and the early nineteenth century was concerned with geological effects of earthquakes, and with effects on buildings. It was noted, for instance, that, in the main, buildings on soft ground were more damaged by earthquakes than those on hard rock. Early in the nineteenth century, earthquake lists were being regularly published, and in 1840 von Hoff published an earthquake catalogue for the whole world.

About the middle of the nineteenth century, the foundations of instrumental seismology were laid when Robert Mallet suggested the

1

setting up of a chain of observatories over the Earth's surface, and Palmieri in Italy devised a 'sismografo elettro-magnetico' capable of detecting earthquakes and of recording some features of the consequent local earth movements such as duration and direction. The history of seismology of this period includes the names of Nöggerath and Schmidt (Germany), who introduced the use of isoseismal lines to estimate the epicentre of an earthquake and the apparent speed of travel of the ensuing disturbance; of Perrey and Montessus de Ballore (France), who compiled notable earthquake records; and of de Rossi (Italy) and Forel (Switzerland), who together produced the Rossi–Forel intensity scale, the first well-known scale for estimating surface effects of earthquakes and determining isoseismal lines.

In 1892, a major step forward was taken in Japan when John Milne (aided by his association with James Ewing and Thomas Gray and later F. Omori) developed seismographs which were sufficiently compact and simple in operation to enable them to be installed and used in many parts of the world. The first identified recording of a distant earthquake was made in 1889 at Potsdam with horizontal pendulums designed by E. von Rebeur Paschwitz, based on earlier work by James Ewing. A large earthquake on 17 April 1889, had been felt in Japan before it was recorded at Potsdam (Fig. 1.1). From this time onwards, instrumental data on earthquakes began to accumulate and seismology developed from a qualitative to a quantitative science. A detailed history of seismometry to 1900 is given by J. Dewey and P. Byerly (1969).

Meanwhile, progress had been taking place on the mathematical 'front'. Studies of wave motion were fashionable among applied mathematicians throughout the nineteenth century, and mathematical theory relevant to seismology was produced. As early as 1828, Cauchy and Poisson determined the equations of motion of a disturbance in a perfectly elastic material, and Poisson showed that there could be two distinct types of waves (the seismological P and S waves) transmitted with different speeds through the interior of such a material. Stokes showed that the P and S waves were of dilatational and rotational types, respectively, and Green studied the reflection and refraction of elastic waves. Later came the work of Kirchhoff, Kelvin and Rayleigh, including Rayleigh's theory of waves on the boundary of a homogeneous elastic substance.

The close of the nineteenth century saw the identification (1900) by R.D. Oldham of the three main types of seismic waves – P, S and surface – on actual records from mechanical seismographs – nearly 70 years after the mathematical theory of P and S waves had been formulated.

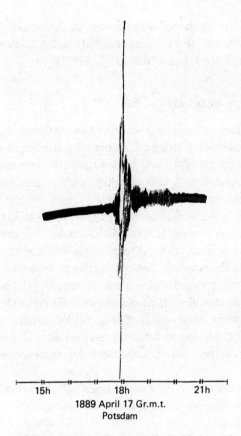

15h   18h   21h
1889 April 17 Gr.m.t.
Potsdam

Fig. 1.1. Seismogram made by a horizontal pendulum in Potsdam, Germany of a Japanese earthquake, 17 April, 1889.

In the first decade of this century, effective seismographs capable of recording three components of short- and long-period waves were constructed by E. Wiechert and B.B. Galitzin. Seismic waves recorded by these instruments immediately aroused the curiosity of mathematicians and geophysicists. In 1904, Lamb attacked the problem of the generation of surface seismic waves. In his Adams Prize Essay of 1911, Love explained the occurrence of a type of surface wave not included in the theory of Rayleigh and made a comprehensive study of vibrations of a compressible gravitating planet.

In 1906, Oldham used recorded seismic waves to show that the Earth has an extensive central core. In the same year, the great San Francisco earthquake occurred, leading to modern ideas about the source of earthquake waves. An International Association of Seismology was founded in 1905 which became, in 1951, the International Association of Seismology and Physics of the Earth's Interior (IASPEI).

1.2 Developments from 1915 to 1960

In this period, seismological data were used to resolve the Earth's internal structure to a remarkable degree. The period started with vague notions about a molten central core and finished with well-determined values of the density, pressure, compressibility, rigidity and gravity throughout practically the whole Earth.

On the instrumental side, there was a notable expansion of a world-wide system of earthquake observatories. By 1920, about 80 stations operated Wiechert type instruments, about 45 had Milne or Milne–Shaw seismographs, and a number had the electromagnetic Galitzin or Galitzin–Wilip instruments which used photographic recording. In the 1930s, H. Benioff designed an extremely successful vertical component seismograph. This short-period instrument provided much higher amplifications of the ground motion than were previously available and made a key contribution to increased detection and measurement precision of seismic waves.

The period saw the rise of the *International Seismological Summary* and the development of an international co-operation that has not been excelled in any other branch of study.

Travel-time tables evolved from crudest beginnings, through the Zöppritz–Turner tables, to the 1940 Jeffreys–Bullen and Gutenberg–Richter tables, in which errors of the order of minutes had been reduced to errors of the order of seconds.

The work of Herglotz and others enabled P and S velocities to be deduced from the travel-time data. These in turn furnished information on compressibility–density and rigidity–density ratios throughout much of the Earth.

In 1914, Gutenberg published his accurate determination of the depth of the boundary of the central core, one of the early results of a long and distinguished career.

Jeffreys became interested in seismology around this time and brought to bear elegant mathematical and statistical methods and a great knowledge of wider geodynamical problems. His attention to scientific method and statistical detail has been one of the main forces through which seismology has attained its present level of precision, and the sample of his results quoted in this book gives some indication of the extent of his contributions (still continuing in 1984).

In 1936, Inge Lehmann produced the first evidence from seismograms of the existence of the Earth's inner core. By 1940, Bullen was able to classify the Earth's interior broadly into a number of shells or regions occupying ranges of depth from the surface to the centre (see § 13.1.5). His construction of Earth Model A was completed around this time, giving close estimates of the interior density and elastic parameters.

Further resolution was made of the Earth's outer layers by near-earthquake studies, by surface-wave studies and, later, by explosion seismology. Starting from the work of A. Mohorovičić in 1909, investigators determined the overall thickness of layers down to the discontinuity that bears his name and, as well, determined the P and S velocities inside the layers. It was also established that crustal thickness is much less under oceans than under continents.

Another important development was the experimental and theoretical investigation, especially by Bridgman, Adams, Williamson and Birch, of the behaviour of matter at pressures and temperatures prevailing in the outer mantle of the Earth.

Following the dislocations of World War II, there was a growth of seismological studies and the total output of results on Earth structure jumped sharply. The stimulus of the International Geophysical Year of 1957–58 led to an increase in the number of first-class seismological observatories, but the mix of seismograph types made comparison of waves between different stations relatively imprecise and non-productive.

Most notable among the achievements at the end of the fifties was a great extension in the spectrum of recorded seismic waves. At one extreme, the instrumentation of seismic prospecting enabled frequencies of order 100 Hz to be measured in ground movements. At the other, new seismographs enabled measurements to be made of surface waves with periods extending up to 10 min and of free oscillations of the whole Earth with periods of about an hour. The spectral gaps between seismic wave vibrations of the order of seconds and daily tidal oscillations of the Earth were at last bridged. In addition, rugged accelerographs were designed that could record the strong ground motions in large earthquakes.

The methods of explosion seismology were greatly developed in the post-war period, particularly in the oil industry. Another development with artificial sources occurred with the advent of nuclear explosions, that changed the face of seismology radically on both the theoretical and observational sides.

1.3 The period since 1960

1.3.1 Seismology and nuclear explosions

With the advent of nuclear explosions, it became possible to extend the seismic explosion method, and all the experimental controls that go with it, to problems of the Earth's deep interior.

Although, so far, seismic applications of nuclear explosions have been mainly secondary to other purposes, a quantity of valuable information has been derived.

The first atom bomb was exploded in New Mexico on 1945 July 16 d 12 h 29 m 21 s (UT) at latitude 33° 40′ 31″ N and longitude 106° 28′ 29″ W from a tower 100 ft above the ground. The origin-time at source was uncertain by 15 s, the time here given being that estimated by Gutenberg from seismic data; this time is considered to be reliable within 2 s. It has become customary to express the energy yielded in a nuclear explosion in terms of equivalent kilotons or megatons of TNT explosive ($1 \text{ kt} = 10^3$ tons; $1 \text{ Mt} = 10^6$ tons). The yield for this explosion was about 19.3 kt.

On 1946 July 24, the first underwater atomic explosion took place 30 m below the ocean surface near Bikini Atoll. This was the first nuclear explosion for which source data were made generally available. It was recorded at eight seismic stations at distances between 69°.0 and 78°.6. Gutenberg and Richter gave P readings for those stations, from which Bullen determined a mean residual of -1.8 ± 0.8 s against the J.B. tables for a surface focus. The negative residuals were attributed to crustal differences between the Bikini and the average continental region. The explosion, meagrely recorded as it was, gave a glimpse of what might be achieved for seismology through nuclear explosions.

In March 1954, it was announced that a hydrogen bomb had been exploded near Bikini. Burke-Gaffney identified a corresponding P wave onset at Riverview Observatory and the identification was confirmed when routine readings arrived from Brisbane. From this start he found readings of seismic waves from four such explosions in routine station bulletins from twelve countries. The readings enabled him and Bullen to compute origin-times which, later, proved to be accurate within 0.0, 0.4, 0.6 and 0.1 s, respectively, for the four explosions.

From the seismic readings of these explosions, Burke-Gaffney and Bullen found negative P residuals, and also found that the P travel times from Bikini to the United States and Australia agreed within less than a second. An unexpected result of some significance was the identification of waves preceding the usual waves recorded after passage through the Earth's core (see § 13.6.5).

In 1956, a series of atomic explosions took place at Maralinga in central Australia and led to the first reliable crustal knowledge in a region where it had not been possible to make inferences from natural earthquakes.

Following an address on seismological aspects of nuclear explosions at the 1957 meeting in Toronto of the IASPEI, the first public release was made of source details of a coming nuclear explosion in Nevada. This was the Rainier explosion of 1957 September 19, which took place 250 m below the surface and was the first underground nuclear explosion. Intense efforts were made by seismologists to record seismic waves from it, but geophysical rewards were not great; the yield was only 1.7 kt with only a small proportion of the energy going into seismic waves.

In 1958, the field of forensic seismology was emphasised when a Geneva Committee of representatives from several countries produced a report on the detection of nuclear explosions, making considerable reference to seismology. Up to the present, underground nuclear explosions have been detonated by the United States, the Soviet Union, the United Kingdom, France (in North Africa and the Pacific Ocean), the People's Republic of China and India.

From 1958 onward, increasing quantities of source data on nuclear explosions were released publicly by the United States and increasing efforts were put into studies of seismological aspects. (Fig. 13.5 shows seismic waves recorded from an underground nuclear explosion at Novaya Zemlya, USSR, recorded in California.) A detailed history of these matters has been written by Bolt (1976).

A summary of recent seismological research on test-ban treaty verification is given in the *Bulletin of the Seismological Society of America*, **72B**, 1982. The main questions discussed involve new high resolution seismographs, estimating explosion yield, application of seismic arrays, wave attenuation in the Earth, and wave amplitude (magnitude) and spectral discriminants.

1.3.2 Standard global recording

By 1958, when the conference of experts met in Geneva to discuss the technical basis for a nuclear test ban treaty, there were in operation only about 700 seismographic stations equipped with seismographs of various types and frequency responses. Few instruments were calibrated so that actual ground motions could not be measured and timing errors of several seconds were common. A special panel set up in the United States in 1959, chaired by L.V. Berkner, recommended the installation of a worldwide standardised seismographic network (WWSSN). Foreign observatories cooperated with the US Coast and Geodetic Survey in installing and operating the new equipment. Each station consisted of six seismographs – three short-period seismographs and three long-period seismographs. Timing and accuracy were maintained by crystal clocks and a calibration pulse was placed daily on each record. Local observatories lent originals of the seismograms to the United States Coast and Geodetic Survey to be copied, and seismologists in any country could request copies for a nominal charge. By 1967, the WWSSN consisted of about 120 stations distributed in 60 countries. Other countries, such as Canada, which did not participate directly in the WWSSN, upgraded their own stations to be compatible with the standardised network. The resulting data provided the basis for significant advances in research in earthquake mechanisms, global tectonics, and the structure of the Earth's interior.

By the 1980s a further upgrading of permanent seismographic stations began, using digital equipment. Among the networks of global digital seismographic stations now operating (see Fig. 1.2) are the seismic research

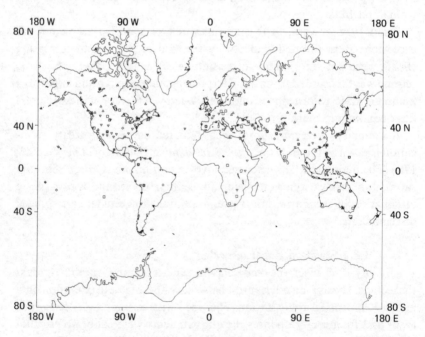

Fig. 1.2. Locations of WWSSN (or equivalent) and digital stations reporting earthquake records in 1983. (Courtesy W. Rinehart, NOAA.)

observatories (SRO) in boreholes 100 m deep, modified high-gain, long-period (surface) observatories (HGLP and ASRO), and digital WWSSN stations (DWWSSN). In addition, a number of gravimeters with digital recording and response to very long wavelengths have been installed worldwide under the International Deployment of Accelerographs (IDA) network. The main thrust is to equip global observatories with seismographs able to record seismic waves over a broad band of frequencies. For example, at the University of California at Berkeley, broadband, three-component analogue recording on magnetic tape was commenced in 1964 and similar digital recording in 1980. Digital signals and magnetic tape storage at the advanced global stations provide more satisfactory measurements of earthquake waves than do photographic recordings and allow rapid input to high-speed computers for estimation of earthquake parameters. Details on the responses of some of these modern instruments are given in chapter 9. Some considerations of present and future trends can be found in the IASPEI Presidential Address 'Seismology in the digital age' given by B.A. Bolt in Hamburg in 1983.

1.3.3 Computers and complexity

A major aim in seismology is to infer the minimum set of properties of the earthquake source and of the Earth which will explain in detail the recorded wave trains. Until the 1960s, this goal was limited severely by the labour needed to evaluate theoretical models and to process the large amounts of recorded seismological data. Applications of high-speed computers cleared the way for major advances in both theoretical work and data handling in seismology.

On the theoretical side, more realistic models of Earth structure that included continental and oceanic boundaries, mountains, alluvial valleys and so on were explored rather than structures with simple geometries. Viscous, non-homogeneous and non-isotropic properties of rocks and soils were included in the analysis where significant. More sophisticated mathematical inverse problems were tackled whereby maximum likelihood estimates of the required parameters (e.g. wave velocity, structural dimensions, density, etc.) were determined from the observations along with resolution and confidence limits. More widespread statistical analyses became possible involving simultaneous analyses of worldwide recordings of seismic waves.

In particular, analysis of the waveforms in terms of the frequency spectra became widespread, and special computing algorithms based on Gauss's 'Fast Fourier Transform' developed by Cooley and Tukey became commonplace. The implication of array and filtering techniques developed in the oil industry were also recognised. Recorded waves from network stations were telemetered to one centre. Large arrays of seismometers were constructed, such as the large-aperture seismic array (LASA) in Montana. This array (closed in 1979) had a circular geometry with a diameter of 200 km and a basic configuration of 21 sub-arrays, each with 25 short-period vertical seismometers. Digital signals from each channel were prefiltered and the combined sub-array signals added after being given appropriate time delays to enhance a particular seismic wave pulse at a specified direction and speed. Such processing by on-line computers, already well-developed in radio astronomy, for example, provided resolution and precision previously unknown in seismology.

Modern mini-computers and microprocessors with peripheral display equipment, allowing the rapid display of wave forms and spectra, are now providing valuable application of graphics methods to seismology. As well, observational seismology is exploiting the storage and retrieval computer facilities that have been long recognised in allied sciences such as meteorology.

1.3.4 Extra-terrestrial seismology

Since 1957 October 4, when the first artificial satellite invaded outer space, the application of seismic methods to the study of extra-terrestrial bodies has been discussed. Space vehicles have carried equipment to the surface of the Moon and Mars to record seismic waves and seismologists on Earth have received telemetered signals. Just as the advent of efficient seismographs late last century led to our present knowledge of the Earth's interior, so will the placing of instruments on other planets lead to much new knowledge of their interiors, as recently reviewed by A.H. Cook (1980). Information may also be expected on many special topics, for example, meteorite impacts.

The experiments will not merely provide knowledge of the Moon and planets directly investigated; they can be expected to have important repercussions on the theories of the Earth's interior, and to influence the course of terrestrial geophysics in a number of ways.

Present knowledge of the interior of the Moon is comparable with, possibly slightly greater than, that of the Earth two decades after seismographs were available to record earthquakes. (See § 13.2.4.)

The mass M', mean radius R', volume and mean density of the Moon are well determined as 7.35×10^{25} g, 1738 km, 2.20×10^{25} cm^3, and 3.34 g/cm^3

Table 1.1. *Lunar seismographic stations*

Station	Coordinates	Duration of Operation
Apollo 12	3.04°S, 23.42°W	19 Nov. 1969–30 Sept. 1977
Apollo 14	3.65°S, 17.48°W	5 Feb. 1971–30 Sept. 1977
Apollo 15	26.08°N, 3.66°E	31 July 1971–30 Sept. 1977
Apollo 16	8.97°S, 15.51°E	21 April 1972–30 Sept. 1977
Apollo 11	0.68°N, 23.45°E	21 July 1969–27 Aug. 1969
Apollo 17	20.16°N, 30.75°E	12 Dec. 1972–30 Sept. 1977

Note: Apollo 11 station did not operate concurrently with other stations. Seismometers for the Lunar Seismic Profiling Experiment and a gravimeter at Apollo 17 station detected only a few major seismic signals.

respectively. The surface value of g is $162 \, \text{cm/s}^2$ and the density ρ varies from $3.28 \, \text{g/cm}^3$ at the surface of the Moon to about $3.40 \, \text{g/cm}^3$ at the centre. Cook gives the moment of inertia as $0.392 \, M'R'^2$, based on lunar librations (forced oscillations) measured by distance observations using laser reflections from the lunar surface.

By 1969, seismographs had been placed at six sites on the moon during the Apollo 12, 14, 15, 16, 11, and 17 missions (Table 1.1). Recording of seismic data ceased in September 1977. The instruments detected between 600 and 3000 moonquakes every year of their operation, although most of the moonquakes were very small. The ground noise is low compared with that of the Earth so that the seismographs could be operated at very high magnifications. Examples of recordings from moonquakes are shown in Fig. 1.3. The long duration seems to arise from a high degree of scattering from fractured rocks with low energy loss in the dry upper part of the Moon.

Because there was more than one station on the Moon, it was possible to use the arrival times of P and S waves at the lunar stations from the moonquakes to determine foci in the same way as is done on the Earth. However, the distribution of lunar seismograph stations was restricted to the front of the Moon, giving considerable uncertainties in locations and travel times.

Moonquakes are of three types (see Fig. 1.3). First, there are the events caused by the impact of lunar modules, booster rockets and meteorites on

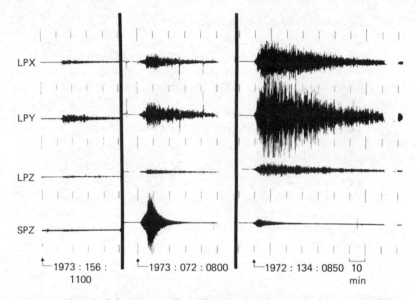

LPX

LPY

LPZ

SPZ

└1973 : 156 :   └1973 : 072 : 0800   └1972 : 134 : 0850 10
     1100   min

Fig. 1.3. Seismograms from three types of moonquakes recorded at the Apollo 16 station. LPX, LPY, and LPZ are the three long-period components and SPZ is the short-period vertical component. The first column shows a deep-focus moonquake; the centre column, a shallow moonquake; the third column shows records from a meteoroid impact on the lunar surface. (Courtesy NASA.)

the surface of the Moon. Second, there are moonquakes centered at shallow depths down to 100 km. These moonquakes are not very common and their locations do not show a structured pattern on the lunar surface. Third, there are relatively deep moonquakes with focal depths of 800 to 1000 km. The cause for the deeper lunar quakes is not known and they do not seem to coincide with any major interior structural boundaries inferred from the travel times of seismic waves.

The mass M'', mean radius R'', mean density and surface gravity value of Mars are 6.42×10^{26} g, 3385 km, 3.9 g/cm^3 and 380 cm/s^2. Observations of the ellipticity of figure lead to the value $0.375\,M''R''^2$ for the moment of inertia.

The landings of Viking 1 on 20 July 1976, and Viking 2 on 3 September 1976, placed the first seismographs on Mars. The seismograph on Viking 1 did not operate, but that on Viking 2 operated as planned for 546 Martian days. Vibrations caused by gusty winds blowing on the space lander hampered detection and identification of seismic onsets, and only one recorded event (on Marsday 80) is considered likely to be a Marsquake.

1.4 The plan of this book

We have in §1.1 indicated that theories of the generation and passage of elastic waves in the Earth are fundamental to seismology. Hence, some knowledge of the mechanics of a deformable body and of wave motion is a necessary preliminary to a more specialised study of seismology. Accordingly, chapters 2 and 3 are concerned with mathematical theories of elasticity, and of vibrations and waves. In both chapters the aim has been to outline the main results relevant to seismological problems. In the interest of terseness, use is made of Cartesian tensors and the summation convention, though it is not necessary for the reader to have previous knowledge of tensors.

In the next five chapters, 4 to 8, wave theory is applied to problems on wave motion in an elastic body, particularly in a sphere and in layered media. Problems treated are those of direct importance to seismology, but much of the theory contained in the book up to chapter 8 will be of interest to students of mathematical physics and applied mechanics. These chapters give the theory of rays of longitudinal and transverse waves, including reflection and refractions, with accompanying discussion on energy and the effect of damping, and introductory surface wave theory, including that of Rayleigh, Love, and Stoneley waves, and the theory of free vibrations of an elastic sphere. In chapter 7, the first basic inverse problem is treated, namely, estimating velocities of internal strata from surface observations of travel-times.

Chapters 9 to 11 are more particularly concerned with the derivation of results special to seismology, with the modern emphasis on resolution, uniqueness, and optimal design of experiments. In chapter 9 the principles of the seismograph are discussed, in chapter 10 methods of analysing instrumental data are described, and in chapter 11 a short account is given of the organisation of earthquake observatories.

In chapter 12, new in this edition, more seismological theory relevant to the determination of properties of structurally complex parts of the Earth's interior is given. Chapter 13 explains how studies of body seismic waves are used to throw light on the structure of planetary interiors. Values of P and S velocities are set down for the whole of the Earth, and are made the basis of a broad classification of the interior of the Earth into regions. The second seismological inverse problem is outlined, namely the determination of the density, pressure, elasticity, and gravity variations in the Earth. Finally, the composition of the Earth's interior is examined to some extent in relation to the seismic data.

Chapter 14 is concerned with observations of long-period surface waves and fundamental free oscillations of the Earth and their inversion to estimate internal properties. Statistical theory is given for the optimal estimation of spectral parameters, and methods for inferring Earth structure from them are discussed.

In chapter 15 the global occurrence and size of earthquakes are considered. The discussion includes reference to related questions such as the quantity of energy released in an earthquake, geographical distribution of earthquake foci, the occurrence of foreshocks and aftershocks, the prediction of earthquake occurrence in space and time, and related statistical theory.

Chapters 16 and 17 are also new and should be helpful to both seismologists and earthquake engineers. Chapter 16 contains an introductory account of the physics and mathematical modelling of the earthquake source.

The final chapter deals with a subject of much recent vigorous work in theoretical seismology: the prediction of ground motion near the source of the seismic waves. A number of empirical results are given that relate to risk estimation and the problem of designing structures which will resist earthquakes.

2

Elasticity theory

The theory of elasticity is concerned with the strain experienced by deformable matter when it is stressed. In the present chapter, the theory is developed so far as is necessary for the basic theory of seismology. It is assumed that media are made up of particles which are sufficiently closely packed for descriptive functions to be continuous and differentiable.

2.1 Analysis of stress

Let P be any particle in the interior of a body, and at a given instant draw through P a unit vector v in any direction (see Fig. 2.1). Let δS be a small plane element of area inside the body, containing P and normal to v. Consider the forces across δS between two separate portions of the body which have a common boundary at δS. In particular, consider the forces exerted on that portion for which v is the outward normal. These forces are

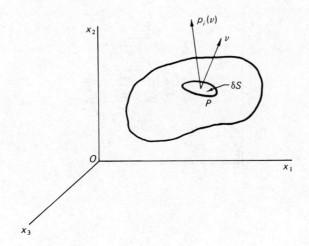

Fig. 2.1. Traction

15

statically equivalent to a single force acting at P together with a couple. This single force is called the *traction* across δS, denoted by the vector $p_j(v)\delta S$.

If δS be indefinitely dimished, the limit of the ratio of the traction across δS to the area δS is called the *stress* at P corresponding to the direction v. In ordinary material this limit is taken to be finite and significant, while the couple (whose moment involves an additional dimension of length) is disregarded.

2.1.1. The stress tensor

The stress at P in general varies with the direction of the normal v to the small surface considered, and is inclined to that normal. We now show that the stress at P across any small plane area containing P is expressible in terms of nine components.

Take two sets of orthogonal axes, called the 1-, 2- and 3-, and the 1′-, 2′- and 3′-axes, respectively. Components of any vector parallel to the axes of the first set will be named the 1-, 2- and 3-components; and similarly with the other set of axes. Consider the matter inside a small tetrahedron (Fig. 2.2) enclosing P, and having the inward normals of three of the faces parallel to the 1-, 2- and 3-axes, respectively, and the outward normal of the fourth face parallel to the 1′-axis. Let the corresponding areas of the four faces be $\delta S_1, \delta S_2, \delta S_3$ and $\delta S_{1'}$. Denote the 2-component of the stress across the face normal to the 1-axis by p_{12}; and similarly with other cases. Let $a_{12'}$ be the cosine of the angle between the 1- and the 2′-axes, etc.

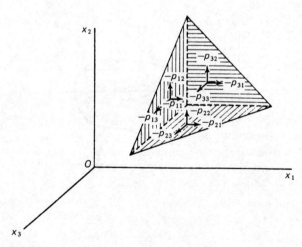

Fig. 2.2. Stress components

By Newton's law, the matter inside the tetrahedron is in equilibrium under the action of body forces (such as gravity), the tractions across the boundary faces, and the reversed mass-accelerations of the constituent particles.

In Fig. 2.2, the stresses on this matter across three of the four boundary faces are shown. Consider resolved parts parallel to say the $2'$-axis. All but the boundary forces are ultimately proportional to the volume of the tetrahedron. Hence, if l be a linear dimension of the tetrahedron, we have

$$p_{1'2'}\delta S_{1'} - p_{11}\delta S_1 a_{12'} - p_{12}\delta S_1 a_{22'} - p_{13}\delta S_1 a_{32'}$$
$$- p_{21}\delta S_2 a_{12'} - p_{22}\delta S_2 a_{22'} - p_{23}\delta S_2 a_{32'}$$
$$- p_{31}\delta S_3 a_{12'} - p_{32}\delta S_3 a_{22'} - p_{33}\delta S_3 a_{32'} = O(l^3),$$

where $O(l^3)$ denotes the same order of magnitude as volume.

Dividing through by $\delta S_{1'}$ and letting $l \to 0$, we have

$$p_{1'2'} = p_{11}a_{11'}a_{12'} + p_{12}a_{11'}a_{22'} + p_{13}a_{11'}a_{32'}$$
$$+ p_{21}a_{21'}a_{12'} + p_{22}a_{21'}a_{22'} + p_{23}a_{21'}a_{32'}$$
$$+ p_{31}a_{31'}a_{12'} + p_{32}a_{31'}a_{22'} + p_{33}a_{31'}a_{32'}$$
$$= \sum_{i=1}^{3} \sum_{j=1}^{3} a_{i1'}a_{j2'}p_{ij}.$$

It is clear that a similar result will hold for the l'-component of stress across a small area normal to the k'-axis, where k and l may each be 1, 2 or 3. It is convenient here to introduce the *summation convention*, whereby if any suffix occurs twice in a single term, it is to be put equal to 1, 2 or 3 in turn and the results added. This convention will be employed regularly in the sequel. (A reader who finds trouble in the use of this adding rule is advised to write out the equations in full when in difficulty.) The relations may thus be written as

$$p_{k'l'} = a_{ik'}a_{jl'}p_{ij}. \tag{1}$$

The reader will perceive that the suffixes i, j on the right-hand side of (1) are *dummy suffixes*, because each may be replaced by any other suffix not already present. (In order to make the summation convention work, the same suffix must never appear more than twice in a single term.)

To summarize, the nine components p_{ij} (associated with any particular point and instant of time) completely determine the stress across any small plane area containing the point. An array p_{ij} which obeys a transformation law of the type (1) is a *cartesian tensor* of the second order. (See Exercise 2.8.1.) This is a simple extension from the tensor of the first order, or *vector*, v_i say ($i = 1, 2, 3$), which has three components and obeys a transformation law of the form

$$v_{k'} = a_{ik'}v_i,\tag{2}$$

and the tensor of zero order, or *scalar*, which is fully determined by a single component whose value is unchanged on transformation of axes. The set of nine elements p_{ij} constitutes the *stress tensor* at P.

2.1.2 Symmetry of the stress tensor

Now consider the matter inside a small parallelepiped which has three edges, PA, PB, PC, passing through P, parallel to the 1-, 2-, 3-axes, of lengths δx_1, δx_2, δx_3, say. In Fig. 2.3 the stresses across the pair of faces perpendicular to Ox_2 are shown.

From Fig. 2.3, the resultant moment about PC of the corresponding tractions is seen (neglecting higher-order terms) to be $-p_{21}\delta x_1 \delta x_3 \cdot \delta x_2$. The resultant moment about PC of the tractions across the pair of faces normal to Ox_1 is found to be $+p_{12}\delta x_2 \delta x_3 \cdot \delta x_1$. It is easy to show that the resultant moment about PC of all other forces (including body forces and reversed mass-accelerations) is of higher order. It follows that $p_{12} = p_{21}$. Similarly $p_{23} = p_{32}$ and $p_{31} = p_{13}$. Hence, for all i,j,

$$p_{ij} = p_{ji}.\tag{3}$$

Thus the stress tensor is symmetrical, and only six of its nine components are independent.

The components p_{ij} $(i \neq j)$ are called *shear components of stress* because of relations like (31) to be given later.

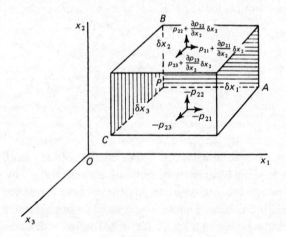

Fig. 2.3. Surface tractions.

2.1.3 Use of the Kronecker delta δ_{ij} and alternating tensor ε_{ijk}

It is convenient to use a special second-order tensor δ_{ij}, the *Kronecker delta*, which is represented by the array

$$\begin{pmatrix} 1 & 0 & 0 \\ 0 & 1 & 0 \\ 0 & 0 & 1 \end{pmatrix}.$$

i.e. $\delta_{ij} = 1$ if $i = j$, and $= 0$ if $i \neq j$. It may be verified that this set of values of δ_{ij} is preserved on transforming the axes. (We use only orthogonal cartesian axes in this chapter.)

The set of direction-cosines a_{ij}, where a_{12} denotes the cosine of the angle between the 1- and the 2-axes, etc., constitutes the tensor δ_{ij}. We note also the *substitution property* of δ_{ij}, namely,

$$\delta_{ij}T_i = T_j, \tag{4}$$

where T_i denotes a tensor of any order which has i as one of the suffixes in its representation; this property is verified on giving to j in (4) the particular values 1, 2, 3.

If in (1) we replace the suffix l' by l, the result is still true (this would correspond to resolving parallel to, say, the 2-axis instead of to the 2′-axis as in §2.1.1). Thus $p_{k'l} = a_{ik'}a_{jl}p_{ij} = a_{ik'}\delta_{jl}p_{ij}$. Using (4) and then replacing l by j, we obtain $p_{k'j} = a_{ik'}p_{ij}$.

We may write this result in the form

$$p_j(\mathbf{v}) = v_i p_{ij}, \tag{5}$$

which gives the stress $p_j(\mathbf{v})$ at P corresponding to the direction v_i, in terms of v_i and the stress tensor p_{ij}. The traction across the small area δS whose normal has the direction v_i is $v_i p_{ij} \delta S$. The relation (5) is, of course, directly obtainable independently of (1).

The 'alternating' tensor ε_{ijk} is useful in elasticity theory, and is used occasionally in seismology. We define the alternating tensor by the array

$\varepsilon_{ijk} = 0$ when any two of i, j, k are equal;

$\quad\ = 1$ when $ijk = 123, 231$ or 312;

$\quad\ = -1$ when $ijk = 321, 132$ or 213.

For a consistent development of vector analysis, the orthogonal reference frames need to be conformable. Right-handed frames in three dimensions are *conformable* with one another; likewise left-handed frames. But a right-handed frame is not conformable with a left-handed frame. The theorem can be easily proved that ε_{ijk} is a Cartesian tensor of third order if, but only if, transformations are restricted to conformable frames. A number of elementary but useful relations (in three 3 dimensions) involving ε_{ijk} can then be derived (see Exercises), such as

$$\varepsilon_{ijk}\varepsilon_{ijk} = 6.$$

We can now define the vector product of two vectors u_i, v_i as $\varepsilon_{ijk}u_jv_k$. (Commonly denoted as $\mathbf{u} \times \mathbf{v}$.) This has components $u_2v_3 - u_3v_2$, $u_3v_1 - u_1v_3$, $u_1v_2 - u_2v_1$. Also, the curl of a vector v_i (commonly denoted as curl $\mathbf{v}$) can be defined as $\varepsilon_{ijk}\partial v_k/\partial x_j$ with components $\partial v_3/\partial x_2 - \partial v_2/\partial x_3$, etc.

Vector products and curls are vectors only in the restricted case where transformations are confined to conformable frames. (That is why books on mathematical physics generally state that they will use right-handed (or left-handed) frames throughout.)

2.1.4 The stress quadric

At a given time, let y_i be the coordinates, referred to P as origin and to axes parallel to the 1-, 2- and 3-axes, of any point of the (central) quadric surface

$$p_{11}y_1^2 + p_{22}y_2^2 + p_{33}y_3^2 + 2p_{23}y_2y_3 + 2p_{31}y_3y_1$$
$$+ 2p_{12}y_1y_2 = \text{constant.}$$

This may, by the summation convention and (3), be written as

$$p_{ij}y_iy_j = \text{constant.} \tag{6a}$$

and, in matrix notation,

$$\mathbf{y}^{\mathsf{T}}\mathbf{Py} = \text{constant,} \tag{6b}$$

where p_{ij} are the elements of $\mathbf{P}$.

Let the coordinates become $y_{i'}$ when axes are taken (still with P at the origin) parallel to the 1'-, 2'- and 3'-axes. Then by (2)

$$y_i = a_{ik'}y_{k'}, \tag{7}$$

and the equation of the quadric becomes

$$p_{ij}a_{ik'}y_{k'}a_{jl'}y_{l'} = \text{constant,}$$

i.e. by (1),

$$p_{k'l'}y_{k'}y_{l'} = \text{constant.}$$

The quadric (6) is the *stress quadric* at P. By (8), the coefficients $p_{k'l'}$ referred to orthogonal Cartesian axes through P give the components of stress across small plane surfaces through P normal to these axes.

We now require the surfaces on which the direction of the traction coincides with the normal v_j, i.e., zero tangential stresses. This condition occurs when

$$p_i(v) = \lambda v_i,$$

or, using the matrix form of p_{ij}, from (5),

$$(p_{ij} - \lambda \delta_{ij})v_j = 0. \tag{8}$$

The three (real) eigenvalues λ and eigenvectors v_j of (8) define the *principal stresses* and *principal axes* of stress at P. Evaluation of these quantities is required in considering seismic source properties (see §16.1.3).

2.1.5 Elastodynamic equations of motion

At any given time t, let x_i be the coordinates of the position of P referred to a fixed set of rectangular cartesian axes, and f_i the acceleration of P; and consider the matter contained inside the parallelepiped shown in Fig. 2.3. Let X_i be the components of body force per unit mass (e.g. gravity) acting on this matter.

The resultant component parallel to the 3-axis of the tractions across the two faces normal to the 2-axis is (neglecting higher-order quantities)

$$-p_{23}\delta x_1 \delta x_3 + \left(p_{23} + \frac{\partial p_{23}}{\partial x_2}\delta x_2\right)\delta x_1 \delta x_3,$$

i.e.

$$\frac{\partial p_{23}}{\partial x_2}\delta x_1 \delta x_2 \delta x_3.$$

Hence the 3-component of the resultant of the tractions across all six faces is

$$\frac{\partial p_{j3}}{\partial x_j}\delta x_1 \delta x_2 \delta x_3.$$

The corresponding equation of motion therefore becomes, on dividing through by $\delta x_1 \delta x_2 \delta x_3$,

$$\rho f_3 = \frac{\partial p_{j3}}{\partial x_j} + \rho X_3,$$

where ρ is the density at x_i at time t. Using (3), we may hence write the three equations of motion as

$$\rho f_i = \frac{\partial p_{ij}}{\partial x_j} + \rho X_i \quad (i = 1, 2, 3). \tag{9}$$

2.2 Infinitesimal strain

Consider the same body again, and take axes as in §2.1.5. Suppose that at the instant under consideration the body has changed configuration such that the displacement of the typical particle P now at x_i has been u_i.

For the present we assume that such displacements are measured from a standard configuration in which the body forces and the stresses are zero at all points. Two special cases are of note (see §5.1): in *plane strain*, u_1, u_2 are functions of x_1, x_2 alone and u_3 is constant; in *antiplane strain*, u_3 only is spacially dependent and $u_1 = u_2 = 0$.

Let Q be the particle at a neighbouring point $x_i + y_i$, where y_i is infinitesimal. Then its displacement must (to sufficient accuracy) have been

$$\partial u_i + \frac{\partial u_i}{\partial x_j} y_j,$$

i.e.

$$u_i - \xi_{ij} y_j + e_{ij} y_j, \tag{11}$$

where

$$\xi_{ij} = \frac{1}{2} \left(\frac{\partial u_j}{\partial x_i} - \frac{\partial u_i}{\partial x_j} \right) \tag{12}$$

and

$$e_{ij} = \frac{1}{2} \left(\frac{\partial u_j}{\partial x_i} + \frac{\partial u_i}{\partial x_j} \right). \tag{13}$$

In elementary problems in elasticity, the deformation gradients $\partial u_i / \partial x_j$ are sufficiently small for their second powers to be neglected in equations containing significant first-order terms; the theory developed on this basis is called *infinitesimal strain theory* and will apply in what follows, except where special mention is made.

In (11), each of the three terms denotes a vector, while each of ξ_{ij}, e_{ij} denotes a tensor of the second order; these results are verified by transforming axes and finding that laws of the forms (2) and (1) are satisfied. Substitution of particular values of i, j shows further that ξ_{ij} is anti-symmetrical (i.e. $\xi_{ij} = - \xi_{ji}$; thus if $i = j$, $\xi_{ij} = 0$) and so has just three independent components, while e_{ij} is symmetrical and has just six independent components.

We leave it as an exercise to the reader to show that if α_{ij}, β_{ij} denote symmetrical and antisymmetrical tensors, respectively, then $\alpha_{ij} \beta_{ij} = 0$. This property will be used several times later.

We now proceed to interpret separately the three terms in (11). The first term u_i is equal to the displacement of P, and so corresponds to a pure translation (of the portion of matter near P) without rotation or deformation. We demonstrate in the following sections that the second term $- \xi_{ij} y_j$ corresponds to a pure rotation without translation or deformation, and the third term $e_{ij}y_j$ to a deformation. In §§2.2.1, 2.2.2, 2.2.3, the origin of axes is taken at the position of P whenever the coordinates y_i are involved.

2.2.1 The rotation tensor

The contribution to the displacement (11) arising from the particular pair of elements ξ_{23} and ξ_{32} of (12) has components $(0, -\xi_{23}y_3, \xi_{23}y_2)$. The associated change in the square of the distance between P, Q must therefore have been

$$y_1^2 + y_2^2 + y_3^2 - \{y_1^2 + (y_2 + \xi_{23}y_3)^2 + (y_3 - \xi_{23}y_2)^2\}, \tag{14}$$

which is zero to the first order in ξ_{23}. But the plane containing PQ and the 1-axis must have been rotated about the 1-axis through an angle

$$\tan^{-1}\left(\frac{y_3}{y_2}\right) - \tan^{-1}\left(\frac{y_3 - \xi_{23}y_2}{y_2 + \xi_{23}y_3}\right), \tag{15}$$

i.e.

$$\tan^{-1}q - \tan^{-1}\left(\frac{q - \xi_{23}}{1 + q\xi_{23}}\right),$$

where

$$q = y_3/y_2, \quad \text{i.e.} \quad \tan^{-1}\xi_{23}.$$

This angle is independent of the coordinates y_i of Q's position relative to P, and is equal to ξ_{23} to sufficient accuracy. The pair of elements ξ_{23} and ξ_{32} is thus associated with a (local) pure rotation (as of a rigid body) about the 1-axis; and similarly for the other two pairs of non-zero elements of ξ_{ij}.

The complete tensor ξ_{ij} thus corresponds to a small pure rotation about some axis through P, and is called the *rotation tensor* at P. The rotation may alternatively be represented as a vector of components $(\xi_{23}, \xi_{31}, \xi_{12})$; the vector whose components are double the latter, namely,

$$\left(\frac{\partial u_3}{\partial x_2} - \frac{\partial u_2}{\partial x_3}, \frac{\partial u_1}{\partial x_3} - \frac{\partial u_3}{\partial x_1}, \frac{\partial u_2}{\partial x_1} - \frac{\partial u_1}{\partial x_2}\right),$$

is the *curl* of the vector u_i. We denote this vector as curl (u_i).

2.2.2 The strain tensor

It will now be shown that the components of e_{ij} as given by (13) are all associated with some internal deformation of the material, i.e. with strain; hence e_{ij} is called the *Cauchy strain tensor* at P.

The contribution to (11) due to the element e_{11} has components of length $(e_{11}y_1, 0, 0)$. This clearly constitutes a strain in which all small lengths parallel to the 1-axis increased by the small fraction e_{11}. Such a strain is an *extension.* The elements e_{11}, e_{22}, e_{33} thus give three mutually perpendicular extensions parallel to the 1-, 2-, 3-axes, respectively.

The contribution to (11) due to the pair of elements e_{23} and e_{32} has components $(0, e_{23}y_3, e_{23}y_2)$. On forming an expression analogous to (14), we find that the corresponding increase to the square of the distance PQ is (to sufficient accuracy)

$$4e_{23}y_2y_3. \tag{16a}$$

This is also the increase to the square of the distance PQ_1, where Q_1 is the projection of Q on the 2–3 plane through P. The proportionate increase in PQ_1^2 is therefore

$$2e_{23}\sin 2\alpha, \tag{16b}$$

where α is the angle between the 2-axis and PQ_1 (see Fig. 2.4). On forming the analogue of (15), we find further that the direction of PQ_1 has turned about the 1-axis through the angle

$$e_{23}(y_2^2 - y_3^2)/(y_2^2 + y_3^2), \quad \text{i.e.} \quad e_{23}\cos 2\alpha. \tag{17}$$

Both (16) and (17) vary with α; thus the elements e_{23}, e_{32} correspond to another type of deformation in the vicinity of P.

From (16) and (17), we can deduce the various properties of this second type of deformation. As an instance of this, let Q be taken to coincide with particles R, S, respectively, initially on the 2-, 3-axes (through P), respectively. By (17), it follows that during the deformation, PR has turned through the (small) angle e_{23}, and PS through $-e_{23}$; thus the angle RPS (initially a right angle) has decreased by the small amount $2e_{23}$ (see Fig. 2.5). By (16), the lengths of PR, PS will, however, not have been changed. (This corresponds to an elementary experiment wherein an initially rectangular prism is distorted without change of side-lengths into a prism whose section is a parallelogram.) A deformation possessing these features is a *shear*, the

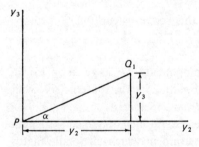
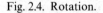

Fig. 2.4. Rotation.

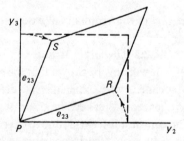

Fig. 2.5. Shear strain.

angle $2e_{23}$ being the *angle of shear*. Similar results of course apply to e_{31}, e_{13} and to e_{12}, e_{21}. We call e_{ij} ($i \neq j$) the *shear components of strain*.

For use in a later section (§2.6), note that (16a) may be derived directly from (10). By (10), the excess, $d(PQ)^2$ say, of the square of the distance between P, Q after strain over that before is given by

$$\mathrm{d}(PQ)^2 = y_i^2 - \left(y_i - \frac{\partial u_i}{\partial x_j} y_j \right)\left(y_i - \frac{\partial u_i}{\partial x_k} y_k \right), \tag{18}$$

where y_i^2 denotes $y_i y_i$, i.e. $y_1^2 + y_2^2 + y_3^2$. As we are here neglecting powers beyond the first in the components of $\partial u_i / \partial x_j$, we may write (18) as

$$\mathrm{d}(PQ)^2 = \frac{\partial u_i}{\partial x_j} y_j y_i + \frac{\partial u_i}{\partial x_k} y_i y_k$$

$$= \left(\frac{\partial u_j}{\partial x_i} + \frac{\partial u_i}{\partial x_j} \right) y_i y_j,$$

on rearranging suffices. Hence, by (13),

$$\mathrm{d}(PQ)^2 = 2e_{ij} y_i y_j. \tag{19}$$

From (19), the result (16a) follows as a particular case.

We now introduce the *strain quadric* at P given by

$$e_{ij} y_i y_j = \text{constant}. \tag{20}$$

By argument similar to that used for the stress quadric, it follows that at any given instant there is one particular set of orthogonal axes through P for which the shear components of strain at P vanish. A knowledge of the directions of these axes, called the *principal axes of strain* at P, and of the corresponding values taken by e_{11}, e_{22}, e_{33}, called the *principal extensions*, is sufficient to determine completely the deformation in the neighbourhood of P. The deformation near a point is thus always expressible as the resultant of simple extensions in three mutually perpendicular directions.

The theory of quadric surfaces gives the further result that $e_{11} + e_{22} + e_{33}$ is invariant for changes in the directions of the axes (the frame remaining orthogonal), i.e. e_{ii} is a scalar. This result may be directly established by noting that $e_{ii} = \partial u_i / \partial x_i$ and applying transformation formulae based on (2).

2.2.3 Cubical dilatation

Consider a small portion of matter, enclosing P, which undergoes deformation. Then the cubical dilatation at P is defined as the limit, as the area of the boundary surface approaches zero, of the proportionate increase in the volume of this matter. If e_{11}, e_{22}, e_{33} are the principal extensions at P

(associated with this deformation), the dilatation θ at P is thus equal to

$$1 - (1 - e_{11})(1 - e_{22})(1 - e_{33}).$$

Thus

$$\theta = e_{11} + e_{22} + e_{33} = e_{ii}, \tag{21}$$

neglecting higher-order terms. By the preceding paragraph, it follows, moreover, that (21) is true even if e_{11}, e_{22}, e_{33} are other than the principal extensions.

From (13) and (21) we have

$$\theta = \frac{\partial u_1}{\partial x_1} + \frac{\partial u_2}{\partial x_2} + \frac{\partial u_3}{\partial x_3} = \frac{\partial u_i}{\partial x_i}. \tag{22}$$

The last expression is the *divergence* of the vector u_i, and denoted as div(u_i).

A (cubical) *compression* is a negative dilatation.

2.2.4 The equation of conservation

We now express mathematically the fact that the mass of any given portion of matter is conserved. Consider the matter inside an indefinitely small parallelepiped S which has the point x_i at one corner and has side-lengths δx_i; let $\delta\tau = \delta x_1 \delta x_2 \delta x_3$. Suppose first that S is fixed in space. The rate of increase of mass within S is then $(\partial\rho/\partial t)\,\delta\tau$, where ρ is the density at x_i at time t. The net rate of inflow of mass across the surface of S is found (by an argument similar to that used in deriving the term $\partial p_{ij}/\partial x_j$ in (9)) to be $-\partial/\partial x_i\{(\rho v_i)\delta\tau\}$, where v_i is the velocity of the particle at x_i at time t. The conservation principle is expressed by equating these two terms. This gives the *conservation equation*

$$\frac{\partial\rho}{\partial t} + \frac{\partial(\rho v_i)}{\partial x_i} = 0. \tag{23}$$

It is convenient for some purposes to express (23) in the form

$$\frac{d\rho}{dt} + \rho\frac{\partial v_i}{\partial x_i} = 0,$$

where the operation

$$\frac{d}{dt} = \frac{\partial}{\partial t} + v_i\frac{\partial}{\partial x_i}$$

denotes 'differentiation following the motion'; the symbol $\partial/\partial t$ is concerned with time changes at an assigned point, i.e. fixed x_i, while d/dt (often denoted D/Dt) is concerned with time changes in the behaviour of an assigned particle or group of particles.

Secondly, suppose that S is not fixed in space, but moves with the particles initially enclosed in it. The conservation principle may then be expressed in the alternative form

$$\frac{d}{dt}(\rho\delta\tau) = 0. \tag{23a}$$

We may note also that the equation (23) is relevant in wider conservation cases, e.g. conservation of energy (§ 3.3.6). The name 'equation of continuity' is commonly used for the equation of conservation in any of its forms.

We note incidentally that the velocity v_i is connected with the displacement u_i by the relation $v_i = du_i/dt$. From this and (23) it may be shown that

$$\frac{\partial v_i}{\partial x_j} = \frac{d}{dt}\left(\frac{\partial u_i}{\partial x_j}\right), \tag{24}$$

neglecting the second-order terms $\{(\partial v_k/\partial x_j)(\partial u_i/\partial x_k)\}$. The formula (24) will be useful in later parts of this chapter; we leave its proof as an exercise to the reader.

2.3 Curvilinear coordinates

Most problems treated in this book will concern plane or spherical geometries. Sometimes, however, more general geometries are encountered in elasticity and elastic wave problems and it is convenient to have a general transformation from Cartesian coordinates to orthogonal curvilinear coordinates.

Let u_1 be a parameter of a curve $x_i = x_i(u_1)$. Then for a displacement dx_i to a neighbouring point along the curve,

$$dx_i = \frac{dx_i}{du_1}du_1,$$

by Taylor's theorem.

Now divide the space into three families of curves, parameters u_1, u_2, u_3, so that through each point there passes one curve of each family. We thus have $x_i = x_i(u_1, u_2, u_3)$ and $\partial x_i/\partial u_1$, etc. are tangent vectors.

The line element of a curve is given, from above, by

$$ds^2 = dx_i dx_i$$

$$= \frac{\partial x_i}{\partial u_\alpha}\frac{\partial x_i}{\partial u_\beta}du_\alpha du_\beta, \tag{25}$$

summed over α and β.

For a mutually orthogonal system, the product of the tangent vectors vanishes ($\alpha \neq \beta$) so that

$$ds^2 = h_1^2 du_1^2 + h_2^2 du_2^2 + h_3^2 du_3^2, \tag{26}$$

where

$$h_1^2 = \frac{\partial x_i}{\partial u_1}\frac{\partial x_i}{\partial u_1}, \text{ etc.}$$

For transformation to spherical polars (r, θ, ϕ), $x_1 = u_1 \sin u_2 \cos u_3$, $x_2 = u_1 \sin u_2 \sin u_3$, $x_3 = u_1 \cos u_2$. Thus, $h_1 = 1$, $h_2 = r$, $h_3 = r \sin \theta$. Similarly, from the elementary definitions, we can derive expressions in orthogonal curvilinear coordinates for the vector operators grad, div, curl and the Laplacian operator $\mathbf{V}^2$ and for strain e_{ij}. These expressions are set down, for example, in Takeuchi and Saito (1972) (see §2.8).

2.4 Perfect elasticity

The occurrence of stress in deformable matter is accompanied by the occurrence of strain. Perfect elasticity is when the components of strain at any point are determined by the components of stress at the point. Thus in perfect elasticity the components of strain are (on the infinitesimal strain theory) homogeneous linear functions of the components of stress, and vice versa. We write

$$p_{kl} = A_{ijkl}e_{ij}, \tag{27}$$

where the A_{ijkl} constitute a set of 81 coefficients. Of these coefficients not more than 36 could be independent, since each of the tensors p_{ij}, e_{ij}, being symmetrical, has only six independent components; thus $A_{ijkl} = A_{jikl} = A_{ijlk}$.

This constitutes a generalisation of *Hooke's Law*, and in many circumstances agrees closely with what happens in practice. We shall assume that the standard equilibrium configuration is stable; this is so in ordinary problems.

2.4.1 Stress–strain relations for a perfectly elastic isotropic material

Let p_{ij} and e_{ij} be the stress and strain tensors at time t at the point x_i in such a material. By *isotropic*, we mean that the elastic behaviour is entirely independent of any particular direction. Then, on account of the linearity of the stress–strain relations and the symmetry associated with isotropy, we may write

$$p_{11} = Ae_{11} + B(e_{22} + e_{33}) + Ce_{23} + D(e_{31} + e_{12}), \tag{28}$$

$$p_{23} = Ee_{11} + F(e_{22} + e_{33}) + Ge_{23} + H(e_{31} + e_{12}), \tag{29}$$

where the coefficients A, B, C, D, E, F, G, H depend only on the particular material and on the thermodynamical conditions, together with two pairs of similar equations involving the same eight coefficients. These coefficients are independent of the directions of the particular axes taken.

Take, in particular, axes (1-, 2-, 3-) coinciding with the principal axes of strain at the point, P say. Then (28) and (29) become

$$p_{11} = Ae_{11} + B(e_{22} + e_{33}), \tag{28a}$$

$$p_{23} = Ee_{11} + F(e_{22} + e_{33}). \tag{29a}$$

Now take a second set of axes (1'-, 2'-, 3'-) obtained by rotating the first reference frame through $\frac{1}{2}\pi$ about the 1-axis; thus the 1'- and 1-axes coincide, the 2'- and 3-axes coincide, and the 3'-axis is opposite to the 2-axis. Since the new axes also coincide with the principal axes of strain at P, we have by (29)

$$p_{2'3'} = Ee_{1'1'} + F(e_{2'2'} + e_{3'3'}),$$

and hence

$$-p_{32} = Ee_{11} + F(e_{33} + e_{22}).$$

Hence, by (29a) and (3), $p_{23} = 0$. We can similarly show that when the axes are principal axes of strain, $p_{31} = 0$ and $p_{12} = 0$.

We have thus shown that for a perfectly elastic isotropic substance, the principal axes of stress at P coincide with the principal axes of strain at P.

Now consider the tensor

$$p_{ij} - B\theta\delta_{ij} - (A - B)e_{ij}.$$

Its components are all zero when referred to principal axes of strain (and hence also of stress) at P; this is obvious for $i \neq j$, and follows by (28a) for the cases $i = j$. Hence, by the tensor transformation rule (1), the components of this tensor are zero whatever the directions of the axes.

Thus for a perfectly elastic isotropic substance, the stress–strain relations are of the form

$$p_{ij} = B\theta\delta_{ij} + (A - B)e_{ij}.$$

We have in effect shown that $C = D = E = F = H = 0$, and that $A - B = G$, so that it follows that the eight coefficients above are expressible in terms of just two *elastic parameters*. The two parameters are commonly taken as λ and μ (the Lamé elastic parameters), where $\lambda = B$, and $\mu = \frac{1}{2}(A - B)$.

Thus in place of (28) and (29), we have

$$p_{11} = \lambda\theta + 2\mu e_{11}, \tag{30}$$

and

$$p_{23} = 2\mu e_{23}. \tag{31}$$

and the full stress–strain relations are

$$p_{ij} = \lambda\theta\delta_{ij} + 2\mu e_{ij}. \tag{32}$$

The use of λ, μ (or corresponding pairs of elastic parameters such as k, μ and E, σ – see §§ 2.4.3, 2.4.4) in the sequel will be understood to imply that the elastic behaviour is isotropic.

It is easy to deduce from (32) that if, in particular, the stress system is such that all components of stress except p_{23} and p_{32} vanish, then all components of strain vanish except e_{23} and e_{32}. Thus such a stress system is associated with a simple shear, and is called a *shearing stress*. A shearing stress is also given by the stress system in which $p_{22} = -p_{33}$ and all other components of p_{ij} are zero. (This may be seen on rotating the 2- and 3-axes through $\frac{1}{4}\pi$ about the 1-axis, and applying the necessary transformation formulae obtained from (1).) We then have

$$e_{22} = -e_{33} = p_{22}/2\mu$$

and all the other e_{ij} zero. The dilatation θ is of course zero.

2.4.2 Equations of motion in terms of displacement
By equations (9) and (32),

$$\rho f_i = \frac{\partial}{\partial x_j}(\lambda\theta\delta_{ij} + 2\mu e_{ij}) + \rho X_i.$$

Hence, by (4) and (13), we have the Navier–Green equations

$$\rho f_i = \frac{\partial}{\partial x_i}(\lambda\theta) + \frac{\partial}{\partial x_j}\left\{\mu\left(\frac{\partial u_j}{\partial x_i} + \frac{\partial u_i}{\partial x_j}\right)\right\} + \rho X_i. \tag{33}$$

If the material is uniform, so that λ and μ are constants,

$$\rho f_i = \lambda\frac{\partial\theta}{\partial x_i} + \mu\frac{\partial}{\partial x_i}\frac{\partial u_j}{\partial x_j} + \mu\nabla^2 u_i + \rho X_i,$$

where ∇^2 is *Laplace's operator*, $\partial^2/\partial x_i^2$. Hence, by (22),

$$\rho\frac{\mathrm{d}^2 u_i}{\mathrm{d}t^2} = (\lambda + \mu)\frac{\partial\theta}{\partial x_i} + \mu\nabla^2 u_i + \rho X_i. \tag{34a}$$

If the displacement u_i and the velocity v_i are always small (this is the case in ordinary problems in elasticity, but not in hydrodynamics), we may by (23) replace $\mathrm{d}/\mathrm{d}t$ by $\partial/\partial t$ on the left-hand side of (34a), for we are already neglecting second powers of components of $\partial u_i/\partial x_j$. In this case, we write

$$\rho\frac{\partial^2 u_i}{\partial t^2} = (\lambda + \mu)\frac{\partial\theta}{\partial x_i} + \mu\nabla^2 u_i + \rho X_i. \tag{34b}$$

Since $\theta = \partial u_i / \partial x_i$, equation (34b) is homogeneous linear in the u_i and X_i. Hence (34b) still holds true if u_i are measured from any equilibrium configuration as standard, and X_i are components of additional body force per unit mass; correspondingly, p_{ij} will be the components of additional stress. This interpretation is of course subject to the requirements of the infinitesimal strain theory and perfect elasticity. Similar remarks apply to all the linear equations in the present theory, and we shall now usually interpret equations in u_i, p_{ij} and X_i in this way. (An exception arises when we consider the strain–energy function (§2.4.5) which is quadratic in the e_{ij}.)

2.4.3 Some perfectly elastic substances

We now consider substances to which the equations (32) apply. By (31), it is seen that the ratio of a shear component of stress to the corresponding shear is independent of the particular axes taken; this ratio, μ, is called the *rigidity*. Perfectly elastic isotropic substances are classified into three types according to the magnitude of μ. If λ and μ are everywhere infinite, so that by (32) the e_{ij} are zero for finite p_{ij}, the substance is an (ideal) *rigid body*; if μ is finite and not zero, the substance is a *perfect solid*; if μ is zero, the substance is a *perfect fluid*.

Observation shows that a fluid at rest closely obeys the conditions at the beginning of §2.4. On putting $\mu = 0$ in (32), if $i \neq j$, we have $p_{ij} = 0$, whatever the directions of the axes. It follows from equation (6) of the stress quadric that at any point of fluid at rest all axes are principal axes, and hence that the stress quadric is a sphere. In this case, the stress across any small plane area is normal to the area. There is now only one independent component of the stress tensor; this component, reversed in sign, is the ordinary fluid pressure of hydrostatics. The term *hydrostatic pressure* is used with any substance when the stress takes this form; a hydrostatic pressure p corresponds to a stress $-p\delta_{ij}$.

Now suppose that during a time increment dt, the material in the vicinity of a particle P of a perfectly elastic isotropic body is subjected to an additional stress in the form of hydrostatic pressure dp. The ratio of this pressure to the compression produced is called the *bulk modulus* or *incompressibility* k at P. We can express k simply in terms of λ and μ. For, by (32),

$$-dp\,\delta_{ij} = \lambda d\theta\,\delta_{ij} + 2\mu de_{ij},$$

contracting (i.e. replacing j by i and doing the summation), we have by (21)

$$-3dp = 3\lambda d\theta + 2\mu d\theta,$$

i.e.

$$k = -\frac{dp}{d\theta} = \lambda + \tfrac{2}{3}\mu. \tag{35}$$

We note incidentally that by (35) and (23*a*)

$$k = \rho\frac{dp}{d\rho}. \tag{36}$$

It is convenient to write equations (32) in terms of k and μ instead of λ and μ; thus

$$p_{ij} = (k - \tfrac{2}{3}\mu)\theta\delta_{ij} + 2\mu e_{ij}. \tag{37}$$

For the particular case $\mu = 0$, (37) degenerates to the *perfect fluid relation*

$$p_{ij} = k\theta\delta_{ij}, \tag{38}$$

i.e. $p = -k\theta$, where $-p = p_{11} = p_{22} = p_{33}$. In particular, the fluid is called *gas* or *liquid* according as k is moderate or very great. An ideal liquid is incompressible and has k infinitely great.

2.4.4 Young's modulus and Poisson's ratio

Suppose that the point P is inside a small right cylinder (of a perfectly elastic isotropic material), that the stress at each end is normal (the magnitude being the same at each end), and that there is no traction across the lateral surface. Then *Young's modulus* E at P is the ratio of the stress at an end to the longitudinal extension of the cylinder; and *Poisson's ratio* σ is the ratio of the lateral contraction to the longitudinal extension.

E and σ are elastic parameters which, like λ and μ, may vary from point to point of the material. They are readily expressible in terms of λ and μ. To show this, take the 1-axis parallel to the cylinder's axis. Then since p_{22}, p_{33} are both zero, we have on adding the three equations like (30) and using (21)

$$p_{11} = (3\lambda + 2\mu)\theta.$$

Using (30) again, we see that p_{11}, e_{11} are connected by

$$p_{11} = \frac{\lambda p_{11}}{3\lambda + 2\mu} + 2\mu e_{11}.$$

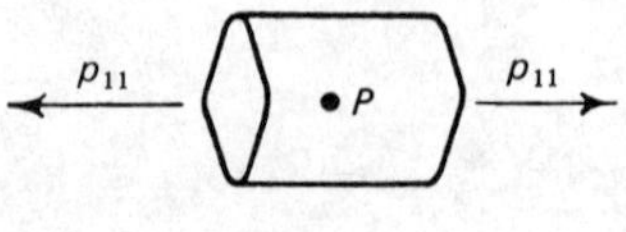

Fig. 2.6. Extension.

Hence

$$E = \frac{p_{11}}{e_{11}} = \frac{\mu(3\lambda + 2\mu)}{\lambda + \mu}. \tag{39}$$

Again from the equations like (30), we have

$$e_{22} = e_{33} = -\lambda\theta/2\mu$$
$$= -\lambda(e_{11} + 2e_{22})/2\mu.$$

Hence

$$\sigma = -\frac{e_{22}}{e_{11}} = \frac{\lambda}{2(\lambda + \mu)}. \tag{40}$$

We can from (39), (40) and (35) express λ, μ, k in terms of E, σ. We obtain

$$\lambda = E\sigma/(1 + \sigma)(1 - 2\sigma), \tag{41}$$

$$\mu = E/2(1 + \sigma), \tag{42}$$

$$k = E/3(1 - 2\sigma). \tag{43}$$

We note also that

$$\lambda/\mu = 2\sigma/(1 - 2\sigma). \tag{44}$$

The relations (32) express p_{ij} in terms of e_{ij}. The parameters E, σ are useful when we wish to express e_{ij} in terms of p_{ij}. Thus, contracting (32) and then using (35), we have

$$p_{ll} = (3\lambda + 2\mu)\theta = 3k\theta.$$

Hence by (32),

$$2\mu e_{ij} = p_{ij} - \lambda\theta\delta_{ij}$$

$$= p_{ij} - \frac{\lambda}{3k}p_{ll}\delta_{ij}.$$

Finally, by (41), (42) and (43)

$$e_{ij} = \frac{1 + \sigma}{E}p_{ij} - \frac{\sigma}{E}p_{ll}\delta_{ij}. \tag{45}$$

Note that, if the standard equilibrium configuration is assumed to be stable, each of μ and k is $\geqslant 0$ (this comes most readily from (56)). By (35) and (39) it follows that $E \geqslant 0$, and from (42) and (43) that $-1 \leqslant \sigma \leqslant \frac{1}{2}$.

2.4.5 Energy in a perfectly elastic body

We consider again the matter, M say, within a small parallelepiped (see Fig. 2.3) which has the particle P at one corner, and take the standard configuration as that in which $p_{ij} = 0$. At time t, let x_i be the coordinates of P's position, δx_i the side-lengths of the parallelepiped, and $\delta\tau = \delta x_1\,\delta x_2\,\delta x_3$

its volume; let u_i, v_i and f_i be the displacement, velocity and acceleration of P. During the ensuing small time dt, let du_i be the additional displacement of P, $dw\,\delta\tau$ the work done on M, $dT\,\delta\tau$ the increase in kinetic energy, and $dQ\,\delta\tau$ the mechanical equivalent of the heat emitted from M. Then, by the first law of thermodynamics,

$$dw = dU + dT + dQ, \tag{46}$$

where $dU\,\delta\tau$ is the increase in the internal energy of M.

The work done on M by the body force, X_i per unit mass, is $\rho\delta\tau\,X_i du_i$. The work done by the 3-components of the tractions across the pair of faces normal to the 2-axis is

$$-p_{23}\delta x_1\,\delta x_3\,du_3 + \left(p_{23}du_3 + \frac{\partial(p_{23}du_3)}{\partial x_2}\delta x_2\right)\delta x_1\,\delta x_3,$$

i.e.

$$\frac{\partial(p_{23}du_3)}{\partial x_2}\delta\tau;$$

and similarly for the work done by other components of traction. Hence

$$dw = \rho X_i du_i + \frac{\partial(p_{ij}du_i)}{\partial x_j}. \tag{47}$$

Also

$$dT\,\delta\tau = d(\tfrac{1}{2}\rho\delta\tau\,v_i^2);$$

hence, by $(23a)$,

$$dT = \rho d(\tfrac{1}{2}v_i^2)$$

$$= \rho v_i \frac{dv_i}{dt}dt$$

$$= \rho f_i du_i. \tag{48}$$

By (47), (48) and (9),

$$dw - dT = p_{ij}\frac{\partial(du_i)}{\partial x_j}. \tag{49}$$

Now by (24),

$$\frac{\partial}{\partial x_j}(du_i) = d\left(\frac{\partial u_i}{\partial x_j}\right) \quad \text{(to sufficient accuracy)}$$

$$= de_{ij} - d\xi_{ij}, \tag{50}$$

by (12) and (13). By (46), (49) and (50), we thus have

$$p_{ij}de_{ij} = dU + dQ, \tag{51}$$

the omitted term $-p_{ij}\mathrm{d}\xi_{ij}$ being seen to be zero since p_{ij} is a symmetrical tensor, and ξ_{ij} antisymmetrical (see the fourth paragraph of §2.2).

When the thermodynamical conditions of the change are isothermal with the absolute temperature ϑ assigned, the state of M is sufficiently described by the values of e_{ij}. By the second law of thermodynamics, $\mathrm{d}Q/\vartheta$, and hence also $\mathrm{d}Q$ (ϑ being here constant), is then a perfect differential in the e_{ij}. By the first law of thermodynamics, $\mathrm{d}U$ is also a perfect differential in the e_{ij}. It then follows from (51) that we may write

$$p_{ij}\mathrm{d}e_{ij} = \mathrm{d}W, \tag{52}$$

where W is a function of the e_{ij}, called the (isothermal) *strain-energy function*. By (52),

$$p_{ij} = \frac{\partial W}{\partial e_{ij}}. \tag{53}$$

Since for perfect elasticity the p_{ij} are (on the infinitesimal strain theory) homogeneous linear functions of the e_{ij}, it follows from (53) that (apart from an additive arbitrary constant, which we shall take as zero) W is a homogeneous quadratic function of the e_{ij}. By (53), we thus have

$$2W = e_{ij}\frac{\partial W}{\partial e_{ij}} = e_{ij}p_{ij}. \tag{54}$$

If, in particular, the elastic behaviour is isotropic, by (54), (32) and (21)

$$2W = e_{ij}(\lambda\theta\delta_{ij} + 2\mu e_{ij})$$
$$= \lambda\theta^2 + 2\mu e_{ij}^2, \tag{55a}$$

λ, μ being isothermal parameters (by e_{ij}^2, we understand $e_{ij}e_{ij}$, the summation convention holding as usual). In terms of k and μ, the isotropic strain-energy function is, by (35), given by

$$2W = k\theta^2 + 2\mu(e_{ij}^2 - \tfrac{1}{3}\theta^2). \tag{55b}$$

Two other types of symmetry deserve attention. First, if there are three orthogonal planes of symmetry at a point we may write (in Love's notation)

$$2W = Ae_{11}^2 + Be_{22}^2 + Ce_{33}^2 + 2Fe_{22}e_{33} + 2Ge_{33}e_{11}$$
$$+ 2He_{11}e_{22} + Le_{23}^2 + Me_{31}^2 + Ne_{12}^2. \tag{56}$$

If there is an axis of symmetry (say x_3), so that the body is transversely isotropic, this form reduces to one with only five coefficients (see Exercise 2.8.5 and equation (64)).

(Note that the coefficients in (56) do not correspond to those in (28) and (29).)

A strain-energy function also exists when the conditions are adiabatic, for in this case $dQ = 0$, and so $p_{ij}de_{ij}$ is again a perfect differential. The strain-energy function in this case agrees with the forms (52), (53) and (54); but in (55) it is necessary to interpret λ, μ, k as adiabatic parameters.

It may be shown (see Jeffreys, *Cartesian Tensors* (CUP, 1952), chapter 8) that the adiabatic and isothermal parameters (indicated here by the suffixes a, i, respectively) are connected by the relations

$$\left.\begin{aligned} \lambda_a &= \lambda_i + \frac{k_i^2\gamma^2\vartheta}{\rho c}, \quad \mu_a = \mu_i, \\ k_a &= k_i\left(1 + \frac{k_i\gamma^2\vartheta}{\rho c}\right) = k_i\left(1 + \gamma_G\gamma\vartheta\right), \end{aligned}\right\} \tag{57}$$

where $\gamma_G = \gamma k_i/\rho c_p$ is Grueneisen's ratio, γ denotes the coefficient of cubical expansion and c the specific heat at constant strain, other symbols being as previously.

In the cases discussed above, the total energy (including elastic strain energy) is conserved. In other cases, $dU + dQ$ will exceed dW, where W is defined by a formula of the type (55). By (51) the energy dissipated per unit volume in time dt is then $p_{ij}de_{ij} - dW$.

2.4.6 Theorems on elastic equilibrium

It will be convenient in this section to make use of *Green's lemma* in the form

$$\iiint \frac{\partial v_i}{\partial x_i}d\tau = \iint v_i v_i\,dS, \tag{58}$$

where v_i is any vector, the triple integral is taken throughout the volume of a body, the double integral is taken over the boundary surface, and v_i are the direction-cosines of the outward normal at a point of this surface. The thermodynamical conditions are such that a strain–energy function exists.

We now prove Kirchhoff's theorem that if either (*a*) the surface displacements or (*b*) the surface tractions of a perfectly elastic body in equilibrium are assigned, then there is at most one solution for the state of strain (and hence also of stress) throughout the body.

The proof is as follows. By (9) and (53), we have, since $f_i = 0$,

$$\frac{\partial}{\partial x_j}\frac{\partial W}{\partial e_{ij}} + \rho X_i = 0. \tag{59}$$

Suppose that u_i', u_i'' are two different sets of displacements each compatible with the assigned conditions, and now introduce $u_i = u_i' - u_i''$. If W, e_{ij}, etc., are now taken to be associated with the u_i as just defined, we deduce from

(59) that

$$\frac{\partial}{\partial x_j}\frac{\partial W}{\partial e_{ij}} = 0.$$

Hence

$$0 = \iiint u_i \frac{\partial}{\partial x_j}\frac{\partial W}{\partial e_{ij}}\,d\tau$$

$$= \iiint \left\{ \frac{\partial}{\partial x_j}\left(u_i\frac{\partial W}{\partial e_{ij}} \right) - \frac{\partial u_i}{\partial x_j}\frac{\partial W}{\partial e_{ij}} \right\}d\tau$$

$$= \iint u_i v_j \frac{\partial W}{\partial e_{ij}}\,dS - \iiint e_{ij}\frac{\partial W}{\partial e_{ij}}\,d\tau,$$

the last line following by use of Green's lemma (58) and the fact that (by (12) and (13)) e_{ij} differs from $\partial u_i/\partial x_j$ by an antisymmetrical tensor (ξ_{ij}). Now when the data (a) are assigned, u_i is zero at all points of the boundary; and when (b) are assigned, $v_j(\partial W/\partial e_{ij})$ is, by (53) and (5), zero at all points of the boundary. Hence in both cases

$$0 = \iiint e_{ij}\frac{\partial W}{\partial e_{ij}}\,d\tau$$

$$= \iiint 2W\,d\tau \tag{60}$$

by (54). Since W cannot be negative at any point (the standard configuration is one of stable equilibrium), it follows from (60) that W is everywhere zero, and hence that the e_{ij} are all zero at all points of the body. From this the theorem follows.

This theorem may be extended to cover the case of a perfectly elastic vibrating body. For example, when there are no variable body forces and the surface is free from traction, the state of stress and strain is uniquely determined throughout the body at all times if the initial displacements and velocities are assigned at all points (see Love, 1949).

A further theorem is that when the surface displacements of a perfectly elastic body and also the body forces (the field of which is conservative) are assigned, the total potential energy in the corresponding equilibrium configuration C is less than that in any other configuration C_1 compatible with the assigned conditions.

To prove this theorem, let u_i now be the displacement at any point of the body in the configuration C, and $u_i + u_i'$ that in C_1; let e_{ij}' correspond to u_i'. Let W be the strain-energy function corresponding to e_{ij}, W' to e_{ij}', and W_1

to $e_{ij} + e'_{ij}$. Since W is a homogeneous quadratic function of the e_{ij}, it follows that

$$W_1 = W + e'_{ij}\frac{\partial W}{\partial e_{ij}} + W'.$$

This step may be shown by writing W in the form $a_{rs}b_r b_s$, where b_r (or b_s) stands for any one of the nine components of e_{ij}; each of the suffixes r, s here runs from 1 to 9; then $W' = a_{rs}b'_r b'_s$, and

$$W_1 = a_{rs}(b_r + b'_r)(b_s + b'_s);$$

it is then necessary to use the result, easily proved, that

$$\partial W/\partial b_r = (a_{rs} + a_{sr})b_s.$$

Hence, by (53),

$$W_1 - W = W' + e'_{ij}p_{ij}$$

$$= W' + p_{ij}\frac{\partial u'_i}{\partial x_j}, \tag{61}$$

since e'_{ij} differs from $\partial u'_i/\partial x_j$ by an antisymmetrical tensor. Also,

$$p_{ij}\frac{\partial u'_i}{\partial x_j} = \frac{\partial}{\partial x_j}(p_{ij}u'_i) - u'_i\frac{\partial p_{ij}}{\partial x_j}$$

$$= \frac{\partial}{\partial x_j}(p_{ij}u'_i) + X_i u'_i \rho, \tag{62}$$

by (9), since $f_i = 0$. Using (61) and (62), and then (58), we have

$$\iiint (W_1 - W)\,d\tau - \iiint X_i u'_i \rho\,d\tau$$

$$= \iiint W'\,d\tau + \iiint \frac{\partial}{\partial x_j}(p_{ij}u'_i)\,d\tau$$

$$= \iiint W'\,d\tau + \iint v_j p_{ij}u'_i\,dS. \tag{63}$$

The surface integral vanishes by virtue of the assigned boundary conditions. Hence, since W' is everywhere positive, the left-hand side of (63) is positive; but this left-hand side consists of the excess in the configuration C_1 over that in C of the sum of the strain energy and the energy associated with the field of X_i. Hence the theorem follows.

2.4.7 Solving problems in elasticity

In elementary equilibrium problems, we are frequently given the boundary conditions and the body forces on a piece of uniform perfectly elastic isotropic matter, and we know from Kirchhoff's theorem that (on the infinitesimal strain theory) there is then at most one solution for the distribution of stress and strain. If therefore we succeed in finding, in terms of x_i, λ and μ (by trial or otherwise), any set of expressions for the e_{ij} which satisfy (32) and the relations (9) with $f_i = 0$, or alternatively which satisfy (34b) with $\partial^2 u_i / \partial t^2 = 0$, and which also satisfy the boundary conditions, then this set of expressions must give the solution. On account of the large number of dependent variables (e_{ij} has six distinct components in general), this method can be expected to succeed only in very simple problems.

The use of the strain-energy function and the minimum-energy theorem leads to alternative and fruitful methods of solving problems. The number of problems in elastostatics and elastodynamics that have been solved exactly is small, and energy methods form the basis of many useful approximations.

In seismology, the main problems are dynamical, being concerned with the passage of disturbances through the Earth, and equations (9) and (34) with non-vanishing accelerations are relevant. These problems are the main concern in this book, and equations (34) have a central place in the theory developed in later chapters.

In the two following sections, we give methods of allowing for departures from the conditions of isotropy and perfect elasticity that we have so far been considering. In §2.7 we comment briefly on the effects of the second-order terms that have so far been neglected.

2.5 Non-isotropic materials and transverse isotropy

A material (still taken to be perfectly elastic) is *anisotropic* (or *aelotropic*) if it deviates from the directionally regular elastic behaviour of an isotropic material. It is then necessary to replace the discussion in §2.4.1 by one starting from more general stress–strain relations (27).

The existence of a strain-energy function in isothermal (or adiabatic) conditions, satisfying the relations (52) and (53), was established in §2.4.5 independently of the isotropic formula (32). It follows from (53) that in these conditions $\partial p_{kl} / \partial e_{ij} = \partial p_{ij} / \partial e_{kl}$, and hence that further relations, of the form $A_{ijkl} = A_{klij}$, hold between the coefficients in (27). From this it follows that in these conditions not more than 21 of the A_{ijkl} can be independent.

Crystals are substances which show some degree of symmetry in their elastic behaviour, but on account of special systematic structure are not completely isotropic. The number of parameters needed to specify the elastic behaviour (under isothermal or adiabatic conditions) of a perfectly elastic crystal of any given type is therefore less than or equal to 21.

Rocks of the Earth contain crystals, but these, in general, will be oriented in all directions at random, so that differences in elastic behaviour in different directions neutralise one another to a large extent. For this reason, in seismology, conditions have been as a rule assumed isotropic. Important exceptions are found, however, in practice (see § 12.1) often associated with shear stress or flow.

One important geophysical case is when elastic moduli for deformation at right angles to the vertical direction are equal but differ from those for deformations parallel to it. Such materials for which the stress–strain relations are symmetric about a fixed axis are *transversely isotropic*. In this case there are five moduli A, C, F, L, N (see equation (56) and § 12.1.1) and for isotropic media

$$A = C = \lambda + 2\mu, \quad L = N = \mu, \quad F = \lambda. \tag{64}$$

2.6 Departures from perfect elasticity due to time effects

When a perfectly elastic (isotropic) body is in equilibrium under an assigned stress distribution (the thermodynamic conditions being also given), the strain distribution is uniquely determined by equations (37). We shall in the following consider cases in which (37) need modification; in particular, cases where the time t enters into the stress–strain relations.

As a preliminary it is convenient to introduce the tensors P_{ij} and E_{ij}, defined by

$$P_{ij} = p_{ij} - \tfrac{1}{3} p_{ll} \delta_{ij}, \tag{65}$$

$$E_{ij} = e_{ij} - \tfrac{1}{3} e_{ll} \delta_{ij}. \tag{66}$$

P_{ij} and E_{ij} are called the *deviatoric* (or *distortional*) *stress tensor* and *deviatoric strain tensor*, respectively. We note that when $i \neq j$, $P_{ij} = p_{ij}$ and $E_{ij} = e_{ij}$.

From (37) we deduce that

$$\tfrac{1}{3} p_{ll} = k\theta. \tag{67}$$

Using (65), (67), (37) and then (66),

$$P_{ij} = 2\mu E_{ij}, \tag{68}$$

for all i, j. The relations (67) and (68) together are equivalent to (37); only five of equations (68) are independent since, by (65) and (66), P_{ii} and E_{ii} are clearly both zero. For perfect elasticity, the relation (67) describes adequately behaviour under a completely symmetrical stress, while (68) describes the effects of any departures from symmetry.

It is easy to deduce from (66) that

$$E_{ij}^2 = e_{ij}^2 - \tfrac{1}{3}\theta^2, \tag{69}$$

and hence by (56) that

$$2W = k\theta^2 + 2\mu E_{ij}^2. \tag{70}$$

The main imperfections of elasticity that are observed arise only under stresses that are not fully symmetrical. Hence in setting up formulae to describe these imperfections, we shall keep (67) unchanged throughout, but shall modify (68) in various ways. All the elastic parameters to be used will be taken to be independent of t.

2.6.1 Fluid viscosity

The simplest mathematical model showing a deviation from perfectly elastic behaviour is an ideal viscous fluid. In this case (67) continues to hold, but in place of (68) we write

$$P_{ij} = 2v\frac{\mathrm{d}}{\mathrm{d}t}E_{ij}, \tag{71}$$

where v is a new parameter denoting the *shear viscosity* of the fluid.

A common unit of viscosity is the *poise* equal to one dyne-s/cm^2. Laboratory experiments show that in general v for liquids decreases with increases in temperature and increases with pressure increases. Values of v are about 10^{14} and 10^{22} poise for glacier ice and limestone, respectively.

Deviatoric stress is here connected with the rate of deviatoric strain in the same way as with the actual deviatoric strain in the case of a perfectly elastic solid.

For the particular case of laminar flow parallel to the 1-axis in planes perpendicular to the 2-axis (Fig. 2.7), we derive from (71), using (24),

$$p_{21} = v\frac{\partial v_1}{\partial x_2} \tag{72}$$

(where v_i denotes velocity), which agrees with a common definition of fluid viscosity.

We notice that if in particular v is zero, (71) reduces to $P_{ij} = 0$, which by (65) and (67) gives the perfect fluid relation (38).

The presence of terms involving $\mathrm{d}/\mathrm{d}t$ in the stress–strain relations is associated with dissipation of energy (see § 3.1.2). By the last paragraph of § 2.4.5, the rate of dissipation of energy per unit volume is given by

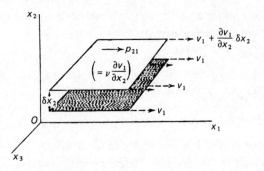

Fig. 2.7. Laminar flow.

$$\Phi = p_{ij}\frac{de_{ij}}{dt} - \frac{dW}{dt},$$

where W in the present case ($\mu = 0$) is, by (70), equal to $\frac{1}{2}k\theta^2$. Thus, by (65), (66),

$$\Phi = (P_{ij} + \tfrac{1}{3}p_{ll}\delta_{ij})\frac{d}{dt}(E_{ij} + \tfrac{1}{3}\theta\delta_{ij}) - \frac{d}{dt}(\tfrac{1}{2}k\theta^2)$$

$$= P_{ij}\frac{d}{dt}E_{ij} + \tfrac{1}{3}p_{ll}\frac{1}{3}\frac{d\theta}{dt}3 - k\theta\frac{d\theta}{dt},$$

since $\delta_{ij}E_{ij} = E_{ii} = 0$, and $\delta_{ij}P_{ij} = p_{ii} = 0$. By (67) and (71) we then have

$$\Phi = P_{ij}\frac{d}{dt}E_{ij}$$

$$= 2v\left(\frac{d}{dt}E_{ij}\right)^2. \tag{73}$$

The reader may be interested to deduce from (9), (71), (24) and the definitions of P_{ij}, E_{ij}, the Navier–Stokes equations of viscous fluid motion, namely,

$$\rho\frac{dv_i}{dt} = \rho X_i - \frac{\partial p}{\partial x_i} + \tfrac{1}{3}v\frac{\partial\theta'}{\partial x_i} + v\nabla^2 v_i,$$

where $p = -\tfrac{1}{3}p_{kk}$, and $\theta' = \partial v_j/\partial x_j$.

2.6.2 Kelvin–Voigt model

The suggestion arises that by suitably combining the perfect elasticity relations (68) with the viscous fluid relations (71), one might obtain a constitutive relation relevant to imperfectly elastic behaviour of a solid. The simplest way of combining (68) and (71) is to write

$$P_{ij} = 2\mu E_{ij} + 2v\frac{dE_{ij}}{dt}. \tag{74}$$

The form (74) corresponds to Kelvin–Voigt or *firmoviscous* behaviour. It has been used to indicate the attenuation in seismic wave transmission but is deficient in not having instantaneous elastic response.

The presence of the term in v implies exponential delay in reaching the full strain under an applied deviatoric stress, and also in recovery of an initial configuration after removal of a stress. It also implies dissipation of energy in accordance with the viscous fluid formula (73).

Next, consider the effect of an applied periodic deviatoric stress of period T, say $P_{ij} = A_{ij}\cos(2\pi t/T)$ (see §3.1.1). If T is small compared with v/μ, we find that E_{ij}, as given by (74), is of the order of $(T/v) A_{ij}$, which is small compared with A_{ij}/μ; hence under rapidly changing stresses, (74) implies an effective rigidity much greater than μ. On the other hand, if the stress periods are long compared with v/μ, the term in v in (74) is unimportant.

The last paragraph brings out the important point that in visco-elastic solids the rigidity may no longer be described by the single parameter μ, as is the case with perfect elasticity; with (74), rigidity depends on v and on the form of applied stress.

A more general and mathematically elegant treatment of linear visco-elastic materials such as (74) involves complex variables and will be postponed until §4.5.

2.6.3 Elastic afterworking

The relations (74) go a little distance towards representing observed elastic afterworking, wherein there is slow movement or *creep*, often of long duration, following application or removal of a deviatoric stress. But with most solids that exhibit elastic afterworking, this is not the whole story; following a sudden change in the deviatoric stress, the creep is preceded by some immediate change of strain not allowed for in (74). We can meet this complication qualitatively by adding to (74) a term in dP_{ij}/dt analogous to $2v dE_{ij}/dt$, obtaining the form

$$P_{ij} + \tau\frac{dP_{ij}}{dt} = 2\mu E_{ij} + 2v\frac{dE_{ij}}{dt}, \tag{75}$$

where τ is a further parameter. The perfect elasticity relation (67) is taken still to hold, in keeping with the observed behaviour of many solids showing elastic afterworking.

To show that (75) meets the requirements of elastic afterworking, suppose that at time $t = 0$ the solid is unstrained and that a constant

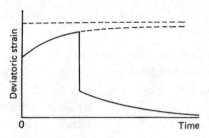

Fig. 2.8. Elastic afterworking.

deviatoric stress is then applied. By (75), we have (the parameters are being taken to be independent of t – §2.6)

$$\int_0^t P_{ij}\,dt + \tau P_{ij} = 2\mu \int_0^t E_{ij}\,dt + 2\nu E_{ij}, \tag{76}$$

so that there is an immediate deviatoric strain equal to $(\tau/2\nu)P_{ij}$. Subsequently, P_{ij} being taken constant, we deduce from (75)

$$E_{ij} = \frac{\tau P_{ij}}{2\nu}\left\{ \left(1 - \frac{\tau'}{\tau}\right)e^{-t/\tau'} + \frac{\tau'}{\tau} \right\}, \tag{77}$$

where we have written τ' for ν/μ. By (77), E_{ij} cannot exceed the value $P_{ij}/2\mu$, no matter how long the stress is applied.

Now suppose the stress is removed. We see by (76) that there is an immediate reduction of amount $(\tau/2\nu)\,P_{ij}$ in the deviatoric strain, and this is followed by an exponential creep to zero strain; the larger the parameter τ' is, the slower is the rate of recovery. Fig. 2.8 shows the main features of the variation of E_{ij} with time, when a constant deviatoric stress is applied and subsequently removed. Note that it is necessary that $\tau' > \tau$ for elastic afterworking.

Under very slowly changing stresses, the terms in τ and ν in (75) are unimportant and the behaviour approximates to that of a perfectly elastic solid of rigidity μ. Under very rapidly changing stresses the terms in τ and ν predominate, and the behaviour approximates to that of a perfectly elastic solid of rigidity ν/τ. Under changing stresses of intermediate periods there is appreciable damping and dissipation of energy. Again the rigidity does not depend on μ alone. It follows that in (77) there is no objection to replacing ν/τ by the symbol μ (a different μ of course from that in (75) and (76), but still having the dimensions of rigidity).

2.6.4 Maxwell model

If we take an extreme form of (75) in which the parameter μ is zero (this is equivalent to taking $\tau' = \infty$), we have the relations

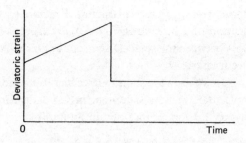

Fig. 2.9. Maxwell model.

$$2v\frac{dE_{ij}}{dt} = P_{ij} + \tau\frac{dP_{ij}}{dt}. \tag{78}$$

A substance whose elastic behaviour is described by (67) and (78) is called the Maxwell or *elasticoviscous* model.

If at time $t = 0$ a constant deviatoric stress P_{ij} is applied to an elasticoviscous substance previously unstrained, we see (using (76) with $\mu = 0$) that a deviatoric strain equal to $(\tau/2v) P_{ij}$ is immediately produced. Subsequently the deviatoric strain increases at the steady rate of $P_{ij}/2v$, as with a viscous fluid, and can become indefinitely great. If at time $t = t_1$ the stress is removed, there is an immediate reduction of amount $(\tau/2v) P_{ij}$ in the deviatoric strain, but now a permanent deviatoric strain of amount $(2v)^{-1}\int_0^{t_1} P_{ij}\,dt$ remains. This is illustrated in Fig. 2.9.

Under deviatoric stresses that change very slowly with the time, the first term on the right-hand side of (78) predominates, so that this case approximates to that of a fluid of viscosity v. If, on the other hand, the stresses are changing very rapidly, the second term predominates, and the Maxwell model approximates to a perfectly elastic solid of rigidity v/τ.

These results are all in qualitative agreement with the observed behaviour of some plastic materials. A substance such as pitch or 'silly putty' can be indefinitely strained under sustained deviatoric stress (i.e. suffers *plastic flow*), and only partially recovers an initial configuration on removal of the stress (i.e. has acquired a *permanent set*).

2.6.5 Strength of a solid

If a solid is subjected to deviatoric stress steadily increasing from zero, the elastic behaviour is usually at first perfectly elastic. This may be followed by a stage in which elastic afterworking becomes evident, but as yet there is no irrecoverable distortion. Ultimately there comes irrecoverable distortion in the form of plastic flow or of fracture. Fracture may occur immediately on application of a high stress, but usually there is a finite range of deviatoric stress between the smallest to produce plastic flow and the smallest to produce fracture.

The question arises as to whether the stage of irrecoverable distortion is determined by any simple function of stress components. A natural function to consider is the scalar P_{ij}^2. From (65) we obtain on multiplying out

$$P_{ij}^2 = \tfrac{1}{3}\{(p_{22} - p_{33})^2 + (p_{33} - p_{11})^2 + (p_{11} - p_{22})^2\}$$
$$+ 2(p_{23}^2 + p_{31}^2 + p_{12}^2), \tag{79}$$

and this expression is in fact proportional to a function whose value is, on the theory of Mises, assumed to determine the onset of plastic flow. There is experimental support for this assumption in the case of metals, and the value of $(3P_{ij}^2)^{\frac{1}{2}}$ at which flow starts has been used as an index of *strength*.

Actually the stage of irrecoverable distortion may depend significantly on the values of other functions in addition to P_{ij}^2, for example, the scalar p_{kk}. On one theory this stage is assumed to be determined by the *stress-difference*, defined as the difference between the greatest and least principal stresses. In geophysics, the difference between results on this theory and those on the Mises theory is unimportant; for we can deduce from (79) that $(P_{ij}^2)^{\frac{1}{2}}$ must lie between 0.71 and 0.82 times the stress-difference, a range of variation of only 15%.

It is important to notice that the terms 'strength' and 'rigidity' are not synonymous. Whereas an elasticoviscous substance has zero strength, we have seen that in certain circumstances it behaves as a perfectly elastic solid with significant rigidity.

2.6.6 Solids and fluids

In §2.3.3, we classified perfectly elastic substances into perfect solids and perfect fluids according to the value of the parameter μ. This classification is too simple for actual substances. It is questionable whether μ is precisely zero for any actual substance; in addition, we have seen that a single parameter μ may be insufficient to describe rigidity; also a substance may show rigidity when μ is zero. In so far as we classify any substance (in a given thermodynamical state) as solid or fluid it is thus necessary to lay down some further convention.

Some writers have taken absence of strength as the criterion that a substance is a fluid, others absence of rigidity; as we have seen these criteria are not equivalent. Since some substances lacking strength (e.g. pitch at ordinary temperatures) can behave as perfectly elastic solids, absence of rigidity is the better criterion. The question then arises as to the detectability of rigidity. It appears from the preceding subsections that rigidity is most likely to be detected under rapidly changing stress. We shall see in chapter 4 that there is a connection between the rigidity of a substance and the transmission of rotational waves through it. Following Jeffreys, we shall therefore call a substance a *solid* if there can be detectable transmission of rotational waves through it; otherwise, a *fluid*. An elasticoviscous substance is thus a solid on this classification.

Before closing the present section, we point out that the question of imperfections has many further complications that we have not considered here. For many purposes in seismology, however, a knowledge of the qualitative effects of imperfections is all that is needed.

2.7 Finite-strain theory

The theory we have so far developed rests on the assumption (see §2.2) that strains are sufficiently small to permit neglect of higher-order terms; but observations indicate that in some cases where the stress is high this assumption cannot hold. It is found, for instance, that if a long, thin, straight, cylindrical rod (the 'Euler column') be subjected to increasing compressive forces at its ends, a state of elastic instability in due course develops. As a result the configuration suddenly changes to one in which the rod is appreciably curved. This behaviour is in discord with the uniqueness theorem of Kirchhoff (§2.46), and since it is found to take place in the effective absence of imperfections of elasticity, implies that second-order terms are important in some problems. Theory which takes these terms into consideration is called *finite-strain theory*.

Whereas in the first-order theory we obtained the equation

$$d(PQ)^2 = 2e_{ij}y_iy_j, \tag{19}$$

where the e_{ij} are given by (13), the second-order theory would by (18) give

$$d(PQ)^2 = \left(\frac{\partial u_j}{\partial x_i} + \frac{\partial u_i}{\partial x_j}\right)y_iy_j - \frac{\partial u_i}{\partial x_j}\frac{\partial u_i}{\partial x_k}y_jy_k,$$

i.e.

$$d(PQ)^2 = 2\varepsilon_{ij}y_iy_j, \tag{80}$$

where

$$\varepsilon_{ij} = \frac{1}{2}\left(\frac{\partial u_j}{\partial x_i} + \frac{\partial u_i}{\partial x_j} - \frac{\partial u_k}{\partial x_i}\frac{\partial u_k}{\partial x_j}\right). \tag{81}$$

Thus any component of the more accurate strain tensor ε_{ij} differs from the corresponding component of e_{ij} by the sum of three second-order terms.

In second-order theory, questions arise such as the influence of rotation on the stress–strain relations, the possible variation of k, μ as the stress increases, and departures from isotropy if an initial stress is high and unsymmetrical. It is evident, therefore, that mathematics of finite strain is more complicated than that of the infinitesimal theory, and in general it is necessary to introduce simplifying assumptions.

In pioneering work, Birch applied finite-strain theory developed by Murnaghan to problems of the Earth's interior, assuming the presence of an initial finite hydrostatic stress $- p\delta_{ij}$ (see §4.7). On certain assumptions, he derived equations of state for internal regions of the Earth. Examples for chemically homogeneous regions are

$$p = - 3k_0\varepsilon(1 - 2\varepsilon)^{\frac{5}{2}}(1 + 2\xi\varepsilon), \tag{82}$$

$$k = k_0(1 - 2\varepsilon)^{\frac{5}{2}}\{1 - 7\varepsilon + 2\xi\varepsilon(2 - 9\varepsilon)\}, \tag{83}$$

$$\frac{dk}{dp} = \frac{12 - 49\varepsilon}{3(1 - 7\varepsilon)}, \tag{84}$$

where ξ is a function of temperature and $3\varepsilon = \varepsilon_{ii}$; in (84), terms in ξ have been neglected.

Further reference to finite strain is made in §4.7.

2.8 Exercises

1   Prove the following elementary properties of Cartesian tensors (C.T.s).
   (a) $\delta_{ii} = 3$.
   (b) The sum or difference of two C.T.s of the same order is a C.T. of that order.
   (c) The product of a scalar and C.T. of any order is a C.T. of that order.
   (d) If u_i, v_i are vectors, then $u_i v_j$ and $\partial u_j/\partial x_i$ are C.T.s of second order.

   [Note that $\partial/\partial x_{k'} = (\partial x_i/\partial x_{k'})\partial/\partial x_i$.]

   (e) If t_{ij} is a C.T. of second order, then t_{ii} is a scalar. (This is an example of tensor *contraction*.)
   (f) The scalar product of two vectors, and the divergence of a vector, are scalars.
      (The scalar product of u_i and v_i is $u_i v_i$; the divergence of v_i (or div **v**) is $\partial v_i/\partial x_i$.)
   (g) If ϕ is a scalar, $\partial\phi/\partial x_i$ (or grad ϕ) is a vector.
   (h) If u_i and v_i are vectors, then $u_j\partial v_i/\partial x_j$ is a vector.
   (i) Any C.T. of the second order can be expressed as the sum of a symmetrical and an antisymmetrical C.T.

   $$t_{ij} = \tfrac{1}{2}(t_{ij} + t_{ji}) + \tfrac{1}{2}(t_{ij} - t_{ji}).$$

(j) If t_{ij} is symmetric, and w_{ij} antisymmetric, $t_{ij}w_{ij} = 0$.

2 (a) If $\xi_{ij} = \frac{1}{2}(\partial v_j/\partial x_i - \partial v_i/\partial x_j)$ and $\xi_i = \varepsilon_{ijk}\partial v_k/\partial x_j$, prove that (a) $\xi_i = \varepsilon_{ijk}\xi_{jk}$; (b) $\xi_{ij} = \frac{1}{2}\varepsilon_{ijk}\xi_k$.

(b) Using the rules of Cartesian tensors, prove that

   (i) $\mathbf{u} \times (\mathbf{v} \times \mathbf{w}) = (\mathbf{u}\cdot\mathbf{w})\mathbf{v} - (\mathbf{u}\cdot\mathbf{v})\mathbf{w}$.

   (ii) curl curl $\mathbf{v} = \text{grad div } \mathbf{v} - \nabla^2\mathbf{v}$ $[\nabla^2 = \partial^2/\partial x_i^2]$

   (iii) div grad $\phi = \nabla^2\phi$.

   (iv) div curl $\mathbf{v} = 0$.

   (v) curl grad $\phi = 0$.

(c) Prove

   (i) $\delta_{ij}\varepsilon_{ijk} = 0$.

   (ii) $\varepsilon_{ijk}\varepsilon_{ljk} = 2\delta_{il}$.

   (iii) $\varepsilon_{ijk}\varepsilon_{klm} = \delta_{il}\delta_{jm} - \delta_{im}\delta_{jl}$.

3 Prove from first principles that, in orthogonal curvilinear coordinates u_i,
   (a) the components of grad ϕ are

$$\frac{1}{h_i}\frac{\partial\phi}{\partial u_i} \quad (i = 1, 2, 3)$$

(b)

$$\text{div } v_i = \frac{1}{h_1 h_2 h_3}\left\{\frac{\partial}{\partial u_1}(v_1 h_2 h_3) + \text{two similar terms}\right\}$$

(c)

$$\nabla^2\phi = \frac{1}{h_1 h_2 h_3}\left\{\frac{\partial}{\partial u_1}\left(\frac{h_2 h_3}{h_1}\frac{\partial\phi}{\partial u_1}\right) + \text{two similar terms}\right\}.$$

4 Derive the elastic equations of motion (9) by an application of Gauss' divergence theorem to the motion of an elementary volume of matter under external forces and internal tractions.

5 Show that the strain-energy function W for a transversely isotropic body (about the x_3 axis) can be written as

$$2W = A(e_{11} + e_{22})^2 + Ce_{33}^2 + 2F(e_{11} + e_{22})e_{33} + L(e_{23}^2 + e_{31}^2) + N(e_{12}^2 - 4e_{11}e_{22}),$$

where A, C, F, L, N are elastic moduli (Love, 1945). When the medium is isotropic derive equations (64).

6 For the case of finite hydrostatic stress (i.e. complete symmetry), show from §2.7 that (22) and (81) yield $\varepsilon_{ii} = \theta - \theta^2/6 = 3\varepsilon$, say.

   From §2.2.3 prove that the density ρ is related to ε by

$$\rho = \rho_0(1 - 2\varepsilon)^{\frac{3}{2}},$$

where the zero suffix relates to the unstrained state.

7 From equation (5), §2.1.3, define the normal traction (a scalar) on a surface with normal v_j as

$$P_N = v_j p_j(v)$$

and the tangential (or shear) traction as P_S where

$$P_S^2 = p_j(v)p_j(v) - P_N^2.$$

Determine the direction v^j for which P_S is a maximum and the maximum value of P_S in terms of the principal stresses p_1, p_2, p_3, where $p_1 > p_2 > p_3$.

8 (a) A Kelvin–Voigt solid is stressed by a constant load. Determine the final strain and the exponential time-constant. On the removal of the load show that there is no permanent strain associated with the deformation.

(b) Discuss briefly the physical basis for a Bingham solid for which

$$P_{ij} = \eta U_{ij} + 2v\frac{dE_{ij}}{dt} \quad (i \neq k),$$

where

$$U_{ij} = \begin{bmatrix} 0 & 1 & 1 \\ 1 & 0 & 1 \\ 1 & 1 & 0 \end{bmatrix}.$$

What is the relation between such a solid and a viscous liquid?

(c) Two systems are subject to Bingham-type behaviour. Let one system be the Earth (the 'prototype'), the other a laboratory experiment (the 'model'). In the Earth $\eta \sim 4 \times 10^9$ c.g.s. units; in the laboratory a length reduction of 10^6 and a density reduction of 5 are required. Determine the values of η and v for the model material. (Note that the ratio of the gravitational accelerations may be taken as unity.)

9 The components of the stress tensor are

$$p_{ij} = \begin{bmatrix} 1 & 1 & 0 \\ 1 & -2 & 0 \\ 0 & 0 & 1 \end{bmatrix}.$$

Calculate the principal stresses and their corresponding directions and determine the maximum tangential traction and the direction of the normal to the surface with this traction.

3

Vibrations and waves

Before we develop seismological aspects of the elastic wave equation (2.34), it is useful here to discuss vibrations and waves to some extent. It is convenient also to set up certain results relevant to the theory of the seismograph.

3.1 Vibrations of systems with one degree of freedom

3.1.1 Simple harmonic motion
The simplest vibrating system is modelled in terms of a single coordinate x which is changing according to

$$\ddot{x} + \omega^2 x = 0, \tag{1}$$

where ω is a real constant, and dots denote differentiations with respect to time t. This is the case of (undamped) simple harmonic motion. The general solution of (1) is

$$x = A \cos \omega t + B \sin \omega t,$$

or

$$x = C \cos (\omega t + \varepsilon), \tag{2}$$

where A, B and C, ε are constants of integration, determinate if the initial values of x and $\dot{x}$ are known. The motion is (time) periodic, the *period* being $2\pi/\omega$; the *frequency* is $\omega/2\pi$; the *amplitude* is the coefficient C; and the *phase constant* is ε.

3.1.2 Damped vibrations
In any vibrating system, friction is present to a greater or less degree; this involves the presence of terms depending on velocities in the equations of motion, and causes *damping* of the motion. We shall consider only cases where the friction forces are proportional to first powers of velocities. The main features of such cases are illustrated in a study of the

51

equation

$$\ddot{x} + 2\lambda\omega\dot{x} + \omega^2 x = 0, \tag{3}$$

which (see chapter 9) corresponds to the free motion of an ideal seismograph, the coefficient λ being an index of damping.

Writing i for $\sqrt{-1}$, we add to (3) i times the equation obtained on replacing x by y in (3), obtaining

$$\ddot{z} + 2\lambda\omega\dot{z} + \omega^2 z = 0, \tag{4}$$

where $z = x + iy$. As a trial solution of (4), we put

$$z = K \exp(i\gamma t),$$

where K and γ are independent of t.

The trial solution satisfies (4) if γ is such that

$$(i\gamma)^2 + 2\lambda\omega i\gamma + \omega^2 = 0,$$

i.e. if

$$i\gamma = -(\lambda \pm [\lambda^2 - 1]^{\frac{1}{2}})\omega. \tag{5}$$

The coefficient K is left undetermined, and the two values of γ in (5) lead to two independent solutions of (4), the coefficients K_1, K_2 say (corresponding to K) being independent and (if initial conditions or their equivalent are not specified) arbitrary. Since (4) is linear in z, we may superpose these two solutions to obtain the general solution of (4), K_1 and K_2 being the two integration constants. We note that K_1, K_2 may be complex numbers.

The general solution of (3) is then deduced from the fact that x is the real part of z. It is clear from (5) that this solution will take different forms according as the damping coefficient λ is less than or greater than unity. We derive when $\lambda < 1$

$$\left.\begin{aligned}
x &= \{A\cos([1 - \lambda^2]^{\frac{1}{2}}\omega t) \\
&\quad + B\sin([1 - \lambda^2]^{\frac{1}{2}}\omega t)\} \exp(-\lambda\omega t), \\
\text{or} \qquad\qquad & \\
x &= C\exp(-\lambda\omega t)\cos([1 - \lambda^2]^{\frac{1}{2}}\omega t + \varepsilon);
\end{aligned}\right\} \tag{6}$$

and when $\lambda > 1$

$$\begin{aligned}
x &= A\exp\{-(\lambda + [\lambda^2 - 1]^{\frac{1}{2}}\omega t\} \\
&\quad + B\exp\{-(\lambda - [\lambda^2 - 1]^{\frac{1}{2}})\omega t\},
\end{aligned} \tag{7}$$

where A, B and C, ε are real constants of integration. When $\lambda = 1$, (5) yields only one distinct value of γ, but the general solution (containing two independent integration constants) may be deduced from either (6) or (7). Thus if, for instance, we expand in power series the cosine and sine terms in the first of (6) and replace $B\omega(1 - \lambda^2)^{\frac{1}{2}}$ by B, we then obtain for the case $\lambda = 1$

$$x = (A + Bt)\exp(-\omega t). \tag{8}$$

The solution (6) corresponds to vibratory motion of period $2\pi/\omega(1 - \lambda^2)^{\frac{1}{2}}$ and amplitude $C \exp(- \lambda\omega t)$; this amplitude diminishes steadily towards zero as time goes on. Neither of the solutions (7) and (8) gives vibratory motion; in both cases there can be at most one zero of x within a finite time, after which x creeps asymptotically to zero. Thus the motion is *periodic* if $\lambda < 1$, *aperiodic* if $\lambda \geq 1$. Note that if λ is only a little less than unity, the amplitude factor has already appreciably diminished by the time one half-period has elapsed, so that the practical result is not greatly different from aperiodicity. In connection with the theory of the seismograph, it is a useful exercise to compare the graphs of x against t for the cases $\lambda = 1$ and $\lambda = 1/\sqrt{2}$; it will be found that the difference in the values of x for these two cases will, if initially small, be always small, compared with the maximum value of x.

3.1.3 Forced vibrations

If a dynamical system like that just considered is further subject to variable external forces, additional terms will need to be included in equations such as (1) and (3). We shall here investigate the equation

$$\ddot{x} + 2\lambda\omega\dot{x} + \omega^2 x = b \cos pt, \tag{9}$$

where b and p are further given constants. The additional term $b \cos pt$ corresponds to the presence of a superimposed periodic force of period $2\pi/p$. Towards solving (9), we introduce y which satisfies the conjugate equation

$$\ddot{y} + 2\lambda\omega\dot{y} + \omega^2 y = b \sin pt,$$

so that, if again $z = x + iy$, we have

$$\ddot{z} + 2\lambda\omega\dot{z} + \omega^2 z = b \exp(ipt). \tag{10}$$

Putting $z = K \exp(ipt)$ as a trial solution of (10) gives

$$(- p^2 + 2i\lambda\omega p + \omega^2)K = b.$$

Thus a particular solution of (10) is

$$z = b \exp(ipt)/(\omega^2 - p^2 + 2i\lambda\omega p).$$

we can rewrite this in the form

$$z = b \exp\{i(pt - \delta)\}/\{(\omega^2 - p^2)^2 + 4\lambda^2\omega^2 p^2)^{\frac{1}{2}},$$

where

$$\delta = \tan^{-1}\{2\lambda\omega p/(\omega^2 - p^2)\}. \tag{11}$$

we then deduce that a particular solution of (9) is

$$x = b \cos(pt - \delta)/\{(\omega^2 - p^2)^2 + 4\lambda^2\omega^2 p^2\}^{\frac{1}{2}}, \tag{12}$$

where the *phase lag* δ is given by (11). The general solution of (9) is given by equating x to the sum of the *complementary function*, i.e. the right-hand side of (6), (7) or (8) obtained in the process of solving (3), and the *particular integral*, i.e. the right-hand side of (12).

It is evident that the complementary function (which has a real exponential factor in all cases) becomes increasingly less important as time goes on. The period of the ultimately dominant term (i.e. the particular integral), namely, $2\pi/p$, is that of the superimposed force. The phase lag δ depends on λ, ω and p. An important feature is that when the damping is slight (i.e. λ small), and p and ω are nearly equal, x as given by (12) can become great; this is the case of *resonance*.

3.1.4 The delta function

At an early stage in the theory of seismic disturbances, a mathematical way to handle impulses must be introduced. The modern procedure is to work with the Dirac delta function $\delta(t)$ and the related Heaviside step function $H(t)$.

These functions are, however, not ordinary ones since they are defined by the equations

$$\int_0^\infty \delta(t)f(t)\,\mathrm{d}t = f(0), \tag{13}$$

and

$$\frac{\mathrm{d}}{\mathrm{d}t}H(t) = \delta(t). \tag{14}$$

Rigorous justification for manipulating with functions of this kind is given by the theory of distributions (see M.J. Lighthill, 1960). We imagine $\delta(t)$ to represent a sequence of bell-shaped impulse functions centred at $t = 0$ which have progressively thinner and taller peaks with the area under the curve always equal to unity. In the limit, the sequence $\delta(t)$ has the 'sifting' property (13). Similarly, the step function can be considered as a limit of ordinary functions with the limiting properties

$$H(t) = \begin{cases} 0, & t < 0 \\ 1, & t \geqslant 0 \end{cases}.$$

Generally, so long as $f(t)$ in (13) is a well-behaved (everywhere differentiable) function, the operations of addition, differentiation, integration by parts and so on can be applied to generalized functions. Some important simple examples are given through the text. Here we note that, for an ordinary function $f(t)$,

$$\frac{\mathrm{d}}{\mathrm{d}t}\{f(t)H(t - t_0)\} = f'(t)H(t - t_0) + f(t)\delta(t - t_0). \tag{15}$$

The application of impulsive ground motion to seismographs can be modelled by the use of $\delta(t)$ and $H(t)$ (see chapter 9) and in chapter 16 we will see how simple earthquake sources can be constructed from them also.

3.1.5 Green's function

A more general dynamical problem in seismology than (9) is the application of an impulse to an elastic system.

We thus examine the inhomogeneous equation

$$p_0(t)\ddot{x} + p_1(t)\dot{x} + p_2(t)x = f(t), \quad t_0 < t < t_1 \tag{16}$$

and the corresponding homogeneous equation

$$p_0(t)\ddot{x} + p_1(t)\dot{x} + p_2(t)x = 0, \quad t_0 < t < t_1. \tag{17}$$

We assume that $p_0(t)$, $p_1(t)$ and $p_2(t)$ are continuous functions of t in (t_0, t_1). Appropriate boundary conditions will be considered later.

It turns out that the solution of the homogeneous equation (17) can be used to construct the solution of the inhomogeneous equation (16). The method is called Lagrange's *variation of parameters*. (See §9.5.3.)

Let $X_1(t)$ and $X_2(t)$ be known linearly independent solutions of (17). But, from the theory of ordinary differential equations, a necessary and sufficient condition for X_1 and X_2 to be linearly independent is

$$W(t) = X_1\dot{X}_2 - \dot{X}_1X_2 \neq 0, \tag{18}$$

where $W(t)$ is the Wronskian of X_1 and X_2.

Now assume that a solution of (16) may be found in the form

$$x(t) = u_1(t)X_1(t) + u_2(t)X_2(t),$$

where $u_1(t)$ and $u_2(t)$ are functions which we must construct. On differentiation,

$$\dot{x}(t) = u_1\dot{X}_1 + u_2\dot{X}_2 + \dot{u}_1X_1 + \dot{u}_2X_2. \tag{19}$$

Let us require that the unknown functions satisfy

$$\dot{u}_1X_1 + \dot{u}_2X_2 = 0, \tag{20}$$

so that (19) has only the first two terms. A second differentiation yields

$$\ddot{x}(t) = u_1\ddot{X}_1 + u_2\ddot{X}_2 + \dot{u}_1\dot{X}_1 + \dot{u}_2\dot{X}_2.$$

Substitution for $\dot{x}$ and $\ddot{x}$ into (16) gives

$$p_0(u_1\ddot{X}_1 + u_2\ddot{X}_2 + \dot{u}_1\dot{X}_1 + \dot{u}_2\dot{X}_2) + p_1(u_1\dot{X}_1 + u_2\dot{X}_2)$$
$$+ p_2(u_1X_1 + u_2X_2) = f.$$

In this equation the coefficient of u_1 is $p_0\ddot{X}_1 + p_1\dot{X}_1 + p_2X_1$, which vanishes because X_1 is a solution of (17). Similarly, the coefficient of u_2

vanishes, and we are left with

$$\dot{u}_1 p_0 \dot{X}_1 + \dot{u}_2 p_0 \dot{X}_2 = f. \tag{21}$$

The condition (18) ensures that (20) and (21) can be solved for $\dot{u}_1$ and $\dot{u}_2$. We find

$$\dot{u}_1 = \frac{-X_2}{X_1 \dot{X}_2 - \dot{X}_1 X_2} \frac{f}{p_0}, \quad \dot{u}_2 = \frac{X_1}{X_1 \dot{X}_2 - \dot{X}_1 X_2} \frac{f}{p_0}. \tag{22}$$

Integration thus gives u_1 and u_2, and hence

$$x(t) = X_1(t) \int^t \frac{-X_2(\sigma)}{W(\sigma)} \frac{f(\sigma)}{p_0(\sigma)} d\sigma + X_2(t) \int^t \frac{X_1(\sigma)}{W(\sigma)} \frac{f(\sigma)}{p_0(\sigma)} d\sigma. \tag{23}$$

The lower limit of each integral is free. In each case a change in the lower limit adds a constant multiple of $X_1(t)$ or $X_2(t)$ to the solution $x(t)$. Therefore, (23) is a solution of (16).

Because two boundary conditions must be accommodated, we add to (23) the terms $AX_1 + BX_2$ which are the general solution of the homogeneous equation (17).

We can best illustrate this general method by finding the solution of

$$\ddot{x} + \omega^2 x = f(t) \quad (\omega \text{ real and constant}, 0 < t < T), \tag{24}$$

which satisfies the boundary conditions $x(0) = 0$, $x(T) = 0$. This equation differs from (9) in having (for simplicity) no damping and a more general forcing function. We take $\cos \omega t$ and $\sin \omega t$ as independent solutions of the auxiliary equation. Then

$$W = \cos \omega t (\omega \cos \omega t) - (-\omega \sin \omega t) \sin \omega t = \omega,$$

so that (23) gives

$$x(t) = \cos \omega t \int_0^t \frac{(-\sin \omega \sigma)}{\omega} f(\sigma) d\sigma + \sin \omega t \int_0^t \frac{\cos \omega \sigma}{\omega} f(\sigma) d\sigma,$$

where arbitrarily the lower end-points are taken as zero. Hence, a particular solution of (24) is

$$x(t) = \frac{1}{\omega} \int_0^t \sin \omega(t - \sigma) f(\sigma) d\sigma. \tag{25}$$

Therefore, adding the solution of the inhomogeneous equation, the complete solution of (24) is

$$x(t) = C \cos \omega t + D \sin \omega t + \frac{1}{\omega} \int_0^t \sin \omega(t - \sigma) f(\sigma) d\sigma.$$

The boundary conditions give $C = 0$ and

$$D \sin \omega T + \frac{1}{\omega} \int_0^t \sin \omega(T - \sigma) f(\sigma) \, d\sigma = 0,$$

so that the required solution is

$$x(t) = -\frac{\sin \omega t}{\omega \sin \omega T} \int_0^T \sin \omega(T - \sigma) f(\sigma) d\sigma$$
$$+ \frac{1}{\omega} \int_0^t \sin \omega(t - \sigma) f(\sigma) d\sigma. \tag{26}$$

If the portion of the first integral from 0 to t is then combined with the second integral, we have

$$x(t) = -\int_0^t \frac{\sin \omega\sigma \sin \omega(T - t)}{\omega \sin \omega T} f(\sigma) d\sigma$$
$$- \int_t^T \frac{\sin \omega t \sin \omega(T - \sigma)}{\omega \sin \omega T} f(\sigma) d\sigma$$

$$= \int_0^T G(t, \sigma) f(\sigma) d\sigma, \tag{27}$$

where

$$G(t, \sigma) = \begin{cases} -\dfrac{\sin \omega\sigma \sin \omega(T - t)}{\omega \sin \omega T} & \text{if} \quad 0 < \sigma < t \\[4mm] -\dfrac{\sin \omega t \sin \omega(T - \sigma)}{\omega \sin \omega T} & \text{if} \quad t < \sigma < T. \end{cases}$$

In the important case when $f(t)$ is the delta-function at $t = \xi$, namely $\delta(t - \xi)$, then (27) becomes

$$x(t) = \begin{cases} -\dfrac{\sin \omega\xi \sin \omega(T - t)}{\omega \sin \omega T} & \text{if} \quad 0 < \xi < t < T, \\[4mm] -\dfrac{\sin \omega t \sin \omega(T - \xi)}{\omega \sin \omega T} & \text{if} \quad 0 < t < \xi < T, \end{cases}$$

$$= G(t, \xi). \tag{28}$$

This result has a very important physical interpretation which is often used in theoretical seismology (see chapter 16). We can see that $G(t, \sigma)$ is the system response to a unit impulse at $t = \sigma$. If this response is known, we can at once construct the response (27) to a more general disturbance $f(t)$. $G(t, \sigma)$ is called the *Green's function* for forced vibrations in this problem.

3.2 Vibrations of systems with more than one degree of freedom

We confine consideration to *holonomic* systems (i.e. systems for which all differential constraint relations can be integrated). Suppose that specification of the configuration of such a system requires knowledge of the value of n coordinates. These coordinates, $q_r (r = 1, 2,..., n)$, may be distances measured from fixed or moving points or lines, or may be angles or other functions of position. They are called *generalised coordinates*, and the system is said to have n *degrees of freedom*.

In investigating vibrations of such a system, we shall not in the remainder of this chapter include terms due to friction in the equations. Allowance for friction involves additional algebra, and the general effects of friction are quite similar to the damping effects found for the particular case of §3.1.2. When the friction is not too great, its presence does not alter the vibratory character of the motion, but entails the presence in the amplitudes of real exponential factors diminishing as t increases, as in (6). The presence of such factors is of course associated with dissipation of energy.

3.2.1 Eigen-vibrations of systems with finite freedom

We shall first take n finite and examine the small vibrations of the system about a stable equilibrium configuration. We take q_r all zero in this configuration. The kinetic energy T is expressible in the form

$$2T = \sum_r \sum_s a_{rs} \dot{q}_r \dot{q}_s, \tag{29}$$

where the coefficients a_{rs} will in general be functions of the q_r, but may, if we neglect terms of higher order than the second, be taken as constants for motion in the vicinity of the equilibrium configuration. In the absence of dissipative forces there exists a potential energy V. Since $\partial V / \partial q_r = 0$ for equilibrium, we may, expanding V in a Taylor series and neglecting terms of higher order than the second, write

$$2V = \sum_r \sum_s b_{rs} q_r q_s, \tag{30}$$

where the coefficients b_{rs} are constants. Without loss of generality, we may take $a_{rs} = a_{sr}$ and $b_{rs} = b_{sr}$. By Lagrange's equations

$$\frac{d}{dt}\left(\frac{\partial T}{\partial \dot{q}_r}\right) - \frac{\partial T}{\partial q_r} = -\frac{\partial V}{\partial q_r} \quad (r = 1, 2,..., n), \tag{31}$$

we then obtain, correct to the first order in the q_r, the n equations of motion

$$\sum_s a_{rs} \ddot{q}_s + \sum_s b_{rs} q_s = 0 \quad (r = 1, 2,..., n). \tag{32}$$

The equilibrium configuration being stable, we can show by taking a trial solution of the form

$$q_r = K_r \exp(i\gamma t) \quad (r = 1, 2, ..., n) \tag{33}$$

that the general solution of (32) is expressible in the real form

or

$$\left.\begin{array}{l} q_r = \sum_s A_{rs} \cos \gamma_s t + \sum_s B_{rs} \sin \gamma_s t \\[2mm] q_r = \sum_s C_{rs} \cos(\gamma_s t + \varepsilon_s) \end{array}\right\} \quad (r = 1, 2, ..., n). \tag{34}$$

In the second of (34), the γ_s, and also the ratios of the C_{rs} for any assigned s, are expressible in terms of the constants a_{rs} and b_{rs} in (32); but for each s one particular member of the corresponding C_{rs} may be taken independently of the a_{rs} and b_{rs}. The $2n$ constants of integration may be taken as the particular n members of C_{rs} that are taken independently of the a_{rs} and b_{rs}, together with the n phase constants ε_s.

The solution (34) may be regarded as a linear combination of n *normal modes* or *eigen-vibrations*, each mode corresponding to a particular value of the suffix s, and having the following properties: (i) the frequency is the same for every coordinate q_r and depends only on the coefficients a_{rs} and b_{rs}; (ii) the phase constant is the same for every q_r but is otherwise arbitrary; (iii) the ratio of the amplitudes of the various q_r is determined by a_{rs} and b_{rs}.

The system if suitably excited can vibrate in any one of the n normal modes; this is merely a matter of arranging the initial conditions so that all but the relevant two integration constants are zero. The frequencies in the n normal modes are called *eigen-frequencies* of the system. Points at which the displacement is permanently zero in a particular normal mode are called *nodes* of the mode.

A further important property of normal modes is that, in any particular mode, the mean values, over a period, of the kinetic and potential energies are equal. To show this, let us write a particular solution (extracted from (34)) corresponding to a normal mode in the form

$$q_r = C_r \cos(\gamma t + \varepsilon) \quad (r = 1, 2, ..., n). \tag{35}$$

On substituting from (35) into (29) and (30), respectively, we find that the corresponding mean kinetic and potential energies over a period are

$$\tfrac{1}{4}\gamma^2 \sum_r \sum_s a_{rs} C_r C_s, \quad \tfrac{1}{4} \sum_r \sum_s b_{rs} C_r C_s. \tag{36}$$

By substituting from (35) into the equations of motion (32), we find that the two expressions in (36) are equal.

3.2.2 Rayleigh's principle

If instead of (35) we substitute the expressions

$$q_r = c_r \cos(\mu t + \varepsilon) \quad (r = 1, 2, ..., n),$$

where c_r and $2\pi/\mu$ are not necessarily the actual amplitudes and period in a normal mode, and then equate the two expressions analogous to (36), we obtain the equation

$$\mu^2 \sum_r \sum_s a_{rs} c_r c_s = \sum_r \sum_s b_{rs} c_r c_s. \tag{37}$$

Rayleigh showed that if the c_r are taken approximately equal to the actual amplitudes C_r in a normal mode, then the value of $2\pi/\mu$ as obtained from (37) differs from the actual period $2\pi/\gamma$ of the mode by a small fraction which is of the order of the squares of the quantities $(C_r - c_r)/C_r$.

It follows that if we have sufficient knowledge in a particular problem to enable us to estimate approximate values of the amplitude ratios of a particular normal mode of vibration, we can then by (37) infer the period of this mode within a fractional error of the order of the squares of the errors in the trial amplitude ratios.

This principle can be made the basis of a rapid determination of the frequency of vibration in the lowest mode (see § 12.3.4).

3.2.3 Particles on an elastic string

In order to illustrate the theory of § 3.2.1, we shall consider vibrations of a light straight flexible elastic string of length l, fixed at both ends and having n particles, each of mass m, attached at equal intervals $h = l/(n + 1)$; the tension S will be taken to be uniform along the string. We shall discuss (in outline) the free vibrations of this system, neglecting gravity and friction, following the initial presence of small lateral coplanar displacements and velocities of the particles.

We take as generalised coordinates the n (lateral) displacements y_r of the particles. It is then easy to derive the equations of small motion of the system in the form

$$m\ddot{y}_r + S\left(\frac{y_r - y_{r-1}}{h} - \frac{y_{r+1} - y_r}{h}\right) = 0 \quad (r = 1, 2, ..., n), \tag{38}$$

where $y_0 = y_{n+1} = 0$. On substituting into (38) a trial solution of the form $y_r = K_r \exp(i\gamma t)$, corresponding to (33), we derive (after some algebra) the general solution of (38) in the form

$$y_r(t) = \sum_{s=1}^{n} \left(A_s \sin\frac{s\pi x_r}{l}\cos\gamma_s t + B_s \sin\frac{s\pi x_r}{l}\sin\gamma_s t\right) \quad (r = 1, 2, ..., n), \tag{39}$$

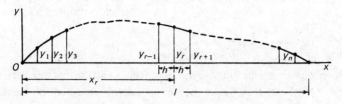

Fig. 3.1. Vibrating string.

where

$$\gamma_s = \left\{ \frac{2S}{mh}\left(1 - \cos\frac{s\pi}{n+1}\right)\right\}^{\frac{1}{2}} \quad (s = 1, 2, ..., n); \tag{40}$$

in these equations, x_r is the distance of the rth particle from the end O of the string (Fig. 3.1), and the A_s and B_s are $2n$ integration constants. There is one normal mode corresponding to each particular value of s, the associated eigen-frequency being $\gamma_s/2\pi$. The phase in the sth mode is determined solely by the particular value of A_s/B_s; this verifies that the phase is the same for each coordinate y_r in a particular mode. We verify also that the ratio of the amplitudes of the y_r in the sth mode is independent of the integration constants.

Particular values of the $2n$ integration constants A_s, B_s may be determined if the initial displacements and velocities are known, for we immediately derive from (39) the $2n$ equations

$$\left. \begin{aligned} y_r(0) &= \sum_s A_s \sin\frac{s\pi x_r}{l} \\ \left(\frac{\mathrm{d}y_r}{\mathrm{d}t}\right)_{t=0} &= \sum_s B_s\gamma_s \sin\frac{s\pi x_r}{l} \end{aligned} \right\} \quad (r = 1, 2, ..., n), \tag{41}$$

which are precisely sufficient for this purpose.

3.2.4 Vibrations of continuous systems

The theory described above may be extended to a vibrating continuous system, the number of degrees of freedom being now infinite. We shall make use of the problem of § 3.2.3 to illustrate this case (again only in outline), letting the number n of the particles increase indefinitely in such a way that the total mass remains finite; we then have the problem of the transverse vibrations of a continuous massive string of length l and uniform line-density, ρ say, fixed at its end-points.

Application of a limit operation to the finite set of ordinary differential equations (38) gives in place of (38) the single partial differential equation

$$\rho \frac{\partial^2 y}{\partial t^2} - S \frac{\partial^2 y}{\partial x^2} = 0, \tag{42}$$

in which x and y are continuous variables replacing the discrete variables x_r and y_r. Note that (42) may be independently derived by setting up expressions for the kinetic and potential energies and then using Hamilton's principle.

Application of the limit operation to the solution (39) gives

$$y(x,t) = \sum_{s=1}^{\infty} \left(A_s \sin \frac{s\pi x}{l} \cos \gamma_s t + B_s \sin \frac{s\pi x}{l} \sin \gamma_s t \right), \tag{43}$$

where, from (40),

$$\frac{\gamma_s}{2\pi} = \frac{s}{2l} \left(\frac{S}{\rho} \right)^{\frac{1}{2}}. \tag{44}$$

We shall later (§ 3.3) show how to derive this solution directly from (42). The solution (43) is seen to be equivalent to the effect of superposing an infinite number of normal modes, each possessing properties as in the case of finite freedom. The eigen-frequencies are given by (44).

As in the case of finite freedom, the values of all the A_s, B_s can be determined if the initial velocities and displacements of all points of the string are known. By (43),

$$\left. \begin{aligned} y(x,0) &= \sum_{s=1}^{\infty} A_s \sin \frac{s\pi x}{l}, \\ \frac{\partial y}{\partial t}(x,0) &= \sum_{s=1}^{\infty} B_s \gamma_s \sin \frac{s\pi x}{l}. \end{aligned} \right\} \tag{45}$$

Using the method of Fourier, we obtain from the first of (45)

$$\int_0^l y(x,0) \sin \frac{r\pi x}{l} dx = \sum_{s=1}^{\infty} A_s \int_0^l \sin \frac{s\pi x}{l} \sin \frac{r\pi x}{l} dx. \tag{46}$$

Since $\int_0^\pi \sin s\xi \sin r\xi \, d\xi$ is equal to zero if $r \neq s$, and equal to $\frac{1}{2}\pi$ if $r = s$, we then obtain

$$\int_0^l y(x,0) \sin \frac{r\pi x}{l} dx = A_r \int_0^l \sin \frac{r\pi x}{l} \sin \frac{r\pi x}{l} dx = \frac{1}{2} A_r l.$$

Hence

$$A_s = \frac{2}{l} \int_0^l y(x,0) \sin \frac{s\pi x}{l} dx. \tag{47}$$

Similarly, using the second of (45) and (44),

$$B_s = \frac{2}{s\pi}\left(\frac{\rho}{S}\right)^{\frac{1}{2}} \int_0^l \frac{\partial y}{\partial t}(x, 0) \sin\frac{s\pi x}{l}\,\mathrm{d}x. \tag{48}$$

3.2.5 Seismological considerations

The normal mode theory discussed in the preceding subsections forms the basis of one line of investigation of earthquake phenomena; for, following the occurrence of an earthquake, the Earth may be considered as a vibrating continuous system described by (infinite) sets of eigenvalues and eigenvectors.

A second line of investigation makes use of the special feature that the initial disturbance in an earthquake is confined to a comparatively small part of the whole region traversed by the ensuing disturbances. In these circumstances, properties of equations analogous to (42) may be developed by a method somewhat different from that so far discussed; this development, which is called wave theory, will be considered in §§ 3.3–3.5.

Vibration theory based on normal mode considerations and wave theory are of course quite complementary or *dual*; either method will lead to the same results, but in particular problems one method may lead to results more readily than the other.

The complementary character of the two methods will be illustrated in a simple example to be discussed in the third paragraph of § 3.3 and in § 3.3.1. As another example, Rayleigh pointed out that Rayleigh surface waves (discussed in chapter 5) are included in Lamb's theory of the vibrations of an elastic sphere (based on the normal mode method); actually Bromwich later deduced from Lamb's theory the equation (5.16) for the velocity of Rayleigh waves. Some further direct applications of normal mode theory may be read in Love (1911).

Normal mode theory is often the direct source of elegant results in seismological theory. An example is the use that has been made by Jeffreys, Lapwood and others of Rayleigh's principle (§ 3.2.2) in the theory of seismic surface waves (see chapter 5). The use of the principle makes possible the estimation of approximate answers to problems that would be very difficult to solve by ordinary methods. (See also chapter 12.)

Another instance is the use that has been made of a reciprocal theorem in the theory of vibrations (see § 4.2.3). Suppose that the application of an impulse at one point P in a dynamical system causes a vibration in which the velocity at some second point Q has the component v in a particular normal mode. The theorem then states that *if the same impulse be instead applied at Q, it will produce at P the same component v in the normal mode in question.* As a corollary, it follows that *if an impulse is applied at a node of any normal mode, then this particular normal mode will be missing from the ensuing vibrations*; if the point of application be near such a node, the amplitude of the normal mode in question will be abnormally small.

Now, as we shall see in chapter 5, the amplitudes of seismic surface waves (which may be regarded as equivalent to superposed normal modes of vibration) decrease exponentially with depth below the Earth's surface. It then follows by the last paragraph that, if an earthquake originates at a greater depth than normal, then the surface waves ensuing should be less marked than in the normal case – in striking agreement with observations. (In fact at an early stage of seismology, the virtual absence of observations of surface waves from some earthquakes was shown by Jeffreys and Stoneley to be crucial evidence of the occurrence of 'deep-focus' earthquakes – see §15.3.2).

3.3 Plane waves

We now proceed to a discussion of the one-dimensional wave equation

$$\frac{\partial^2 y}{\partial t^2} = c^2 \frac{\partial^2 y}{\partial x^2}, \tag{49}$$

where c is an assigned constant, t denotes the time, x is a rectangular Cartesian coordinate, and y is a function of x and t which represents a disturbance of some physical quantity; in the seismological applications, y will denote a component of displacement.

We can readily show (by changing to new independent variables $x - ct$, $x + ct$, respectively) that the general solution of (49) is

$$y(x, t) = f(x - ct) + F(x + ct), \tag{50}$$

where the forms of f and F are restricted only by initial and boundary conditions. If t is increased by any value, Δt say, and simultaneously x by $c\Delta t$, the value of $f(x - ct)$ is clearly not altered; hence the first term on the right-hand side of (50) represents (see Fig. 3.2) a disturbance advancing unchanged in form in the positive x-direction with speed $c\Delta t/\Delta t$, i.e. c. At any instant, y depends only on x, and so is the same at all points of any plane normal to the x-axis; thus $f(x - ct)$ represents an advancing *plane wave*. The term $F(x + ct)$ clearly represents a plane wave proceeding in the negative direction.

The discussion illustrates that waveforms given by equation (50) such as

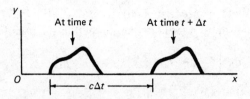

Fig. 3.2. Travelling pulse.

water waves in a stream or plane waves propagate without change of form, i.e. they leave no 'wake' behind them. This behaviour should be compared with that given below for cylindrical and spherical waves.

The equation (42) obtained for the vibrating continuous string is a particular case of (49), the speed c being $(S/\rho)^{\frac{1}{2}}$. The solution (43) must therefore be deducible from the form (50). The deduction is as follows. The particular boundary condition $y(0, t) = 0$ when applied to (50) gives $0 = f(-ct) + F(ct)$ for all t; hence $f(\xi) = -F(-\xi)$, and so $y(x, t) = F(x + ct) - F(-x + ct)$. The other particular boundary condition $y(l, t) = 0$ then gives $0 = F(l + ct) - F(-l + ct)$ for all t; hence $F(\xi + 2l) = F(\xi)$, which shows that F is a periodic function of period $2l$. It follows from Fourier theory that $F(\xi)$ is then expressible in the form

$$F(\xi) = \frac{1}{2} \sum_{s=1}^{\infty} A_s \sin \frac{s\pi\xi}{l} - \frac{1}{2} \sum_{s=0}^{\infty} B_s \cos \frac{s\pi\xi}{l},$$

where the A_s, B_s are constants (the fractions $\frac{1}{2}$ and the minus sign have been included with a view to making the A_s, B_s correspond exactly to the A_s, B_s in (43)). Hence

$$y(x, t) = \frac{1}{2} \sum_{s=1}^{\infty} A_s \left\{ \sin\left(\frac{s\pi x}{l} + \frac{s\pi ct}{l} \right) + \sin\left(\frac{s\pi x}{l} - \frac{s\pi ct}{l} \right) \right\}$$
$$- \frac{1}{2} \sum_{s=0}^{\infty} B_s \left\{ \cos\left(\frac{s\pi x}{l} + \frac{s\pi ct}{l} \right) - \cos\left(\frac{s\pi x}{l} - \frac{s\pi ct}{l} \right) \right\}$$
$$= \sum_{s=1}^{\infty} \left(A_s \sin \frac{s\pi x}{l} \cos \frac{s\pi ct}{l} + B_s \sin \frac{s\pi x}{l} \sin \frac{s\pi ct}{l} \right),$$

which agrees with (43).

3.3.1 Fourier's integral theorem and spectra

According to this theorem, a function $f(x)$ (subject to very few restrictions, which we shall ignore) may be expressed in the form

$$f(x) = \frac{1}{\pi} \int_0^{\infty} d\kappa \int_{-\infty}^{\infty} f(\eta) \cos\{\kappa(x - \eta)\} \, d\eta \qquad (51)$$

(see Whittaker and Watson, *Modern Analysis*, §9.7 (CUP, 1927)). This enables us to write the solution (50) of equation (49) in the useful form

$$y(x, t) = \frac{1}{\pi} \int_0^\infty d\kappa \int_{-\infty}^\infty d\eta \, [f(\eta) \cos \{\kappa(x - ct - \eta)\}$$
$$+ F(\eta) \cos \{\kappa(x + ct - \eta)\}]. \tag{52}$$

We notice that (52) indicates the connection between the equation (49) and the normal modes discussed in § 3.2.1, and the solution (43) for the vibrating continuous string may be constructed using (52). The first boundary condition $y(0, t) = 0$ is satisfied if the expression

$$\{f(\eta) + F(\eta)\} \cos \kappa\eta \cos \kappa ct - \{f(\eta) - F(\eta)\} \sin \kappa\eta \sin \kappa ct$$

is zero for all t; this will be the case (i) if $\cos \kappa\eta = 0$ and $f(\eta) = F(\eta)$, or (ii) if $\sin \kappa\eta = 0$ and $f(\eta) = -F(\eta)$. The second boundary condition, $y(l, t) = 0$, is then satisfied (i) if $2F(\eta) \sin \kappa l \sin \kappa\eta \cos \kappa ct$, or (ii) if $2F(\eta) \sin \kappa l \cos \kappa\eta \sin \kappa ct$, is zero for all t; i.e. if

$$\sin \kappa l = 0. \tag{53}$$

The equation (53) restricts values of κ to the discrete set $\kappa_s = s\pi/l$, where s is an integer, and implies that κ is not a continuous variable; from (i) $\cos \kappa\eta = 0$ or from (ii) $\sin \kappa\eta = 0$, it is implied that η also is discontinuous. Thus due to the boundary conditions in this problem, the form (52) must degenerate from an integral to an infinite series; it is in fact easy (superposing results given separately by (i) and (ii)) now to construct the form (43) previously obtained. We note that the *characteristic equation* (53) yields a discrete set of values of κ and hence also of the eigen-frequencies. This type of result, due to the special boundary conditions, is sometimes described as a discrete *spectrum* of frequencies; in the particular problem the frequencies are given by equation (44).

In seismology, we shall be particularly concerned with the case when the form (52) does not degenerate to a series. In this case, κ being a continuous variable, there exists a continuous set of values of the eigen-frequencies, i.e. a *continuous spectrum*.

3.3.2 Simple harmonic plane wave

The form (52) may be regarded as the result of superposing elementary waves of the form

$$y = A \cos \{\kappa(x - ct) + \varepsilon\}, \tag{54}$$

in which A, κ, ε vary from element to element (together with waves of similar form travelling in the opposite direction). The equation (54), which

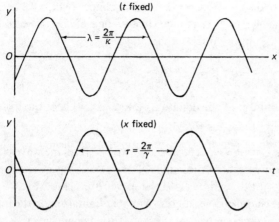

Fig. 3.3. Simple harmonic wave.

represents a *simple harmonic* (or *sinusoidal*) *advancing plane wave*, is thus fundamental in wave theory; we easily verify of course that (54) is a particular solution of (49). If we write (54) as

$$y = A \cos\left\{2\pi\left(\frac{x}{\lambda} - \frac{t}{\tau}\right) + \varepsilon\right\}, \tag{55}$$

we see that at any given point y oscillates with (time) *period* τ; while at any instant, y is a periodic function of x, the periodic distance being λ, the *wavelength*. We obviously have the relations

$$\lambda = 2\pi/\kappa, \quad \tau = 2\pi/\kappa c, \quad \lambda/\tau = c, \quad f = 1/\tau, \tag{56}$$

f being the *frequency* and κ the *wave number*. For mathematical convenience, we shall use instead of (55) the relation

$$y = A \exp\{i(\kappa(x - ct) + \varepsilon)\}, \tag{57}$$

where we understand that the real part is to be taken when physical interpretation is being made. We shall frequently write ω or γ for κc; then $2\pi/\omega$ is the period. Also, we write $\omega = 2\pi f$, with ω in radians and f in hertz.

3.3.3 Vector waves. Polarisation

Many physical problems are concerned with the transmission through some medium of a disturbance that is vectorial in character. In such a case we may meet a set of three equations each of the form (49), namely,

$$\frac{\partial^2 u_i}{\partial t^2} = c^2 \frac{\partial^2 u_i}{\partial x_1^2} \quad (i = 1, 2, 3), \tag{58}$$

where the u_i are the components of the disturbance (which we here take to be a displacement), and x_1 is a particular one of the three rectangular coordinates x_i. Suppose further that

$$\frac{\partial u_i}{\partial x_2} = \frac{\partial u_i}{\partial x_3} = 0 \quad (i = 1, 2, 3). \tag{59}$$

The general solution of the equations (58) and (59) is of the form

$$u_i = f_i(x_1 - ct) + F_i(x_1 + ct) \quad (i = 1, 2, 3), \tag{60}$$

and corresponds to the superposition of two vector plane waves travelling in the positive and negative directions of the 1-axis, respectively, with speed c. On account of the relations (59) the waves are plane waves.

It commonly happens that there is a further limitation on the freedom of variation of the components u_i of the transmitted disturbance; when this is so the disturbance is said to be *polarised*. If, for instance, the u_i constitute a *solenoidal vector* (i.e. div(u_i) vanishes at all times and places), it follows from (60) and (2.22) that $\partial u_1/\partial x_1$ is zero at all times and places; thus only the components u_2, u_3 are associated with the wave transmission, i.e. the disturbance is in this case restricted to being always at right angles to the 1-axis.

Consider the particular case of polarisation given by

$$u_i = A_i \cos\{\kappa(x_1 - ct) + \varepsilon_i\} \quad (i = 1, 2, 3),$$

with $A_1 = 0$. We derive from this

$$\frac{u_2^2}{A_2^2} + \frac{u_3^2}{A_3^2} - \frac{2u_2 u_3}{A_2 A_3}\cos(\varepsilon_3 - \varepsilon_2) = \sin^2(\varepsilon_3 - \varepsilon_2),$$

which shows that in this case the particles of the medium will in general describe ellipses in planes normal to the direction of propagation. In this case the disturbance is *elliptically polarised*; the particles will describe ellipses in senses corresponding to the right-hand or left-hand screw law according as the phase difference $\varepsilon_3 - \varepsilon_2$ of the two components u_2 and u_3 lies between 0 and π or between 0 and $-\pi$; the former case is illustrated in

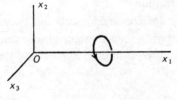

Fig. 3.4. Circular polarisation.

Fig. 3.4. If it should happen that $|\varepsilon_3 - \varepsilon_2| = \frac{1}{2}\pi$ and $A_2 = A_3$, the ellipses are circles and the disturbance is *circularly polarised*. If the ellipses degenerate to straight lines (which will all be parallel to a particular plane through the 1-axis), the disturbance is *plane polarised*; this will be the case for instance when one of A_2 or A_3 is zero, or when $\varepsilon_2 = \varepsilon_3$.

3.3.4 Standing waves

If we superpose a pair of simple harmonic plane waves of the same amplitude A, wave-length $2\pi/\kappa$ and speed c, travelling in opposite directions, the resultant disturbance is given by

$$y = 2A \cos(\kappa x + \varepsilon) \cos(\kappa ct + \varepsilon'), \tag{61}$$

where ε, ε' depend on the phase constants of the constituent waves. The equation (61) corresponds to what is called a *standing wave*, since there is no resultant progressive wave motion. It will be noticed that, in (61), x and t occur in separate factors, and that the disturbance is permanently zero at points where $\kappa x + \varepsilon = (r + \frac{1}{2})\pi$, r being any integer, and is a maximum where $\kappa x + \varepsilon = r\pi$; the former points are *nodes* and the latter *antinodes*.

The disturbance corresponds to the superposition of a pair of simple harmonic waves. In more general cases, a standing wave may take the form

$$y = X(x)T(t), \tag{62}$$

where X and T are any functions of x and t respectively. It is easy to show that general plane wave motion may be regarded as the result of superposing standing waves of the form (62). In fact, a further method of solving the wave equation (49) is to start by substituting the trial form (62) into (49).

In general the result of superposing a set of standing waves is not itself a standing wave, since the sum of a number of terms like that in (62) is not in general expressible as a product of two factors containing x alone and t alone.

3.3.5 Dispersion of waves

The preceding discussion of wave theory was developed from (49) in which c is a constant; and we saw that the disturbance is transmitted in the direction of the x-axis without change of form.

In problems where the equations of motion of a disturbance are more complicated than (49), substitution of the form (57) leads to a solution of the form (52), with the qualification that c varies with κ. A simple illustration is provided by modifying the vibrating string problem of §3.2.4 so that each element of the string is now subject to an additional applied force

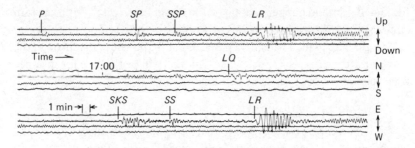

Fig. 3.5. Three components of ground motion recorded by a WWSSN intermediate-period seismograph at Berkeley 88° from a New Ireland earthquake on 7 May 1972. The symbols LR and LQ denote Rayleigh and Love waves, respectively.

proportional to the displacement. The equation of motion replacing (42) is found to be

$$\rho \frac{\partial^2 y}{\partial t^2} = S \frac{\partial^2 y}{\partial x^2} - hy, \tag{63}$$

where h is a new constant. The trial substitution (57) clearly satisfies

$$\rho c^2 \kappa^2 = S\kappa^2 + h, \tag{64}$$

and we can readily build up a solution of the form (52); but on account of (64), we see that c is now a function of κ. Another illustration will be found in §5.4 on Love waves.

Such dependence of c on κ (and thus by (56) on the wave-length and period) implies that the shape of the disturbance will in general continually change as time goes on, since each simple harmonic constituent represented in (52) will now travel with a wave velocity special to itself. If the initial disturbance is confined to a finite range of values of x and the medium is unlimited, it follows that as time goes on there will be a continual spreading out of the disturbance into trains of waves. This phenomenon, called *dispersion*, is treated below. An example of dispersed *LR* seismic waves is given in Fig. 3.5.

3.3.5.1 Generally in seismology, we are considering not the superposition of two waves, as in §3.3.4, but infinitely many. We use the Fourier integral (52) in the form

$$y(x, t) = \int_0^\infty F(\kappa) \exp \{i[\kappa x - \gamma(\kappa)t]\} \, d\kappa, \tag{65}$$

to allow a thorough analysis of a dispersive system. It should be noted that a similar integral, as in (50), for waves travelling in the negative x-direction should be added for a general solution of (49), but this is omitted because we are here concerned with asymptotic properties of the general form.

We write the phase in (65) as

$$\psi(\kappa) = \kappa x/t - \gamma(\kappa),$$

so that

$$y(x,t) = \int_0^\infty F(\kappa)\exp\left[it\psi(\kappa)\right]d\kappa. \tag{66}$$

Physically, interference of individual sinusoidal components will tend to give mutual cancellation when the waves are out of phase and mutual reinforcement when they are in phase. For large t, the main contribution to the wave form may therefore be expected when the phase $\psi(\kappa)$ is stationary, i.e., when

$$d\gamma/d\kappa = x/t = C, \quad \text{say.} \tag{67}$$

The interpretation is that, as t becomes large, the gaze of an observer follows a wave group at a fixed velocity x/t called the *group velocity*. The discussion suggests further that because, for linear systems, energy cannot transfer from one part of the spectrum to another, the wave energy in the dispersed waves is also transported with the group velocity x/t. A full analysis is given in the book by James Lighthill. An elementary evaluation of (66) for dispersed waves and diffraction is given in §3.7.

3.3.5.2 Summing up the foregoing discussion, we have that the predominant effect at time t (sufficiently large) and place x arises in general from the cluster with wave-lengths near $2\pi/\kappa$, where κ is the solution of (67). It follows that as time goes on, the original disturbance is continually sorting itself out into groups of simple harmonic waves, each group being associated with a particular wave-length, and travelling forward with its group velocity

$$C(\kappa) = d\gamma/d\kappa = d(\kappa c)/d\kappa. \tag{68}$$

We note incidentally from (68) that when there is dispersion the group velocity C is in general different from the wave velocity c and

$$c + \kappa dc/d\kappa = C. \tag{69}$$

3.3.5.3 For the modified vibrating string problem referred to earlier in §3.3.5, we find from (64) and (68) that the group velocity is

$$C = S/c\rho = S\kappa/\gamma\rho. \tag{70}$$

3.3.6 Energy in plane wave motion

At any instant, the kinetic and potential energies of the vibrating system (or medium) between the planes $x = x'$ and $x = x''$, due to the displacements associated with a wave of the simple type described at the start of §3.3, are expressible in the forms

$$\tfrac{1}{2}a \int_{x'}^{x''} \left(\frac{\partial y}{\partial t}\right)^2 dx, \quad \tfrac{1}{2}b \int_{x'}^{x''} \left(\frac{\partial y}{\partial x}\right)^2 dx, \tag{71}$$

respectively, where a, b are constants. (As an exercise, this may be verified for the case of the vibrating continuous string, §3.2.4.) By Hamilton's principle, we may then deduce that

$$a\frac{\partial^2 y}{\partial t^2} = b\frac{\partial^2 y}{\partial x^2}, \tag{72}$$

giving the wave velocity c equal to $(b/a)^{\frac{1}{2}}$.

If we substitute $f(x - ct)$ for y in the expressions (71), we find that the results are equal. It follows that in an advancing (or receding) plane wave, the energy at any instant is half kinetic and half potential. Note the correspondence between this conclusion and the result on mean energies stated in the last paragraph of §3.2.1.

If in (71) we replace y by ky, where k is any constant, we see that the energies are multiplied by k^2. Thus for sets of waves of similar shape which differ only in the magnitudes of the amplitudes, the energies are proportional to the squares of corresponding amplitudes.

In some problems, it is convenient to make use of the *energy density, w* say, i.e. the energy per unit volume in the medium. In the absence of energy dissipation, we can then write down a conservation equation analogous to (2.23), namely,

$$\frac{\partial w}{\partial t} + \frac{\partial}{\partial x_i}(wv_i) = 0, \tag{73}$$

where v_i is the velocity of energy transmission.

The one-dimensional form of (73) is

$$\frac{\partial w}{\partial t} + \frac{\partial}{\partial x}(wv) = 0, \tag{74}$$

where w is now the energy per unit length in the direction of propagation of the wave.

As an instance of the use of this equation, we can verify that when the wave disturbance is transmitted unchanged in form, energy is conveyed through the medium with the wave velocity c. To demonstrate this, we have by (71) for the case in question

$$w = \tfrac{1}{2}a\left(\frac{\partial y}{\partial t}\right)^2 + \tfrac{1}{2}b\left(\frac{\partial y}{\partial x}\right)^2. \tag{75}$$

Hence

$$\frac{\partial w}{\partial t} = a\frac{\partial y}{\partial t}\frac{\partial^2 y}{\partial t^2} + b\frac{\partial y}{\partial x}\frac{\partial^2 y}{\partial t}$$

$$= b\frac{\partial y}{\partial t}\frac{\partial^2 y}{\partial x^2} + b\frac{\partial y}{\partial x}\frac{\partial^2 y}{\partial x \partial t} \quad \text{(by 72)}$$

$$= -\frac{\partial}{\partial x}\left(-b\frac{\partial y}{\partial t}\frac{\partial y}{\partial x}\right).$$

Thus by (74) the speed of energy flow is $-\{b(\partial y/\partial t)(\partial y/\partial x)\}/w$; on putting $y = f(x - ct)$ in this expression, we find that this speed is equal to c.

When there is dispersion, the expression (75) for the energy density needs modification. For instance, in the case of the modified string problem considered in §3.3.5, we have in place of (75)

$$w = \tfrac{1}{2}\rho\left(\frac{\partial y}{\partial t}\right)^2 + \tfrac{1}{2}S\left(\frac{\partial y}{\partial x}\right)^2 + \tfrac{1}{2}hy^2. \tag{76}$$

By (63), (74) and (76), it is then easy to show that the local speed of transmission of energy at time t is $\{-S(\partial y/\partial t)(\partial y/\partial x)\}/w$, which is similar to the corresponding result in the last paragraph.

3.3.7 Propagation of plane waves in a general direction

We have previously considered the propagation of plane waves in the direction of one of the axes of reference. If instead the direction of propagation has direction-cosines $l_i\,(i = 1, 2, 3)$, we obtain from (50), (54) and (57) by transforming axes the relevant corresponding equations

$$y = f(l_i x_i - ct) + F(l_i x_i + ct), \tag{77}$$

$$y = A\cos\{\kappa(l_i x_i - ct) + \varepsilon\}, \tag{78}$$

$$y = A\exp\{i(\kappa(l_i x_i - ct) + \varepsilon)\}, \tag{79}$$

the summation convention being understood.

3.4 The wave equation

There are numerous equations associated with wave propagation. One of these, *the* (classical) wave equation, is of fundamental importance, namely,

$$\frac{\partial^2 y}{\partial t^2} = c^2 \mathbf{V}^2 y, \qquad (80)$$

where c is constant and $\mathbf{V}^2 \equiv \partial^2/\partial x_i^2$. Equation (49) is the one-dimensional form of (80).

It will be seen on substitution that each of (77), (78), (79) is a particular solution of (80) for all directions l_i. Thus a linear combination of plane waves of any form travelling in any direction with the same speed c and not changing in form constitutes a solution of (80).

3.4.1 Case of spherical symmetry

It is easy to derive the most general solution of (80) subject to the restriction that y is symmetrical about some centre O. If r denotes the distance from O, we can deduce from (80) that in this case

$$\frac{\partial^2 (ry)}{\partial t^2} = c^2 \frac{\partial^2 (ry)}{\partial r^2}, \qquad (81)$$

of which the general solution is seen by comparison with (49) and (50) to be of the form

$$y = r^{-1}\{f(r - ct) + F(r + ct)\}. \qquad (82)$$

The result (82) differs from the one-dimensional result (50) only in that r replaces x and that there is an additional amplitude factor r^{-1}. The solution (82) thus corresponds to the transmission of a *spherical wave*, the value of y at any fixed instant being the same at all points on the surface of any sphere of centre O. In many physical applications, the term in F is not relevant; we are then left with the solution $y = r^{-1}f(r - ct)$, which corresponds to a spherical wave advancing outward with speed c, the amplitude being inversely proportional to the distance from P. We note that there is at any instant a spherical *wave front* and, like plane waves, spherical waves do not leave a wake behind them.

3.4.2 General solution

A form of the general solution of the wave equation (80) has been obtained by Kirchhoff.

The solution gives an expression for $y_P(t)$, the value of y at time t at any point P, in terms of the circumstances existing at certain previous instants at points of any arbitrary closed surface S surrounding P. For convenience we use square brackets here to denote that a function is to be evaluated at a point Q of S at the instant $t - r/c$, where r is the distance PQ. The result is then that

$$y_P(t) = \frac{1}{4\pi} \int \int \left\{ \frac{1}{cr} \frac{\partial r}{\partial v} \left[\frac{\partial y}{\partial t} \right] - [y] \frac{\partial}{\partial v} \left(\frac{1}{r} \right) + \frac{1}{r} \left[\frac{\partial y}{\partial v} \right] \right\} dS, \tag{83}$$

where the integration is taken over the surface S, and δv is an element of the outward normal at a point of S.

For any point Q of S, it is seen that the instant at which the square-bracket functions are to be evaluated precedes the instant t by PQ/c, which is the time a disturbance would take to travel in a straight line from Q to P with speed c. In the right-hand side of (83), y occurs only inside square brackets, and it follows that $y_P(t)$ is fully determined as the resultant of appropriate effects travelling with speed c from all points of S towards P. This representation is used by J.A. Hudson (1980).

It is now seen further that any disturbed small region may be regarded as a *secondary source* transmitting a secondary disturbance in all directions. This corresponds to *Huygens's principle* in optics, and the subsequent superposition of effects at any point P. The results of course apply also to elastic waves, and not merely to optics.

3.4.3 Ray theory

We return now to the propagation of waves in three dimensions, and introduce the concept of *rays*.

Consider the transmission of waves through a homogeneous medium, following an initial disturbance that is symmetrical about a centre O and is confined to within a small sphere surrounding O. If c is the speed of wave transmission, it follows from (82) that after the lapse of time t, the disturbance will be confined to the vicinity of the surface of a sphere of centre O and radius ct. Now consider what happens as time goes on within a cone of small solid angle, or *pencil*, whose vertex is at O.

Looking at the matter from the point of view of Huygens's principle given in § 3.4.2, we see that all the secondary disturbances must be interfering with one another to such an extent that a major part of the whole disturbance is being transmitted with speed c in the direction of the cone's axis. It follows that the circumstances are essentially equivalent to rectilinear propagation of the disturbance along *rays* as in geometrical optics, each ray proceeding normally outwards from the wave front at any instant.

When, as in the case of the Earth, the properties of the medium vary from point to point, the wave velocity is a function of position and not a constant. In this case, the wave front at time t after an initially confined disturbance will not in general be a sphere, even if the disturbance is symmetrical about a centre. But use may still be made of the concept of rays, the paths of which cut wave fronts at right angles; the rays will now, of course, be curved.

If in the vicinity of such a region there is a surface of discontinuity (the radius of curvature of which is sufficiently large), on opposite sides of which the wave velocity has different but constant values, plane wave theory may be used to determine the character of the reflected and refracted waves. This process is carried out in chapter 6 for elastic waves, and it is shown that laws hold which are analogous to the laws of reflection and refraction in geometrical optics.

In all cases, it may be shown that the rays obey *Fermat's principle*, according to which among feasible paths joining any two assigned points the actual ray has stationary travel-time (see § 3.9, exercise 2).

The connection of energy transmission with the ray paths will be discussed in chapter 8 for the case of elastic waves. It will be shown (§ 8.4.1) that in a medium of continuously varying properties, the energy is in general transmitted along the rays with negligible loss on the way.

Suppose that a wave front is described by the phase relation

$$\tau(x_i) = t. \tag{84}$$

Then, as a generalization of (77), we seek a series solution of (80), called the *ray series*, given by

$$y(x_i, t) = \sum_{k=0}^{\infty} y_k(x_i) F_k(\tau(x_i) - t), \tag{85}$$

where F_k are usually complex functions. In standard ray theory, the phase function $\tau(x_i)$ is taken as real and depends on the spacial coordinates only. For a given t, equation (84) clearly gives the position of the wave phase; the surfaces, $\tau(x_i) = $ constant, define wave fronts and the normals to them define the rays. Further mathematical development is given in § 7.1.1.

3.5 Two-dimensional wave motion

The two-dimensional form of the wave equation (72) is

$$\frac{\partial^2 y}{\partial t^2} = c^2 \left(\frac{\partial^2 y}{\partial x_1^2} + \frac{\partial^2 y}{\partial x_2^2} \right). \tag{86}$$

We consider here the solution of (86) for the case in which y is symmetrical about a centre O. Changing to polar coordinates (r, θ), we obtain from (84) for the case of symmetry

$$\frac{\partial^2 y}{\partial t^2} = c^2 \left(\frac{\partial^2 y}{\partial r^2} + \frac{1}{r} \frac{\partial y}{\partial r} \right).$$

The solution of this equation involves Bessel functions. In accordance with a method indicated in §3.3.4, we take as a trial solution $y = R(r)T(t)$, where functions of r and t are separated. Substituting gives

$$\frac{1}{c^2 T}\frac{d^2 T}{dt^2} = \frac{1}{R}\frac{d^2 R}{dr^2} + \frac{1}{rR}\frac{dR}{dr}.$$

The left-hand side is independent of r, and the right-hand side of t; hence both sides are constant, equal to $-\kappa^2$ say. Hence

$$T = C\cos(\kappa ct + \varepsilon), \tag{87}$$

where C, ε are constants, and

$$r^2\frac{d^2 R}{dr^2} + r\frac{dR}{dr} + \kappa^2 r^2 R = 0. \tag{88}$$

The equation (88) is Bessel's equation of zero order for R in terms of κr, and its solution is

$$R = AJ_0(\kappa r) + BY_0(\kappa r), \tag{89}$$

where A, B are constants,

$$J_0(x) = 1 - \frac{x^2}{2^2} + \frac{x^4}{2^2 4^2} - \frac{x^6}{2^2 4^2 6^2} + \cdots, \tag{90}$$

and

$$Y_0(x) = J_0(x)\log x + \frac{x^2}{2^2} - \frac{x^4}{2^2 4^2}(1 + \tfrac{1}{2}) + \frac{x^4}{2^2 4^2 6^2}(1 + \tfrac{1}{2} + \tfrac{1}{2}) - \cdots; \tag{91}$$

$J_0(x)$ and $Y_0(x)$ are Bessel's functions of zero order of the first and second kinds, respectively. By (87) and (89), a solution of (85) is therefore

$$y = \{AJ_0(\kappa r) + BY_0(\kappa r)\}\cos(\kappa ct + \varepsilon); \tag{92}$$

and a more general solution can be obtained by superposing solutions of the form (92) with different values of κ.

It may be shown (see Whittaker and Watson, *Modern Analysis* (C.U.P., 1927), §§17.5, 17.6) that for large values of x

$$J_0(x) \approx \left(\frac{2}{\pi x}\right)^{\frac{1}{2}}\cos(x - \tfrac{1}{4}\pi)$$

and

$$Y_0(x) \approx \left(\frac{2}{\pi x}\right)^{\frac{1}{2}}\sin(x - \tfrac{1}{4}\pi),$$

the neglected terms being of the order of $x^{-\frac{3}{2}}$. It follows that at large distances from an initially confined disturbance, the amplitude is inversely proportional to the square root of the distance from the centre. This property is important in the theory of the transmission of seismic surface waves (see §8.7).

The more complicated behaviour of cylindrical seismic waves compared with both plane and spherical waves should be noted. Consider a cylindrical wave propagating with velocity c from the origin of coordinates. Then the displacement at an observation point x, y at time t depends on contributions from all points within the range ct of the origin. In contrast to plane waves, however (see Fig. 3.2), the contributions from different points are proportional to the factor $(c^2t^2 - R^2)^{-\frac{1}{2}}$ where R is the distance between the observation point and the origin. As shown in Fig. 3.6, the shape of the initial displacement or 'first motion' of a struck membrane travels with a

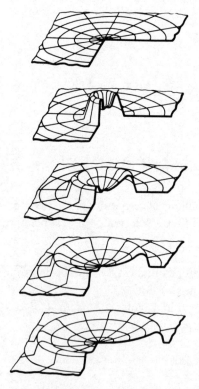

Fig. 3.6. Change of cylindrical wave shape on a struck membrane. Initial displacements are zero, and the centre portion is started upwards.

velocity c as with plane waves, but the shape of the wave changes as it spreads out leaving a residual wake after it. These properties agree with the general rule that waves in an odd number of dimensions leave no residual wake behind, whereas waves with even dimensions do so. Seismic plane waves do not change shape or size as they travel, while spherical seismic waves do not change shape but attenuate proportionally to the factor $1/r$. By contrast, cylindrical seismic waves, such as surface waves, change both their shape and size leaving residual displacements in their tails.

3.6 Scattering

Deflections of a portion of wave energy occur when elastic waves encounter an obstacle or a region in which the elastic properties of the medium differ from the values outside the region. Theoretically, it is usual to consider the difference between the actual wave and the wave present when the scattering obstacle is absent. In this sense, a seismic wave incident on an anomalous region in the Earth will produce, in addition to an undisturbed plane wave, an interfering *scattered wave* that will spread out from the obstacle in all directions.

Three cases are generally considered, but there is overlap between them and some difference in usage and emphasis (see § 3.7). First, if the scattering obstacles are very small compared with the wave-length of the incident waves, the scattered waves spread outwards in all directions and there exist no sharp-edged shadows. If seismic waves pass through regions where there are many small scatterers, the scattered waves will interfere with each other, giving rise to coherent or incoherent wave trains, depending on whether the scatterers are regularly spaced or not. At any point, the scattered wave can be constructed from the sum of the relative phases and amplitudes of individual components determined by the distribution of the sizes of the scatterers and their locations. An explanation of the coda of seismic waves travelling in the Earth's crust has been given in these terms.

If a seismic wave is incident on an obstacle that is very large compared with the wave-length, the assumptions of geometrical optics or ray theory are approximately valid (see § 6.5). In this case, the scattered waves can be thought of as reflected and refracted waves at the boundary. High-frequency seismic waves incident on the Earth's core are an example. In the third case, when the incident wave-length is of the same order as the curvature of the object, there may be a number of interesting interference effects. A realistic treatment requires solution of the appropriate wave equation or at least a specially adapted approximation, such as that provided by variational methods (see §13.2). A mathematical analysis of elastic wave scattering for relevant seismological problems in the Earth is beyond the scope of this text and the reader is referred to the books by W.L. Pilant and B.L. Moiseiwitsch for references. (See also chapter 12.)

3.7 Diffraction

When the scattering surface has a curvature large compared with the curvature of the incident wave front, the reflected or refracted wave front will be sharply curved, and ray theory ceases to be fully applicable. When this is the case, the waves are usually said to be *diffracted* rather than scattered. For instance, if a screen is placed in front of an advancing beam of plane light waves, there is some light transmitted to that part of the region on the far side of the screen that would remain dark on pure ray theory. At the edge of the shadow, the wave amplitude oscillates before dropping to zero well inside the shadow. Elastic waves show similar effects. For example, dilatational elastic waves propagating in the interior of a spherical shell of the Earth set up variable stresses on the internal boundary and these act on the matter near the boundary and so produce diffraction effects in the fringing region.

Diffraction is also important close to the centre of an initially confined disturbance, where conditions assumed for ray theory fail to hold. Thus, for instance, full wave theory is needed for an adequate description of the wave in the region close to the origin of an earthquake.

It is important, however, to note that even when diffraction effects are significant, the time of arrival at any point in the medium of the first (high frequency) part of any transmitted disturbance is always given by Fermat's principle, suitably applied. This is a consequence of the general solution described in §3.4.2. Ray theory is of much avail in many (but not all) problems in seismology; much use of it will be made in later chapters.

Diffracted waves can be analysed as in dispersion (equation (65)) by stationary phase $\psi(\kappa)$. Let stationary values of the group velocity C occur near κ_0 so that the Taylor expansion gives

$$\psi(\kappa) = \psi(\kappa_0) + (\kappa - \kappa_0)\psi'(\kappa_0)$$
$$+ \tfrac{1}{6}(\kappa - \kappa_0)^3 \psi'''(\kappa_0) + O(\kappa - \kappa_0)^4.$$

Thus, near its stationary point the integral (66) becomes

$$y(x,t) \simeq \int F(\kappa_0)\exp\left\{it[\psi(\kappa_0) + (\kappa - \kappa_0)\psi'(\kappa_0)\right.$$
$$\left. + \tfrac{1}{6}(\kappa - \kappa_0)^3 \psi'''(\kappa_0)]\right\}\,d\kappa. \tag{93}$$

In (93), substitute

$$s = (\kappa - \kappa_0)[\tfrac{1}{2}t\psi'''(\kappa_0)]^{\frac{1}{3}}$$

so that the cubic term is $\frac{1}{3}is^3$. Next, put

$$X = [\tfrac{1}{2}t\psi'''(\kappa_0)]^{-\frac{1}{3}}t\psi'(\kappa_0).$$

so that the linear term becomes isX. We thus find

$$y(x,t) \simeq 2\pi F(\kappa_0)\{\exp[it\psi(\kappa_0)]\}[\tfrac{1}{2}t\psi'''(\kappa_0)]^{-\frac{1}{3}}\mathrm{Ai}(X), \tag{94}$$

where the Airy function is defined as

$$\mathrm{Ai}(X) = (2\pi)^{-1}\int_{-\infty}^{\infty}\exp[i(sX + \tfrac{1}{3}s^3)]\,ds. \tag{95}$$

This function has been widely tabulated and suitable asymptotic forms are available. The slowly decaying wave-like form of $\mathrm{Ai}(X)$ for $X < 0$, and exponentially descending form for $X > 0$, are shown in Fig. 3.7. The envelope of a diffracted or dispersed seismic wave train indeed often resembles $\mathrm{Ai}(X)$ with a dispersed wavy portion terminated abruptly. The wave packet or group corresponding to a stationary value of $C(\kappa)$ develops relatively large amplitudes and is called the *Airy phase* in seismology (see Fig. 10.1).

The amplitude of diffracted waves, period T s, at an angular distance $\Delta°$ into the shadow of the large spherical core of the Earth (see § 13.1) can be shown approximately to be (Scholte, 1956),

$$A = A_0\exp(-0.20\Delta/T^{\frac{1}{3}}). \tag{96}$$

The depth of penetration is a function of frequency with the highest frequency diffracted waves being extinguished quickly.

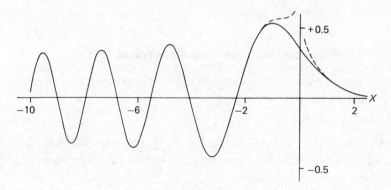

Fig. 3.7. Full line; the $\mathrm{Ai}(X)$ solution of Airy's equation. Broken lines; asymptotic forms for $X > 0$ and $X < 0$.

Elastic wave diffraction can also occur without the presence of sharp physical boundaries. An example is the diffraction at the optical caustic produced by light through a spherical lens and given an early mathematical treatment by G.B. Airy. Here we define an *elastic wave caustic* as a boundary between a region with an interference wave pattern between two separate trains of waves and a neighbouring region with no waves. Near a caustic, straightforward ray theory breaks down, but the problem can be 'healed' (in the terminology of M.J. Lighthill) by the use of a single mathematical concept: the Airy integral. Near cusps of travel-time curves, the seismic rays run together and the locus of points which separate regions without rays and regions twice covered by rays is the caustic curve or surface.

The caustic corresponds to $X = 0$ in the graph of the Airy integral (Fig. 3.7). For $X < 0$, groups of waves of similar wave number beat together while, for $X > 0$, the wave amplitude falls exponentially. The asymptotic form of $\mathrm{Ai}(X)$ is

$$\mathrm{Ai}(X) \approx \tfrac{1}{2}\pi^{-\frac{1}{2}}X^{-\frac{1}{4}}\exp\left(-\tfrac{2}{3}X^{\frac{3}{2}}\right) \quad \text{when } X > 0. \tag{97}$$

The dashed lines in Fig. 3.7 allow a comparison of the degree of approximation in the asymptotic theory to the smooth behaviour of the Airy integral. While ray theory predicts an infinite amplitude on the caustic surface, the Airy theory covers this 'wound' with a finite transition.

In 1939, Jeffreys was the first to apply Airy's theory of waves at a caustic to a seismological case. Near the caustic that occurs for seismic waves through the Earth's core at about $\Delta_c = 142°$, from (97), the amplitude A of the waves in the shadow decreases approximately exponentially with Δ as

$$A = A_0 \exp\left(-250(\Delta_c - \Delta)^{\frac{3}{2}}/T\right), \tag{98}$$

where T is in seconds.

3.8 Helmholtz and Sturm–Liouville equations

After separation of variables in the wave equation (58), using (62), we first obtain the *Helmholtz equation*

$$\nabla^2 X + k^2 X = 0, \tag{99}$$

where permissible values of k are fixed by the imposed boundary conditions. For spherical polar coordinates r, θ, ϕ the equation (99) may be further separated by substituting

$$X(r, \theta, \phi) = R(r)\, S(\theta, \phi). \tag{100}$$

The process yields

$$\frac{1}{R}\frac{d}{dr}\left(r^2\frac{dR}{dr}\right) + k^2r^2 = -\frac{1}{S}\left\{\frac{1}{\sin\theta}\frac{\partial}{\partial\theta}\left(\sin\theta\frac{\partial S}{\partial\theta}\right) + \frac{1}{\sin^2\theta}\frac{\partial^2 S}{\partial\phi^2}\right\}$$

and, because the left-hand side is a function of r only and the right-hand side is a function of θ and ϕ only, both sides may be put constant α so that

$$\frac{d}{dr}\left(r^2\frac{dR}{dr}\right) + (k^2r^2 - \alpha)R = 0 \tag{101}$$

and

$$\frac{1}{\sin\theta}\frac{\partial}{\partial\theta}\left(\sin\theta\frac{\partial S}{\partial\theta}\right) + \frac{1}{\sin^2\theta}\frac{\partial^2 S}{\partial\phi^2} + \alpha S = 0. \tag{102}$$

Next, put

$$S(\theta, \phi) = \Theta(\theta)\Phi(\phi) \tag{103}$$

so that

$$\frac{d^2\Phi}{d\phi^2} + \beta\Phi = 0 \tag{104}$$

and

$$\frac{1}{\sin\theta}\frac{d}{d\theta}\left(\sin\theta\frac{d\Theta}{d\theta}\right) + \left(\alpha - \frac{\beta}{\sin^2\theta}\right)\Theta = 0, \tag{105}$$

where β is the constant of separation.

Equation (104) has a single-valued solution within the range $0 \leqslant \phi < 2\pi$ if $\beta = m^2$ where m is an integer, in which case

$$\Phi(\phi) = A\exp im\phi + B\exp(-im\phi), \tag{106}$$

where A and B are constants. Let $\mu = \cos\theta$ and write $\Theta(\theta) = P(\mu)$ so that (105) becomes

$$\frac{d}{d\mu}\left\{(1 - \mu^2)\frac{dP}{d\mu}\right\} + \left(\alpha - \frac{m^2}{1 - \mu^2}\right)P = 0. \tag{107}$$

In general the solutions of (107) are infinite at $\mu = \pm 1$. However if $\alpha = l(l + 1)$, where l is a positive integer or zero, one of its solutions is bounded for $-1 \leqslant \mu \leqslant 1$. This bounded solution is denoted by $P_l^m(\mu)$, called an associated Legendre function; it has the form of a polynomial of order $l - |m|$ in μ times the function $(1 - \mu^2)^{\frac{1}{2}|m|}$.

The equations (101), (104) and (107) are all of Sturm–Liouville type,

$$\frac{d}{dx}\left\{p(x)\frac{dy}{dx}\right\} + \{q(x) + \lambda r(x)\}y = 0, \tag{108}$$

where λ is a constant and $p(x)$, $q(x)$ are algebraic functions having a finite number of zeros. These functions must not have singular points in the range of definition $a < x < b$. Equations of Sturm–Liouville type include Legendre's equation, Airy's equation, Bessel's equation and others of great importance in seismology. Vibrating elastic bodies such as strings, membranes, and spheres, as we have seen above, also give rise to Sturm–Liouville equations and a well-developed theory is available to analyse the eigen-vibrations.

3.9 Exercises

1   (a) Write the equation for a progressive surface seismic wave moving along the negative x-axis with the following characteristics: amplitude, 0.2 cm; period 5 s; wave-length 3 km.

(b) A wire 1 m long has a mass m of 0.010 kg/m and is under a tension F of 10 newton. If the wire is rigidly held at both ends and is set into vibration, find the frequencies of the fundamental and first two overtones.

2   In the calculus of variations, the condition for

$$I = \int_{t_0}^{t_1} f(\dot{x}, x, t)\,dt$$

to be stationary in the case of fixed end-points is Euler's equation

$$\frac{\partial f}{\partial x} - \frac{d}{dt}\frac{\partial f}{\partial \dot{x}} = 0.$$

Consider the transmission time from $(0,0)$ to (x_1, z_1) of a wave whose velocity is proportional to the distance from a fixed plane, $v = c(z_0 + z)$, say, where c and z_0 are constants.

Show that

$$T = \int_0^{x_1} \frac{(1 + p^2)^{\frac{1}{2}}}{c(z_0 + z)},$$

where $p = dz/dx$.

Hence prove, by Fermat's principle, that the time and distance traversed by the ray from surface to surface are

$$T = \frac{2}{c}\tanh^{-1}\sin e, \quad X = 2z_0 \tan e,$$

where e is the angle of emergence at the surface (see §6.1).

3   A uniform string of length l is stretched in tension between two fixed points. A point at distance b from one end is plucked aside a small distance β and then released from rest. Show that the transverse displacement of a point on the string at distance x from one end at time t is given by

$$y = \frac{2\beta l^2}{\pi^2 b(l - b)}\sum_{s=1}^{\infty}\frac{1}{s^2}\sin\frac{s\pi b}{l}\sin\frac{s\pi x}{l}\cos\frac{s\pi ct}{l}.$$

Note that (a) the harmonics diminish rapidly, like $1/s^2$, (b) the nearer β is to the end, the greater the amplitude (What similarities are there with seismic waves produced by an analogous source at depth in the Earth?), (c) there is a reciprocal relation between the motion at x and source at b, (d) if $b = l/s$ (i.e. at the node of the sth mode), then the mode is not excited.

4 A string of length $l + l'$ is stretched with tension P between two fixed points. The length l has mass m per unit length, and the other m' per unit length. Prove that the period τ of transverse vibrations is given by

$$\frac{\tan\left(\dfrac{2\pi l}{\tau}\left(\dfrac{m}{p}\right)^{\frac{1}{2}}\right)}{\tan\left(\dfrac{2\pi l}{\tau}\left(\dfrac{m'}{p}\right)^{\frac{1}{2}}\right)} + \left(\dfrac{m}{m'}\right)^{\frac{1}{2}} = 0.$$

Compare the eigen-periods with those of a homogeneous string. The problem is related to that of the eigen-vibrations of a heterogeneous Earth model.

5 A weightless string with three particles of equal mass m distributed uniformly is stretched with tension P between two fixed points (see § 3.2.3). If y_1, y_2, y_3 are the particle displacements from equilibrium positions, prove that the kinetic and potential energies of the system are

$$T = \tfrac{1}{2}m(\dot{y}_1^2 + \dot{y}_2^2 + \dot{y}_3^2)$$

and

$$V = \frac{1}{2}\frac{P}{a}\{y_1^2 + (y_2 - y_1)^2 + (y_3 - y_2)^2 + y_3^2\}.$$

Show (a) by ordinary theory (see § 3.2.2) and (b) by Rayleigh's principle that the squared eigenfrequencies are 2μ, $(2 \pm \sqrt{2})\mu$, where $\mu = P/ma$.

[Hint. For symmetric vibration, frequency ω, assume $y_1 = \beta y_2 = y_3$, form Rayleigh's quotient by equating mean energies, and differentiate to find stationary values.]

If the string is assumed to take parabolic shape, show that Rayleigh's principle gives the gravest frequency as about 0.588μ.

6 Consider the wave packet

$$\phi(x, t) = \int_{-\infty}^{\infty} a(\kappa)\exp\{2\pi i(\kappa x - \gamma t)\}\,d\kappa,$$

where κ is the wave number.

Investigate the case

$$a(\kappa) = \begin{cases} 1 & \text{if} \quad |\kappa - \kappa_0| < \kappa_1 \\ 0 & \text{otherwise} \quad (\kappa_0,\, \kappa_1 \text{ constants}). \end{cases}$$

Assume that only the first two terms of the Taylor expansion of γ in terms of κ are required and show that at time t the disturbance is

$$\phi(x, t) = \frac{\sin\{2\pi(x - ct)\kappa_1\}}{\pi(x - ct)}\exp 2\pi i(\kappa_0 x - \gamma_0 t),$$

where $c = (dy/d\kappa)_0$. With what speed does the packet move as a whole? [Note the connection between the rectangular function and the sinc or diffraction function.]

7   Show from (97) that diffracted PKP waves with 1 s period might be expected to be observed up to 3° distance before the PKP caustic, while 10 s waves should be observable back to about 132° (see §13.1.4).

8   Show that the Airy integral (95) is a solution of the Airy equation

$$\frac{d^2y}{dX^2} - Xy = 0.$$

Show that this equation is of Sturm–Liouville type and calculate the two independent series solutions at the origin.

9   Derive the Sturm–Liouville equation (108) by finding the stationary condition (Euler–Lagrange) for the integral

$$I = \int_a^b \tfrac{1}{2}\{p(x)\left(\frac{dy}{dx}\right)^2 - q(x)y^2\} \, dx,$$

subject to the condition

$$\int_a^b r(x)y^2 \, dx = \text{constant}.$$

10   Consider a harmonic spherical acoustic wave diverging from a point source. Show that the pressure field can be defined by a scalar $\phi \propto \exp i\omega t$ and construct the solution for the displacement u at radius r

$$\rho u = \frac{A}{r^2} \cos \omega(p - r/c) - \frac{A\omega}{rc} \sin \omega(t - r/c),$$

where ρ is the density and A a constant.

4

Body elastic waves

We now investigate some features of the transmission of a disturbance through an infinite medium. To begin with, we assume the material to be homogeneous, isotropic and perfectly elastic, and use infinitesimal strain theory. We take the undisturbed configuration as standard, and we ignore the effects of possible fluctuations in external forces during the passage of the disturbance. Later we consider how far the simple theory based on these assumptions needs to be amended for transmission of seismic waves through the Earth.

4.1 P and S waves

In the circumstances just described, the equations of motion of the disturbance are ($2.34b$) with $X_i = 0$, i.e.

$$\rho \frac{\partial^2 u_i}{\partial t^2} = (\lambda + \mu)\frac{\partial \theta}{\partial x_i} + \mu \nabla^2 u_i. \tag{1}$$

If we differentiate both sides of (1) with respect to x_i (this involves, in accordance with the summation convention, adding the results of separate differentiations for $i = 1, 2, 3$), we obtain using (2.22)

$$\rho \frac{\partial^2 \theta}{\partial t^2} = (\lambda + 2\mu)\nabla^2 \theta = (k + \tfrac{4}{3}\mu)\nabla^2 \theta, \tag{2}$$

since $\lambda = k - \tfrac{2}{3}\mu$ by (2.35). If we apply the operation curl to both sides of (1), we obtain

$$\rho \frac{\partial^2}{\partial t^2}\mathrm{curl}(u_i) = (\lambda + \mu)\mathrm{curl}\left(\frac{\partial \theta}{\partial x_i}\right) + \mu \nabla^2 \mathrm{curl}(u_i).$$

But a typical component of curl $(\partial \theta / \partial x_i)$ is of the form

$$\left(\frac{\partial}{\partial x_i}\frac{\partial \theta}{\partial x_j} - \frac{\partial}{\partial x_j}\frac{\partial \theta}{\partial x_i}\right),$$

which is zero. Hence

$$\rho \frac{\partial^2}{\partial t^2} \text{curl}(u_i) = \mu \nabla^2 \text{curl}(u_i). \tag{3}$$

The equation (2) and (3) are scalar and vector wave equations, respectively. By (2), a *dilatational* (or *irrotational*) disturbance θ may be transmitted through the substance with speed α, where

$$\alpha = \left(\frac{\lambda + 2\mu}{\rho} \right)^{\frac{1}{2}} = \left(\frac{k + \frac{4}{3}\mu}{\rho} \right)^{\frac{1}{2}}. \tag{4}$$

By (3), a *rotational* (or *equivoluminal*) disturbance may be transmitted with speed β, where

$$\beta = \left(\frac{\mu}{\rho} \right)^{\frac{1}{2}}. \tag{5}$$

We notice that $\beta < \alpha$. In seismology, the two types of waves are called, respectively, the *primary*, *P*, or 'push' waves, and the *secondary*, *S*, or 'shake' waves. In the case we have taken, both speeds depend only on the elastic parameters and the density of the substance, and there is no dispersion. In transmission of seismic waves through the Earth, the elastic parameters λ, μ, k are taken as adiabatic ones.

If, in particular, μ is zero, then by (5) β is zero. Thus rotational waves are not transmitted through a substance of zero rigidity, a result already referred to in §2.6.6 in connection with discrimination between fluid and solid states.

Note that if k/ρ is infinite and μ/ρ finite, only rotational waves are transmitted with finite speed. This result is the basis of the old 'incompressible solid ether' theory of light and electromagnetism.

4.1.1 Case of plane waves

Sufficiently distant from the source of an initially confined disturbance, the waves may be regarded as plane. This approximation is relevant to many seismological problems, for the distance of a station recording the local seismic displacements is often great compared with the dimensions of the initially disturbed region. In this case, called the 'far field', the displacements associated with the *P* and *S* waves are in effect *longitudinal* and *transverse*, respectively. The transverse character of the *S* waves is indicated by the theory of §3.3.3, since div curl (u_i) is readily seen to be zero.

The theory of plane waves may be set up independently of (2) and (3) by making a trial substitution (cf. (3.57)) of the form

$$u_i = A_i \exp\{i\kappa(l_j x_j - ct)\}, \tag{6}$$

where $l_j^2 = 1$, in equations (1). On eliminating the three A_i, it is found that the form (6) is a possible solution of (1) if and only if the square of the speed c obeys a cubic equation whose roots are α^2, β^2, β^2, where α, β are given by (4), (5). The speeds α, β are found as before to be associated with longitudinal and transverse waves, respectively; it follows that the types of waves described in the last paragraph are the only possible types of body plane waves.

It emerges also that the two types P and S are independent of each other, and further that the latter may be plane polarised. In seismology, when an S wave is polarised so that all particles of the substance move horizontally during its passage, it is denoted SH; when the particles all move in vertical planes containing the direction of propagation, the wave is denoted SV.

4.1.2 Poisson's relation

For many solids, and in particular for most rocks of the Earth, values of the two elastic parameters λ and μ are not greatly different. Poisson's relation neglects any difference in the values of λ and μ and may be expressed by any one of the following equations

$$\lambda = \mu, \quad k = \tfrac{5}{3}\mu, \quad \sigma = \tfrac{1}{4}, \quad \alpha = \sqrt{3}\,\beta, \quad \alpha^2 = 3\mu/\rho. \tag{7}$$

In some problems, it is convenient to assume that Poisson's relation holds in order to reduce the algebra.

4.2 Inclusion of the seismic source in infinite media

4.2.1 Spherical source

We discuss now the form of waves to be expected following the occurrence of an initial disturbance inside a confined region. We first consider two particular cases of initial disturbances which, though simple, serve as a basis for investigating movements in the Earth following an earthquake.

Suppose that a dilatational wave is generated by a sudden explosion within a sphere of radius a and centre O (inside the given medium) with a pressure symmetrical with respect to O. More precisely, let r denote the distance from O, t the time, and $H(t)$ Heaviside's unit function ($H(t) = 0$ if $t < 0$; -1 if $t > 0$). Then we suppose there is zero displacement at all points for $t < 0$, and that the sphere $r = a$ is acted on from inside by a symmetrical pressure $AH(t)$, where A is constant. It follows from the theory of chapter 3 that at any point r outside the sphere, the displacement remains zero until the time t_1, where $t_1 = (r - a)/\alpha$.

For $t > t_1$, Jeffreys has shown (assuming Poisson's relation (7) to hold) that there is a radial displacement u_r given by

$$u_r = \frac{Aa^3}{4\mu r^2}\left[1 + \left\{\left(\frac{r}{a} - \frac{1}{2}\right)\sqrt{2}\sin\left(\frac{2\sqrt{2}\,\alpha t'}{3a}\right)\right.\right.$$
$$\left.\left. - \cos\left(\frac{2\sqrt{2}\,\alpha t'}{3a}\right)\right\}\exp\left(-\frac{2\alpha t'}{3a}\right)\right],$$
(8)

where $t' = t - t_1$. If the point is in the far field, so that terms in r^{-1} dominate over terms in r^{-2}, and t' is not too great, (8) gives

$$u_r \approx \frac{\sqrt{2}\,Aa^2}{4\mu r}\sin\left(\frac{2\sqrt{2}\,\alpha t'}{3a}\right)\exp\left(-\frac{2\alpha t'}{3a}\right).$$
(9)

When $t' = 0$, the value of the exponential factor in (9) is unity. When the sine term is passing through zero for the first time after $t' = 0$, t' is equal to about $3.4a/\alpha$, and the value of the exponential factor has been reduced to about one-tenth. After this time, the right-hand side of (8) is always small compared with its first maximum. Thus the displacement at any point for which r/α is appreciable is essentially a single swing from the zero position, followed by a rapid approach to nearly zero.

Suppose next that a rotational wave is generated by the action of a tangential impulse over the sphere $r = a$, the impulse being symmetrical about a diameter. More precisely, let r, θ, ϕ be spherical polar coordinates referred to O as origin, and let the stress across the sphere $r = a$ be $-AH(t)$ $\sin\theta$ in the direction of ϕ increasing. There will then be no displacement at a point r outside the sphere until the instant t_2, where $t_2 = (r - a)/\beta$. For $t > t_2$, Jeffreys obtained an expression analogous to (8) for the relevant displacement component u_ϕ. When r/a is appreciable and $t - t_2, = t''$ say, is not too great, the approximate form is

$$u_\phi \approx \frac{2Aa^2}{\sqrt{3}\,\mu r}\sin\left(\frac{\sqrt{3}\,\beta t''}{2a}\right)\exp\left(-\frac{3\beta t''}{2a}\right).$$
(10)

This result is similar in form to (9), but the numerical values of the coefficients are a little different. The value of t'' when the sine term in (10) is passing through zero for the first time after $t'' = 0$ is now $3.6a/\beta$. The exponential factor in (10) diminishes even more rapidly than that in (9). Note that both waves attenuate like r^{-1} in the far field.

On account of the independence of the transmission of P and S waves, we

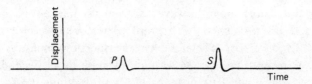

Fig. 4.1. Seismogram from a distant point source in an infinite medium.

may superpose effects, and we then have conditions corresponding to a simple earthquake source sending out both P and S waves through an unbounded homogeneous medium. In these circumstances, the 'synthetic seismogram' of the displacement of a distant particle would be of the form in Fig. 4.1. Essential features are the absence of trains of waves and a quiescent interval between the arrivals of the P and S pulses. (See also §5.5.3 and §16.3.3.)

4.2.2 Green's function representation for point sources

The problem is to solve the wave equation with the addition of a body force term appropriate to the physics of the source. First, generalising from the scalar forms (2), we consider the solution, at the point x_i in an unbounded medium, of the equation

$$\frac{\partial^2 \theta}{\partial t^2}(x_i, t) - c^2 \nabla^2 \theta(x_i, t) = \delta(x_i)\delta(t) \quad (i = 1, 2, 3). \tag{11}$$

From equation (3.28), a solution of a similar inhomogeneous equation is known in terms of Green's function $G(t, \xi)$ for an impulsive force applied at time $t = \xi$. The source, of course, will not be localised at the origin of coordinates but, in spherical coordinates, at $\mathbf{r}_0$.

We therefore seek the solution of

$$\gamma^2 G(\mathbf{r}, \mathbf{r}_0) + c^2 \nabla^2 G(\mathbf{r}, \mathbf{r}_0) = -\delta(\mathbf{r} - \mathbf{r}_0), \tag{12}$$

where G is the spacial factor for a wave from a unit point source at $\mathbf{r}_0$.

For a periodic simple source $\exp(-i\gamma t)$ it can be shown (see, for example, Morse and Feshbach, chapter 7) that the general form of $G(\mathbf{r}, \mathbf{r}_0)$ is

$$G(\mathbf{r}, \mathbf{r}_0) = \frac{1}{4\pi c^2 R} \exp i\kappa R, \tag{13}$$

where $R = |\mathbf{r} - \mathbf{r}_0|$ and $\kappa = \gamma/c$ (cf. equation (9)).

More generally, for the source defined by $\delta(\mathbf{r} - \mathbf{r}_0)\delta(t - \xi)$, the solution of the inhomogeneous scalar wave equation is

$$G(\mathbf{r}, t) = \frac{1}{4\pi c^2 R} \delta\left(t - \xi - \frac{R}{c}\right). \tag{14}$$

The first point to note about the wave field from a simple seismic source is, from (13) and (14), its symmetry with regard to source $\mathbf{r}$ and receiver $\mathbf{r}_0$, i.e. it depends only on $\mathbf{r} - \mathbf{r}_0$. Therefore, Green's function conforms to the principle of reciprocity (see §4.2.3 and Exercise 4.9.4.) Secondly, the solution is, like (9) and (10), in terms of a retarded time $(t - R/c)$. The geometrical spreading is inversely as distance R but, because the second factor in (13) and (14) usually varies, the latter dominates the wave shape recorded at the receiver. Finally, we note that a more general solution of (11) requires the addition of any scalar solution, $\chi(\mathbf{r})$ say, of the homogeneous equation $\nabla^2 \chi + \kappa^2 \chi = 0$. If a surface bounds the medium, then these complementary functions may be needed to satisfy the boundary conditions.

For an extended source, superposition by integration over the volume allows appropriate solutions to be constructed from the above solutions. In more physically realistic elastodynamic problems in which both compression and shear motion at the source are considered along with source directionality (see §16.3), the necessary vector solutions can be found from the scalar solutions by reconstruction from elastic potentials (see e.g. §5.1). Mathematical results along these lines are now extensive and the reader is referred to Aki and Richards, for example, for more detail.

4.2.3 Reciprocity theorem

A powerful result used in studies of the seismic source is a simple reciprocal relation between displacements and forces at the source and receiver.

Consider the equilibrium of an elastic body, as defined in §2.1.5, under surface tractions P_i and body forces F_i. For initial simplification, let the forces be harmonic so that the resulting displacements are

$$u_i = u_i(x_1, x_2, x_3) \exp i\gamma t.$$

Then (2.9) becomes

$$-\gamma^2 \rho u_i = \frac{\partial p_{ij}}{\partial x_j} + F_i. \tag{15}$$

Denote a second force and traction system by primed symbols with corresponding displacement u_i'. We first form an expression I by integration of F_i over the volume V and P_i over the surface S, using (2.5), and multiplying by u_i', so that

$$I = \int_V \left\{ \gamma^2 \rho u_i u_i' + \frac{\partial p_{ij}}{\partial x_j} u_i' \right\} dV - \int_S v_j p_{ji} u_i' \, dS.$$

Now introduce the derivative of the product $p_{ij}u'_i$ in the first term and use the divergence theorem in the second term. We obtain

$$I = \gamma^2 \int_V \rho u_i u'_i \, dV - \int_V p_{ji} \frac{\partial u'_i}{\partial x_j} \, dV. \tag{16}$$

Similarly, for the second force system, we can derive an expression I' with primes interchanged on u_i, p_{ji}. The reader will find it easy to prove the equality of the two right-hand integrals in I and I' by writing the integrand in (16) as $\frac{1}{2} p_{ji} e'_{ji}$ and noting the symmetry of the stress–strain relations.

Equating the two expressions I and I' for the same frequency, and using (15), we find the reciprocal equation, called Betti's theorem,

$$\int_V F_i u'_i \, dV + \int_S P_i u'_i \, dS = \int_V F'_i u_i \, dV + \int_S P'_i u_i \, dS. \tag{17}$$

There can be many variations of the general conditions on S and of the body forces. Note also that the result can be easily generalised for non-oscillating forces and that initial conditions have not been specified. The theorem, therefore, holds when the primed and unprimed systems apply to different times. The important result is that, if the elastic displacements due to a unit force are known at one point, it is possible to write down displacements at any other place (see §4.8 and §16.3). Thus, Lamb's calculation (§5.5.3) of the displacement at a point beneath the free surface of a half-space, produced by an applied force at the free surface, can, by means of the reciprocity theorem, be used to write down the surface displacements due to a buried source. A more advanced treatment is developed in Hudson (1980).

4.3 Form of ground motion in an earthquake

In an actual earthquake, the displacement of any point of the Earth's surface is markedly different from that pictured in Fig. 4.1 for a medium with no boundaries. Fig. 4.2 shows the component of the relative ground motion recorded on a seismograph following a small shallow earthquake. The arrows indicate the arrival-times of the P and S movements. Notice that there is a train of waves following both the first P movement and the first S movement, and that there is no interval of quiescence.

We now discuss as possible causes of these oscillatory movements (apart from the effects of the free surface): (i) the existence of initial conditions more complicated than those taken in §4.2; (ii) fluctuations in the local

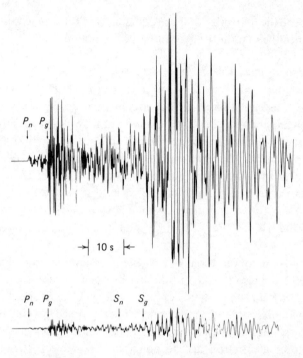

Fig. 4.2. Vertical component seismograms from an earthquake magnitude 5.9 on 3 September 1978 in the Swabian Jura, Germany. The recording was at the Graefenberg Array at $\Delta = 220$ km. The upper seismogram is the broad-band instrument recording; the lower record is the recording after pass-band filtering to enhance local earthquake phases. (Courtesy E. Wielandt and D. Seidl.)

gravity during the passage of a disturbance (we ignored the term ρX_i in forming (1)); (iii) imperfections of elasticity; (iv) departures from homogeneity within the rocks; (v) temperature effects; (vi) finite strain.

In regard to (i), the initial conditions taken in §4.2 are of course simpler than those in a usual earthquake. The initial dislocation of rock causing a tectonic earthquake would normally be spread over a finite interval of time, the 'rise time', and if there were several distinct movements of comparable size during this interval, it would follow from superposition considerations that there would be several distinct P and S pulses at a distant point (see §16.3).

The intervals between these pulses should, however, be independent of the distance of the point from the source of the disturbances, and this gives a means of deciding whether (i) is a cause of the oscillatory movements, called the wave 'coda'. There is in fact considerable evidence that a proportion of earthquakes do show successive energy release. The existence of such multiple earthquakes was first put beyond doubt by Stoneley as a result of a statistical study of observatory readings of the Mongolian earthquake of 1931 August 10. Stoneley showed that there were two shocks of comparable intensity separated by an interval of about 33 s, and possibly other shocks in between.

Nevertheless, small to moderate earthquakes are often not significantly multiple, the part of the initial disturbance which sends out the main energy being confined to a short interval of time. Conditions approximate to the occurrence of an initial impulsive displacement (cf. § 4.2) with a rise time of a few seconds at most. The cause of oscillatory motions observed at the surface for many seconds is therefore not always found in complicated initial conditions. In the near field, the dynamics of the moving dislocation over a ruptured surface becomes very important (cf. § 16.3) and the stationary point model ('the focus') cannot explain the observed wave motions.

We shall show that none of (ii), (iii), (v) or (vi) can be an adequate cause of dispersion of the disturbance as it travels outwards. Indeed the evidence is that the main causes of the coda must be sought in the layering of the half-space and (iv) – heterogeneity of the Earth.

We note first that, in the theory of §§ 4.1, 4.2, the implications of the presence of the Earth's outer surface were ignored. It turns out (cf. chapter 5) that surface waves along this outer boundary do exist and are important; also they suffer dispersion. But their speed is always less than the relevant speeds of the P and S body waves, and hence they do not affect the earlier movements at any given place. Thus surface waves in an earthquake cannot cause the early oscillatory movements.

Note next that these oscillatory movements are observed at points at considerable distances from the source of an earthquake. This indicates that the Earth must be markedly heterogeneous at certain depths below the outside surface. We shall describe in chapter 13 a composite lithosphere in the Earth of 100 km thickness, inside which there are quite considerable changes of property. The existence of such structural heterogeneities is likely to account for much of the observed dispersion of body waves, particularly at higher frequencies. It should be noted, however, that strong repeated reflections of body waves between the Earth's outer surface and interfaces between the layers do not give a fully satisfactory explanation, since there is evidence that these interfaces are not sufficiently sharp. Jeffreys pointed out that for steeply incident waves the reflections at such interfaces would be insignificant (see § 8.4.1), while less steeply incident P

waves at the outer surface would mostly be strongly and steeply reflected into S waves (see §6.4), leaving the observed strong P wave coda still unexplained. Jeffreys further suggested that the observed distribution of amplitudes in near earthquakes is substantially influenced by diffraction at irregular interfaces in the Earth's crust. More recently, K. Aki and others have published observations that demonstrate the importance of wave scattering from crustal irregularities (see §17.5).

We now proceed to discuss the effects of gravity fluctuations and of imperfections of elasticity on the transmission of body waves. Finally, we shall consider the influence of temperature conditions, and the possible significance of finite-strain theory.

4.4 The effect of gravity fluctuations

In (1), we excluded the term ρX_i of (2.34). But this term will not be zero if there are changes in the gravity value at any point x_i as an elastic wave passes by. We shall use the suffix 0 to denote values of certain scalars in the undisturbed configuration. Then on the Newtonian theory of attraction, X_i is expressible in the form $-\partial/\partial x_i (V - V_0)$, where V is the gravitational potential at x_i. Hence in place of (1), we have

$$\rho \frac{\partial^2 u_i}{\partial t^2} = (\lambda + \mu) \frac{\partial \theta}{\partial x_i} + \mu \nabla^2 u_i - \rho \frac{\partial}{\partial x_i}(V - V_0). \tag{18}$$

Now V, V_0 satisfy Poisson's equation, i.e.

$$\nabla^2 V = 4\pi G \rho, \quad \nabla^2 V_0 = 4\pi G \rho_0, \tag{19}$$

where G is the constant of gravitation. From (2.26), we deduce that

$$\rho - \rho_0 = -\rho\theta, \tag{20}$$

approximately. By (19) and (20),

$$\nabla^2(V - V_0) = -4\pi G \rho\theta. \tag{21}$$

Differentiating (18) with respect to x_i and using (21), we have in place of (2), correct to the first order in the displacements,

$$\rho \frac{\partial^2 \theta}{\partial t^2} = (\lambda + 2\mu)\nabla^2 \theta + 4\pi G \rho^2 \theta;$$

i.e.

$$\frac{\partial^2 \theta}{\partial t^2} = \alpha^2 \nabla^2 \theta + \omega^2 \theta, \tag{22}$$

where

$$\omega^2 = 4\pi G \rho. \tag{23}$$

The equation (22) was solved by Jeffreys (for the case $\lambda = \mu$), using the first set of initial conditions in § 4.2. As before, there is no disturbance at the point r until the instant t_1. For $t > t_1$, the expression (8) for the displacement needs to be modified chiefly in the following ways: (i) the term unity inside the square brackets in (8) needs to be replaced by $1 + \frac{1}{2}\omega^2 t t'$; (ii) the exponential term inside the square brackets needs to be multiplied by $1 + \varepsilon$, where ε is of the order of $\omega^2 a t/\alpha$.

Representative values for typical P and S waves are $\rho = 5\,\mathrm{g/cm^3}$, $a/\alpha = 1\,\mathrm{s}$; also, $G = 6.7 \times 10^{-8}$ c.g.s. units. It then follows that

$$\tfrac{1}{2}\omega^2 t t' \approx 2 \times 10^{-6} t'(t_1 + t'),$$

which is small compared with unity until t' is of the order of several hundred seconds. But at a time as great as this (t' denotes the time that has elapsed after the onset of the first pulse at a particular place), the displacement associated with direct P waves would in actual earthquake conditions be unobservable. Thus the modification (i) is normally insignificant. The modification (ii) is also insignificant, since on the above data $\omega^2 a t/\alpha = 4 \times 10^{-6} t$.

If we apply the operator curl to (18) we obtain equation (3) unaltered, showing that gravity has no effect in the propagation of S waves. This is, of course, because a purely rotational disturbance causes no disturbance in density.

We conclude, therefore, that neglect of gravity effects is unimportant in problems on seismological body waves. (But compare § 5.6.3.) In particular, these effects cannot account for the observed dispersion of these waves.

4.5 The effects of elastic imperfections

In obtaining the equations (1), we used the perfect elasticity stress–strain relations (2.32). We now consider the significance of deviations from (2.32) of the type set down in § 2.5.

4.5.1 Constitutive laws for anelasticity

We wish to generalise the stress–strain relation to include time rates of change of stress and strain. The main mathematical approaches use either differential forms or integral forms. In the first, for example, Jeffreys replaced μ in the perfect elasticity equations by an operator containing d/dt. He found for plane distortional waves that elastic afterworking (see § 2.6.3) causes attenuation of a P pulse but it remains sharp and no dispersion is introduced. A more general analytical theory is in terms of an integral representation and we will summarise it here.

The fundamental relations (2.67) and (2.68) may be generalised to contain a time variable, so that

$$p_{ll}(\mathbf{x}, t) = 3ke_{ll}(\mathbf{x}, \tau) \qquad -\infty < \tau < t \tag{24}$$

and

$$P_{ij}(\mathbf{x}, t) = 2\mu E_{ij}(\mathbf{x}, \tau) \qquad -\infty < \tau < t. \tag{25}$$

Because present values of stress do not depend on future strains and vice versa, the forms must be defined non-retroactively or *causally*. Under appropriate conditions we can also write down the inverse relations to (24) and (25).

The similarity of (24) and (25) suggests that for simplicity we write

$$p(t) = Ke(\tau), \tag{26}$$

and consider the response to a step change in strain $AH(t)$

$$p(t) = AKH(\tau) \qquad -\infty < \tau < t$$
$$= A\psi(t), \tag{27}$$

where $\psi(t)$ is the *relaxation* function, zero for $t < 0$ and continuous and non-decreasing for $t > 0$.

The strain can be changed by a series of small steps at successive times. This procedure suggests addition over the total time and the extension of (27) to the convolution

$$p(t) = \int_0^t \psi(t - \tau)\, de(\tau) \qquad t > 0$$
$$= \int_0^t \psi(t - \tau)\dot{e}(\tau)\, d\tau. \tag{28}$$

We obtain (27) by noting that if $e = AH(t)$, $\dot{e} = A\delta(t)$.

For perfect elasticity

$$\psi(t) = mH(t), \tag{29}$$

where m is an elasticity modulus. At $t = 0$ in general there will be an elastic response of the anelastic model defined by

$$\psi(0+) - \psi(0-) = m_0, \tag{30}$$

where m_0 is the instantaneous elastic modulus. Integration of (28) by parts gives, for $t > 0$, $\psi(0-) = 0$,

$$p(t) = m_0 e(t) + \int_0^t \dot{\psi}(t - \tau)e(\tau)\, d\tau, \tag{31}$$

where the instantaneous response is given by the first term and the creep or afterworking by the second.

Now returning to the tensor forms (24) and (25) we may write the generalised visco-elastic forms as

$$p_{ll}(\mathbf{x}, t) = 3k_0 e_{ll}(\mathbf{x}, t) + 3 \int_0^t \dot{R}_k(t - \tau) e_{ll}(\mathbf{x}, \tau) d\tau, \tag{32}$$

and

$$P_{ij}(\mathbf{x}, t) = 2\mu_0 E_{ij}(\mathbf{x}, t) + 2 \int_0^t \dot{R}_\mu(t - \tau) E_{ij}(\mathbf{x}, \tau) d\tau, \tag{33}$$

where R_k, R_μ are the dilatational and shear relaxation functions, and k_0, μ_0 are related to the Lamé moduli.

The equations of motion for a homogeneous medium (1) now become $(X_i = 0)$, with $R_\lambda = R_k - \frac{2}{3} R_\mu$,

$$\rho \frac{\partial^2 u_i}{\partial t^2} = (\lambda_0 + \mu_0) \frac{\partial \theta}{\partial x_i} + \mu_0 \nabla^2 u_i$$
$$+ \int_0^t \{ [\dot{R}_\lambda(t - \tau) + \dot{R}_\mu(t - \tau)] \frac{\partial}{\partial x_i} \theta(\mathbf{x}, \tau)$$
$$+ \dot{R}_\mu(t - \tau) \nabla^2 u_i(\mathbf{x}, \tau) \} d\tau. \tag{34}$$

As in §4.1 we can show that there are just two wave speeds α_0, β_0 given by (4) and (5) with substitution of the instantaneous moduli. This is because there is zero contribution from the integral in (34) on the wave front as long as the relaxation functions remain bounded at $t = 0$. Further discussion is given by Hudson (1980) and Brennan and Smylie (1981).

4.5.2 Linear models and the Jeffreys power law

Treatment of wave motion in anelastic materials in the general case is not elementary but useful seismological results can be obtained for some linear models. Such models can be considered physically as systems of elastic springs and viscous dash-pots connected in series or parallel.

For the spring we may write

$$p = me, \tag{35}$$

and for the dash-pot (see (2.71) for three dimensional case)

$$p = \eta \dot{e} \tag{36}$$

with corresponding relaxation function $\eta \delta(t)$.

A spring and dash-pot in series (see Fig. 4.3) gives equation (2.78) or a Maxwell model. In this case the tension is the same in both elements and the total extension is the sum of that in each element. Therefore, differentiating,

$$\dot{e} = \dot{e}_S + \dot{e}_D$$
$$= \dot{p}/m + p/\eta. \tag{37}$$

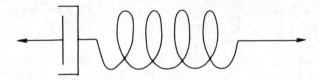

Fig. 4.3. The Maxwell element. Dash-pot and spring in series.

It is left as an exercise to the reader to verify that the Maxwell relaxation function is

$$\psi(t) = m \exp(-mt/\eta) \qquad t > 0. \tag{38}$$

The inverse relation for strain as a function of stress gives the corresponding *creep* function (i.e. response to constant force)

$$\phi(t) = 1/m + t/\eta, \qquad t > 0 \tag{39}$$

Parallel connection of the elements gives the Kelvin–Voigt model (see (2.74)). In this case the total stress is the sum of the two separate stresses

$$p = p_S + p_D$$
$$= me + \eta \dot{e}. \tag{40}$$

Here, simply,

$$\psi(t) = mH(t) + \eta\delta(t) \tag{41}$$

and

$$\phi(t) = [1 - \exp(-mt/\eta)]/m \qquad t > 0. \tag{42}$$

Another class of model of anelastic behaviour is suggested by experimental studies. Rather than fitting complicated systems of springs and dashpots, we can fit the strain increase to a power or logarithm of time. Consider, for example,

$$\phi(t) = 1/m_0 + At^\alpha, \qquad t > 0, \tag{43}$$

where m_0, A and α are constants ($0 < \alpha < 1$). Note that $\alpha = 1$ yields Maxwell elasticoviscosity.

Jeffreys in 1958 proposed a composite form for anelastic behaviour in the Earth

$$\phi(t) = 1/m_0 + q[(1 + at)^\alpha - 1]/\alpha, \tag{44}$$

where q and a are positive constants and $0 \leqslant \alpha \leqslant 1$. A value $\alpha = 0.25$ gives an effective model for many seismic wave propagation problems. Note that when at is small $\phi(t)$ is again the Maxwell form.

4.5.3 Damping of harmonic waves. The quality factor Q

Consider sinusoidal oscillations of a homogeneous visco-elastic medium. We may write

$$e_{ij} = \bar{e}_{ij}\exp(-i\gamma t) \tag{45}$$

and similarly for p_{ij}. By (32) and (33), for a disturbance starting at $t = -T$,

$$\bar{p}_{ij} = \lambda_0 \delta_{ij}\bar{e}_{ll} + 2\mu_0\bar{e}_{ij}$$
$$+ \int_{-T}^{t} [\dot{R}_\lambda(t-\tau)\exp i\gamma(t-\tau)\delta_{ij}\bar{e}_{ll}$$
$$+ 2\dot{R}_\mu(t-\tau)\exp i\gamma(t-\tau)\bar{e}_{ij}]\,d\tau. \tag{46}$$

In the limit as $T \to \infty$, equation (46) becomes

$$\bar{p}_{ij} = (\lambda_0 + \lambda_1(\gamma))\delta_{ij}\bar{e}_{ll} + 2(\mu_0 + \mu_1(\gamma))\bar{e}_{ij}, \tag{47}$$

where λ_1, μ_1 are the one-sided Fourier transforms of $\dot{R}_\lambda$ and $\dot{R}_\mu$.

This stress–strain relation is identical with (2.32) with the real perfectly elastic moduli replaced by the complex moduli $\bar{\lambda} = \lambda_0 + \lambda_1(\gamma)$, $\bar{\mu} = \mu_0 + \mu_1(\gamma)$. Hereafter we will interpret $\bar{p}_{ij}$, $\bar{e}_{ij}$ as complex functions.

By equation (2.52) the energy dissipated per unit volume in time dt is given by

$$dW/dt = p_{ij}\dot{e}_{ij} \tag{48}$$
$$= -\tfrac{1}{2}\gamma\mathscr{I}\{\bar{k}|\bar{\theta}|^2 + 2\bar{\mu}|\bar{E}_{ij}|^2\}, \tag{49}$$

from (2.70).

At this stage we introduce the important seismological quantity $Q(\gamma)$ called the quality factor. Its inverse Q^{-1} is the ratio of the energy lost in one harmonic cycle to 2π times the purely elastic energy stored in the oscillation.

We may then write the damping factors for purely dilatational and deviatoric waves as

$$Q_k^{-1} = -\frac{\tfrac{1}{2}\mathscr{I}\bar{k}|\bar{\theta}|^2}{\tfrac{1}{2}k_0|\bar{\theta}|^2}, \tag{50}$$

and

$$Q_\mu^{-1} = \frac{\tfrac{1}{2}\mathscr{I}\bar{\mu}|\bar{E}_{ij}|^2}{\tfrac{1}{2}\mu_0|\bar{E}_{ij}|^2}. \tag{51}$$

In either case, the form is

$$Q^{-1}(\gamma) = -\mathscr{I}(\bar{m})/m_0 = -\mathscr{I}(m_1(\gamma))/m_0 \tag{52}$$

where m_0 is the instantaneous modulus given in §4.5.1. It is convenient in terms of the Fourier transforms of the relaxation functions, to write

$$m_1(\gamma) = \int_0^\infty \dot{\psi}(t) \exp(i\gamma t) \, dt. \tag{53}$$

We can then demonstrate two important properties of $Q^{-1}(\gamma)$ for well-behaved relaxation functions: from (52) and (53), $Q^{-1}(\gamma) \to 0$ as $\gamma \to 0$ and as $\gamma \to \infty$. (In some treatments, m_0 is replaced in (52), with some disadvantage, by $\mathcal{R}(\bar{m})$.)

It is now possible to calculate the explicit forms of Q for the various anelastic models. As an illustration, for waves through an elasticoviscous or Maxwell medium for which, from (38),

$$\dot{\psi}(t) = -m_0^2 \{\exp(-m_0 t/\eta)\}/\eta \quad (t > 0),$$

integration of (53) and then substitution in (52) yields

$$Q^{-1} = \frac{\eta\gamma/m_0}{1 + (\eta\gamma/m_0)^2}. \tag{54}$$

As expected, as frequency both increases and decreases, $Q^{-1}(\gamma)$ becomes small with a maximum at $\gamma = m_0/\eta$. Other formulae have been derived and the reader is referred to the book by J.A. Hudson (1980) for an evaluation of $Q(\gamma)$ for the Jeffreys law.

4.6 Thermodynamical conditions

The thermodynamical conditions during the transmission of seismic waves will be approximately adiabatic, and so, as pointed out in §4.1, we take the elastic parameters λ, μ, k to be the adiabatic ones. It is of interest, however, to compare the speeds of P and S waves under adiabatic and isothermal conditions.

As in §2.4.5, we let γ denote the coefficient of cubical expansion of the medium, c the specific heat at constant strain, and ϑ the absolute temperature. Then, to sufficient accuracy, it follows from (2.57) and (4) that the speed of P waves would in isothermal conditions be given by

$$\left\{ k\left(1 - \frac{k\gamma^2\vartheta}{\rho c}\right) + \tfrac{4}{3}\mu \right\}^{\frac{1}{2}} \rho^{-\frac{1}{2}}, \tag{55}$$

where k is the adiabatic incompressibility. For an ordinary solid, the difference in the speeds given by (4) and (55) is only about 1 per cent, but it might approach 10 per cent in the Earth's deep interior where ϑ and k/ρ have greater values than at the surface.

Since the rigidity μ is the same in adiabatic and isothermal conditions, the speed of S waves is precisely the same in both cases.

4.7 Finite-strain effects

The high stresses reached in the Earth's interior involve finite strains. But there is the simplifying feature that the strengths of materials in the Earth set limits of the order of 10^9dynes/cm^2 to the stress-differences that can be sustained; any cause tending to raise the stress-difference beyond the limit results in fracture (and hence an earthquake) or plastic flow. On the other hand, the mean, $-p$ say, of the principal stresses, p_1, p_2, p_3, say, is already (see § 13.5) 10^{10}dynes/cm^2 at a depth of 35 km, and steadily increases to values exceeding 10^{12}dynes/cm^2 in the deeper interior. Thus throughout most of the Earth, $p_1 \approx p_2 \approx p_3$, and the stress is effectively of the form $-p\delta_{ij}$ and representable in terms of a hydrostatic pressure p.

The stresses accompanying seismic wave transmission may therefore to close accuracy be treated as additional infinitesimal stresses superposed on a finite hydrostatic stress. In these circumstances, it is permissible to ignore effects of anisotropy which would enter significantly if the initial stress were not hydrostatic. The initial stress also causes ρ, k and μ to vary in otherwise homogeneous regions of the Earth, but the effect on wave transmission is negligible when ρ, k and μ change only by small proportions inside one wave-length. Hence theory based on (4) and (5) remains serviceable in general.

Finite-strain theory based on (2.81) to (2.86) is of importance in applying the results of seismological investigations to questions of the Earth's internal composition. In particular, useful results on the variation of ρ, k and μ with p have been obtained by Birch and others. Parameters additional to those in the first-order theory have to be introduced, and data from high-pressure experiments have to be brought to bear.

The case of seismic waves superimposed on a media first subjected to static finite strain has received some study. Under realistic restrictions, the appropriate differential equation is similar to (2.9) for an inhomogeneous, anisotropic elastic medium, except that, for example, A_{ijkl} does not have full symmetry (see equation (2.27)). Progress towards usable solutions then depends upon selection of simplifying assumptions. The analysis is quite complicated in general, but for homogeneous pre-strain the mathematical methods, such as plane-wave expansions and slowness surfaces, used for waves in anisotropic media apply. In this case, there is a P wave slowness surface and two slowness surfaces for S waves which are, however, in general coupled. The shapes of the nodal surfaces on which there is zero wave amplitude, of interest for fault-plane solutions (see § 16.2.1), also may become more complicated for certain pre-strains. Further analysis has been given by K. Walton.

4.8 Case of spherical waves

Transformation to spherical coordinates (r, θ, ϕ), by the formula of § 2.3, of the tensor equation of motion (2.9) and stress–strain relation (2.32) was shown by A.E.H. Love to produce the elastodynamic equations

$$\left(\frac{\partial^2}{\partial t^2} - \alpha^2 \nabla^2 \right) \operatorname{div} \mathbf{u} = 0, \tag{56}$$

$$\left(\frac{\partial^2}{\partial t^2} - \beta^2 \nabla^2 \right) (r \operatorname{curl}_r \mathbf{u}) = 0, \tag{57}$$

and

$$\frac{\partial^2 u_r}{\partial t^2} = \alpha^2 \frac{\partial}{\partial r} \operatorname{div} \mathbf{u} - \beta^2 \operatorname{curl}_r \operatorname{curl} \mathbf{u}, \tag{58}$$

where α and β are given by (4) and (5) and curl_r denotes the r component of the rotation.

As in electromagnetic theory, we may then use the method of Helmholtz to decompose the displacement $\mathbf{u}$ into a scalar (dilatational) and vector (shear) potential

$$\mathbf{u} = \operatorname{grad} \Phi \pm \operatorname{curl} \mathbf{A}, \tag{59}$$

where we may impose the solenoidal condition $\operatorname{div} \mathbf{A} = 0$. The decomposition (59) is not unique because $\mathbf{A}$ is arbitrary to an additive term $\operatorname{grad} \chi$, where χ is some scalar function fixed by the boundary conditions.

Substitution of (59) in (56), (57) and (58) yields equations which are clearly satisfied if

$$\left(\frac{\partial^2}{\partial t^2} - \alpha^2 \nabla^2 \right) \Phi = 0, \tag{60}$$

$$\left(\frac{\partial^2}{\partial t^2} - \beta^2 \nabla^2 \right) (r A_r) = 0 \tag{61}$$

and

$$\left(\frac{\partial^2}{\partial t^2} - \beta^2 \nabla^2 \right) (r \operatorname{curl}_r \mathbf{A}) = 0. \tag{62}$$

In deriving (62) we use the vector identity

$$r\operatorname{curl}_r\operatorname{curl}\mathbf{a} = \frac{1}{r}\frac{\partial}{\partial r}(r^2\operatorname{div}\mathbf{a}) - \nabla^2(r\,a_r).$$

Again clearly the construction of the wave displacement from the potential solutions of the above equations is not unique; the question is discussed in the book by J.A. Hudson.

It is now straightforward to show that equations (60), (61), and (62) describe spherical waves of P, SH and SV type, respectively. In the first case, simply, $\operatorname{div}\mathbf{u} = \nabla^2\Phi = 0$ and the wave is dilatational. In the second case, $\Phi = 0$, $\operatorname{curl}_r\mathbf{A} = 0$ so that $u_r = 0$. Introduce the arbitrary Hertzian function M, where

$$u_\theta = \frac{1}{\sin\theta}\frac{\partial M}{\partial\phi}, \quad u_\phi = -\frac{\partial M}{\partial\theta}. \tag{63}$$

Then, from (61),

$$\left(\frac{\partial^2}{\partial t^2} - \beta^2\nabla^2\right)M = 0 \tag{64}$$

which is the wave equation for S waves with motion tangential to the surface of the sphere. The third case is left as an exercise (see §4.9.7).

The wave equations (60), (61) and (62) can be analysed by separation of variables (see §3.8). Consider spherical P wave displacements given by (60). If

$$\Phi = U(r)F(\theta,\phi)\exp i\omega t,$$

separation gives Legendre's associated equation in θ, the simple harmonic equation in both ϕ and t, and a form of Bessel's equation (see §3.5) in r. Explicitly, the latter equation is

$$\frac{d^2U}{dr^2} + \frac{2}{r}\frac{dU}{dr} + \left\{\frac{\omega^2}{\alpha^2} - \frac{l(l+1)}{r^2}\right\}U = 0, \tag{65}$$

where l and ω are separation constants.

The general solution is thus a constant multiple of

$$\Phi = \left(\frac{\omega r}{\alpha}\right)^{-\frac{1}{2}} H_{l+\frac{1}{2}}\left(\frac{\omega r}{\alpha}\right) P_l^m(\cos\theta)\exp\{i(m\phi \pm \omega t)\}, \tag{66}$$

where m is a third separation constant, $P_l^m(\cos\theta)$ is a Legendre associated function (a tesseral harmonic) and $H_{l+\frac{1}{2}}(\omega r/\alpha)$ is Hankel's Bessel function of the first or second kind. The components of $\mathbf{u}$ come from (59).

4.9 Exercises

1   Show that because the equations (2.78) are formally derived from the elastic afterworking equations the general conclusions of §4.2.1 still hold if the medium is of Maxwell type. In particular, there is no displacement at a given point until the instant of onset as given by perfect elasticity theory; there is a sharp onset (with slightly reduced amplitude); and there is no dispersion.

2   Using a suitable Green's function, solve the one-dimensional scalar wave equation in an infinite medium with no body forces for the initial conditions

$$u(x, 0) = U(x),$$
$$\dot{u}(x, 0) = V(x).$$

3   Consider a pressure acoustic wave propagating outwards from a small harmonically vibrating source. If the flow of the medium density ρ away from the source at time t is $S \exp(-i\gamma t)$, show from the pressure wave equation that the pressure, fluid velocity and mean square energy are, respectively,

$$p = -\frac{i\gamma\rho}{4\pi r} S \exp\{i\kappa(r - ct)\},$$

$$u_r = \frac{p}{\rho c}\left(1 + \frac{ic}{\gamma r}\right),$$

$$W = \rho\left(\frac{1}{4\pi r^2}\right)^2 |S|^2 [(\kappa r)^2 + \tfrac{1}{2}].$$

Show that (a) in the *far field* (i.e. $r \gg 2\pi/\kappa$), $u_r = p/\rho c$ and the velocity is in phase with the pressure, and (b) in the *near field* ($r \ll 2\pi/\kappa$) there is a velocity component out of phase with the pressure near the source giving the second term in the third equation above.
[Hint: For acoustic waves

$$\rho\frac{\partial \mathbf{u}}{\partial t} = -\operatorname{grad} p \quad \text{and} \quad \frac{\partial p}{\partial t} = -\operatorname{div} \mathbf{u}.]$$

4   A taut infinite string under tension P is subjected to a transverse force F, distributed uniformly along the string from $x = 0$ to $x = b$, and applied for $0 < t < \tau$. Show that the appropriate point force to obtain the Green's function is $F\delta(x - x_0)$ for $0 < t < \tau$ and zero for $t < 0, t > \tau$. Hence prove the Green's function is

$$G(x, x_0, t) = \begin{cases} 0, & ct < |x - x_0|, \\ ct - |x - x_0|, & |x - x_0| \ll t < |x - x_0| + c\tau, \\ c\tau, & |x - x_0| + c\tau < ct, \end{cases}$$

apart from a factor $F/2p$. Draw the wave for the times $t = 2\tau/3, 5\tau/3$.
    By integration of G above as a function of x_0 from 0 to b, show that for

$x > b$, the transverse wave displacement for the distributed load is

$$y(x,t) = F(ct + b - x)^2, \quad ct < x < ct + b, \quad x - b < ct < x,$$
$$= 2Fbc\tau, \quad ct - c\tau > x, \quad x + c\tau < ct,$$

if $c\tau > b$. Derive solutions also for the interval $x < b$ and then draw the wave form.

5 By analogy with the relaxation function $\psi(t)$ defined in §4.5.1, derive the inverse relations

$$e(t) = \frac{P(t)}{m_0} + \int_0^t \dot{\phi}(t - \tau)\sigma(\tau)\,d\tau \quad (t > 0),$$

where $\phi(t)$ is the *creep function*, zero for $t < 0$ and continuous and non-decreasing for $t > 0$.

Show that $\dot{\phi}(t) \to$ constant as $t \to \infty$ and interpret this property in terms of 'elastic afterworking'.

What is the form of the above equation when there is no resistance to shear stresses but resistance to dilatation?

6 Prove from the definitions of the relaxation function $\psi(t)$ and the creep function $\phi(t)$ that

$$\int_0^t \phi(t - \tau)\psi(\tau)\,d\tau = t \quad (t > 0).$$

Then show that, given $\psi(t)$, the Fourier transform of $\phi(t)$ can be written down.

7 Show that the equation (62) corresponds to spherical waves of SV type by introduction of the arbitrary Hertzian function

$$A_\theta = \frac{1}{\sin\theta}\frac{\partial M}{\partial\phi}, \quad A_\phi = -\frac{\partial M}{\partial\theta}.$$

Prove that the case corresponds to $\text{div}\,\mathbf{u} = 0$ and $\text{curl}_r\,\mathbf{u} = 0$, i.e. the component of rotation normal to the spherical wavefront is zero.

8 Given the bulk modulus and rigidity modulus of a rock is 2×10^{11} dynes/cm^2 and the density is $2.5\,\text{g/cm}^3$, find the velocity of longitudinal and transverse waves and Poisson's ratio.

5

Surface elastic waves and eigen-vibrations of a sphere

We now proceed to examine the transmission of waves over the boundary surface of an elastic medium. Except where explicitly stated (as in §5.7), we take the medium to be perfectly elastic and isotropic, ignore effects of fluctuations in the external forces, and use the infinitesimal strain theory. In chapter 4, the effects of departures from ideal conditions for the case of seismic body waves were examined in some detail, and it was shown that these effects are usually very small. We expect, therefore, that the theory of surface waves based on similar assumptions will likewise apply with close accuracy to actual seismic surface waves.

We will consider at length Rayleigh and Love waves which are important in seismology. These two types of waves are transmitted along the free surface of the Earth in appropriate circumstances; but it is convenient first to examine a more general problem.

5.1 Waves guided along a plane boundary

Let M and M' be two homogeneous, perfectly elastic media in welded contact, separated by a plane horizontal boundary and extending to indefinitely great distances from the boundary, M' being above M (the boundary is taken horizontal to enable us later to use the symbols SH and SV – see §4.1.1). We use the usual symbols ρ, α, β, etc. to denote properties of the medium M, and dashed symbols for M'. As a reference system we take a set of orthogonal axes $Ox_1x_2x_3$, the origin O being any point of the boundary, and Ox_3 pointing normally into M' (see Fig. 5.1).

Consider a wave travelling in the direction Ox_1 in such a manner that (a) the disturbance is confined to the neighbourhood of the boundary, and (b) at any instant all particles in any line parallel to Ox_2 have equal displacements. On account of (a), the wave is an *internal wave*; and on account of (b), the case we have taken is analogous to the plane waves described in chapter 3.

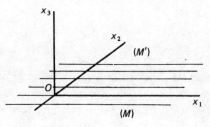

Fig. 5.1. Plane boundary.

It follows from (*b*) that all partial derivatives with respect to the coordinate x_2 are zero. Functions, ϕ and ψ say, of x_1, x_3 and t then exist such that the components u_1 and u_3 of the displacement at any point may be expressed in the form (see §4.8),

$$u_1 = \frac{\partial \phi}{\partial x_1} + \frac{\partial \psi}{\partial x_3}, \quad u_3 = \frac{\partial \phi}{\partial x_3} - \frac{\partial \psi}{\partial x_1}. \tag{1}$$

From (1) we deduce

$$\nabla^2 \phi = \theta \tag{2}$$

and

$$\nabla^2 \psi = \frac{\partial u_1}{\partial x_3} - \frac{\partial u_3}{\partial x_1}. \tag{3}$$

The equations (2) and (3) illustrate the reason for the scalar functions ϕ and ψ. We thereby separate the dilatational and the rotational disturbances associated with the components u_1, u_3 (plane strain). The component u_2 is of course associated with purely distortional movement (antiplane strain). Thus ϕ is associated with P waves, ψ with SV waves, and u_2 with SH waves. (A similar decomposition for spherical elastic waves is given in §4.8.) Our assumptions are that the u_i within the medium M satisfy (4.1). This will be the case if ϕ, ψ and u_2 satisfy

$$\left. \begin{array}{l} \dfrac{\partial^2 \phi}{\partial t^2} = \alpha^2 \nabla^2 \phi, \\[2mm] \dfrac{\partial^2 \psi}{\partial t^2} = \beta^2 \nabla^2 \psi, \end{array} \right\} \tag{4a}$$

$$\frac{\partial^2 u_2}{\partial t^2} = \beta^2 \nabla^2 u_2 \tag{4b}$$

in M; and similar relations in M', with α, β replaced by α', β'. This step follows from a consideration of (2), (3) and the various equations of §4.1.

Towards solving the equation (4), we put

$$\left.\begin{aligned}\phi &= f(x_3)\exp\{i\kappa(x_1 - ct)\},\\ \psi &= g(x_3)\exp\{i\kappa(x_1 - ct)\},\end{aligned}\right\} \tag{5a}$$

$$u_2 = h(x_3)\exp\{i\kappa(x_1 - ct)\} \tag{5b}$$

in M; and similar relations in M', the symbols f, g, h being replaced by f', g', h'. This will lead us to a particular solution corresponding to a group of simple harmonic waves of wave-length $2\pi/\kappa$ travelling forward with speed c.

It is convenient to introduce r, s, r', s', where

$$r = \left(\frac{c^2}{\alpha^2} - 1\right)^{\frac{1}{2}}, \quad s = \left(\frac{c^2}{\beta^2} - 1\right)^{\frac{1}{2}},$$

$$r' = \left(\frac{c^2}{\alpha'^2} - 1\right)^{\frac{1}{2}}, \quad s' = \left(\frac{c^2}{\beta'^2} - 1\right)^{\frac{1}{2}}, \tag{6}$$

the positive value of the square root being taken in each case.

Substituting from (5b) into (4b), we have the differential equation for SH motion for the uniform half-space M'

$$\frac{d^2 h'}{dx_3^2} + \kappa^2 \left(\frac{c^2}{\beta'^2} - 1\right) h' = 0, \tag{7a}$$

of which the solution (β' constant) is

$$h'(x_3) = C' \exp(-i\kappa s' x_3) + F' \exp(i\kappa s' x_3), \tag{7b}$$

where C', F' are constants of integration. For the waves to be essentially trapped, $h'(x_3)$ must diminish with increasing distance from the boundary. This will be the case if $h'(x_3)$ contains an exponential factor in which the exponent is real and negative. Hence s', and similarly s, r and r', are taken to be imaginary. Furthermore constants corresponding to C' must vanish in M', and likewise those corresponding to F' in M.

The form of the solution in M is therefore

$$\left.\begin{aligned}\phi &= A\exp\{i\kappa(-rx_3 + x_1 - ct)\},\\ \psi &= B\exp\{i\kappa(-sx_3 + x_1 - ct)\},\end{aligned}\right\} \tag{8a}$$

$$u_2 = C\exp\{i\kappa(-sx_3 + x_1 - ct)\}; \tag{8b}$$

and in M'

$$\left.\begin{aligned}\phi &= D'\exp\{i\kappa(r'x_3 + x_1 - ct)\},\\ \psi &= E'\exp\{i\kappa(s'x_3 + x_1 - ct)\},\end{aligned}\right\} \tag{9a}$$

$$u_2 = F'\exp\{i\kappa(s'x_3 + x_1 - ct)\}, \tag{9b}$$

where A, B, C, D', E', F' are constants, and r, s, r', s' are all positive imaginaries.

We now apply the boundary conditions, which are (i) the displacement at, and (ii) the stress across the boundary surface between M and M' are continuous at all times and places.

From (1), the conditions (i) give

$$\left.\begin{array}{l} A - Bs = D' + E's', \\ - Ar - B = D'r' - E', \end{array}\right\} \tag{10a}$$

$$C = F'. \tag{10b}$$

The components of stress involved in (ii) are p_{31}, p_{32}, p_{33}. By (2.32) and (1), these components are for M

$$p_{33} = \lambda\theta + 2\mu\frac{\partial u_3}{\partial x_3} = \lambda\mathbf{V}^2\phi + 2\mu\left(\frac{\partial^2\phi}{\partial x_3^2} - \frac{\partial^2\psi}{\partial x_3\partial x_1}\right), \tag{11}$$

$$p_{31} = \mu\left(\frac{\partial u_3}{\partial x_1} + \frac{\partial u_1}{\partial x_3}\right) = \mu\left(2\frac{\partial^2\phi}{\partial x_3\partial x_1} - \frac{\partial^2\psi}{\partial x_1^2} + \frac{\partial^2\psi}{\partial x_3^2}\right), \tag{12}$$

$$p_{32} = \mu\frac{\partial u_2}{\partial x_3}, \tag{13}$$

and similarly for M'. On substituting from (8) and (9) into (11), (12) and (13), and using (4.4) and (4.5) we find that the conditions (ii) give

$$\left.\begin{array}{l} [\{\alpha^2(1 + r^2) - 2\beta^2\}A + 2\beta^2 sB]\rho \\ \quad = [\{\alpha'^2(1 + r'^2) - 2\beta'^2\}D' - 2\beta'^2 s'E']\rho', \\ [-2rA - (1 - s^2)B]\beta^2\rho \\ \quad = [2r'D' - (1 - s'^2)E']\beta'^2\rho', \\ -sC\beta^2\rho = s'F'\beta'^2\rho'. \end{array}\right\} \tag{14b}$$

Because s and s' are both positive imaginaries, (10b) and (14b) give that C and F' are both zero. Thus there is no propagation of the component u_2 of displacement, i.e. in this problem there are no internal boundary SH waves.

The detailed solution for the components u_1 and u_3 depends on (10a) and (14a), and involves much algebra in the general case. In the following sections certain special cases will be considered. Further discussion of the general problem may be found in Aki and Richards (1980).

5.2 Rayleigh waves

The case of the above problem in which the plane boundary is a free surface (so that M' is replaced by a vacuum) was first examined by Lord Rayleigh. The corresponding waves are called *Rayleigh waves*.

As before, there are no SH surface waves.

For the other components of displacement, the absence of stress over the free surface enables us to replace the right-hand sides of (14a) by zero, giving

$$(s^2 - 1)A + 2sB = 0$$

and

$$2rA + (1 - s^2)B = 0. \tag{15}$$

Eliminating A and B, we obtain

$$(1 - s^2)^2 = -4rs,$$

i.e., by (6),

$$R(\kappa)/\kappa^4 = \left(2 - \frac{c^2}{\beta^2}\right)^2 - 4\left(1 - \frac{c^2}{\alpha^2}\right)^{\frac{1}{2}}\left(1 - \frac{c^2}{\beta^2}\right)^{\frac{1}{2}} = 0, \tag{16a}$$

where $R(c)$ is Rayleigh's function. On rationalising this equation and removing the factor c^2/β^2, we have

$$\frac{c^6}{\beta^6} - 8\frac{c^4}{\beta^4} + c^2\left(\frac{24}{\beta^2} - \frac{16}{\alpha^2}\right) - 16\left(1 - \frac{\beta^2}{\alpha^2}\right) = 0, \tag{16b}$$

from which c may be determined.

If we substitute into the left-hand side of (16b) the values $c = \beta$ and $c = 0$, we obtain unity and $-16(1 - \beta^2/\alpha^2)$, respectively; this last expression is negative, since $\beta < \alpha$. Hence (16b) has a real root of c lying between 0 and β; from (6), such a root makes both r and s imaginary, thus satisfying the restriction on r and s stated in §5.1.

Examination of the complete range of solutions for (16b) for values of Poisson's ratio $-1 < \sigma < 0.5$ yields conditions for real and complex roots. Complex (conjugate) roots correspond to waves, called *leaky* or *evanescent modes*, that decay with distance along the surface and radiate energy into the half-space (see also §12.3.2). Thus, given suitable generating conditions, Rayleigh waves persist along a free boundary of a homogeneous perfectly elastic half-space. The waves are polarised so that the particles of the medium move in vertical planes parallel to the direction of the wave motion (see Fig. 3.5). Note that the other roots correspond in general to P waves incident on the free surface producing only reflected S waves.

In (8) and (9), κ may have any real positive value, and solutions for particular values of κ may be superposed (cf. §§3.3.1, 3.3.2) to give more general solutions. But since in the present problem the velocity c is by (16) determined independently of κ, it follows that there will be no dispersion of a general wave form in the present conditions.

When Poisson's relation (§4.1.2) holds, (16) yields three real values of c^2/β^2, namely, 4, $(2 + 2/\sqrt{3})$ and $(2 - 2/\sqrt{3})$. The first two of these values are both greater than 3 and thus make both r and s real, so that there

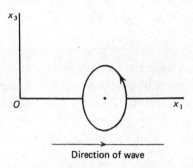

Fig. 5.2. Rayleigh wave particle motion (half-space).

could be no corresponding surface wave solutions. The third value leads to the results

$$c = 0.92\beta, \quad r = 0.85i, \quad s = 0.39i. \tag{17}$$

The first of the equations (17) shows that the speed of Rayleigh waves in a homogeneous, isotropic, perfectly elastic half-space is 0.92 of the speed of S body waves in the medium. The relevant expressions for u_1 and u_3 corresponding to simple harmonic waves of wave-length $2\pi/\kappa$ are found from (17) using (1), (8a) and (15), and taking real parts. The results are

$$\left.\begin{aligned} u_1 &= a\{\exp(0.85\kappa x_3) - 0.58\exp(0.39\kappa x_3)\}\sin\{\kappa(x_1 - ct)\}, \\ u_3 &= a\{-0.85\exp(0.85\kappa x_3) + 1.47\exp(0.39\kappa x_3)\}\cos\{\kappa(x_1 - ct)\}, \end{aligned}\right\} \tag{18}$$

where a is a constant related to A and κ. Putting $x_3 = 0$, we see that during the passage of the disturbance, a surface particle describes an ellipse given by

$$u_1 = 0.42a\sin\eta, \quad u_3 = 0.62a\cos\eta, \tag{19}$$

where η is a parameter which decreases as time increases. The ellipse is therefore described in a 'retrograde' fashion (see Fig. 5.2) and the maximum displacement parallel to the direction of transmission is about two-thirds of that in the vertical direction, in this case.

5.3 Stoneley waves

It turns out that real internal elastic waves involving the scalars ϕ and ψ may also propagate as trapped interface waves along the boundary between the two homogeneous half-spaces M and M' (see Fig. 5.1). The existence theorem was first proved by Stoneley in 1924. We need to examine the roots of the 4×4 determinant of the coefficients of A, B, D', E' in equations (10a) and (14a).

The behaviour of the roots of the Stoneley determinant depends upon the elastic properties of the media (see Exercise 5.8.4). For a Poisson's ratio $\sigma = 0.25$ in M and M', there is a range of real solutions; the corresponding Stoneley waves are not dispersive, with the Stoneley wave velocity lying between β' ($\beta' < \beta$) and the Rayleigh wave velocity in M. Particle motions are in general retrograde and there is an exponential decay in amplitude away from the boundary.

Under certain conditions there are complex roots of the quartic equation in c. Generally these solutions correspond to 'pseudo-waves' for which the amplitudes die off more rapidly with distance of propagation than they do in purely real internal waves. Further discussion is given in the book by Pilant.

Obviously, Stoneley waves and Rayleigh waves are closely related, with limiting transitions occurring if M is replaced by a layer of finite width. Stoneley waves have been experimentally observed along the interface between a liquid layer and an elastic half-space. The liquid allows wave receivers to be placed at various points in the fluid layer (see § 12.4). In this case, equations (10a) and (14a) simplify and the Stoneley equation reduces to the form given in Exercise 5.8.4. There is always a real root in this case.

5.4 Love waves

Surface waves of the SH type are observed to occur on the Earth's surface (see Fig. 3.5). The reason is that actual conditions in the Earth differ in essential respects from those set down in the problem of § 5.1.

Consider first the presence of a homogeneous layer over the half-space.

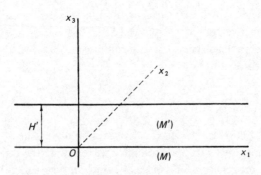

Fig. 5.3. Layer over half-space.

Suppose that this layer M' is bounded above by a plane (horizontal) free surface, and below by a parallel boundary at a distance H' from the free surface. The latter boundary separates M' from the half-space M, the contact being, as before, welded. A.E.H. Love showed that in these circumstances, waves of the SH type can travel along the free surface.

It is sufficient to consider the component u_2 of the displacement. We take an origin in the plane boundary between M and M', and directions of axes as in § 5.1 (see Fig. 5.3). The essential difference between the analysis of § 5.1 and that now required is that the displacement in M' may no longer diminish with distance from the boundary between M and M'. Thus for the medium M' we preserve the full solution obtained using (7), and replace (9b) by

$$u_2 = C' \exp\{i\kappa(-s'x_3 + x_1 - ct)\} + F' \exp\{i\kappa(s'x_3 + x_1 - ct)\},$$
(20)

where s' is now not necessarily imaginary. For M we still have (8b) with s imaginary.

The boundary conditions are (i) and (ii) of § 5.1, together with the requirement (iii) that there shall be no stress across the free surface $x_3 = H'$. These three conditions give, using (8b), (13) and (20),

$$C = C' + F',$$
(21a)

$$\mu s C = \mu' s'(C' - F'),$$
(21b)

$$C' \exp(-i\kappa s'H') = F' \exp(i\kappa s'H').$$
(21c)

Eliminating C, C' and F' from (21a, b, c), we obtain Love's function

$$L(c, \kappa) = \mu i s + \mu' s' \tan(\kappa s'H') = 0.$$
(22)

SH surface (Love) waves can therefore exist if (22) can be satisfied. Substitution for s and s' from (6) into (22) gives the equation for the velocity c of Love waves, namely,

$$\mu\left(1 - \frac{c^2}{\beta^2}\right)^{\frac{1}{2}} - \mu'\left(\frac{c^2}{\beta'^2} - 1\right)^{\frac{1}{2}} \tan\left\{\kappa H'\left(\frac{c^2}{\beta'^2} - 1\right)^{\frac{1}{2}}\right\} = 0.$$
(23)

The requirement that s should be imaginary and hence, by (22), s' real is, by (6), satisfied if $\beta' < c < \beta$. Thus for Love waves to be possible, the velocity of S waves in the lower medium M must be greater than that in M', and the wave velocities c of the Love waves must lie between the two velocities β and β' of the S body waves.

The equation (23) shows that c is dependent on wave number κ, and is not a constant, so that in the present boundary conditions there will be dispersion of a general wave form. We see from (23) that if κ is small, $c \to \beta$;

i.e. the velocities of the longer Love waves approach the velocity of S waves in M. Also if κ is large, $c \to \beta'$, i.e. the velocities of the very short waves approach the velocity of S waves in the upper medium M'. Note that for $\beta < \beta'$, (22) has no positive real roots and there are no propagating surface waves. If complex wave numbers κ are considered, (22) shows that we have another case of leaky modes.

5.4.1 Nodal planes

Now change the origin of axes to a point in the free surface (i.e. the upper boundary of M') and reverse the direction of the 3-axis. We find on using (20) and (21c) that the displacement in M' corresponding to a given κ is given by the real part of

$$2C' \cos(\kappa s' x_3) \exp\{i\kappa(-s'H' + x_1 - ct)\}. \tag{24}$$

The presence of the real cosine factor containing x_3 shows that the displacement is zero for all values of x_1, x_2 and t, if $\kappa s' x_3$ takes any of the particular values $\frac{1}{2}\pi$, $\frac{3}{2}\pi$, $\frac{5}{2}\pi$,.... The corresponding planes (which are parallel to the free surface) are nodal planes (cf. § 3.2.1), the wave numbers κ are *eigenvalues* and the waves correspond to *modes* of the vibrations.

For values of κ for which $\kappa s' H' < \frac{1}{2}\pi$, there will be no nodal planes; for values of κ for which $\frac{1}{2}\pi < \kappa s' H' < \frac{3}{2}\pi$, one nodal plane; and so on. It follows that the cases of no, one, two, etc., nodal planes will be associated with ranges of values of κ which correspond to different branches of the tangent function in (22) or (23) and hence to different modes.

In practice in seismology, recorded surface waves of both Love and Rayleigh type consist of the superposition of different modes. Usually, however, the fundamental mode (no nodal planes) and first and second higher modes predominate (see Fig. 3.5).

5.4.2 Dispersion curves

Let C be the group velocity associated with any particular values of κ; then, by (3.69),

$$\begin{aligned}
\frac{C}{\beta'} &= \frac{1}{\beta'} \frac{\mathrm{d}(\kappa c)}{\mathrm{d}\kappa} \\
&= \frac{c}{\beta'} + \kappa H' \frac{\mathrm{d}(c/\beta')}{\mathrm{d}(\kappa H')}.
\end{aligned} \tag{25}$$

If we assign values of μ/μ' and ρ/ρ', we can, using (23) and (4.5) draw up a corresponding table connecting values of c/β' and $\kappa H'$ for the case of any particular Love wave mode. Using this table and (25), we can then find values of C/β' in terms of c/β' and also in terms of $\kappa H'$. The curves connecting these values are called *dispersion curves*; their forms depend on the values taken for μ/μ' and ρ/ρ'.

In seismology, various periods $(2\pi/\kappa c)$ may be measured from an earthquake record and the corresponding group velocities determined from a knowledge of the time, phase properties and location of the earthquake source. Phase velocities can be determined between recording points from phase spectral differences (see § 5.6.6 and § 14.3.3). If the waves are Love waves, the dispersion curves constructed as described then enable us to estimate H', if β, μ/μ' and ρ/ρ' be presumed known. This method has been usefully employed in the study of the Earth's crustal structure, and will be referred to again in chapters 12 and 13. Examples of dispersion curves constructed by Oliver from observed surface waves are given in Fig. 11.1.

Jeffreys first showed, using (23) and (3.6), that with materials similar to those in the Earth's crust there exists a minimum group velocity of Love waves, which is less than β'. Waves with maxima or minima group velocities are called *Airy phases* and § 3.6 is then relevant.

5.4.3 The differential equation for continuously varying media

Stratification is, of course, unnecessary for Love waves to exist. Let the half-space M now have a free surface at $x_3 = 0$ (take Ox_3 into the medium), and continuously varying elastic properties. In this case, instead of the second order differential equation (7a) for Love waves, the equation of motion contains gradients in β. The starting point is (2.33) rather than (2.34b). With $\theta = 0$, $X_i = 0$, we obtain

$$\rho \frac{\partial^2 u_2}{\partial t^2} = \frac{\partial}{\partial x_3} \left\{ \mu \frac{\partial u_2}{\partial x_3} \right\} + \frac{\partial}{\partial x_1} \left\{ \mu \frac{\partial u_2}{\partial x_1} \right\}, \tag{26}$$

where ρ and μ are functions of x_3.

As for Love waves in a homogeneous layer (cf. equation (5b)) put

$$u_2 = h(x_3) \exp\{i\kappa(x_1 - ct)\}.$$

Thus, if primes denote differentiation with respect to x_3,

$$\mu h'' + \mu' h' - \kappa^2 \rho(\beta^2 - c^2)h = 0. \tag{27}$$

This equation (cf. equation (7a)) is similar to Bessel's equation; for arbitrary gradients, it is best solved numerically as an eigen-problem, i.e. find κ such that the eigenvector h has the correct surface wave form for a given phase velocity c. It is convenient to write $y_1 = h, y_2 = \mu \partial h / \partial x_3$ because this transformation on (26) yields

$$-\rho \omega^2 h = \frac{\partial}{\partial x_3} \left\{ \mu \frac{\partial h}{\partial x_3} \right\} - \kappa^2 \mu h. \tag{28}$$

We immediately obtain the equivalent simultaneous equations for numerical integration

$$y_1' = h' = y_2/\mu,$$

and

$$y_2' = (\kappa^2\mu - \rho\omega^2)y_1. \tag{29}$$

Such equations may be solved by a Runge–Kutta integration formula and have the advantage that y_1 and y_2 are proportional to displacement and stress, respectively.

5.5 Surface waves in the presence of multiple layers and sources

The persistence of surface waves following an actual earthquake is evidence of heterogeneity within the Earth, and makes it desirable to study the theory of surface waves in layered media. The simplest case, for which matrix theory is well suited, is a vertical stack of plane-parallel homogeneous layers. Complications such as lateral inhomogeneities and dipping layers require more advanced mathematical treatment (cf. chapter 12).

5.5.1 Rayleigh waves for a single surface layer

The characteristics of Rayleigh wave dispersion for the fundamental and first and second higher modes in a uniform elastic layer M' overlying a uniform elastic half-space M are exhibited in Fig. 5.4. The computations were done by Mooney and Bolt (1966).

The computed dependent variables are phase velocity c, group velocity C, and the ratio of horizontal to vertical surface displacement S_x/S_z. The latter represents the relative amplitude of horizontal to vertical ground displacement at the free surface. It is an observable quantity and sometimes, in experimental work, a diagnostic one. A multiplying factor of -1 is understood for all ratios; the factor corresponds to retrograde motion and a 90° phase shift between the two components. Positive values of the ratio, S_x/S_z, correspond to retrograde particle motion and negative values to progressive or prograde motion. Points where the curve crosses the horizontal axis yield purely vertical surface displacement. Points where the curve approaches infinity yield purely horizontal displacement.

In the medium M, ϕ and ψ are given by (8a) as before, with r and s imaginary but, in M', additional expressions in (9a) of the form

$$A' \exp\{i\kappa(-r'x_3 + x_1 - ct)\} \quad \text{and}$$
$$B' \exp\{i\kappa(-s'x_3 + x_1 - ct)\}$$

are needed (r', s' real). The boundary conditions in §5.1 and §5.2 now lead to six equations in A, B, A', B', D', E'. The Rayleigh wave frequency equation is obtained by setting the 6×6 determinant equal to zero.

For the single surface layer, it is possible to reduce the frequency equation to the form

$$\Delta(c, \kappa) = 0, \tag{30}$$

where Δ is a 3×3 determinant and κ is the wave number. Real solutions of this equation exist under the condition that $1 \leqslant c/\beta' \leqslant \beta/\beta'$ which entails that $\beta' < \beta$. When $c = \beta'$ or β, there are limiting cases but no computational difficulty arises at these points. For all modes except the fundamental, the upper end-point ($c = \beta$) represents the long-period cutoff of unattenuated propagation. No restriction applies to α' or α beyond the physical requirement that Poisson's ratio in both media must lie in the range 0.0 to 0.5; in particular, real solutions exist for either $\alpha' < \beta$ or $\alpha' > \beta$. Complex solutions to the period equation correspond to leaky modes.

To obtain a solution for equation (30), choose a value for the period τ, estimate a trial value for c (and therefore for κ), and evaluate the determinant Δ together with its partial derivatives. An improved estimate for c can then be obtained by the second-order Newton–Raphson iteration method.

On automatically incrementing τ by $\Delta\tau$, the partial derivatives can be used to estimate a new trial value for c. The evaluation of cutoff points for the higher modes was carried out by solving equation (30) for κ (hence τ) subject to the condition that $c = \beta$. Amplitude ratios are obtained from the boundary conditions by substitution of the solutions of equation (30).

Group velocities are often computed by taking finite differences of the phase velocity values for substitution in the relation (3.69). The numerical differentiation may lead to significant errors, particularly in regions of rapid change in phase velocity, near the higher-mode cutoffs, and near the beginning and end of the range of computation.

Alternative methods have been devised for computing more accurate values of group velocity, for example using Rayleigh's principle (cf. §12.3.4). For the problem considered here, however, differentiation may be carried out analytically at the outset.

To obtain a value for $dc/d\kappa$, differentiation of equation (30) gives

$$\left(\frac{\partial\Delta}{\partial c}\right)dc + \left(\frac{\partial\Delta}{\partial\kappa}\right)d\kappa = 0, \tag{31}$$

and substitution into (3.69) or (25) then yields

$$C = c - \kappa \left(\frac{\partial \Delta}{\partial \kappa} \bigg/ \frac{\partial \Delta}{\partial c} \right). \tag{32}$$

The partial derivatives are obtainable by differentiating successive columns in the determinant Δ.

An interesting application of surface wave dispersion methods is the determination of the properties of unconsolidated alluvium. In this case, simple artificial sources can be used to provide data on alluvial thickness

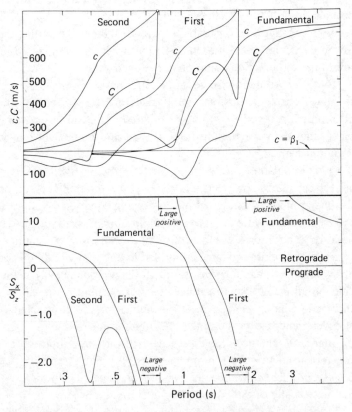

Fig. 5.4. Dispersion curves for the fundamental and two first high modes of Rayleigh waves travelling in the alluvial layer, thickness 100 metres, over shale. Bottom half shows ratio of horizontal to vertical particle motion as a function of period. (Courtesy H.M. Mooney and B.A. Bolt, 1966.)

and physical properties averaged over large distances. This application involves a very large shear velocity ratio and an unusually large
value for Poisson's ratio in the upper layer (0.40 to 0.45). As a specific
example for which shear velocity data are available, the underlying bedrock
was chosen to be the Pierre Shale in the Rocky Mountain area; the
parameters (in km/s) are $\alpha' = 0.57$, $\alpha = 2.6$, $\beta' = 0.20$, $\beta = 0.80$, and the
densities are $\rho' = 1.70$, $\rho = 2.20$ g/cm^3. The results for other choices would
not be greatly different. Thickness is taken arbitrarily as $H' = 100$ metres.

Fig. 5.4 shows the complicated dispersion pattern to be expected. The
fundamental mode group velocity exhibits two steeply rising limbs
separated by an inflection. The first and second higher Rayleigh modes
produce no less than three group-velocity minima each. The dominant
group velocity minimum for all three modes is unusually broad, and the
group-velocity maximum for the first higher mode is unusually narrow. The
Airy phase from the fundamental mode travels very slowly and will appear
much later than the major part of the dispersed wave train.

The combination of high shear velocity ratio and high Poisson's ratio
yields an example of *prograde motion in the fundamental mode*. The
prograde motion is confined to the narrow period range 1.05 to 1.90
seconds, so that the periods at which it would be observed depend very
strongly upon thickness of the alluvial layer. All of the modes pass very
steeply through the region of dominantly vertical motion, hence we may
conclude that most of the surface displacement will be horizontal except at
the shorter periods. (See §17.4.4).

5.5.2 Matrix theory. Love and Rayleigh waves

Matrix notation and theory are now universally used in engineering and seismology to handle efficiently elastic vibrations of layered
systems. The reader is referred to Pestel and Leckie (1963) for the
elastomechanical development.

As an introduction, consider a vibrating longitudinal system of masses
m_i and springs with constant k_i in series (see e.g., Fig. 17.3). The
displacement of the point i from its rest position is x_i and the internal force is
N_i, say, in the spring. Then the *state vector* $\mathbf{z}_i = (x_i, N_i)^T$. Let superscripts L
and R denote conditions just to the left and right of m_i. Then, for
equilibrium of spring k_i,

$$N_i^L = N_{i-1}^R = k_i(x_i - x_{i-1}). \tag{33}$$

We may rewrite (33) as

$$\mathbf{z}_i^L = E_i \mathbf{z}_{i-1}^R, \tag{34}$$

where

$$E_i = \begin{bmatrix} 1 & 1/k_i \\ 0 & 1 \end{bmatrix}.$$

The equation (34) thus transfers the state vector from the right to the left of the spring. Matrices of type $\mathbf{E}$ are therefore called *transfer* or *propagator* matrices.

A vibrating layered elastic system, modelled as above, can be reduced to a set of first-order differential equations such as (29), and transfer matrix theory can be used to obtain numerical solutions. Thus, in matrix form, the equation with constant coefficients

$$\frac{d\mathbf{B}(z)}{dz} = \mathbf{AB}(z) \tag{35}$$

has a solution of form

$$\mathbf{B}(z) = \exp\left[(z - z_0)\mathbf{A}\right]\mathbf{B}(z_0), \tag{36}$$

where $\mathbf{B}(z_0)$ is the initial state matrix and the exponential is the propagator matrix in this case. The basic seismological requirements of matrix evaluation are described in Gantmacher (1959).

Of the alternative general methods of solving (34), a straightforward algorithm is Picard recursion. We note that (36) is equivalent to

$$\mathbf{B}(z) = \mathbf{B}(z_0) + \int_{z_0}^{z} \mathbf{A}(\xi)\mathbf{B}(\xi)d\xi. \tag{37}$$

Because the integral equation contains the initial condition, it is suitable for iteration through the z space. When $\mathbf{B}_i$ represents the ith approximation for $\mathbf{B}$ we have

$$\mathbf{B}_{i+1}(z, z_0) = \mathbf{I} + \int_{z_0}^{z} \mathbf{A}(\xi)\mathbf{B}_i(\xi, z_0)d\xi, \tag{38}$$

and

$$\mathbf{B}_0(z, z_0) = \mathbf{I}.$$

Repeated recursion, in fact, yields

$$\mathbf{B}(z, z_0) = \left(\mathbf{I} + \int_{z_0}^{z} \mathbf{A}(\xi)d\xi + \int_{z_0}^{z} \mathbf{A}(\xi_1)\left[\int_{z_0}^{\xi_1} \mathbf{A}(\xi)d\xi\right]d\xi_1 + \cdots\right)\mathbf{I}.$$

The expression within the outer parentheses, called the matrisant, usually converges (Picard's theorem) and, moreover, for $\mathbf{A}$ constant becomes the exponential series. The matrisant is thus the required propagator matrix $\mathbf{E}(z, z_0)$.

In the seismological problem of transferring the state vector of displacements and stresses from layer zero successively to layer n, with reference to the above mass–spring system, we recognise that the transfer matrices satisfy, for continuous systems, a chain rule,

$$\mathbf{E}(z, z_0) = \mathbf{E}(z, \xi)\mathbf{E}(\xi, z_0). \tag{39}$$

Then for n layers defined by $z_0 \leqslant z_1 \leqslant \cdots \leqslant z_n$,

$$\mathbf{E}(z_n, z_0) = \prod_{i=1}^{n} \mathbf{E}(z_i, z_{i-1}), \tag{40}$$

and for thin or constant property layers, and $z_{i-1} \leqslant \xi_i$,

$$\mathbf{E}(z_n, z_0) \approx \prod_{i=1}^{n} \exp\left[(z_i - z_{i-1})\mathbf{A}(\xi_i)\right]. \tag{41}$$

The method just described is equivalent to the matrix method of Haskell (1953) where the matrix $\mathbf{E}_k$ is called the layer propagator matrix in the kth layer. We next set down the layer matrices for seismic surface waves.

First, the eigenvalues of the Love wave equation (7a) are, for a homogeneous isotropic layer,

$$v^2 = \kappa^2 - \frac{\omega^2 \rho}{\mu}. \tag{42}$$

Elements of the layer matrix of the Love wave for layer thickness h are then, by substitution in (36), and Sylvester's expansion (see §5.8.14)

$$\begin{aligned}
E_{11}(h) &= E_{22}(h) = 1 + 2\,\sinh^2(vh/2) = \cosh vh, \\
E_{12}(h) &= (1/\mu)(\sinh vh)/v, \\
E_{21}(h) &= v^2\mu(\sinh vh)/v.
\end{aligned} \tag{43}$$

We see in (43) that the elements of the layer matrix are always real even if v is imaginary. For transversely isotropic SH motion (see §2.5) we may put $\mu = L$ and $\kappa^2 = N\kappa^2/L$ in (42) and (43).

The eigenvalues of the Rayleigh wave equations of motion in an anisotropic medium (see §12.1.2) may be complex and although the layer matrix is always real by definition, computations may be complicated. In an isotropic body, however, v^2 is always real and the layer matrix simplifies. The eigenvalues of (4a) are then given by

$$\left.\begin{aligned}
v_\alpha^2 &= \kappa^2 - \frac{\omega^2 \rho}{\lambda + 2\mu}, \\
v_\beta^2 &= \kappa^2 - \frac{\omega^2 \rho}{\mu},
\end{aligned}\right\} \tag{44}$$

and the elements of the layer matrix by

$$E_{11}(h) = E_{22}(h) = 1 + \frac{2\mu}{\omega^2\rho}\left[-(\kappa^2 + v_\beta^2)\sinh^2\frac{v_\alpha h}{2} + 2\kappa^2\sinh^2\frac{v_\beta h}{2}\right],$$

$$E_{12}(h) = \frac{1}{\omega^2\rho}\left[-v_\alpha^2\frac{\sinh v_\alpha h}{v_\alpha} + \kappa^2\frac{\sinh v_\beta h}{v_\beta}\right],$$

$$E_{13}(h) = -E_{42}(h) = \frac{\kappa\mu}{\omega^2\rho}\left[2v_\alpha^2\frac{\sinh v_\alpha h}{v_\alpha} - (\kappa^2 + v_\beta^2)\frac{\sinh v_\beta h}{v_\beta}\right],$$

$$E_{14}(h) = -E_{32}(h) = \frac{2\kappa}{\omega^2\rho}\left(\sinh^2\frac{v_\alpha h}{2} - \sinh^2\frac{v_\beta h}{2}\right),$$

$$E_{21}(h) = \frac{\mu^2}{\omega^2\rho}\left[-(\kappa^2 + v_\beta^2)^2\frac{\sinh v_\alpha h}{v_\alpha} + 4\kappa^2 v_\beta^2\frac{\sinh v_\beta h}{v_\beta}\right],$$

$$E_{23}(h) = -E_{41}(h) = 2\mu^2(\kappa^2 + v_\beta^2)E_{14}(h),$$

$$E_{24}(h) = -E_{31}(h) = \frac{\kappa\mu}{\omega^2\rho}\left[(\kappa^2 + v_\beta^2)\frac{\sinh v_\alpha h}{v_\alpha} - 2v_\beta^2\frac{\sinh v_\beta h}{v_\beta}\right],$$

$$E_{33}(h) = E_{44}(h) = 1 + \frac{2\mu}{\omega^2\rho}\left[2\kappa^2\sinh^2\frac{v_\alpha h}{2} - (\kappa^2 + v_\beta^2)\sinh^2\frac{v_\beta h}{2}\right],$$

$$E_{34}(h) = \frac{1}{\omega^2\rho}\left(\kappa^2\frac{\sinh v_\alpha h}{v_\alpha} - v_\beta^2\frac{\sinh v_\beta h}{v_\beta}\right),$$

$$E_{43}(h) = \frac{\mu^2}{\omega^2\rho}\left[4\kappa^2 v_\alpha^2\frac{\sinh v_\alpha h}{v_\alpha} - (\kappa^2 + v_\beta^2)^2\frac{\sinh v_\beta h}{v_\beta}\right].$$

$$\tag{45}$$

In a liquid layer the elements are

$$E_{11}(h) = E_{22}(h) = 1 + 2\sinh^2\frac{v_\alpha h}{2},$$

$$E_{12}(h) = -\frac{v_\alpha^2}{\omega^2\rho}\frac{\sinh v_\alpha h}{v_\alpha},$$

$$E_{21}(h) = -\omega^2\rho\frac{\sinh v_\alpha h}{v_\alpha}.$$

$$\tag{46}$$

5.5.3 Lamb's problem

So far surface wave motions have been assumed without enquiry as to their source. It is instructive to consider as a starting problem the two-dimensional case of an impulsive source concentrated along a line on the surface of an elastic half-space. H. Lamb first solved this problem in a classic paper in 1904 and extended it to a surface point source in three dimensions. Lamb also considered the effect on the wave-forms generated of burying the source. He went on to compute the first synthetic seismogram for vertical displacements far from the source (the 'far field').

Let, at first, the applied normal surface traction downwards be

$$[p_{33}]_0 = -Z(\kappa)\exp\{i\kappa(x_1 - ct)\}. \tag{47}$$

Then from (15) we obtain

$$A = \frac{2\kappa^2 - k^2}{R(\kappa)} \cdot \frac{Z}{\mu}, B = -\frac{2\kappa^2 r}{R(\kappa)} \cdot \frac{Z}{\mu},$$

where $R(\kappa)$ is the Rayleigh function and k is ω/β.

Values for the scalars ϕ and ψ in (1) can now be obtained by back substitution in (8a) and for the displacements from (1). For the tangential component of displacement we find

$$[u_1]_0 = -\frac{i\kappa(2\kappa^2 - k^2 - 2rs\kappa^2)}{R(\kappa)}\frac{Z}{\mu}\exp\{i\kappa(x_1 - ct)\}. \tag{48}$$

and, for the vertical displacement,

$$[u_3]_0 = \frac{\kappa r k^2}{R(\kappa)}\frac{Z}{\mu}\exp\{i\kappa(x_1 - ct)\}. \tag{49}$$

We now replace the uniform surface pressure by a line sinusoidal pressure source at $x_1 = 0$ of magnitude $Q/2\pi$. Integration over all wave numbers yields for the tangential displacement at $(x_1, 0)$

$$[u_1]_0 = \frac{iQ\exp(-i\omega t)}{2\pi\mu}\int_{-\infty}^{\infty}\frac{\kappa(2\kappa^2 - k^2 - 2rs\kappa^2)}{R(\kappa)}\exp(i\kappa x_1)d\kappa \tag{50}$$

and a similar expression for the vertical component

$$[u_3]_0 = \frac{Q\exp(-i\omega t)}{2\pi\mu}\int_{-\infty}^{\infty}\frac{\kappa r k^2}{R(\kappa)}\exp(i\kappa x_1)d\kappa. \tag{51}$$

If Q is subsequently treated as a delta function $Q\delta(t)$, the displacement can be determined for an impulse. Finally, synthetic seismograms can be constructed for more complicated sources by superposition of similar Green's function solutions (see §3.1.2).

Note that the displacements are given as improper integrals because the Rayleigh function denominator vanishes at two wave numbers $\pm\kappa$, say, that are the roots of (16) corresponding to free Rayleigh waves travelling to left and right.

Evaluation of the above integrals is not elementary. Using real variable analysis, Cauchy Principal Values must be evaluated using numerical or other means. Analysis is more satisfactory, however, with complex variables and Cauchy integration theory. It is found that the integrand is many-valued with various singularities. In particular, there are six singularities on the real axis, four branch points corresponding to P and S motion and two, called the *Rayleigh poles*, to the real roots of Rayleigh's equation (16). For a detailed modern treatment, see the book by Hudson (1980).

5.6 Normal oscillations of an elastic sphere

An instructive introductory consideration of the vibrations of an elastic sphere is to consider the case of a homogeneous sphere. The equations of motion for spherical waves (4.60), (4.61), (4.62) are then directly applicable. (Thermodynamical conditions are ignored.)

The first and third of these equations define an eigen-problem in which there is elastic dilatation, curl $u_r = 0$ and spherical surfaces are deformed. This type of vibration is called *spheroidal*. For a body the size of the Earth there is an added complication that the dilatation perturbs the gravity field significantly and a further second-order differential equation for this interaction is required.

Equation (4.61) as shown by (4.64) describes an eigen-vibration with $u_r = 0$, div $\mathbf{u} = 0$, so there is no gravitational perturbation. Motion is tangential to the surface of the sphere and therefore sets up a twist about a diameter of the sphere. The motion is thus called *torsional* (or less descriptively *toroidal*).

The first step to provide a suitable numerical algorithm for the computations of eigenvalues and -vectors for the vibrations of the non-homogeneous Earth is to include radial variation of the elastic constants. A summary of the mathematical characterisation is given below and in § 14.2. Details of the mathematics can be found in Lapwood and Usami (1981).

5.6.1 The basic equations.

Let r, θ, ϕ be spherical polar coordinates referred to the centre of a spherically symmetrical and non-rotating Earth model of radius a. During an oscillation, let u, v, w be the components of the displacement from the undisturbed configuration in the directions of r, θ, ϕ increasing, respectively. Let λ and μ denote the Lamé parameters and ρ and k the density and incompressibility. Zero suffixes will relate to the undisturbed configuration. Allowance needs to be made for the presence of initial stress which is taken, as usual, to be a hydrostatic pressure p_0 such that $dp_0/dr = -g_0\rho_0$. Let p_{rr}, $p_{r\theta}$, etc., and e_{rr}, $e_{r\theta}$, etc., denote the spherical polar components of additional stress and strain.

The strain components and the dilatation Δ are found to be connected with u, v, w by

$$
e_{rr} = \frac{\partial u}{\partial r}, \quad e_{\theta\theta} = \frac{1}{r}\frac{\partial v}{\partial \theta} + \frac{u}{r}, \quad e_{\phi\phi} = \frac{1}{r\sin\theta}\frac{\partial w}{\partial \phi} + \frac{v}{r}\cot\theta + \frac{u}{r},
$$

$$
2e_{r\theta} = \frac{\partial v}{\partial r} - \frac{v}{r} + \frac{1}{r}\frac{\partial u}{\partial \theta}, \quad 2e_{r\phi} = \frac{1}{r\sin\theta}\frac{\partial u}{\partial \phi} + \frac{\partial w}{\partial r} - \frac{w}{r},
$$

$$
2e_{\theta\phi} = \frac{1}{r}\frac{\partial w}{\partial \theta} - \frac{w}{r}\cot\theta + \frac{1}{r\sin\theta}\frac{\partial v}{\partial \phi},
$$
$$\tag{52}$$

$$
\Delta = \frac{\partial u}{\partial r} + \frac{2u}{r} + \frac{1}{r\sin\theta}\frac{\partial}{\partial \theta}(v\sin\theta) + \frac{1}{r\sin\theta}\frac{\partial w}{\partial \phi}. \tag{53}
$$

The stress–strain relations take the form

$$
p_{rr} = p_1 + 2\mu e_{rr}, \quad p_{\theta\theta} = p_1 + 2\mu e_{\theta\theta}, \quad p_{\phi\phi} = p_1 + 2\mu e_{\phi\phi},
$$

$$
p_{r\theta} = 2\mu e_{r\theta}, \quad p_{r\phi} = 2\mu e_{r\phi}, \quad p_{\theta\phi} = 2\mu e_{\phi\theta},
$$
$$\tag{54}$$

where $\quad p_1 = -(p_0 + ug_0\rho_0) + \lambda\Delta.$

We shall consider only free oscillations, so that gravity will be the only external force entering the equations. During an oscillation let ψ be the gravitational potential associated with the excess density distribution $\rho - \rho_0$ and the accompanying surface displacements. The components of body force per unit mass are then expressible as

$$
g_0 - \partial\psi/\partial r, \quad -r^{-1}\partial\psi/\partial\theta, \quad -(r\sin\theta)^{-1}\partial\psi/\partial\phi.
$$

The equations of motion (2.9) are found (putting $f_i = \partial^2 u_i/\partial t^2$ – see §2.4.2) to take the form

$$
\rho\frac{\partial^2 u}{\partial t^2} = \rho g_0 - \rho\frac{\partial\psi}{\partial r} + \frac{\partial p_{rr}}{\partial r} + \frac{1}{r}\frac{\partial p_{r\theta}}{\partial \theta} + \frac{1}{r\sin\theta}\frac{\partial p_{r\phi}}{\partial \phi}
$$

$$
\quad + r^{-1}(2p_{rr} - p_{\theta\theta} - p_{\phi\phi} + p_{r\theta}\cot\phi),
$$

$$
\rho\frac{\partial^2 v}{\partial t^2} = -\frac{\rho}{r}\frac{\partial\psi}{\partial \theta} + \frac{\partial p_{r\theta}}{\partial r} + \frac{1}{r}\frac{\partial p_{\theta\theta}}{\partial \theta} + \frac{1}{r\sin\theta}\frac{\partial p_{\theta\phi}}{\partial \phi}
$$
$$\tag{55}$$

$$
\quad + r^{-1}\{(p_{\theta\theta} - p_{\phi\phi})\cot\theta + 3p_{r\theta}\},
$$

$$
\rho\frac{\partial^2 w}{\partial t^2} = -\frac{\rho}{r\sin\theta}\frac{\partial\psi}{\partial \phi} + \frac{\partial p_{r\phi}}{\partial r} + \frac{1}{r}\frac{\partial p_{\theta\phi}}{\partial \theta} + \frac{1}{r\sin\theta}\frac{\partial p_{\phi\phi}}{\partial \phi}
$$

$$
\quad + r^{-1}(3p_{r\phi} + 2p_{\theta\phi}\cot\theta).
$$

The equation of continuity is found to yield

$$
\rho - \rho_0 = -(\rho_0\Delta + ud\rho_0/dr) \tag{56}
$$

(cf. (4.20)), and (4.21) gives

$$\nabla^2 \psi = -4\pi G(\rho_0 \Delta + u \mathrm{d}\rho_0/\mathrm{d}r). \tag{57}$$

On substituting from (52), (53) and (54) into (55) and (57) a set of second-order differential equations between the variables u, v, w and ψ is derived. These equations can be solved when λ, μ and ρ_0 are given as functions of r, and when suitable boundary conditions are prescribed.

The boundary conditions include regularity at the origin, and, at the deformed surface at all times, vanishing of the stress-components and equality of the internal and external gravitational potentials and their gradients. Other boundary conditions may have to be set down to allow for discontinuities in λ, μ and ρ_0 as functions of r in the models considered.

A discussion of the numerical integration is given in §14.2.

5.6.2 Torsional (toroidal) modes

Equation (4.64) describes the motion of purely *SH* waves. For radially symmetrical standing waves, rather than travelling spherical waves, we may separate r and t from the angular variables and obtain an equation similar to (4.65) where ω is the eigen-frequency. This relation gives the radial distribution of potential (and displacement) for torsional modes of oscillation of a perfectly elastic and isotropic (solid) sphere which is homogeneous throughout (i.e. μ, ρ = constant, all r). The family of torsional modes is generated as l takes successive integer values $l = 1, 2, \ldots$. In the case of a radially inhomogeneous sphere (i.e. μ and ρ are functions of r) the equation requires an additional term (cf. Alterman, Jarosch and Pekeris, 1959) and becomes

$$\mu\left\{\frac{\mathrm{d}^2 W}{\mathrm{d}r^2} + \frac{2}{r}\frac{\mathrm{d}W}{\mathrm{d}r}\right\} + \frac{\mathrm{d}\mu}{\mathrm{d}r}\left\{\frac{\mathrm{d}W}{\mathrm{d}r} - \frac{W}{r}\right\} + \left\{\omega^2\rho - \frac{l(l+1)}{r^2}\mu\right\}W = 0.$$

$$\tag{58}$$

The eigen-frequency corresponding to each mode number l will be that value of ω which allows a real solution W of (58) to be found which satisfies the boundary conditions. For a solid sphere, these conditions will be the vanishing of the stresses on the outside of the sphere, and the vanishing of the displacements as the centre of the sphere $(r = 0)$ is approached. For a planet with a liquid core, such as the Earth and probably Venus, the second condition is replaced by the vanishing of the stresses at the boundary between the solid mantle and liquid core; motion is confined to the solid mantle.

From (4.63), it is evident that the components of displacements in the torsional modes are

$$
\left.
\begin{aligned}
u_r &= 0, \\
u_\theta &= \frac{W(r)}{\sin\theta}\frac{\partial}{\partial\varphi}Y_l^m(\theta,\varphi)\exp i\omega t, \\
u_\varphi &= -W(r)\frac{\partial}{\partial\theta}Y_l^m(\theta,\varphi)\exp i\omega t,
\end{aligned}
\right\}
\tag{59}
$$

where $Y_l^m(\theta,\phi) = P_l^m(\cos\theta)\exp im\phi$. Components of this type near the Earth's surface are represented by the Love-type surface waves.

5.6.3 Spheroidal and radial modes

The components of displacements in the spheroidal modes may be expressed (see §4.8) as

$$
\left.
\begin{aligned}
u_r &= U(r)Y_l^m(\theta,\varphi)\exp i\omega t, \\
u_\theta &= V(r)\frac{\partial}{\partial\theta}Y_l^m(\theta,\varphi)\exp i\omega t, \\
u_\varphi &= \frac{V(r)}{\sin\theta}\frac{\partial}{\partial\varphi}Y_l^m(\theta,\varphi)\exp i\omega t.
\end{aligned}
\right\}
\tag{60}
$$

The radial displacement factors $U(r)$ and $V(r)$ are the solutions of two second-order differential equations, corresponding to (58) above, which may be derived from (4.60) and (4.62). In this case, curl$_r$ **u**, which is the radial component of rotation, vanishes so that there is no *SH* component of motion. The vertical and horizontal motions at the surface are appropriate to Rayleigh-type surface waves.

Substitution from (60) shows that div **u** $\neq 0$ in this case so there will be elastic dilatation. Deformations of the spherical boundaries of the equilibrium configuration will further perturb the gravity field. (The eigen-period for the gravest spheroidal oscillation of a homogeneous Earth model is 48.9 min without consideration of gravity, 44.3 min with gravity, a difference of 9.4 per cent; for the less massive Moon the difference is 0.64 per cent. For the Earth the effect of gravity falls to 0.15 per cent for modes of vibration with $l = 100$). These perturbations may be represented by

$$
p = P(r)\,Y_l^m(\theta,\varphi),
\tag{61}
$$

where $P(r)$ is the solution derived from equation (57). For an inhomogeneous sphere with radial symmetry, the eigenfunctions must therefore be computed, from a linear system of six first-order equations with variable coefficients.

There is a special case of spheroidal oscillations in which there is no *SV* motion; it arises formally when the order l of the spherical harmonic $Y_l(\theta, \phi)$ is zero. When gravity and rotation are neglected, an analytic form for the solution can be derived easily from equation (4.65). A solution for a purely radial pulsation which is non-singular at the Earth's centre is

$$U = B\left(\frac{n \cos nr}{r} - \frac{\sin nr}{r^2}\right), \tag{62}$$

where $n = \omega/\alpha$, and B is a constant.

Now, at the free surface ($r = R$), the radial stress vanishes so that, by (54),

$$p_{rr} = \left(k - \frac{2}{3}\mu\right)\text{div }\mathbf{u} + 2\mu\frac{du_r}{dr} = 0. \tag{63a}$$

Substitutions in (63a) from (62) using (4.4), (4.5), and the equation $u_r = \exp i\omega t \,(dU/dr)$ from (4.59), yields, for $r = R$,

$$nR/\tan nR = (1 - n^2R^2\alpha^2/4\beta^2). \tag{63b}$$

This transcendental equation provides a useful check on numerical integration methods used to determine the eigen-frequencies of non-homogeneous models. If the gravity is taken into account as a body force, (63b) holds with the modification $n^2 = (\omega^2 + 4A)/\alpha^2$, where the surface gravity is AR.

5.6.4 Geometrical description of the oscillations

After a great earthquake, such as the great Alaskan earthquake of 29 March 1964, there is at first a transient elastic-wave pulse which spreads in all directions through the body of the planet. A few seconds after 20 min following the energy release, the diametrical point on the surface of the Earth, the anticentre, is affected by the first longitudinal motion. At some later time, the complex system of travelling waves through the body (P and S waves, etc.) and along the surface (Love and Rayleigh waves, etc.) of the planet form a stationary interference pattern of standing waves. The appropriate description of these standing waves is given by the system of free oscillations (eigen-vibrations) of the sphere.

The geometrical appearance of the vibrations can be derived from the expressions (59) and (60). Taking first the torsional oscillations u_θ and u_ϕ will vanish in two distinct ways: at the zeros of the radial function $W(r)$ and at the zeros of the associated Legendre function. At the zeros of $W(r)$, equations (59) show that the displacements u_θ and u_ϕ vanish together so that at these radii there will be internal nodal spherical surfaces; one such surface occurs in the first overtone of the oscillation of order l, and so on.

The angular harmonics form a network of nodal latitude and longitude lines over any spherical surface. The displacement u_θ vanishes at the latitudes (relative to the source as pole) defined by $(\partial/\partial\varphi) Y_l^m(\theta, \varphi) = 0$ and u_φ at the meridians defined by $(\partial/\partial\theta) Y_l^m(\theta, \varphi) = 0$.

There is a corresponding geometrical description for the spheroidal modes described by (60). Again, a tesseral network of nodal lines exists on any spherical surface, and there will be internal surfaces on which $U(r)$ or $V(r)$ vanish (not, in general, together). It should be noted that if a seismic recorder on the surface happens to be on a nodal line of a particular oscillation, then it will not record that mode.

The nodal lines become more closely spaced as the mode number increases. In Fig. 5.5, a few examples of the modes are sketched. In the gravest torsional mode for the degenerate case $m = 0$, from (59),

$$u_\theta = \frac{W(r)}{\sin\theta} P_2(\cos\theta) \exp i\omega t = \frac{1}{2} \frac{W(r)}{\sin\theta} (3\cos^2\theta - 1) \exp i\omega t,$$

$$u_\varphi = -W(r) P_2'(\cos\theta) \exp i\omega t = 3W(r) \sin\theta \cos\theta \exp i\omega t. \tag{64}$$

There is thus a change of sign in u_φ at the equator ($\theta = 90°$); the northern

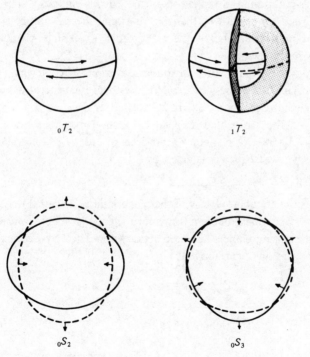

Fig. 5.5. Geometry of displacements in the modes $_0T_2$, $_1T_2$, $_0S_2$, $_0S_3$.

hemisphere components of displacement along the latitudes are in one sense, the southern hemisphere components are in the other.

In the gravest spheroidal mode ($l = 2$) there are, from (60), two nodal latitude lines on which $u_r = 0$ at a colatitude (relative to the source) of $\cos^{-1}$ $(1/\sqrt{3})$ north and south.

A particular notation has come to be used generally for the eigen-vibrations. In terms of the integers l, m of the associated Legendre functions, the torsional and spheroidal modes are denoted by $_nT_l^m$ and $_nS_l^m$, respectively. The prefix n denotes the order of the overtone which for $_nT_l$ is the number of internal nodal surfaces $W(r) = 0$ and for $_nS_l^m$ for a homogeneous sphere is the number of internal surfaces on which $U(r)$ (cf. equation (60)) vanishes. The forms of these modes are illustrated in Fig. 5.5.

5.6.5 Effects of rotation and ellipticity. Terrestrial spectroscopy

Some of the first frequency spectral analyses of the Earth's eigen-vibrations, made from recordings following the 1960 Chilean earthquake, showed significant fine structure in some of the graver modes of vibration. For a time, two possible explanations were considered: perturbation from the Earth's rotation, and the effects of the Earth's ellipticity. Theoretical work followed which showed that the rotational effect would be the more prominent and would be of the right order to explain the observed frequency multiplets.

The rotation of a planet introduces Coriolis-force terms into the equations of motion of the elastic sphere; they cause, in general, dependence between eigen-frequencies and the azimuthal angle φ so that the upper index m of the associated Legendre function becomes an essential parameter. A mode of planetary vibrations of order l will contain $2l + 1$ multiplets whose frequency is given by

$$_n\omega_l^m = {}_n\omega_l^0 + m_n\beta_i\,\Omega, \tag{65}$$

where Ω is the angular velocity of the planet and $_n\beta_l$ is called the *splitting parameter* which takes different forms for the two classes of vibration. The theory for rotating planets has been set down in detail by Ledoux (1958). In the notation given above it is

$$_n\beta_l = \begin{cases} \dfrac{\displaystyle\int_0^R \rho(2U_nV_n + V_n^2)r^2\,\mathrm{d}r}{\displaystyle\int_0^R \rho(U_n^2 + l(l+1)V_n^2)r^2\,\mathrm{d}r} & \text{for spheroidal,} \quad (66a) \\[6pt] 1/l(l+1) & \text{for torsional modes.} \quad (66b) \end{cases}$$

The phenomenon is the elastodynamic analogy of the splitting of atomic spectral lines by a magnetic field discovered by Zeeman in 1896.

Two further consequences of this dynamical theory are of general interest. First, the nodal lines on the surface, say, of a rotating planet will not be stationary with respect to a seismological observatory but will rotate westward with a velocity which is a function of l and n for spheroidal oscillations and of l only for torsional oscillations. An observatory on Earth would lie on a $_0S_2$ nodal line every 60 h, approximately, and on a $_0T_2$ nodal line every 144 h. The rate of drift approaches zero as l becomes large; for example, for $_0T_{12}$ the period of nodal precession on Earth is about 3740 h.

Secondly, relative amplitudes of the components of the multiplets are a function of the source characteristics. Hence, it may be possible to infer properties of the source from observed amplitudes. Pekeris, Alterman and Jarosch suggested in 1961 that the name *terrestrial spectroscopy* is appropriate to this type of study.

The Earth's ellipticity contributes to the fine structure of multiplets but the effect is relatively small. Flattening introduces some slight coupling between the motions of torsional and spheroidal type. In the case of torsional oscillations of oblate planets the $_nT_l^m$ vibration splits into $(l+1)$ modes with a difference in extreme frequencies which reaches only 0.7 per cent for the multiplet $l = 2$ compared with 2 per cent maximum for Coriolis splitting.

Viscosity will introduce damping of eigen-vibrations and broadening of lines in the frequency spectra (see § 14.2.3). Dissipation cannot be measured for most graver modes unless some correction for the Earth's rotation is applied. Both the Coriolis and the flattening effects rapidly become inconsequential as the modal order increases.

5.6.6 Duality with travelling waves

The geometrical description of spherical P waves is given by equation (4.66) and there will be a similar expression for S motions. The asymptotic approximation for $P_l(\cos \theta)$ for $l > 1$ is

$$P_l(\cos \theta) \sim \left(\frac{2}{\pi l \sin \theta} \right)^{\frac{1}{2}} \cos \left((l + \tfrac{1}{2})\theta - \frac{\pi}{4} \right). \tag{67}$$

Thus, the wave displacement at the surface of the Earth contains a phase factor

$$\exp i \left\{ (l + \tfrac{1}{2})\theta - \frac{\pi}{4} - \omega t \right\}.$$

Now, interpreting the standing wave at $r = R$ as a travelling wave like

(3.57), the phase velocity is

$$c = \lambda/\tau = \omega R/(l + \tfrac{1}{2}).$$ (68)

This formula is of considerable importance in seismology and was obtained in another way by Jeans in 1923. For $l > 20$ the particle motion in the core of the Earth becomes unimportant, and for $l > 100$ there is little motion below 500 km depth. For even higher order modes, the alternative representation in terms of travelling surface waves is more usually adopted. From the theoretical and observational analysis, (68) is valid for $l \geqslant 20$ with an accuracy well within the numerical precision used in most work. Since $C = c - \lambda dc/d\lambda$, the group velocity can be expressed conveniently, using (68), as

$$C = c + (l + 1/2)dc/dl.$$ (69)

In 1961, Bolt and Dorman derived from the computed phase velocities the following approximate empirical relation between the wave-length, the phase velocity c in a gravitating sphere, and the corresponding velocity c_h for a layered half-space with no gravity

$$c \sim c_h(1 + 0.000\ 16\tau).$$ (70)

This equation allows the estimation of c within 1 per cent for $100 < \tau < 300$ s from values computed for a flat Earth model.

5.7 Seismic waves in linear visco-elastic media

5.7.1 Equation of motion. The correspondence principle

We have already considered (§ 4.5) the general effects of elastic imperfections on harmonic waves. In particular for linear constitutive relations we were led to consider complex elastic moduli. For linear visco-elastic continuum mechanics, the starting point is again the equation (2.9) connecting stress p_{ij} with small displacements u_i. The stress–strain equations (2.32) of ordinary infinitesimal strain theory are replaced by the appropriate complex form.

Consider sinusoidal vibrations of such a medium with frequency ω. We then put

$$p_{ij} = \mathscr{R}[\bar{p}_{ij}\exp i\omega t],$$ (71)

$$e_{ij} = \mathscr{R}[\bar{e}_{ij}\exp i\omega t],$$ (72)

where bars denote functions of spacial coordinates only. (Body forces may be assumed to vary like $\exp i\omega t$ also). It follows that equations (2.9) and

(2.13) become

$$- \rho \omega^2 \bar{u}_i = \frac{\partial \bar{p}_{ij}}{\partial x_j} + \rho \bar{X}_i \quad (i = 1, 2, 3) \tag{73}$$

and

$$\bar{e}_{ij} = \frac{1}{2} \left(\frac{\partial \bar{u}_j}{\partial x_i} + \frac{\partial \bar{u}_i}{\partial x_j} \right). \tag{74}$$

In place of the isotropic perfectly elastic moduli we define from (2.67) and (2.68) the complex deviatoric and dilatational moduli $M(\omega)$ and $K(\omega)$.

$$\bar{p}_{ll} = 3K\bar{e}_{ll} \tag{75}$$

and

$$\bar{P}_{ij} = 2M\bar{e}_{ij}. \tag{76}$$

The relation between K and M and the complex compliances follows from the developments in §4.5.1.

It has been pointed out, for example, by Bland (1960), that the one-to-one similarity evident above, between perfectly elastic and linear visco-elastic vibration problems, leads to a useful *correspondence principle*. In brief: the solution for a dynamic problem for a visco-elastic material can be obtained from the solution of the corresponding elastic problem by applying a Fourier transform to the elastic solution, replacing elastic moduli by complex moduli, and inverting the transform.

In the next sections the correspondence in the seismic wave case is immediately clear.

5.7.2 Damped seismic waves

Consider the case of spherically symmetric waves given by (3.80). Application of the correspondence principle to the formulae in §4.1 and §4.8 yields.

$$\text{div}\,(\bar{u}_i) = \frac{A}{r} \exp \left\{ \pm i\omega \left(\frac{\rho}{K + \frac{4}{3}M} \right)^{\frac{1}{2}} r \right\} \tag{77}$$

and

$$\text{curl}\,(\bar{u}_i) = \frac{A}{r} \exp \left\{ \pm i\omega \left(\frac{\rho}{M} \right)^{\frac{1}{2}} r \right\}, \tag{78}$$

where A is a constant. These results, when multiplied by $\exp i\omega t$, represent damped P and S waves diverging from a point source. The velocities are

$$\alpha = \{ \mathscr{R}[3\rho/(3K + 4M)]^{\frac{1}{2}} \}^{-1} \quad \text{and} \quad \beta = \{ \mathscr{R}(\rho/M)^{\frac{1}{2}} \}^{-1}.$$

The two attenuations are (see (14.20))

$$Q_P^{-1} = -2\mathscr{I}[3\rho/(3K+4M)]^{\frac{1}{2}},$$ (79)

and

$$Q_S^{-1} = -2\mathscr{I}(\rho/M)^{\frac{1}{2}}.$$ (80)

It is important to note that, in general, since K and M are functions of frequency, the P and S waves in a linear visco-elastic solid are both dissipative and dispersive. A discussion of the wave speeds as functions of the dilatational and deviatoric quality factors Q_k and Q_μ is given by Kennett (1983). See also Exercise 5.8.13. The application of the correspondence principle to Love and Rayleigh waves is left as an exercise for the reader.

5.7.3 Damped oscillations of a visco-elastic sphere

From §5.6.1 we can apply the correspondence principle and introduce appropriate complex elastic moduli to write down the equations of motion for the eigenvibrations of a visco-elastic sphere. In 1965, Jeffreys warned that eigen-spectra for an inelastic Earth differed from those of the perfectly elastic one. The effect is geophysically significant, with eigen-frequency observations differing from the predicted elastic case by 8 s (0.25 per cent) for the fundamental spheroidal mode $_0S_2$ and 12 s for $_0T_2$. Consequently, all Earth models based on solutions of the eigen-vibration inverse problem that do not incorporate anelasticity are suspect (see §13.5.3).

It is sufficient here, in order to establish the link with previous theory, to work through only the simplest case: radial oscillations of a uniform, homogeneous, sphere.

Rewrite the eigenvalue equation (63) as

$$nR \cot nR = 1 - \frac{k + \frac{4}{3}\mu}{4\mu}(nR)^2.$$ (81)

By the correspondence principle, we now can replace $k + \frac{4}{3}\mu$ by the complex moduli to yield

$$n^2 = 3\rho\omega^2/(3K+4M)$$ (82)

so that the wave frequency is complex in general.

For numerical simplicity assume a Maxwell model with Poisson's condition i.e. $3K = 5M$ (equation (4.7)). Then (82) becomes for spring modulus m,

$$\omega = n\left(\frac{3m}{\rho}\right)^{\frac{1}{2}}\left(1 + \frac{m}{i\omega\eta}\right)^{-\frac{1}{2}}$$ (83)

from (4.37). For $_0S_0$, $m/\omega\eta \ll 1$, so that, on expansion,

$$\omega \sim n\left(\frac{3m}{\rho}\right)^{\frac{1}{2}} + \frac{i\omega}{2\eta}. \tag{84}$$

On substitution into the equations of §5.6.3, the resulting radial displacement is

$$u_r = B\left(\frac{n\cos nr}{r} - \frac{\sin nr}{r^2}\right)\exp\left(-\frac{m}{2\eta}t\right)\cos n\left(\frac{3m}{\rho}\right)^{\frac{1}{2}}t. \tag{85}$$

We note that the damping term has $Q = \omega\eta/m$ and the displacement is reduced by $1/e$ in time $2\eta/m$ independently of the frequency and radius of the sphere in this case.

5.8 Exercises

1   The fundamental frequency of an open organ pipe 100 cm long is 180 vibrations per second. What is the velocity of sound in the pipe? What is the frequency of the second possible overtone of that open pipe? Make a diagram showing the loops and nodes in this problem. What is the analogy in the Earth?

2   Consider the way that the wave motions in surface Love waves decrease with depth in the Earth.

Would an earthquake with focus at depth 110 km produce (compared with the fundamental mode) significant first higher mode Love waves with period 30 s? At what depths would there be little second higher mode energy in Love waves of 30 s period?

3   Using the theory of §5.4.1, prove that the nth higher mode Love wave exists only for frequencies $\omega > \omega_n$, where

$$\omega_n = \frac{n\pi\beta'}{H'}\bigg/\left(1 - \frac{\beta'^2}{\beta^2}\right)^{1/2}.$$

Show that this *cutoff frequency* for the first higher mode is 0.08 Hz for a continental crust with $\beta' = 3.5$ km/s, $\beta = 4.5$ km/s and $H' = 35$ km.

4   Consider a Stoneley wave propagating along the plane interface between a liquid M_1 and a perfectly elastic solid half-space M_2. Let the axis of x be along the interface and the axis of z be into the solid. Assume elastic potentials of the form

$$\phi = f_1(z)\exp[i\kappa(x - ct)] \text{ in } M_1,$$

and

$$\begin{cases} \phi = f_2(z)\exp[i\kappa(x - ct)] \\ \psi = f_3(z)\exp[i\kappa(x - ct)] \end{cases}, \text{ in } M_2.$$

From the equations of motion determine the proper forms for the amplitude functions $f(z)$ for a Stoneley wave and calculate the displacements (u_1, w_1) and (u_2, w_2). By substitution of the displacements u, w in the three boundary conditions, at $z = 0$, determine three equations of condition for the amplitude terms. Show, in particular, that the vanishing of the shear stress at the surface of the solid yields an equation of the form

$$2B\kappa r_2 + (s_2^2 + \kappa^2)iC = 0,$$

where B, C are constants, and

$$r_2^2 = \kappa^2(1 - c^2/\alpha_2^2)$$
$$s_2^2 = \kappa^2(1 - c^2/\beta_2^2).$$

Write down the 3×3 determinant which must be satisfied if a Stoneley wave is to propagate and then show that the equation for its velocity is

$$\left(\frac{2\beta_2^2}{c^2} - 1\right) - 4\left(\frac{\beta_2^2}{c^2}\right)\left(\frac{\beta_2^2}{c^2} - \frac{\beta_2^2}{\alpha_2^2}\right)^{\frac{1}{2}}\left(\frac{\beta_2^2}{c^2} - 1\right)^{\frac{1}{2}}$$
$$+ \frac{\rho_1}{\rho_2}\left(\frac{\beta_2^2}{c^2} - \frac{\beta_2^2}{\alpha_2^2}\right)^{\frac{1}{2}} \Big/ \left(\frac{\beta_2^2}{c^2} - \frac{\beta_2^2}{\alpha_1^2}\right)^{\frac{1}{2}} = 0.$$

Is the wave dispersive?

5   As with Lamb's problem, consider a sinusoidal transverse line source applied in the x_2 direction on the free surface of an elastic layer, thickness H', welded to an elastic half-space (Fig. 5.3).

Show that the Love wave displacement at $P(x_1, x_3, t)$ is proportional to

$$\int_{-\infty}^{\infty} [\mu i s \sin \kappa s'(H' - x_3) + \mu' s' \cos \kappa s'(H' - x_3)] \exp i\kappa(x - ct) d\kappa / L(c, \kappa),$$

where $L(c, \kappa)$ is Love's function. Plot the Love poles on the complex wave number plane and define the types of singularity present.

6   State how the Rayleigh pole arises in Lamb's problem.

Calculate the residue at the Rayleigh pole $\zeta = \kappa$ in the integral

$$\int_{-\infty}^{\infty} \frac{\zeta(2\zeta^2 - k^2 - 2rs\kappa^2) \exp i\zeta x}{R(\zeta)} d\zeta,$$

where $R(\zeta)$ is Rayleigh's function.

Show that the addition of the residues at the poles $\zeta = \pm \kappa$ gives rise to a standing wave $A \cos \kappa x$, when A depends upon the elastic constants.

7   In the Rayleigh wave problem (§ 5.2) prove that the general condition for three real roots of Rayleigh's equation is

$$(10 - 12x)^3 > (45x - 28)^2,$$

where $x = 1 - \beta^2/\alpha^2 = 1/2\,(1 - \sigma)$.

Hence show that if $\sigma > 0.264$ there is only one real root and two complex conjugate roots. Further show that if κ is real, the amplitude of either the P or SV component of the wave increases away from the boundary so that the wave is a leaky mode.

8   A steel sphere, 20 cm in diameter, is available, and Young's modulus for steel is 20×10^{11} dyne/cm^2. Assuming that Poisson's ratio is 1/3, compute the frequency of the two gravest radial spheroidal oscillations ($_0S_0$ and $_1S_0$). How would you excite and observe these oscillations?

9   Consider Love waves propagating in a homogeneous crustal layer with $\beta = 3.5$ km/s and thickness 30 km over a homogeneous mantle with $\beta = 4.5$ km/s. Calculate the low-frequency cutoff frequencies for the first and second higher-mode Love waves.

Consider Rayleigh waves propagating in a homogeneous half-space and assume the source is a point force in both time and space. Using the Betti reciprocal relation (see § 4.2.3) examine the efficiency for exciting Rayleigh waves of sources at different depths. How does the excitation of the surface wave depend upon the orientation of the force? How does the frequency spectrum of a Rayleigh wave depend upon the source depth?

10   Consider the free oscillations of the Earth described geometrically in chapter 5. Sketch the surface pattern of nodal lines and displacements for modes $_0T_3$, $_1T_4$ and $_0S_4$. Draw the internal nodal pattern for $_2S_0$, $_2S_2$ and $_2T_2$.

Can properties of the inner core be inferred from frequency measurements of the modes $_0T_2$, $_1T_2$, $_0S_{40}$?

11   Because the Earth's free oscillations are damped, their amplitudes slowly decrease with time after a great earthquake. Calculate (see chapter 14) by what percentage the amplitudes of $_0T_2$, $_0S_2$ and $_0S_0$ decrease after 20 hours.

12   Consider plane waves in linear visco-elastic medium, i.e. take λ, μ complex and the wave number κ complex:

$$\kappa = \kappa_0 + i\kappa_1.$$

For a plane wave

$$\mathbf{u} = \mathbf{a}\exp(-\kappa_1 \cdot \mathbf{x})\exp i(\kappa_0 \cdot \mathbf{x} - \omega t).$$

Show that planes of constant phase are given by $\kappa_0 \cdot \mathbf{x} = $ constant, and planes of constant amplitude are $\kappa_1 \cdot \mathbf{x} = $ constant.

If $\mathbf{a} = A\kappa$ (A complex), deduce that

$$\mathcal{R}\mathbf{u} = |A|\exp(-\kappa_1 \cdot \mathbf{x})\{\boldsymbol{\xi}_0 \cos(\kappa_0 \cdot \mathbf{x} - \omega t + \varepsilon')$$
$$- \boldsymbol{\xi}_1 \sin(\kappa_0 \cdot \mathbf{x} - \omega t + \varepsilon')\}$$

where $\boldsymbol{\xi}_0$ and $\boldsymbol{\xi}_1$ are real, $\boldsymbol{\xi}_0 + i\boldsymbol{\xi}_1 = \kappa\exp(-ia)$, and $\varepsilon' = \arg A + a$.

Hence deduce that the P wave is no longer longitudinal, but is elliptically polarised in the plane of κ_0 and κ_1.

13   Show that the S wave speed β^* in a linear visco-elastic medium, modulus M, is

$$\beta^{*2}(\omega) = \beta^2\{1 + \mathcal{R}\{M(\omega)\}/\mu - i\,\text{sgn}(\omega)Q_\mu^{-1}(\omega)\},$$

where β and μ are the corresponding perfectly elastic values; and

$$Q_\mu^{-1}(\omega) = -\mathcal{I}\{M(\omega)\}/\mu.$$

14   Derive (43) using Sylvester's *interpolation formula* for the function of a square matrix with roots λ_κ, i.e.

$$f(\mathbf{A}) = \sum_{k=1}^{n} f(\lambda_k) \prod_{j \neq k} (\mathbf{A} - \lambda_j\mathbf{I}) / \prod_{j \neq k} (\lambda_k - \lambda_j).$$

6

Reflection and Refraction
of elastic waves

6.1 Formulation

As in § 5.1, we consider two homogeneous media, M and M', in welded contact, separated by a plane boundary. We now investigate the laws of reflection and refraction for plane waves incident through a medium M towards this boundary, which we take to be horizontal. We adopt a frame of reference similar to that in § 5.1; the origin O is any point of the boundary, Ox_3 points normally into the medium M', and Ox_1, Ox_2 are respectively perpendicular to and parallel to the line in which an incident wave front cuts the boundary. To begin with, we consider waves of simple harmonic form.

The equations of motion are again the equations (4.1). Partial derivatives with respect to x_2 are again zero, and the general part of the analysis of § 5.1, including the boundary conditions, is relevant. The principal change is that we must refer back to the more general solutions corresponding to (5.7), instead of using the particular solutions selected in § 5.1 to fit the special requirements of surface waves. The result is that an additional term has to be added to each of the equations (5.8); the form of the solution now obtained for the medium M is accordingly as in equations (2) below.

6.1.1 Laws of reflection and refraction

It follows as in § 5.1 that the relevant solutions of the wave equations contain no terms in x_2, and that x_1 always enters only as a factor $\exp\{i\kappa(x_1 - ct)\}$. This factor (see, for example, (2) and (3) below) will be common to every term, whether associated with the incident or with reflected or refracted waves. However, note that in the present problems the c in this factor is not the actual wave velocity, but the velocity of advance of the line in which a (plane) wave front cuts the plane boundary surface. An analogous interpretation holds for κ.

In a particular problem, let e be the (acute) angle between the normal to a

140

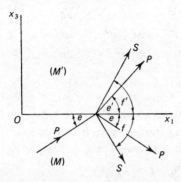

Fig. 6.1. Reflection and refraction at a plane interface.

wave front and the x_1-axis for any particular one of the waves involved (incident, reflected or refracted), and let i be the complement of e; we shall throughout this book take e, i to lie in the range from 0 to $\frac{1}{2}\pi$. Let v be the corresponding wave velocity. It follows from the property of c stated in the last paragraph that $v = c \cos e$. Hence we have

$$\frac{\sin i}{v} = \frac{\cos e}{v} = \text{constant} \tag{1}$$

for all the waves involved. (This relation may be alternatively inferred from a consideration of (3.79); or it may be deduced, using the calculus of variations, from Fermat's principle (§ 3.4.3).)

The relation (1) is analogous to laws of reflection and refraction applying to rays in geometrical optics. The elastic case is more general, however, since there are liable to be waves of P as well as S type (which alone occurs in optics), both reflected and refracted.

With an incident wave, i is called the *angle of incidence* and e the *angle of emergence*. In the sections that follow, we shall often be concerned with two different directions of wave velocity in the same medium, associated respectively with P and S reflections or refractions arising from a given incident wave. In such cases, we shall (see Fig. 6.1) denote the angles between the normals to wave fronts and the x_1-axis by e for P waves and by f for S waves in the medium M (i.e. for incident or reflected waves), and by e' and f' in the medium M' (i.e. for refracted waves).

6.1.2 General equations for the two media

In accordance with our preliminary remarks, a solution for the medium M which replaces the less general solution (5.8) has the form

$$\left.\begin{aligned}
\phi &= A_0 \exp\{i\kappa(x_3\tan e + x_1 - ct)\} + A\exp\{i\kappa(-x_3\tan e + x_1 - ct)\}, \\
\psi &= B_0 \exp\{i\kappa(x_3\tan f + x_1 - ct)\} + B\exp\{i\kappa(-x_3\tan f + x_1 - ct)\}, \\
u_2 &= C_0 \exp\{i\kappa(x_3\tan f + x_1 - ct)\} + C\exp\{i\kappa(-x_3\tan f + x_1 - ct)\},
\end{aligned}\right\}$$

$$(2)$$

The coefficients $\tan e$ and $\tan f$ which replace the r and s of (5.8) are determined using (3.70). The first terms on the right-hand sides of (2) correspond to incident waves, and the second terms to reflected waves. With separate incident P, SV and SH waves, we have, respectively,

$$B_0 = C_0 = 0, \quad C_0 = A_0 = 0, \quad A_0 = B_0 = 0.$$

For M', there are, of course, no terms corresponding to the terms containing A, B, C in (2), since in M' the waves will all be moving away from the boundary. Hence for M' we may (corresponding to (5.8), replacing D', E', F' by A', B', C') write

$$\left.\begin{aligned}
\phi &= A' \exp\{i\kappa(x_3\tan e' + x_1 - ct)\}, \\
\psi &= B' \exp\{i\kappa(x_3\tan f' + x_1 - ct)\}, \\
u_2 &= C' \exp\{i\kappa(x_3\tan f' + x_1 - ct)\}.
\end{aligned}\right\}$$

$$(3)$$

Relations between the coefficients A, B, A_0, etc., are determined from the boundary conditions corresponding to continuity of displacement and stress across the boundary at all times and places. From such formulae, transmission and energy coefficients (see chapter 8) and phase-shifts can be computed as a function of e (see e.g. Pilant (1979) chapter 12).

6.2 Special cases

6.2.1 Case of incident SH waves

When the incident wave is of the SH type, we have $A_0 = B_0 = 0$; also $\mu \neq 0$. The boundary conditions which involve ϕ and ψ then show that A, B, A' and B' have no dependence on C_0, and the relevant solution is

$$\begin{aligned}
u_2 = C_0 \exp\{i\kappa(x_3\tan f + x_1 - ct)\} \\
+ C\exp\{i\kappa(-x_3\tan f + x_1 - ct)\}
\end{aligned}$$

$$(4)$$

for the medium M, and

$$u_2 = C' \exp\{i\kappa(x_3\tan f' + x_1 - ct)\}$$

$$(5)$$

for M'.

If $\mu' \neq 0$, continuity of displacement at the boundary then gives

$$C_0 + C = C',$$

$$(6)$$

and continuity of the stress component p_{32} gives, by (5.13),

$$(C_0 - C)\mu \tan f = C'\mu' \tan f'. \tag{7}$$

From (6) and (7) we derive

$$\frac{C}{\mu \tan f - \mu' \tan f'} = \frac{C'}{2\mu \tan f} = \frac{C_0}{\mu \tan f + \mu' \tan f'}. \tag{8}$$

Also, by (1), we have

$$\frac{\cos f}{v} = \frac{\cos f'}{v'}, \tag{9}$$

where v, v' are the velocities of S waves in M, M'. In the case of normal incidence ($f = f' = \tfrac{1}{2}\pi$), the formula (8) fails to determine the ratios of C and C' to C_0; but using (9) and taking limits, we derive for this case

$$\frac{C}{\mu v' - \mu' v} = \frac{C'}{2\mu v'} = \frac{C_0}{\mu v' + \mu' v}. \tag{10}$$

By (8), (9) and (10) we can determine C and C' in terms of C_0 and f, and hence determine the amplitude ratios of the incident, reflected and refracted waves for any angle of emergence and elastic medium.

The above solution shows that when the incident wave is SH type, then the reflected and refracted waves can only be SH. Notice that the coefficient C in (8) can be zero only for an angle of emergence which satisfies (9) and the equation $\mu \tan f - \mu' \tan f' = 0$. Apart from this possible value of f, there will always be a reflected wave. It may happen, moreover (depending on the values of μ, μ', ρ, ρ'), that the expression $\mu \tan f - \mu' \tan f'$ has no zero in the range $0 \leqslant f \leqslant \tfrac{1}{2}\pi$, in which case there is a reflected wave for every angle of incidence.

The question of the existence of a refracted wave conveying energy from the boundary into M' is determined by equation (9). For $v' > v$, $\cos f'$ exceeds unity for a range of real values of f. In this case, $\tan f'$ is imaginary and (5) takes the form

$$u_2 = C' \exp(-bx_3) \exp\{i\kappa(x_1 - ct)\}, \tag{11}$$

where b is a positive real number. This gives inside M', instead of a refracted wave, a surface movement whose amplitude diminishes exponentially with increasing distance from the boundary, and which travels along the boundary with the velocity of advance of the line in which the incident wave front cuts the boundary. Such evanescent waves are sometimes called *inhomogeneous* plane waves (see Exercise 6.6.7).

When $\tan f'$ is imaginary, it follows from (8) that C/C_0 is equal to the ratio of two conjugate complex numbers. This entails that $|C/C_0| = 1$, and hence $C = C_0 \exp(-i\delta)$, where δ is a real number depending on f, μ/μ' and v/v'. Hence in this case, there is *total reflection* of the incident wave, the amplitude of the reflected wave is equal to that of the incident wave, and there is a change of phase given by δ. For the angle of emergence $\cos^{-1}(v/v')$ which gives f' a zero value, there is total reflection with zero change of phase.

If $\mu' = 0$, i.e. if M' is a fluid or vacuum, there is free slipping along the boundary and the condition which gives (6) no longer holds; also (5) is now irrelevant. But since p_{32} is now zero at the boundary, the left-hand side of (7) is zero, and $C = C_0$ in place of (6). The solution is therefore given entirely by (4) with C replaced by C_0. In this case there is complete reflection without change of type or phase.

6.2.2 *P* wave incident against a free plane boundary

We next consider a *P* wave incident through the medium *M* against a free plane boundary, the angle of emergence being e. Thus in (2) and (3), we put $B_0 = C_0 = 0$, $A' = B' = C' = 0$. Hence the relevant equations are

$$
\begin{aligned}
\phi &= A \exp\{i\kappa(-x_3 \tan e + x_1 - ct)\} \\
&\quad + A_0 \exp\{i\kappa(x_3 \tan e + x_1 - ct)\}, \\
\psi &= B \exp\{i\kappa(-x_3 \tan f + x_1 - ct)\}, \\
u_2 &= C \exp\{i\kappa(-x_3 \tan f + x_1 - ct)\},
\end{aligned}
\tag{12}
$$

holding in the medium *M*.

Let us assume for simplicity that Poisson's relation holds in *M*. Since α is the velocity associated with ϕ, and β with ψ and u_2, we then have by (1) and (4.7)

$$
\cos^2 e = 3 \cos^2 f.
\tag{13}
$$

The boundary conditions are that the stress components p_{31}, p_{32}, p_{33} vanish at the boundary. The condition involving p_{32} gives by (5.13) that C is zero. Thus there are no reflected *SH* waves in the present case.

The boundary conditions involving p_{33}, p_{31} give, using (5.11), (5.12) and (4.7),

$$
(A_0 + A)(1 + 3\tan^2 e) + 2B \tan f = 0,
$$
$$
(A_0 - A)2 \tan e + B(\tan^2 f - 1) = 0.
$$

From these and (13) we derive

$$\frac{A}{4\tan e \tan f - (1 + 3\tan^2 e)^2} = \frac{B}{-4\tan e(1 + 3\tan^2 e)}$$

$$= \frac{A_0}{4\tan e \tan f + (1 + 3\tan^2 e)^2}. \quad (14)$$

By (13) and (14) we can determine A and B in terms of A_0 and e, and hence determine amplitude ratios of the incident and two reflected waves for any angle of incidence.

For normal incidence we have $e = f = \frac{1}{2}\pi$, and for grazing incidence $e = 0$. It is readily deduced from (13) and (14) that B is zero in both these cases, but in no other case. Thus there exists a reflected disturbance of SV type for all angles of incidence except zero and $\frac{1}{2}\pi$.

The coefficient vanishes if $(1 + 3\tan^2 e)^2 = 4\tan e \tan f$. Using (13), we find that this equation has two relevant roots, namely, $e = 12°.8, 30°$. For these angles of emergence, there are accordingly no reflected waves of P type. Between them, A/A_0 rises to a maximum of only 0.058 (at $e = 25°$). Thus over a fair range of values of e there is very little reflection in the P type. Jeffreys has pointed out that a small departure from Poisson's relation (e.g. when $\alpha = 1.8\,\beta$) would prevent A from vanishing at all; nevertheless, the ratio A/A_0 would remain small over a similar range of values of e to that just indicated. Indeed, for $2° < e < 63°$, at least half of the reflected energy is in the SV type; this result is relevant to an argument used in §4.3.

6.2.3 *SV* wave incident against a free plane boundary

The solution for this case is similar to that in §6.2.2. In (2) and (3), put $A_0 = C_0 = 0, A' = B' = C' = 0$. We find again that C is zero for all angles of incidence, and hence that there are no reflected SH waves. Taking the angle of emergence of the incident wave as f, we have

$$\phi = A \exp\{i\kappa(-x_3 \tan e + x_1 - ct)\}, \quad (15)$$

$$\psi = B \exp\{i\kappa(-x_3 \tan f + x_1 - ct)\} + B_0 \exp\{i\kappa(x_3 \tan f + x_1 - ct)\}. \quad (16)$$

Assuming Poisson's relation, we have (13) again, and A and B are determined in terms of B_0 and f by the equations

$$\frac{A}{4\tan f(1 + 3\tan^2 e)} = \frac{B}{4\tan e \tan f - (1 + 3\tan^2 e)^2}$$

$$= \frac{B_0}{4\tan e \tan f + (1 + 3\tan^2 e)^2}. \quad (17)$$

The resemblance between the forms (14) and (17) is an instance of a reciprocal theorem proved by Gutenberg.

We find, similarly to the case of §6.2.2, that B is zero if $e = 12°.8$ or $30°$, i.e. (by (13)) if $f = 55°.7$ or $60°$. For a range of angles between these angles of emergence there is little reflection in the SV type.

There is the complication with incident SV waves that e is imaginary if $0 \leqslant f < \cos^{-1}(1/\sqrt{3}) = 54°.7$. For this range of values of f, (15) takes a form analogous to the right-hand side of (11), and we find (cf. the case described in §6.2.1) that there is then complete reflection in the SV type, with a change of phase in general.

6.3 Curved boundaries and head waves

In seismology, reflection and refraction problems often involve effects of curvature of boundaries or wavefronts. In many circumstances, such as large distances from the source or short wave-lengths, the plane-wave theory (Snell's law) is sufficiently accurate for most purposes. All the special cases described in §§6.2.1–.2.3 are important in seismology, and are illustrative of types of solution arising from the equations set up in §6.2. Sometimes, special cases corresponding to specific properties of the two media M and M' need to be considered with values corresponding to particular discontinuities within the Earth, described in chapters 12 and 13.

In the cases so far considered, the incident wave was taken to be a single simple harmonic wave. For more complicated incident waves, results are obtained by superposition of the effects of the separate harmonic constituents, provided there is no dispersion. The law of reflection and refraction continues to be given by (1), and formulae such as (8), (14) and (17) continue to hold, coefficients such as A_0, A, B, etc., being interpreted as amplitudes of the displacement in the incident and derived waves.

New theoretical considerations arise when wave fronts are incident on boundaries with significantly different curvature (see §6.5). The simple energy partition involved with plane-wave incidence on a plane boundary does not occur and, mathematically, a match of boundary conditions is no longer elementary. Energy will, however, still be reflected and refracted but wave energy will also flow along the boundary itself. We usually say that the seismic wave is reflected, refracted and scattered. If the radius of curvature is large, the reflected wave may be represented as a reflected plane wave plus a scattered wave, which measures the degree of deviation from the planar geometry.

An important case occurs when a spherical seismic wave leaves from a source in medium M (see Fig. 6.1) and is incident on a plane boundary Ox_1 at the *critical angle* f_c, such that, from (1), $f_c = \cos^{-1} v/v'$. (A common seismological circumstance is shown in Fig. 7.10 for Pn and Sn waves for which there is a linear travel-time relation given by (7.81)).

After critical reflection at the interface, the ray description has a wave travelling parallel to the interface in M' but ray theory does not provide an adequate physical or quantitative explanation for the observed waves. These waves travelling along the boundary are called *diffraction waves*, *head waves* or *conical waves* (see Exercise 6.6.6), and their description requires appropriate full-wave diffraction theory found, for example, in the works by Cagniard (1962), Berry and West (1966) and Kennett (1983). Surface motions from head waves are often small because, as first shown by Jeffreys, the amplitude attenuates with distance along the interface like the decay in the Airy integral (see §3.8).

6.4 Refraction of dispersed waves

We consider the refraction of waves, subject to dispersion, across an interface separating two homogeneous media M and M'. Stoneley pointed out that the direction of the refracted wave front is determined by the ratio of the velocities of wave propagation, while transmission of energy in the two media follows the law of group velocity.

The following proof of this result is due to Jeffreys. Suppose a disturbance issues at time $t = 0$ from point O in medium M, and consider the effect at time t at a point O' in medium M'. For ease of exposition, the proof is given for the two-dimensional case; it is not difficult to extend the proof to the three-dimensional case.

Consider (see Fig. 6.2) a pencil of rays which all pass from O to O' in time t, the pencil cutting off a length ds of the interface near the point P say, and let the wave periods for the rays in this pencil all lie between $2\pi/\gamma$ and

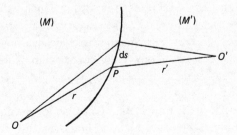

Fig. 6.2. Refraction at a curved boundary.

$2\pi/(\gamma + d\gamma)$. Let $OP = r$, $PO' = r'$, and let the wave and group velocities in the two media corresponding to the period $2\pi/\gamma$ be c, c', and C, C', respectively.

By analogy with the form (3.65) (but taking γ as argument, instead of κ), it is seen that the displacement at O' at time t will be of the form

$$\iint f(\gamma)g(\theta)\exp\{i\gamma(rc^{-1} + r'c'^{-1} - t)\}\,d\gamma\,ds, \tag{18}$$

where $f(\gamma)$ is analogous to $F(\kappa)$ in (3.65) and $g(\theta)$ is a function of direction. The functions f and g will, like the function $F(\kappa)$, in general be slowly varying functions. It may then be shown (cf. § 3.3.5) that the predominant contribution to the integral (18) will arise from values of s and γ near those for which the integrand is stationary. The conditions for this (neglecting variation of f and g) are

$$\frac{1}{c}\frac{dr}{ds} + \frac{1}{c'}\frac{dr'}{ds} = 0, \tag{19}$$

$$r\frac{d(\gamma c^{-1})}{d\gamma} + r'\frac{d(\gamma c'^{-1})}{d\gamma} - t = 0. \tag{20}$$

Since dr/ds and dr'/ds are opposite in sign but equal in magnitude to the sines of the angles which the pencil makes with the normal to the interface at P on the two sides, equation (19) is equivalent to the usual law of refraction (1). Thus paths of the rays are determined by the wave velocities. Because $\gamma = \kappa c$, equation (20) is by (3.68) equivalent to $t = r/C + r'/C'$. This shows that the travel-time of the predominant disturbance is determined by the group velocities.

6.5 Scattered seismic waves. Matrix theory

It is usual nowadays to formulate seismic reflection/refraction problems in terms of matrices in order to budget more easily the multiple partitioning of wave energy. As § 6.1.2 shows, the algebraic manipulation is elementary, but soon becomes unwieldy as attested by many erroneous published transmission coefficients. Perhaps the most succinct nomenclature for the matrix forms introduces physically keyed symbols for wave amplitudes at the outset, rather than the uncorrelative terms A, B, C, D etc. used in previous sections.

As an illustration, consider the important case of plane seismic waves incident on the plane boundary between a solid homogeneous isotropic half-space M_1 and a liquid homogeneous half-space M_2. (The case

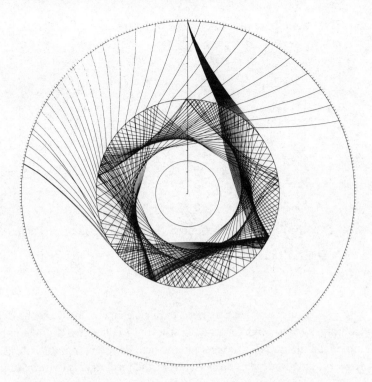

Fig. 6.3. Computer plot of $P7KP$ waves reflected inside the core seven times before reaching the surface. (Courtesy C. Chapman.)

corresponds to body waves incident through the Earth on the boundary of the liquid core. See §10.6 and Fig. 6.3.)

The notation and ray configuration is an extension of that in Fig. 6.1. There are in general three incident $(\grave{P}_1, \grave{S}_1, \grave{P}_2)$ rays, three reflected (scattered) rays $(\acute{P}_1, \acute{S}_1, \acute{P}_2)$ and two refracted rays $(\acute{P}_1, \acute{P}_2)$. The subscripts indicate the upper or lower half-space and the primes the direction of the rays. In this handy form we can now rewrite (2) as

$$\phi_1 = \grave{P}_1 \exp\{i\kappa(x_3 \tan e_1 + x_1 - ct) \\ + \acute{P}_1 \exp\{i\kappa(-x_3 \tan e_1 + x_1 - ct)\}, \qquad (21)$$

and so on.

Substitution of the potentials into the appropriate boundary conditions (see §5.1), namely (i) continuity of normal displacements u_3, and (ii) continuity of stress components p_{31}, p_{33}, yields three equations of type (6) and (7) that can be written in matrix form as

$$
\begin{bmatrix}
\tan e_1 & 1 & \tan e_2 \\
2\tan e_1 & 2 - c^2/\beta_1^2 & 0 \\
\lambda_1 c^2/\alpha_1^2 + 2\mu_1 \tan^2 e_1 & 2\mu_1 \tan f_1 & -\lambda_2 c^2/\alpha_2^2
\end{bmatrix}
\begin{bmatrix}
\acute{P}_1 \\
\acute{S}_1 \\
\acute{P}_2
\end{bmatrix}
$$

$$
=
\begin{bmatrix}
\tan e_1 & -1 & \tan e_2 \\
2\tan e_1 & c^2/\beta_1^2 - 2 & 0 \\
-\lambda_1 c^2/\alpha_1^2 - 2\mu_1 \tan^2 e_1 & 2\mu_1 \tan f_1 & \lambda_2 c^2/\alpha_2^2
\end{bmatrix}
\begin{bmatrix}
\grave{P}_1 \\
\grave{S}_1 \\
\grave{P}_2
\end{bmatrix}. \tag{22}
$$

Verification of the algebra is left as an exercise for the reader. The solution involves matrix inversion. Write (22) for brevity as

$$\mathbf{ND} = \mathbf{MS} \tag{23}$$

where $\mathbf{D}$ is the *incident matrix* and $\mathbf{S}$ the *scattering matrix*. Then

$$\mathbf{S} = \mathbf{RD}, \tag{24}$$

where

$$\mathbf{R} = \mathbf{M}^{-1}\mathbf{N}.$$

For 3×3 matrices, such as here, explicit elements of $\mathbf{R}$ can be derived algebraically, although in practice they are calculated by a computer. Numerical methods of inversion are usually more reliable. A general property can be inferred, however, from the form of (24) by rewriting it as

$$
\begin{bmatrix}
\acute{P}_1 \\
\acute{S}_1 \\
\acute{P}_2
\end{bmatrix}
=
\begin{bmatrix}
R_{11} & R_{12} & R_{13} \\
R_{21} & R_{22} & R_{23} \\
R_{31} & R_{32} & R_{33}
\end{bmatrix}
\begin{bmatrix}
\grave{P}_1 \\
\grave{S}_1 \\
\grave{P}_2
\end{bmatrix}. \tag{25}
$$

Identification of the individual types of ray thus shows that the elements R_{ij} are the scattering coefficients for the various rays.

It is helpful to put

$$
\mathbf{R} =
\begin{bmatrix}
\grave{P}_1 \acute{P}_1 & \grave{P}_1 \acute{S}_1 & \grave{P}_1 \acute{P}_2 \\
\grave{S}_1 \acute{P}_1 & \grave{S}_1 \acute{S}_1 & \grave{S}_1 \acute{P}_2 \\
\grave{P}_2 \acute{P}_1 & \grave{P}_2 \acute{S}_1 & \grave{P}_2 \acute{P}_2
\end{bmatrix}. \tag{26}
$$

Subscripts can be omitted. The solid–solid case gives 4×4 matrices.

Strictly, as in Fresnel's theory in optics, curved boundaries impose wave frequency dependence on the scattering coefficients R_{ij}. The effect is most important for body waves with periods greater than a few seconds with near-grazing angles, and with proximity to turning points of the rays (see § 7.3). Such conditions arise, for example, for $PmKP$ and $SmKS$ core waves (see Fig. 10.6). Adequate corrections to the coefficients can be made simply by replacing the trigonometric functions by generalized functions. One treatment, due to P. Richards (see Ben-Menahem and Singh, 1981, chapter

8), replaces $\cos i$ by generalised cosines $C^{(1)}$ and $C^{(2)}$, for upgoing and downgoing waves, respectively. The form is

$$C^{(1,2)}(r, p, \omega) = (1 - p^2 v^2 / r^2)\left\{ e^{\pm \pi i/6}\frac{H^{(1,2)}_{2/3}(\omega\tau)}{H^{(1,2)}_{1/3}(\omega\tau)} \right\}, \tag{27}$$

where p is the ray parameter (see (7.12)), $H(x)$ is a Hankel function (see (4.66)), and τ is given by (7.63). Note that, from (7.1), $\cos i = (1 - p^2 v^2 / r^2)^{\frac{1}{2}}$, and $C^{(1,2)} \to \cos i$ as $|\omega\tau| \to \infty$.

The reader can then easily show, using the theory of §6.1.2, that for a P wave incident on a boundary between two homogeneous fluids, the reflection coefficient is

$$R = A/A_0 = \frac{\rho_2\alpha_2 C_1^{(2)} - \rho_1\alpha_1 C_2^{(2)}}{\rho_2\alpha_2 C_1^{(1)} + \rho_1\alpha_1 C_2^{(2)}} \tag{28}$$

in the usual notation.

6.6 Exercises

1 Consider, as an approximation, P rays in an Earth of uniform velocity, radius r and centre O. Consider a source F on a diameter such that $OF = \frac{1}{10}r$. Consider stationary rays of PP type from F to a receiving station S on the surface, such that FOS is $90°$. Prove that three paths are possible (i.e. pP, PP and $P'P'$ types) in theory, such that the angle of incidence i at S is given by a cubic equation in i.

2 Water has an index of refraction of 1.33. At what angle must a beam of light strike the water surface in order that the reflected beam may be plane-polarised? Could the same effect occur when a beam of seismic S waves strikes the Mohorovičić discontinuity? [Hint: The tangent of the angle of incidence for complete polarisation equals the refraction index of the reflector.)

3 An elastic P wave, travelling in the x, z plane, is emergent at an angle e on the horizontal free surface $z = 0$ of a homogeneous perfectly elastic medium.

   Show that, in general, there are a reflected P and a reflected S wave, but if $e = 30°$ and $\lambda = \mu$, there is no reflected P wave.

4 In terms of the seismological notation introduced in §10.6 show that equation (26) can be written as

$$\mathbf{R} = \begin{bmatrix} PcP & PcS & PKP \\ ScP & ScS & SKP \\ PKP & PKS & PKKP \end{bmatrix},$$

   for the specific problem of P and S waves scattered at the boundary of the solid mantle and liquid core.

5 Let A_1, B_1 be the amplitude functions of an incident and a reflected wave

respectively. The *coefficient of reflection R* is defined to be the ratio $R = |B_1/A_1|$.

Two semi-infinite bars are joined to form an infinite rod. Their Young's moduli are E_1 and E_2 and densities ρ_1, ρ_2. Find the reflection coefficient R and the phase change on reflection, when harmonic waves in the first medium meet the join in the bars.

6 Consider a line source at S in medium M distance h above a plane boundary with medium M', emitting cylindrical wave fronts at equal intervals of time. Draw fronts of the incident, reflected, and refracted waves for the case of a pure P wave source when $\alpha = 6.0$, $\beta = 3.0$, $\alpha' = 8.0$ and $\beta' = 4.0$, all in km/s.

By Huygens principle, points on the boundary defined by the advancing refracted cylindrical fronts of P and S waves in M' are energy sources. Hence draw the plane wave fronts in M corresponding to the conical (head) waves Pn and Sn.

7 Consider a plane wave propagating with complex slowness p and direction $\mathbf{m}$ in a linear visco-elastic medium with displacement

$$u = A \exp(-\mathbf{a} \cdot \mathbf{x}) \exp[i(\mathbf{m} \cdot \mathbf{x}/c - t)],$$

where $\mathbf{a}$ is the direction of maximum attenuation.

Show that

$$\mathcal{R}(p^2) = m^2/c^2 - a^2, \quad \text{and} \quad \mathcal{I}(p^2) = 2(am/c)\cos\gamma$$

where $\cos\gamma = \mathbf{a} \cdot \mathbf{m}/am$.

Prove that, for elastic media, $a = 0$ or $\gamma = \pi/2$. Further prove that, for anelastic media, $0 < \gamma < \pi/2$ corresponds to inhomogeneous plane waves.

7

Seismic rays in a spherically stratified Earth model

We next consider transmission of body seismic waves in the interior of an Earth model assumed to be spherical and symmetrical about its centre in all its properties. We ignore diffraction effects and investigate properties of seismic rays. Each member of the families of rays has its end-points on the outer surface of the sphere, and is refracted through any surface of discontinuity encountered. The P or S ray character will be taken to be assigned at all levels in any one family, and, on account of the symmetry about O, it is sufficient to consider rays which all lie in the same plane through O and have their deepest points all on the same radius.

7.1 The parameter p of a seismic ray

We may regard the Earth model as composed of an indefinitely large number of thin homogeneous concentric spherical shells. Consider a portion $PP'P''$ of a seismic ray, where P, P', P'' are points on three consecutive boundaries between these shells. Let v', v'' be the ray speeds along PP', $P'P''$, respectively, and let the angles i', j', i'' be as in Fig. 7.1.

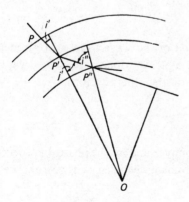

Fig. 7.1. Snell's law refraction.

153

Then, by (6.1),

$$\frac{OP' \sin i'}{v'} = \frac{OP' \sin j'}{v''}.$$

But $OP' \sin j'$ is equal to $OP'' \sin i''$, because both equal the perpendicular from O to $P'P''$. It follows that

$$p = \frac{r \sin i}{v}, \tag{1}$$

where r is the distance from O to any point P of the model, v is the ray speed at P, i is the (acute) angle between OP and the direction of the ray at P, and p is constant along the entire ray.

The value of p, called the *parameter* of the ray, will differ from member to member of a given family. The ray parameter is of great importance in what follows.

7.1.1 Rays in inhomogeneous media. The eikonal equation

We now consider elastic waves in inhomogeneous perfectly elastic and isotropic media. In this case (2.34) yields, with zero body force,

$$\rho f_i = \frac{\partial}{\partial x_i}(\lambda \theta) + \frac{\partial}{\partial x_j}\left\{ \mu \left(\frac{\partial u_j}{\partial x_i} + \frac{\partial u_i}{\partial x_j} \right) \right\}, \tag{2}$$

where u_i is the particle displacement vector and the elastic parameters and derivatives are continuous in x_i. In vector notation, (2) becomes

$$\rho \frac{\partial^2 \mathbf{u}}{\partial t^2} = (\lambda + \mu) \nabla(\nabla \cdot \mathbf{u}) + \mu \nabla^2 \mathbf{u} + \nabla \lambda (\nabla \cdot \mathbf{u})$$

$$+ \nabla \mu \mathbf{x}(\nabla \mathbf{x} \mathbf{u}) + 2(\nabla \mu \cdot \nabla)\mathbf{u}. \tag{3}$$

Let the equation of a wave front be

$$t = \tau(x_i). \tag{4}$$

This form defines the *position* of the wave front at a given *time* or, alternately, the *curve* for a ray as travel time varies. Now, we try as a solution (see §3.4.3) a *ray series* in the separated form

$$\mathbf{u}(x_i, t) = \sum_{k=0}^{\infty} \mathbf{u}_k(x_i) F_k(t - \tau), \tag{5}$$

where $\mathbf{u}$ and F_k may be real or complex and τ is the phase. In order to allow the order of discontinuity of F_k to change with k on a wave front, we assume that

$$F_k'(\xi) = F_{k-1}(\xi). \tag{6}$$

In a wide range of practical cases **u** can be taken as real but, for waves near a caustic and supercritical reflection, a complex displacement vector is needed for analysis.

For high-frequency monochromatic waves, a convenient form is

$$F_k(\xi) = (-i\omega)^{-k-\gamma} \exp(-i\omega\xi), \tag{7}$$

so that

$$\mathbf{u}(x_i, t) = \exp\{-i\omega(t - \tau(x_i))\} \sum_{k=0}^{\infty} (-i\omega)^{-k-\gamma} \mathbf{u}_k(x_i), \tag{5'}$$

where γ is a constant parameter.

Equations for τ and u_k may be found by substitution of (5') in (3) and equating coefficients. The resulting recurrent systems of equations ($k = 0, 1, 2, \ldots$) can then be solved for τ and u_k. For $k = 0$, we obtain three equations

$$\rho\mathbf{u}_0 = (\lambda + \mu)\nabla\tau(\mathbf{u}_0 \cdot \nabla\tau) + \mu(\nabla\tau)^2 \mathbf{u}_0. \tag{8}$$

The system has a non-trivial solution if its determinant vanishes. The reader may easily show that this occurs when

$$(\nabla\tau)^2 = \frac{\partial\tau}{\partial x_i}\frac{\partial\tau}{\partial x_i} = c^{-2}, \tag{9}$$

where $c = \alpha, \beta$, the seismic velocities.

This first-order partial differential equation is called the *phase* or *eikonal* equation, from the Greek word meaning 'image'. The equation is often used as a starting point for ray theory in optics, acoustics and so on because $\tau(x_i) = $ constant defines the position of the wave front at a particular time (see equation (4)). It also provides one way of numerically computing the coordinates of points along a specified ray $x_i = x_i(\tau)$ and the ray parameter along it.

Consider the problem of ray tracing (usually now done by means of a small computer program). Write

$$p_i = \partial\tau/\partial x_i,$$

so that the *slowness vector* components p_1, p_2, p_3 replace $\nabla\tau$. We note that, from differential geometry, in general p_i will be normal to the wave front and tangent to the ray, and

$$p_i p_i = c^{-2}. \tag{10}$$

From the theory of partial differential equations, the *characteristic* waves of the eikonal equation provide a way of constructing the set of wave fronts and orthogonal rays. It can be shown that for (9) the characteristic

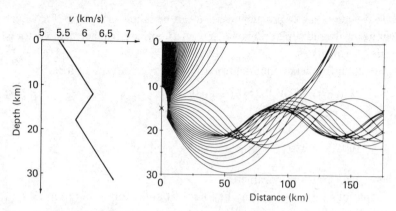

Fig. 7.2. Rays from a seismic source in a low-velocity layer in the crust. (Courtesy V. Červený *et al.*, 1977.)

curves are defined by

$$\frac{\mathrm{d}x_i}{\mathrm{d}\tau} = c^2 p_i, \quad \frac{\mathrm{d}p_i}{\mathrm{d}\tau} = -\frac{1}{c}\frac{\partial c}{\partial x_i}, \qquad i = 1, 2, 3 \tag{11}$$

We thus have six linear differential equations and given initial values $\tau = \tau_0$, $x_i = (x_i)_0$, $p_i = (p_i)_0$. An appropriate numerical integration method, such as Runge–Kutta, then gives x_i and p_i (see Exercise 7.6). Much elegant work along these lines has been done by Červený, Chapman and others and details may be found in the monograph, Červený *et al.* (1977). An example of the method for rays from a source in a low-velocity layer is given in Fig. 7.2.

7.2 Relations between p, Δ, T for a given family of rays

In subsequent discussions, we refer to the values which r, v take at the outer surface of the model and also at the deepest point of the ray whose parameter is p. Suffixes 0 and p, respectively, denote such values.

7.2.1 The relation $p = \mathrm{d}T/\mathrm{d}\Delta$

Consider a family of rays of the type described at the start of this chapter. Let Δ be the angle subtended at O by the whole length of the ray whose parameter is p, and let T be the time of travel along this ray. Let $p + \mathrm{d}p$, $\Delta + \mathrm{d}\Delta$, $T + \mathrm{d}T$ be the corresponding values for an adjacent ray. Then if P_0, Q_0 be adjacent end-points of the two rays and P_0N be normal to the ray through Q_0, we have (see Fig. 7.3)

$$\sin i_0 = \frac{NQ_0}{P_0Q_0} = \frac{v_0 \cdot \frac{1}{2}\mathrm{d}T}{r_0 \cdot \frac{1}{2}\mathrm{d}\Delta}.$$

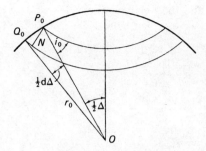

Fig. 7.3. Ray parameter.

Hence, by (1),

$$p = \frac{dT}{d\Delta}. \tag{12}$$

Thus the ray parameter p corresponds to the reciprocal of the ray speed or the slowness at radius r_0, a result first proved by Benndorf.

(In connection with an argument in §8.1, note that the formula (12) is true also for rays which start from some assigned point F in the interior of the Earth model. This is easily seen.) Equation (12) is called Benndorf's relation.

7.2.2 Some integral expressions for T, Δ

Let P be any point of a ray whose parameter is p, let the polar coordinates of P be (r, θ) as indicated in Fig. 7.4, and let the arc-length P_0P be s. Then, by (1),

$$\frac{r^2 \, d\theta}{v \, ds} = p. \tag{13}$$

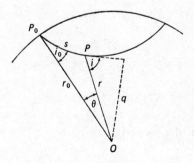

Fig. 7.4. Ray curvature.

Eliminating ds between (13) and the polar relation

$$(ds)^2 = (dr)^2 + r^2(d\theta)^2, \tag{14}$$

we obtain

$$\frac{r^4}{p^2 v^2} = \left(\frac{dr}{d\theta}\right)^2 + r^2, \tag{15}$$

i.e.

$$d\theta = \pm p r^{-1}(\eta^2 - p^2)^{-\frac{1}{2}} dr, \tag{16}$$

where we have introduced the key parameter η, defined by

$$\eta = r/v. \tag{17}$$

At the bottom of the ray $\eta = p$, by (1).

Integrating (16) between P_0 and the deepest point of the ray, we obtain a useful expression for Δ, namely,

$$\tfrac{1}{2}\Delta = p \int_{r_p}^{r_0} r^{-1}(\eta^2 - p^2)^{-\frac{1}{2}} dr. \tag{18}$$

If we eliminate $d\theta$ between (13) and (14) we obtain

$$\frac{p^2 v^2}{r^2} = r^2 \left(\frac{d\theta}{ds}\right)^2 = 1 - \left(\frac{dr}{ds}\right)^2,$$

i.e.

$$ds = \pm r v^{-1}(\eta^2 - p^2)^{-\frac{1}{2}} dr. \tag{19}$$

Because the time T is equal to the integral $\int ds/v$ taken along the ray, we obtain from (19) a formula for T, analogous to the formula for Δ, namely,

$$\tfrac{1}{2}T = \int_{r_p}^{r_0} \eta^2 r^{-1}(\eta^2 - p^2)^{-\frac{1}{2}} dr. \tag{20}$$

The following connection between T and Δ is useful. From (20),

$$T = 2\int_{r_p}^{r_0} \{p^2 r^{-1}(\eta^2 - p^2)^{-\frac{1}{2}} + r^{-1}(\eta^2 - p^2)^{\frac{1}{2}}\} dr.$$

Hence, by (18),

$$T = p\Delta + 2\int_{r_p}^{r_0} r^{-1}(\eta^2 - p^2)^{\frac{1}{2}} dr. \tag{21}$$

Also, because $dr/d\theta$ is zero at the deepest point of the ray, we have

$$p = \eta_p = r_p/v_p. \tag{22}$$

7.2.3 The functions ξ and ζ.

Let ξ and ζ be defined by

$$\zeta = \frac{d \ln v}{d \ln r} = \frac{r}{v}\frac{dv}{dr}, \quad \xi = \frac{2}{1-\zeta} = \frac{2 d \ln r}{d \ln \eta}. \tag{23}$$

These functions of r are important because the simple condition $\zeta < 1$, or $\xi > 0$ (or $d\eta/dr > 0$), determines whether there can exist a ray of the family with its lowest point L at a given level r. This follows since, by (70) and (22), the downward curvature of a ray at L is $v^{-1}dv/dr$, while the curvature of the level surface through r is r^{-1}. Values of r at which ζ and ξ pass through the values 1 and ∞, respectively, are associated with discontinuous changes in the shapes of rays of a family; correspondingly, divergent integrals appear in the ray analysis.

7.2.4 Expressions for $d\Delta/dp$ and dT/dp

By (18), (20) and (23),

$$\Delta = \int_{\eta_p}^{\eta_0} p \xi \eta^{-1}(\eta^2 - p^2)^{-\frac{1}{2}} d\eta, \tag{24}$$

$$T = \int_{\eta_p}^{\eta_0} \xi \eta(\eta^2 - p^2)^{-\frac{1}{2}} d\eta. \tag{25}$$

Integrating (24) and (25) by parts and using (22) gives

$$\Delta = [\xi \cos^{-1}(p/\eta)]_{\eta_p}^{\eta_0} - \int_{\xi_p}^{\xi_0} \cos^{-1}(p/\eta) d\xi$$

$$= \xi_0 \cos^{-1}(p/\eta_0) - \int_{\xi_p}^{\xi_0} \cos^{-1}(p/\eta) d\xi, \tag{26}$$

$$T = \xi_0(\eta_0^2 - p^2)^{\frac{1}{2}} - \int_{\xi_p}^{\xi_0} (\eta^2 - p^2)^{\frac{1}{2}} d\xi. \tag{27}$$

Differentiating (26) and (27) with respect to p, we have

$$\frac{d\Delta}{dp} = -\xi_0(\eta_0^2 - p^2)^{-\frac{1}{2}} + \int_{\xi_p}^{\xi_0} (\eta^2 - p^2)^{-\frac{1}{2}} d\xi = -X + Y, \text{ say}, \tag{28}$$

$$\frac{dT}{dp} = -p\xi_0(\eta_0^2 - p^2)^{-\frac{1}{2}} + \int_{\xi_p}^{\xi_0} p(\eta^2 - p^2)^{-\frac{1}{2}} d\xi. \tag{29}$$

It will be noticed that (28) and (29) entail (12). The formulae (26) to (29) assume that η, ξ and their derivatives are differentiable in the range $r_p \leqslant r \leqslant r_0$, and that ζ does not become equal to unity.

7.3 Relations between Δ and T, corresponding to assigned variations of v with r

Consider an Earth model such that at most points of its interior, v is slowly increasing as r decreases, and the rate of change of v is changing only slowly; and that these circumstances hold near the outer surface. By (26) it follows that for such behaviour of v which we shall call 'ordinary', ζ is negative and moderate in value, and $d\zeta/dr$ and $d\xi/dr$ small; in particular ζ_0 is negative; also $d\eta/dr$ is positive.

We next examine the changes in Δ and T for a given family of rays as the angle of incidence i_0 at the outer surface decreases steadily between $\frac{1}{2}\pi$ and zero. Simultaneously, by virtue of the relation

$$p = \eta_0 \sin i_0, \tag{30}$$

it follows that p decreases steadily between η_0 and zero. When $i_0 = \frac{1}{2}\pi$, Δ and T are of course both zero.

7.3.1 Case 1

Since ζ_0 is negative and p is equal to η_0 when Δ is zero, it follows by (23) and (28) that for small values of Δ, X is large and positive. On the other hand, Y is found to be small for rays that do not penetrate deeply, owing to the smallness of $d\xi/dr$ and the smallness of the relevant range of integration. Thus for small Δ, $d\Delta/dp$ is large and negative. Hence, by (12), $d^2T/d\Delta^2$ is negative, tending to 0 for small Δ; and hence (assuming a power series expansion to exist)

$$T = \eta_0\Delta - a\Delta^3, \tag{31}$$

approximately, where a is a positive constant.

A useful alternative to (31) comes directly from (28). If Y is small compared with X, we have approximately

$$\frac{d\Delta}{dp} = -\xi_0(\eta_0^2 - p^2)^{-\frac{1}{2}}.$$

Since $p = \eta_0$ when $\Delta = 0$, this gives

$$\Delta = \xi_0 \cos^{-1}(p/\eta_0);$$

and hence by (12)

$$T = \eta_0\xi_0 \sin(\Delta/\xi_0). \tag{32}$$

Also, from (29),

$$T = \xi_0(\eta_0^2 - p^2)^{\frac{1}{2}}.$$

The formula (32) is exact when $Y = 0$, i.e. by (28), when ξ is constant

$(= \zeta_0)$, and of course agrees with (31) so far as the third power in an expansion in powers of Δ. It is specially useful because, by (23), it corresponds to a velocity distribution given by the simple law, due to Mohorovičić,

$$v = ar^b, \tag{33}$$

where a and b are constants (actually $b = \zeta_0$).

7.3.2 Case 2

When i_0 is zero (normal incidence), p is zero by (30). By (12) it then follows that for rays that pass near the Earth's centre, for which by (1) p is a little greater than zero,

$$T = a - b(\pi - \Delta)^2, \tag{34}$$

approximately, where a and b are positive constants.

7.3.3 Case 3

Now suppose that at a certain level there is a marked increase in the rate of increase of velocity with depth. More precisely, suppose that (i) $X > Y$, i.e. $d\Delta/dp < 0$, for rays whose deepest points are just above the level in question; (ii) the behaviour of v is ordinary just above this level, while, just below, the variation of v is such that ζ decreases rapidly as the depth increases, becoming rapidly large (and negative); (iii) as the depth further increases v soon resumes ordinary behaviour.

If the rate of change of ζ in connection with (ii) and (iii) is sufficiently great, it is evident from (28) that Y will exceed X for a range of values of p. Accordingly, $d\Delta/dp$ will become positive for this range, and circumstances will be as illustrated in Fig. 7.5(a). The associated features of the (T, Δ) relation may then be inferred after the manner indicated in Fig. 7.5(b) and (c); in (a), (b) and (c), the points A, A', A'' etc. correspond. The features include the presence of cusps at B'' and C'', triplication of the curve over a

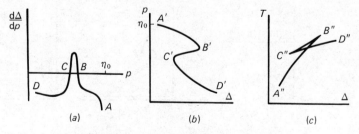

Fig. 7.5. Slowness and travel-time curves.

range of values of Δ, and upward concavity of the branch $B''C''$ of the travel-time curve (c).

7.3.4 Case 4

If the velocity variation in the vicinity of another level is similar to that described in §7.3.3, except that the rate of change of ζ falls just a little short of the amount needed to make Y exceed X, $d\Delta/dp$ nearly but not quite reaches the zero value. In this case, circumstances are as illustrated in Fig. 7.6. There is now no triplication, but the (T, Δ) curve has a considerable curvature in the vicinity of the appropriate value of Δ.

7.3.5 Case 5

The case in which there is a discontinuity surface across which v increases by a finite amount from above to below, the behaviour on both sides of the surface being ordinary, may be deduced from the case of §7.3.3. The forms of the corresponding curves are illustrated in Fig. 7.7. In the graph (b), Δ has a finite reduction from B' to C', owing to the finite decrease in r/v across the discontinuity surface, which entails a finite decrease in p; since r_p is changing continuously in the circumstances under consideration,

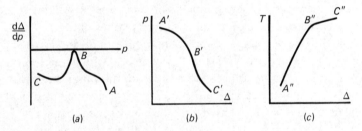

Fig. 7.6. Slowness and travel-time curves.

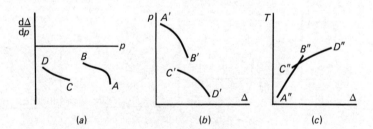

Fig. 7.7. Further slowness and travel-time curves.

it follows from (18) that Δ has a finite decrease. (Values of p, and of i_0, in the gap correspond to rays that are totally reflected upwards at the discontinuity surface. These rays are not included in the above treatment, since $dr/d\theta$ is discontinuous at the deepest points of these rays. It is easy to discuss these rays separately once the theory of this chapter is set up.) Features of the (T, Δ) curve, Fig. 7.7(c), include a discontinuity from B'' to C''; and the gradient at C'' is less than that at B''.

7.3.6 Case 6
Next suppose that at another level the velocity decreases with increase of depth. More precisely, suppose that (i) $X > Y$, i.e. $d\Delta/dp < 0$, for rays whose deepest points are just above this level; (ii) the behaviour of v is ordinary just above this level, while, just below, a decrease of velocity sets in which causes ζ to increase with depth for some distance; (iii) as the depth further increases, ζ soon reaches a maximum and then rapidly decreases until the velocity resumes ordinary behaviour.

If the rate of decrease of velocity is such that ζ reaches the value $+1$, the integral in (28) diverges, and it is evident that there is a discontinuity in Δ as p continuously decreases. At the level, L_1 say, where ζ first passes through the value unity, $d\eta/dr$ is zero and decreasing (as the depth increases); just below this level η will therefore increase until, corresponding to (iii), ζ returns through the value $+1$; subsequently, if the further behaviour of v is ordinary, η will decrease in value until in due course it regains at some lower level, L_2 say, its value at L_1. On considering changes in the parameter p, i.e. η_p, it is evident from (22) that no ray of the family can have its deepest point between the levels L_1 and L_2, and that r_p, Δ and T have discontinuous changes.

On account of the divergence of the integral in (28) in this case, a full analytical discussion would require some modification of the formulae (28) and (29). Here we shall examine (in §7.3.8) only the extreme case of discontinuity in the velocity.

7.3.7 Case 7
If the velocity variation in the vicinity of some level is as in the case of §7.3.6, except that the changes are not quite sufficient for ζ to reach the value $+1$, the effect is merely that $d\Delta/dp$ (which remains negative) is somewhat increased in magnitude for a range of values of p. It is easy to show, similarly to §7.3.4, that this case entails a diminution in the curvature of the (T, Δ) curve for a range of values of Δ. The curve may become nearly straight.

7.3.8 Case 8

We now discuss the case in which there is a discontinuity surface, $r = r_1$ say, across which v decreases by a finite amount from above or below. We now need to separate the contributions to $d\Delta/dp$ arising from the regions above and below this surface; let the suffix a indicate values of variables just above, and b values just below. For rays that penetrate below the level $r = r_1$ we now have that $d\Delta/dp = -X + Y$, where

$$X = \xi_0(\eta_0^2 - p^2)^{-\frac{1}{2}} - \xi_a(\eta_a^2 - p^2)^{-\frac{1}{2}} + \xi_b(\eta_b^2 - p^2)^{-\frac{1}{2}}, \tag{35}$$

$$Y = \left(\int_{\xi_1}^{\zeta_0} + \int_{\xi_p}^{\xi_1} \right)(\eta^2 - p^2)^{-\frac{1}{2}}\,d\xi. \tag{36}$$

Since $v_b < v_a$, we have $\eta_b > \eta_a$; if below the level $r = r_1$ the behaviour of v is ordinary for sufficient distance, η decreases as the depth increases until at a lower level, $r = r_2$ say, η regains the value η_a. It follows that the given discontinuity in velocity entails a discontinuous increase in Y of amount

$$\int_{\xi_2}^{\xi_1} (\eta^2 - \eta_a^2)^{-\frac{1}{2}}\,d\xi. \tag{37}$$

The integral (37) is convergent and its value is moderate. On the other hand, the second term on the right-hand side of (35) is large for the highest rays which penetrate below the level $r = r_1$ (since p is nearly equal to η_a for these rays), while the first and third terms of (35) are moderate. It follows that as p decreases steadily through the value η_a, $d\Delta/dp$ changes suddenly to a large positive value. As p further decreases, the second term in (35) rapidly diminishes to a moderate value, and so $d\Delta/dp$ soon returns to a value comparable with its value for rays just above the level $r = r_1$. If the behaviour of v below the level $r = r_1$ continues to be ordinary for a sufficiently large increase in depth, $d\Delta/dp$ will in due course become negative again. This follows because for rays of the family which pass near the centre of the model, p is small (and positive), and must by (34) obey the approximate relation

$$p = 2b(\pi - \Delta),$$

where $b > 0$.

The case described is illustrated in Fig. 7.8. In the corresponding form of the (T, Δ) relation, it will be noticed that there is a gap in the values of Δ, which would correspond to a 'shadow zone' on the surface of the Earth model. Also, as p continually decreases, the subsequent values of Δ pass through a minimum value at cusp D'', analogously to the minimum deviation of a ray of light through a prism or lens.

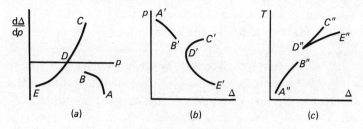

Fig 7.8. Further slowness and travel-time curves.

For points near D in Fig. 7.8(a) we may write

$$\frac{\mathrm{d}\Delta}{\mathrm{d}p} \approx \alpha(p - \beta),$$

where α is the gradient and β the value of p at D. This relation gives

$$\Delta \approx \tfrac{1}{2}\alpha(p - \beta)^2 + \gamma,$$

where γ is constant; and using (2) we deduce that for points near D'' in Fig. 7.8(c),

$$T - T_1 = a(\Delta - \Delta_1) \pm b(\Delta - \Delta_1)^{\frac{3}{2}}, \tag{38}$$

where a and b are constants, and T_1, Δ_1 are the values of T, Δ at the point D''. Thus, near D'', the (T, Δ) curve approximates to a semicubical parabola.

7.3.9 Case 9

The cases considered in §§ 7.3.1–7.3.8 cover the principal features of the (T, Δ) relations that are met with in the actual Earth (see chapters 12 and 13).

Notice from Figs. 7.5–7.8 that $\mathrm{d}^2 T/\mathrm{d}\Delta^2$ is mostly negative; correspondingly by (12), $\mathrm{d}p/\mathrm{d}\Delta$ is then negative; then by (30), $\mathrm{d}\Delta/\mathrm{d}i_0$ is negative and $\mathrm{d}\Delta/\mathrm{d}e_0$ positive, where $e_0 = \tfrac{1}{2}\pi - i_0$. But $\mathrm{d}^2 T/\mathrm{d}\Delta^2$ may sometimes be positive, as, for example, in the branch $D''C''$ in Fig. 7.8(c); in such cases $\mathrm{d}\Delta/\mathrm{d}e_0$ is negative.

7.4 Derivation of P and S velocity distributions from (T, Δ) relations

From readings of seismograms, empirical tables may be derived that connect values of T and Δ for various families of seismic rays (details will be given in chapter 10). The theory in § 7.3 then enables us to infer features of the P and S velocity distributions in the Earth. We now set up theory by which the detailed P and S velocity distributions can be found.

Assume that the relation between T and Δ is known for a family of rays of prescribed type in a spherically symmetrical Earth model. By (12), Δ is thus a known function of p, and vice versa.

7.4.1 Herglotz–Wiechert–Bateman inversion

We assume further that η decreases monotonely as r decreases in the range $r_0 \geqslant r > r'$, where r' is a particular value of r less than r_0; dashes denote values of variables at the level $r = r'$.

Then by (18), provided $p > \eta'$,

$$\Delta = \int_{\eta_p}^{\eta_0} 2pr^{-1}(\eta^2 - p^2)^{-\frac{1}{2}} \frac{\mathrm{d}r}{\mathrm{d}\eta} \mathrm{d}\eta. \tag{39}$$

The equation (39) is a form of Abel's integral equation or Radon's transform whose solution determines η as a function of r. The equation was first investigated in seismological contexts by Herglotz, Wiechert and Bateman. The following mode of solution is due to Rasch.

Let r_1 be a value of r such that $r_0 \geqslant r_1 > r'$, and let the suffix 1 be used to denote values of variables (e.g. v_1, η_1) at the level r_1. Incidentally, $\eta_1 > \eta'$. Also, let Δ_1 be the value of Δ for the ray whose deepest point is at the level r_1.

Apply the operation

$$\int_{\eta_1}^{\eta_0} \mathrm{d}p(p^2 - \eta_1^2)^{-\frac{1}{2}}$$

to both sides of (39); this integration ranges over rays of the family from the highest, for which $p = \eta_0$, down to the ray for which $p = \eta_1$. We obtain

$$\int_{\eta_1}^{\eta_0} \Delta(p^2 - \eta_1^2)^{-\frac{1}{2}} \mathrm{d}p = \int_{\eta_1}^{\eta_0} \mathrm{d}p \int_{\eta_p}^{\eta_0} 2pr^{-1}\{(p^2 - \eta_1^2)(\eta^2 - p^2)\}^{-\frac{1}{2}} \frac{\mathrm{d}r}{\mathrm{d}\eta} \mathrm{d}\eta$$

$$= \int_{\eta_1}^{\eta_0} \mathrm{d}\eta \int_{\eta_1}^{\eta} 2pr^{-1}\{(p^2 - \eta_1^2)(\eta^2 - p^2)\}^{-\frac{1}{2}} \frac{\mathrm{d}r}{\mathrm{d}\eta} \mathrm{d}p.$$

The last step, in which the order of integration is changed, is seen from the relevant (triangular) domain of integration in the graph of η against p. Integrating by parts on the left side, and integrating with regard to p on the right, we obtain, since $\eta > \eta_1$.

$$\left[\Delta \cosh^{-1}\left(\frac{p_1}{\eta_1}\right) \right]_{\eta_1}^{\eta_0} - \int_{\eta_1}^{\eta_0} \frac{\mathrm{d}\Delta}{\mathrm{d}p} \cosh^{-1}\left(\frac{p}{\eta_1}\right) \mathrm{d}p = \int_{\eta_1}^{\eta_0} \pi r^{-1} \frac{\mathrm{d}r}{\mathrm{d}\eta} \mathrm{d}\eta.$$

The first term on the left side of this expression is zero, for Δ is zero when $p = \eta_0$, while $\cosh^{-1}(p/\eta_1)$ is zero when $p = \eta_1$. Hence

$$\int_0^{\Delta_1} \cosh^{-1}\left(\frac{p}{\eta_1}\right) \mathrm{d}\Delta = \pi \ln\left(\frac{r_0}{r_1}\right). \tag{40}$$

However, p is given as a function of Δ. Also η_1 is the known value of $dT/d\Delta$ at Δ_1. Hence r_1 is found from (40) in terms of Δ_1 and so in terms of η_1 or r_1/v_1. This process can be carried out for any r_1 in the range $r_0 \geqslant r_1 > r'$. We have thus derived a direct method of finding v as a function of r in the range $r_0 \geqslant r > r'$.

If we use (T, Δ) data in this way for various ray families, we can find the P and S velocities at all points down to the level at which the η concerned first starts to increase with increase of depth.

For all the cases discussed in §7.3, excepting those in §§7.3.6 and 7.3.8, η decreases as the depth increases, and so the corresponding departures of v from ordinary behaviour will not affect the success of the method. It must be noted that when there is a loop in the (T, Δ) curve, as in §7.3.3, $d\Delta$ must be reckoned negative on the upper branch in the use of (40).

The method fails to find v below any level at which velocity decreases with depth as discussed in §§7.3.6 or 7.3.8; in the former of these cases $d\eta/dr$ changes sign and reaches a turning value, while in the latter η suffers a discontinuous increase as the depth increases.

The method can easily be adapted to meet the case of §7.3.8. Suppose there is a sudden (finite) increase in η from above to below the level $r = r'$, and that η decreases monotonely between this level and a lower level $r = r''$. The procedure determines v as a function of r for $r_0 \geqslant r > r'$. Using (18) and (20) with the lower limit of integration replaced by r', it is then possible to compute, for rays which go below the level $r = r'$, the contributions to T and Δ for the portions of the path above this level. By subtraction from values in the (assumed known) (T, Δ) table, a table of values of T' against Δ' can then be drawn up, where T' and Δ' apply to the sphere $r = r'$. The velocities for $r' > r > r''$ can then be determined as before. This process is called 'stripping the Earth' and leads to fairly precisely determined velocities throughout most of the Earth.

The theory of the case of §7.3.6 is more troublesome. This case may arise in the outermost 200 km of the Earth and also near a depth of 2900 km.

7.4.2 Bullen's method

From (28), we have

$$Y = d\Delta/dp + \xi_0(\eta_0^2 - p^2)^{-\frac{1}{2}} = d\{\Delta - \xi_0 \cos^{-1}(p/\eta_0)\}/dp, \qquad (41)$$

from which values of Y in terms of p can be deduced for a family of rays when the (T, Δ) relation is given. Also by (28), we have the integral equation

$$Y = \int_{\xi_p}^{\xi_0} (\eta^2 - p^2)^{-\frac{1}{2}} d\xi, \qquad (42)$$

which can be solved to yield

$$\int_0^{Y_1} (p^2 - \eta_1^2)^{\frac{1}{2}} \, dY = -\tfrac{1}{2}\pi(\xi_1 - \xi_0). \tag{43}$$

Hence we can take the velocity law (33), or $\xi = \xi_0$, as a first approximation to a sought velocity distribution, and use (43) to compute corrections $\xi - \xi_0$ to ξ_0 at various depths, Y being a known function of p. The method is effective over wide ranges of depth in parts of the Earth, and is specially powerful where (32) gives a moderately good approximation to the (T, Δ) values.

7.4.3 Linear inverse method

In modern notation, much use is made in seismological inverse theory of a functional that maps a set of functions $f(x)$ over the real numbers, called the function or model space, into the set of real numbers. In what follows, we deal mainly with real functions. Functionals may take the form of matrix, differential, and integral operators, although in this book the main emphasis is on the latter case.

For example, the length s of the curve $y = y(x)$ from x_1 to x_2 is

$$s = \int_{x_1}^{x_2} \left[1 + \left(\frac{dy}{dx}\right)^2\right]^{\frac{1}{2}} dx. \tag{44}$$

Here the functional is the mapping (an integral operator) to a length (a real number), which exists for every curve. Note that the mapping is not one-to-one, since for a given s there is no unique curve $y = y(x)$.

A linear functional ϕ has the property

$$\phi\{af(x) + bg(x)\} = a\phi\{f(x)\} + b\phi\{g(x)\}, \tag{45}$$

where a, b are constants and $f(x), g(x)$ are functions. We note that s is not a linear functional of y in (44); for example, mass M and moment of inertia I are linear functionals of density $\rho(x)$ in § 13.4.1. Truncation after the term in δx shows that the error is of order $(\delta x)^2$.

Assume that the functional I connecting the Earth model to the seismological observations has integral form and the model function is $m = m(x)$. Explicitly we write

$$I(z) = \int_a^b F(z, m) \, dx, \tag{46}$$

where F is an integrable real function. Often $F(z, m)$ can be factored so that (46) takes the form

$$I(z) = \int_a^b G(z, m)m(x) \, dx, \tag{47}$$

where $G(z, m)$ is the *kernel* of the integral equation. The kernel acts as a weighting or smoothing function on the $m(x)$. The inverse feature of (47) is that $I(z)$ is given and $m(x)$ is unknown. For each particular datum $I(z)$ there is a corresponding kernel G so that $G(z, m)$ is called the *data kernel*.

Where possible we would like to work with linear functionals. Suppose a trial model $m_0(x)$ is chosen and we consider a neighbouring model $m(x)$ such that $m = m_0 + \delta m$. Then, from (46) and Taylor's expansion,

$$\delta I = \int_a^b \left(\frac{\partial F}{\partial M_0} \right) \delta m \, dx, \quad \text{to first order,} \tag{48}$$

$$= \int_a^b G(z, m_0) \delta m(x) \, dx, \tag{49}$$

where the kernel is obtained by differentiation. Note that (49) is a linear functional in $\delta m(x)$. δI is the residual between the observed datum (appropriate to the improved model m) and the calculated datum (appropriate to the assumed model m_0).

The small change in m in (49) is called, in the calculus of variations (see §12.3.4), the *first variation* of m or the Fréchet differential. For continuous functions $\delta m(x)$, the exact meaning of (49) requires a treatment using Hilbert spaces. For estimation, it is usually sufficient to consider a finite number of discrete variations to the model at specific values of the independent variable, $x = x_1, x_2, ..., x_m$, say.

It is preferable always to consider a dimensionless perturbation to the model so that we replace $\delta m/m$ by $\delta m'$ and mG by G' and then drop the primes. In this case we simply write (49) as

$$\delta I = \int_a^b G(z, m_0) \delta m(x_1) \, dx + \int_a^b G(z, m_0) \delta m(x_2) \, dx + \cdots$$

$$= \delta m(x_1) \int_{\Delta x_1} G(z, m_0) \, dx + \delta m(x_2) \int_{\Delta x_2} G(z, m_0) \, dx + \cdots, \tag{50}$$

where each perturbation is treated as a narrow rectangular pulse. The idea is illustrated in §13.5.3 for the elementary density problem. Or, replacing z by N data points,

$$\delta I_i = \sum_{j=1}^M G_{ij} m_j, \tag{51}$$

where $m_j = \delta m(x_j)$, and $i = 1, 2, ..., N$ denotes the N different data involved.

It is important to note that in many seismological contexts, $I(z)$ in (46) is not a linear functional so that the kernel must be evaluated anew for each starting model and the estimated solution is dependent upon the initial assumption.

Consider in this formalism the procedures developed by Backus and Gilbert (1970) for determining $v(r)$ from $T(\Delta)$. If $v(r)$ were known, T could be calculated from (20) and (21) (the direct problem). The seismological inverse problem is to use a set of values of T to determine $v(r)$. In this case, the functional given by equation (20) is non-linear, but this situation can be remedied by considering residual times δT (observed times minus times predicted by a trial $v(r)$) and seeking perturbations δv to $v(r)$.

Then, approximately,

$$\delta T(p) = \int_{r_p}^{r_0} G(r, v, p) \delta v(r) \mathrm{d}r, \tag{52}$$

where

$$G(r, v, p) = -2\eta v^{-2} (\eta^2 - p^2)^{-\frac{1}{2}}. \tag{53}$$

The correction term in (52) is small except near $\eta = p$. Here, r_p is the radius to the point where the ray bottoms. Now, in (52), the functional is linear in the perturbation $\delta v(r)$. For each of n rays there are n distinct p and r_p, in general. For a numerical estimate, we discretise $\delta v(r)$ into m segments, δv_j for $r_{j-1} < r < r_j, j = 1, 2, \ldots, m$. Thus, (52) becomes

$$\delta T_i = \sum_{j=1}^{m} A_{ij} \delta v_j \tag{54}$$

where

$$A_{ij} = -2 \int_{r_{j-1}}^{r_j} \frac{\eta \mathrm{d}r}{v^2 (\eta^2 - p^2)^{\frac{1}{2}}}. \tag{55}$$

The solution for the velocity perturbation is thus reduced to a linear algebra problem (see §10.4.2).

7.4.4 Inversion for low velocity layers

The inverse problem of determining $v(r)$ from $T(\Delta)$ in polar coordinates can be reduced to an equivalent Cartesian one for a half-plane $Y \geqslant 0$ by the transformations

$$X = r_0 \Delta / v(r_0) \tag{56}$$

$$Y = r_0 v^{-1}(r) \ln (r_0/r) \tag{57}$$

$$V(Y) = \frac{v\{r_0 \exp [-v(r_0) Y / r_0]\}}{v(r_0) \exp [-v(r_0) Y / r_0]}. \tag{58}$$

Substitution in (18) and (20) gives

$$X(p) = \int_0^{Y(p)} p[u^2(y) - p^2]^{-\frac{1}{2}} \mathrm{d}y, \tag{59}$$

$$T(p) = \int_0^{Y(p)} u^2(y)[u^2(y) - p^2]^{-\frac{1}{2}} dy, \tag{60}$$

where $u(y) = V^{-1}(y)$ is wave slowness, $2X(p)$ is epicentral distance, and $2T(p)$ is the travel-time along a ray with parameter p.

Normally, the velocity function $V(y)$ and its derivative are continuous. There may, however, be a finite number of low-velocity layers (i.e. gaps in the $T(X)$, $p(X)$ curves).

Let the ith low-velocity layer begin at depth y_{1i} and end at y_{2i} so that $u(y) > u(y_i)$ for $y_{1i} < y < y_{2i}$.

It then may be proved, as shown by Gerver and Markushevitch, that the solution of the above inverse problem is given by

$$Y(p) = 2\pi^{-1} \int_p^1 X(q)[q^2 - p^2]^{-\frac{1}{2}} dq$$

$$+ \sum_i 2\pi^{-1} \int_{p_{1i}}^{p_{2i}} \tan^{-1}\{[u^2(y) - q_i^2]/[q_i^2 - p^2]\}^{\frac{1}{2}} dy. \tag{61}$$

Note that the first integral is the Herglotz–Wiechert form with $p = 1$ at $X = 0$, and the second is the contribution from the low velocity layers. The equation (61) gives a solution that bounds all velocity–depth curves consistent with the observed travel-time curves. Finally, consider the transformed ray parameter $p = dT(X)/dX$. Then

$$T(X) = \int_0^X p(x) dx.$$

Integration by parts in terms of p rather than X yields (cf. equation (21))

$$T(X) = pX(p) + \int_p^1 X(q) dq. \tag{62}$$

This key equation connects the travel-times with the wave slownesses.

7.4.5 The tau (τ) method
Rewrite (21) as

$$\tau(p) = T(p) - p\Delta(p) = 2 \int_{r_p}^{r_0} r^{-1}(\eta^2 - p^2)^{\frac{1}{2}} dr \tag{63}$$

or

$$\tau(p) = T(p) - pX(p) = \int_0^{Y(p)} (u^2(y) - p^2)^{\frac{1}{2}} dy. \tag{64}$$

The basic role of $\tau(p)$ in connecting travel times and velocity depth curves has been developed by a number of workers, particularly Gerver and

Markushevitch. It has been employed to avoid the direct use of the Wiechert–Herglotz integral inversion and to estimate uncertainties in $v(r)$ directly from the travel-time observations.

First, note that $\tau(p)$ is a monotonely decreasing function of the ray parameter (in contrast to (T, Δ) curves, see Fig. 7.6). It is continuous everywhere, except at a finite number of points $p = q_i$, corresponding to low-velocity layers where there is a negative jump

$$\sigma_i = \int (u^2(y) - q_i^2)^{\frac{1}{2}} dy \tag{65}$$

integrated over the layer width.

To allow for a number of such layers, we require a modified $\tau(p)$ function

$$\ddot{\tau}(p) = \tau(p) - \sum_i \sigma_i, \quad \text{for } q_i > p. \tag{66}$$

$$= \int_p^1 X(q) dq, \quad \text{from (62).} \tag{67}$$

In the absence of low-velocity layers, $\tau(p)$ is simply the sum of source–receiver distances.

Secondly, because, from (67), the derivative $\tau'(p) = -X(p)$, it follows that $X(p)$, and hence from (64) $T(p)$, can be calculated from $\tau(p)$ except at a finite number of points. Geometrically, $\tau(p)$ is the time intercept of the tangent of the travel-time curve at $X(p)$, $T(p)$.

Thirdly, (64) may be written as a first-order differential equation

$$\tau = p\tau' + T(-\tau'). \tag{68}$$

In a more usual notation, (68) is Clairaut's equation

$$y - xy' = f(y') \tag{69}$$

where f is a given function. Differentiation gives

$$y''[f'(y') + x] = 0.$$

Thus either $y'' = 0$ and the general solution is the one-parameter family of straight lines $y = ax + f(a)$; or, from the second factor and (68), there is a singular solution (no arbitrary constants), that can be shown to be the envelope of the general solution (see Exercise 7.7.8).

In the travel-time application, this singular solution is the key to the stable estimation of $\tau(p)$ from the travel-time curve. The envelope of $\tau(p)$ can be computed from $T(X)$ without numerically differentiating the travel-time curve (in practice, consisting of scattered points) or using array measurements of slowness p.

The construction is as follows. Assume the curve $T(X)$ is a single branch, convex up or down. (The detection and evaluation of loops from scattered data needs special treatment.) In graphical terms, we plot $\tau(p_k, X)$ from (64) for a given p_k. The extreme value $\tau(p_k, X_k)$ in this interval defines the correct X_k for this particular p_k (i.e., when $d\tau/dX = 0$, from (64), $p_k = dT/dX$ at $X = X_k$). The algorithm continues by steps δp through the appropriate range $p_1 < p_k < p_2$, thereby constructing from well-resolved extrema the slowness p versus distance X curve for the given travel-time branch and hence, from (64), the $\tau(p)$ curve itself. The reader should prove that this algorithm yields the singular solution to Clairaut's equation.

Statistical account may be taken of the scatter in $T(X)$ data points. We select a band δX in the neighbourhood of the extrema in the $\tau(p_k, X)$ curve for which τ is approximately constant. Estimates of the variance of $\tau(p)$ can then be found under various assumptions of sample distribution. Weaknesses in the algorithm arise from the tradeoff of step-size δp against consequent interval size δX. Smaller δp entails greater resolution of the $\tau(p)$ curve but decreased precision in $\tau(p)$ estimates. Nevertheless, use of the dispersion band-width at the extremum does provide uncertainty bands for the velocity curves after inversion.

7.5 Special velocity distributions

We now consider some specially useful model velocity distributions.

7.5.1 Curvature of a seismic ray

Let ρ be the upward radius of curvature at the point P of a seismic ray (see Fig. 7.3). Then $\rho = - r\,dr/dq$, where q is the perpendicular from O to the tangent at P to the ray. Since, by (1), $q = pv$, we then have

$$\rho = -\frac{r\,dr}{p\,dv}. \tag{70}$$

It is an immediate consequence of (70) that for a family of rays of given (i.e. P or S) type, the curvature at any level is proportional to p. It follows further from (1) that $\rho \sin a$ is constant at a fixed level for a given family of rays, being equal to $- v\,dr/dv$.

From the formula for curvature in polar coordinates, or independently, using (1), (15) and (70), it follows that

$$\rho \sin^3 i = -\left\{ \frac{d^2}{d\theta^2}\left(\frac{1}{r}\right) + \frac{1}{r} \right\}^{-1}. \tag{71}$$

7.5.2 Rays in a homogeneous medium

The particular case of (70) where v is constant for a range of values of r gives $\rho = \infty$, and hence (of course) that the rays are rectilinear in the corresponding region. Thus, for a homogeneous Earth model, we have by (23) $\zeta = 0$, and thence, putting $\xi_0 = 2$ in (32), the (T, Δ) relation

$$T = 2\frac{r_0}{v_0}\sin\frac{\Delta}{2}; \tag{72}$$

the relation (72) is, of course, readily established directly.

7.5.3 Circular rays; the law $v = a - br^2$

By (70), it follows that if v satisfies the law

$$v = a - br^2, \tag{73}$$

the rays are circles of radii $(2pb)^{-1}$. The law (73) has frequently been used in approximate investigations concerning the region of the Earth just below the outer layers. The corresponding relation between T and Δ is most conveniently expressed in parametric form, the parameter used for this purpose being the angle i_0 used earlier in this chapter. In Fig. 7.9 we have drawn a particular ray of the series whose parameter is p say. K is the centre of the (circular) path of this ray. In the triangle OP_0K, we have $OP_0 = r_0$, $KP_0 = (2pb)^{-1}$, the angle $KOP_0 = \frac{1}{2}\Delta$, and the angle $OP_0K = \frac{1}{2}\pi + i_0$; hence the angle $OKP_0 = \frac{1}{2}\pi - \frac{1}{2}\Delta - i_0$. Therefore

$$2pb\sin\tfrac{1}{2}\Delta = r_0^{-1}\cos\left(\tfrac{1}{2}\Delta + i_0\right).$$

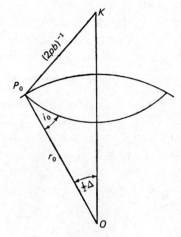

Fig. 7.9. Circular ray.

But by (30)

$$p = r_0 v_0^{-1} \sin i_0.$$

Eliminating p and using (73), we obtain

$$\Delta = 2 \tan^{-1}(\cot i_0 / \lambda), \tag{74}$$

where

$$1 + \lambda = 2a/v_0. \tag{75}$$

By (12), (1) and (74),

$$dT = p\, d\Delta$$

$$= -\frac{r_0 \sin i_0}{v_0} \frac{2\lambda}{\lambda^2 + \cot^2 i_0} \operatorname{cosec}^2 i_0 \, di_0.$$

Hence

$$T = \frac{r_0}{v_0(\lambda^2 - 1)^{\frac{1}{2}}} \ln \frac{\lambda + (\lambda^2 - 1)^{\frac{1}{2}} \cos i_0}{\lambda - (\lambda^2 - 1)^{\frac{1}{2}} \cos i_0}. \tag{76}$$

By (74) and (76)

$$T = \frac{2r_0}{v_0(\lambda^2 - 1)^{\frac{1}{2}}} \ln \left[(\lambda^2 - 1)^{\frac{1}{2}} \sin \tfrac{1}{2}\Delta + \{ (\lambda^2 - 1)\sin^2 \tfrac{1}{2}\Delta + 1 \}^{\frac{1}{2}} \right]. \tag{77}$$

Equations (75) and (77) connect T with Δ.

7.5.4 Mohorovičić's law $v = ar^b$

This law is the relation (33), already established in §7.3.1. It is specially powerful because of the simple form (32) of the corresponding (T, Δ) relation. It includes the case of constant v (§7.5.2), but is more flexible through having the two adjustable parameters η_0 and ξ_0, and in fact provides a close approximation to the actual velocity variation over many ranges of depth in the Earth. Further, where the approximation is not adequate, corrections can be derived after the manner indicated in §7.4.1. A limitation is that it entails $v \to \infty$ as $r \to 0$ (assuming ζ_0 negative), so that the law is not serviceable where r/r_0 is small.

An important analysis of ray theory was made by Slichter in 1932, containing information supplementary to that given above. Other significant contributions are discussed in Hron (1972) and Kennett (1983).

7.6 Theory of travel-times in near earthquakes

7.6.1 Special form of the (T, Δ) relation for near earthquakes

For the model Earth described in §7.1, the travel-time T and the epicentral distance Δ in the case of a surface focus are, by (17) and (21), connected by the relation

$$T = p\Delta + 2 \int_{r_p}^{r_0} \left(\frac{1}{v^2} - \frac{p^2}{r^2} \right)^{\frac{1}{2}} dr, \tag{78}$$

where v is the ray speed at distance r from the centre of the model, p is the parameter of the ray, and the suffixes 0 and p are used to denote values at the outer surface and at the lowest point of the ray, respectively. By (1), introducing $z = r_0 - r$, we may rewrite (78) in the form

$$T = \frac{r_p\Delta}{v_p} + 2 \int_0^{z_p} \left\{ \frac{1}{v^2} - \left(\frac{r_p}{r} \right)^2 \frac{1}{v_p^2} \right\}^{\frac{1}{2}} dz. \tag{79}$$

If the focus instead of being at the outer surface is at the depth h below this level, we need to replace the operation $2 \int_0^{z_p}$ in (79) by $(\int_0^{z_p} + \int_h^{z_p})$; this is seen on considering separately the portions of the ray on either side of the lowest point. (We are here considering rays which leave the focus in a downward direction.)

In the case of near earthquakes, a simplification arises from the fact that the rays do not penetrate to a great depth, so that $r_0 - r_p$ is small compared with r_0. In this case, we have approximately

$$T = \frac{r_0\Delta}{v_p} + \left(\int_0^{z_p} + \int_h^{z_p} \right) (v^{-2} - v_p^{-2})^{\frac{1}{2}} dz. \tag{80}$$

In specific near-earthquake problems, it is convenient to write (80) in the form

$$T = r_0\Delta/v_p + a_p, \tag{81}$$

noting that a_p depends only on the focal depth h and the velocity distribution down to the depth z_p.

It has already been pointed out (§4.3) that the extent of oscillatory moment in seismograms is evidence of marked heterogeneity in the Earth. This conclusion makes it desirable to consider theory for a layered crust.

7.6.2 Application to a layered crustal structure

We shall now apply (80) and (81) to the case of a model crust consisting of a limited number m of homogeneous layers whose boundaries are parallel to the outer surface (see Fig. 7.10). Consider rays of the same type (either P or S) which issue from a focus and are refracted without change of type through each boundary crossed. (We see in chapter 8 that a good part of the energy is carried along these rays if the changes in the properties from layer to layer are not too great.)

Assume that the epicentre has been determined independently and take

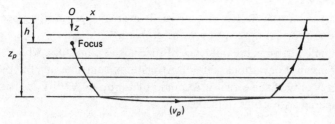

Fig. 7.10. Refracted ray in a layered crust.

as known the distances $r_0\Delta$ to n stations which record near-earthquake phases.

First, consider a family of rays which all have their lowest points in the same layer. We shall, on account of the smallness of the curvature of the boundaries, take as zero the depth penetrated by any ray below the lowest boundary it crosses. Then v_p and a_p are the same for all members of the family, and, by (81), T is proportional to Δ, except for the term a_p which is called the *apparent delay in starting* for the particular pulse.

From (81) we may form equations of condition to determine values of v_p and a_p for this family. A common practice is to plot arrival-times against the Δ for all phases read on seismograms at the various stations, and to attribute to the same phase those points which associate with a particular branch of the graph.

We take a tentative estimate of the origin-time as zero, and let τ be the actual origin-time (referred to this zero). For a particular station, let t be the measured, and $t + \varepsilon$ the actual, arrival-time for the phase in question. The actual travel-time T is $t + \varepsilon - \tau$. Hence by (81)

$$(a_p + \tau) + (r_0\Delta)v_p^{-1} - t = \varepsilon. \tag{82}$$

For the n stations, there are n equations like (82).

Since t is known, and $r_0\Delta$ is assumed known, we thus have n equations of condition to determine $(a_p + \tau)$ and v_p. The corresponding least squares solution is the solution of the pair of equations

$$n(a_p + \tau) + r_0(\Sigma\Delta)v_p^{-1} - \Sigma t = 0, \tag{83}$$

$$(\Sigma\Delta)(a_p + \tau) + r_0(\Sigma\Delta^2)v_p^{-1} - \Sigma(t\Delta) = 0. \tag{84}$$

When allowance is made for continuous variation of v with z inside a layer, this usually takes the form of a linear variation. In place of a single parameter v_p, two parameters are then required to represent the velocity in the layer. It is not difficult to extend the theory to cover this case.

Parameters that are introduced to deal with departures from horizontal layering likewise involve increased uncertainties. In practice when such departures are suspected, and especially in seismic prospecting (see § 7.6.5), there are special procedures in the gathering of data.

Finally, there is the influence of the Earth's curvature, which theoretically involves deviations of the (T, Δ) relations from linearity. But by (31) the effect is only of the order of Δ^3, and at least up to $\Delta = 10°$ is insignificant compared with other influences.

7.6.3 Error, resolution and network design

The above theory involves linear equations so that the statistical treatment follows the linear inverse method of § 7.4.2. Variances of layer velocities and thickness can be calculated from the covariance matrix (10.26). The variance of one observation is obtained from the sum of the squares of the residuals divided by the number of degrees of freedom remaining after estimation.

It is also valuable to calculate the resolution and information matrices (10.23) and (10.25) when working with near earthquakes, quarry blasts, or other artificial sources. Many apparent discrepancies in published estimates of local crustal structure arise from lack of the resolution claimed from the observing network and seismic sources used.

Recent seismological research aims at designing the configuration of seismographic networks, whether permanent or temporary, in a way that provides non-redundant information and maximum resolution. Design principles follow from analysis using the theory of § 7.4.2. For example, we can define an optimal network geometry as that which produces the smallest uncertainties in the unknown parameters, for example, coordinates of a hypocentre. In the latter example, one approximate rule might be that uncertainties should be smallest when rays to the stations leave the focus with a uniform special distribution.

Further discussion of optimal experimental design is given by Lee and Stewart (1981). Geological complexities often make parameterisation extensive and linear analysis difficult.

7.6.4 Determination of layer thicknesses

We continue to consider the m-layered model crustal structure of § 7.6.2. We now let the subscript p relate to the pth layer, R_p say, from the top, and $(m + 1)$ to the region R_{m+1} just below the bottom layer R_m. Let H_p be the thickness of $R_p(p = 1, \ldots, m)$, and let h be the depth of the earthquake focus below the outer surface.

We shall assume that $v_p < v_{p+1}$ $(p = 1, \ldots, m)$. The theory of § 7.1 shows that every R_p for which $s < p \leqslant m + 1$ can then be the lowest point of some family of rays issuing from a focus inside R_s. The presence of significant reduction of velocity with depth inside the crust, adds difficulties to the problem of determining crustal structure.

Consider first the case of an earthquake with focus inside the top layer P_1. Then, by (80) and (81).

$$a_p = 2 \sum_{q=1}^{p-1} \alpha_{pq} H_q - \alpha_{p1} h \quad (p = 2, \ldots, m+1), \tag{85}$$

where

$$\alpha_{pq} = (v_q^{-2} - v_p^{-2})^{\frac{1}{2}} \quad (p = 2, \ldots, m+1; q = 1, \ldots, m; p \neq q). \tag{86}$$

Equations like (83) and (84) can be formed for m phases each corresponding to a family of rays of given type (P or S) with lowest points in $R_2, \ldots, R_{m+1}$, respectively. These equations yield values of v_p and $(a_p + \tau)$ for $p = 2, \ldots, m+1$. The values of the v_p enable the *delay–depth coefficients* α_{pq} to be determined. With (85), we then have in effect $2m$ equations involving the $(2m + 2)$ quantities a_p $(p = 2, \ldots, m+1)$, τ, h and H_p $(p = 1, \ldots, m)$.

Additional information comes from observations of arrival-times associated with the family of rays which leave the focus in an upward direction and lie entirely in R_1. For these rays (81) has to be replaced by

$$T = \{(r_0 \Delta) + h^2\}^{\frac{1}{2}} v_1^{-1} \tag{87}$$

which, however, reduces approximately to a particular case of (81), namely,

$$T \approx r_0 \Delta / v_1 \tag{88}$$

when Δ is sufficiently large for squares of $(h/r_0\Delta)$ to be neglected. Since $T = t - \tau + \varepsilon$, (88) can be written as

$$\tau + (r_0 \Delta) v_1^{-1} - t = \varepsilon, \tag{89}$$

corresponding to (82). Through (89), with the use of stations not too close to the epicentre, we can determine values of v_1 and τ by least squares.

If there are sufficient reliable observations at small Δ, the more accurate (T, Δ) relation (87) may in theory be used to determine h as well as v_1 and τ. There are then sufficient equations to determine all the unknowns, including the thicknesses H_p.

In practice, uncertainties in the estimated thicknesses H_p are often so great in a near-earthquake investigation that no significance can be attached to the result. The reason for this is that P and S waves pass vertically through the entire crust in not more than a few seconds, so that appreciable changes in the H_p, may be accompanied by only slight changes in the a_p as given by (85).

The foregoing discussion shows that the uncertainties attaching to determinations of thicknesses of crustal layers and focal depths within the crust by near-earthquake studies are in general much greater than those attaching to the velocities.

7.6.5 Use of artificial sources. Seismic prospecting

Natural earthquakes can be simulated by explosions which take place at or near the surface of the Earth, and send into the interior seismic waves which can be recorded on emerging at the surface. Elaborate techniques have been developed both in designing artificial seismic sources and in instrumental recording in the search for mineral resources.

The seismic survey method has also been increasingly applied to determine the Earth's structure down to and below the Mohorovičić discontinuity. In addition, seismic waves from underground nuclear explosions (see chapter 1) have been used to increase knowledge of the deeper interior.

Artificial sources have several advantages over natural earthquakes in unravelling the internal structure of the Earth. First, the focal depth is effectively zero, thereby removing a source of uncertainty (see § 10.1). Secondly, the location of the epicentre is known to high precision. Thirdly, the origin-time is usually precisely known. Fourthly, advance knowledge may be available of the approximate origin-time, enabling portable instruments to be set up in strategic places and enabling recorders to run more rapidly than would be feasible in ordinary seismograph operation. In practice, it proves possible to detect changes in the local ground displacement taking place during intervals of the order of 0.001 s. Fifthly, explosions can be used to determine crystal structure in regions where natural earthquakes are scanty. All the theory of the preceding subsections is relevant to the explosion method, with the simplification that, in general, τ, h and certain other parameters no longer have to be included.

Two basic practical methods are used in seismic prospecting – the reflection and the refraction method. *Refraction shooting* is designed to determine seismic speeds and configurations of layers in regions where the speed does not decrease with depth. Field seismographs, or other recording instruments, are set up at distances from an explosion source that are several times the depth to the lowest layer being investigated, and theory based on that of § 7.6.2 is used. In *reflection shooting*, the recording instruments are close to the source, the aim being to record near-vertical reflections from the boundaries of the layer investigated. Periods of the recorded waves are commonly of the order 0.01–0.1 sec. Attention is given to isolating signals from background effects including scattered waves and surface waves, and to securing adequate magnification. Various filtering procedures (§9.8), including the use of arrays of seismographs, are brought to bear, and magnetic-tape recording is commonly employed. Another recent productive technique in studying crustal structure uses a vibration mechanism mounted on a truck as the source of seismic wave energy. The vibrations impart vertical forces up to 30 tonnes over $2m^2$ of ground surface. The forces are approximately sinusoidal but with a slowly varying frequency from about 8 to 32 Hz over a 20 s interval. Recorded signals from a seismometer array are then correlated by computer with the input signals.

Reflection profiling at sea has been developed with great success using 'airguns' as the seismic source (see Fig. 7.11). The gun is towed at a depth of a few metres and delivers an air pressure pulse to the water of order 100 bar metres.

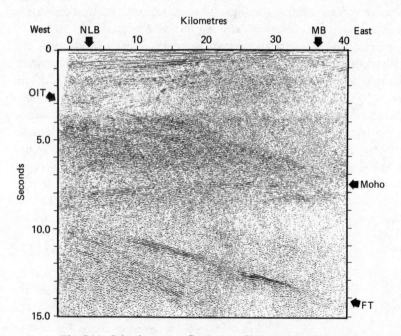

Fig. 7.11. Seismic wave reflection profile (unmigrated) using an airgun array in the Caledonian Forehand north of the Isle of Lewis, Scotland. NLB denotes North Lewis Basin; MB Minch Basin. Strong structural reflectors occur in the lower crust along the Outer Isles Thrust (OIT) and, cutting the Mohorovičić discontinuity, in the upper mantle along the Flannan Thrust (FT). To convert time to depth in kilometres multiply by 3 (in the crust) and by 4 (in the mantle). (Courtesy Brewer *et al.*, *Nature*, 305, 1983.)

Methods of *stacking* and *migration* of observed seismic waves occupy a central place in mapping a cross-section of dipping crustal layers. Stacking to enhance the signal-to-noise ratio involves the addition of signals from appropriate sets of seismic records at neighbouring stations from repeated or neighbouring seismic sources. In migration, the apparent position of a reflection point is transformed to the true position. The same aim applies as in 'stripping the Earth' in global seismology (see chapter 13): the surface wavefield (seismic record section) is theoretically propagated back by *downwards continuation* to the reflecting structure.

The books by Dobrin (1976) and Grant and West (1965) give a useful account of the methods used. By suitably disposing sources and arrays, the shapes of sloping and curved boundaries can be elucidated. Seismic methods have located thin ore bodies 20 m wide at depths of order 200 m.

The methods of seismic exploration, such as the use of arrays of seismometers to increase the signal-to-noise ratio, have also been extensively applied to problems of detecting nuclear explosions (§1.3.1).

Seismic prospecting has been used also to determine ice thicknesses, the first experiments having been made in the Austrian Alps. Of special interest, ice thicknesses have been determined in Antarctica on traverses totalling over 20 000 km. The results indicate that the rock surface in eastern Antarctica is mainly above sea level, while western Antarctica is more of archipelago type. This result is supported in inferences from earthquake surface-wave observations. The maximum ice thickness so far found exceeds 4 km. The values found for the P velocities in Antarctic ice range from 3.8 to 4.0 km/s, the lower value being found in coastal regions, and the higher near the Pole of Relative Inaccessibility. The S velocity is about 1.95 km/s. The methods have also yielded P velocities in rocks below the ice ranging from 4.7 to 6.3 km/s.

Seismic methods have also been used to measure the thicknesses and elastic properties of floating ice in both the Arctic and Antarctic.

7.7 Exercises

1   Consider an Earth model with two homogeneous shells. The P wave velocity in the core is 8.0 km/s and in the mantle 10.6 km/s. Let a family of P rays be incident on the core. Using Snell's law, trace these rays for PKP and $PKKP$ paths for at least five incident rays separated by $r/10$ where r is the core radius. From the envelopes of the rays locate the caustics. If the incident family of rays represents points on a plane wave of uniform amplitude, sketch the amplitudes of the PKP and $PKKP$ wave fronts.

2   Suppose the gradient p of the travel-time curve remained constant for a range of distances Δ. Can the Herglotz–Wiechert equation be integrated

in this range and what is the implication for the corresponding velocity v as a function of r?

3   Consider the case when the (T, Δ) travel-time equation for P phases is given by $T = 700\Delta - 120\Delta^3, 0.03 \lesssim \Delta \lesssim 1$, where T is in seconds and Δ in radians.

(a) What is the travel-time of a P phase for epicentral distance $\Delta = 30°$?

(b) Consider the Herglotz–Wiechert inversion equation (40) in §7.4.1. For $\Delta_1 \lesssim 0.5$, the integrand can be approximated by a circle of radius Δ_1 (centred on the origin) in the upper-right quadrant of the $\cosh^{-1}(p/p_1)$–Δ plane. Using a micro-computer, solve the inversion equation for $\Delta_1 = 0.05$ (0.05) 0.20 to obtain a velocity-depth profile.

(c) Explain the lower limit on $\Delta = 0.03$ given in the (T, Δ) equation above.

4   Show how Euler's equation arises as a condition that integrals of a certain type should be stationary.

A seismic ray AB is a curve in the Earth's interior terminated at the points A and B on the outside surface. If v is the speed of travel at the point P of the ray, and r the distance of P from the Earth's centre O, prove that the time of travel T along the ray and the angle Δ subtended at O by the whole ray are given by

$$T = 2 \int \frac{r \, dr}{v(r^2 - p^2)^{\frac{1}{2}}}, \qquad \Delta = 2 \int \frac{p \, dr}{r(r^2 - p^2)^{\frac{1}{2}}},$$

where p is the perpendicular from O to the tangent at P to the ray. Given that T is stationary when Δ is assigned, prove that $p = av$, where a depends only on Δ.

5   Derive formula (81) directly by neglecting the curvature of the Earth and assuming horizontal stratification from the outset. Take Ox horizontal and in the plane containing a ray whose focal depth is h, and Oz vertically downwards. Let i be the (acute) angle between the ray and the vertical at any point of its path. [Hint: Show $T = \int v_p^{-1} \csc^2 i \, dx = r_0 \Delta/v_p + \int \cot i/v_p \, dz$, where $r_0 \Delta$ is the epicentral distance measured as an arc-length.]

6   The characteristics of the eikonal equation are given by

$$\dot{x}_i = v^2 p_i, \quad \dot{p}_i = -\partial \ln v/\partial x_i, \quad i = 1, 2, 3$$

where a dot denotes differentiation with respect to τ.

By substituting

$$F(x_i, \dot{x}_i, \tau) = [v(x_i)]^{-1} (\dot{x}_k \dot{x}_k)^{1/2},$$

and using the summation convention on k, show that the three second-order characteristic equations

$$\frac{d}{d\tau}(v^{-2} \dot{x}_i) = -\frac{\partial \ln v}{\partial x_i}, \quad i = 1, 2, 3$$

reduce to the Euler equations

$$\frac{d}{d\tau}\left(\frac{\partial F}{\partial \dot{x}_i}\right) - \frac{\partial F}{\partial x_i} = 0, \quad i = 1, 2, 3.$$

Explain how these yield the extremals (rays) of the Fermat functional

$$T = \int v^{-1}(x_i)ds = \int F(x_i, \dot{x}_i, \tau)d\tau,$$

integrated along the ray.

7   For polar angular coordinates $0 \leqslant \phi \leqslant 2\pi, 0 < \delta < \pi$, show that the system

$$p_1 = c^{-1}\cos\phi\sin\delta, \quad p_2 = c^{-1}\sin\phi\sin\delta, \quad p_3 = c^{-1}\cos\delta$$

satisfies the eikonal equation, and that the characteristic equations are

$dx_1/d\tau = c\cos\phi\sin\delta$

$dx_2/d\tau = c\sin\phi\sin\delta$

$dx_3/d\tau = c\cos\delta$

$d\phi/d\tau = (\sin\phi \, dc/dx_1 - \cos\phi \, dc/dx_2)/\sin\delta$

$d\delta/d\tau = -(\cos\phi \, dc/dx_1 + \sin\phi \, dc/dx_2)\cos\delta + \sin\delta \, dc/dx_3.$

8   Show that the Clairaut equation

$$y = xy' + (y')^2$$

has a general solution

$$y = cx + c^2$$

and a singular solution

$$4y = -x^2.$$

Show that the parabola is the envelope of a family of straight lines.

8

Amplitudes of the surface motion due to seismic waves in a spherically stratified Earth model

8.1 Energy considerations

We continue theory relevant to symmetric Earth models. As in chapter 7, let r denote the distance of any point from the centre, v the speed of a particular wave at this point and $\eta = r/v$ (equation 7.17); again the suffix 0 denotes values which variables take at the outer surface of the model.

We now investigate effects at the outer surface of the model following the occurrence of a disturbance initially confined to a small region; for simplicity we treat this region as a point F (*the focus*). We assume that energy propagates in all directions symmetrically from F, and then investigate the amplitudes produced at the outer surface by the body elastic waves. We assume that conditions permit us to make use of the ray theory in chapter 7, and that discontinuity surfaces inside the model are sufficiently distant from F to permit use of plane-wave theory of reflection and refraction set up in chapter 6.

We first give a simplified theory which ignores any energy losses that a particular wave may suffer between F and points of the outer surface of the model. The derivation of the necessary correcting factors arising from such energy losses will be discussed in §8.4.

8.1.1 Energy per unit area of wave front in an emerging wave

Consider a wave of a particular type (e.g. dilatational) which leaves F and travels to the outer surface in the same type. Let I denote the energy in this type emitted per unit solid angle from F, and let e_1 be the angle which any ray leaving F makes with the level surface of the model through F. Then the energy conveyed along all the rays for which e lies in a particular range, e_1 to $e_1 + de_1$, is $2\pi I \cos e_1 de_1$ (see Fig. 8.1).

Now the area of the outer surface of the model containing points whose angular distances from F lie between Δ and $\Delta + d\Delta$ is

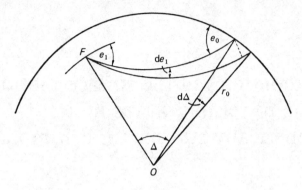

Fig. 8.1. Energy in a ray bundle.

$$2\pi r_0^2 \sin\Delta\,|d\Delta|.$$

(The modulus of $d\Delta$ is used in this formula because $d\Delta$ and de_1 can sometimes be of opposite signs – see §7.3.9.) It follows that the area of a wave front that emerges at the outer surface between these distances is $2\pi r_0^2 \sin\Delta \sin e_0 |d\Delta|$, where e_0 is the angle of emergence. Therefore, the energy $E(\Delta)$ per unit area of the portion of the wave front emerging at the angle e_0 is

$$E(\Delta) = \frac{I}{r_0^2 \sin\Delta \sin e_0} \frac{\cos e_1}{} \left| \frac{de_1}{d\Delta} \right|, \tag{1}$$

where the range e_1 to $e_1 + de_1$ corresponds to the range Δ to $\Delta + d\Delta$.

If T is the travel-time from F for distance Δ, we have by (7.1) and (7.12)

$$\eta_1 \cos e_1 = \eta_0 \cos e_0 = dT/d\Delta, \tag{2}$$

where η_1 and η_0 are the values of η at F and at the surface. By (2) we have

$$-\eta_1 \sin e_1 \frac{de_1}{d\Delta} = \frac{d^2 T}{d\Delta^2}.$$

Hence

$$\frac{\cos e_1}{\sin e_0} \left| \frac{de_1}{d\Delta} \right| = \frac{\cot e_1}{\eta_1 \sin e_0} \left| \frac{d^2 T}{d\Delta^2} \right|$$

$$= \frac{\eta_0}{\eta_1} (\eta_1^2 \tan^2 e_0 - \eta_0^2 \sin^2 e_0)^{-\frac{1}{2}} \left| \frac{d^2 T}{d\Delta^2} \right|, \tag{3}$$

using (2) again. Because we shall consider reflections at the outer surface of the model, it is convenient to denote the angle of emergence at the outer surface by the symbol e instead of e_0. We may then by (1) and (3) express

$E(\Delta)$ in the form

$$E(\Delta) = \frac{I\eta_0}{r_0^2 \eta_1 \sin \Delta} (\eta_1^2 \tan^2 e - \eta_0^2 \sin^2 e)^{-\frac{1}{2}} \left| \frac{d^2 T}{d\Delta^2} \right|. \tag{4}$$

8.1.2 Relation between energy and amplitude

Consider an advancing P or S wave, and let w be the energy per unit volume over a region such that the material can be treated as uniform within it (we exclude complications due to diffraction). Then by (3.73)

$$\frac{\partial w}{\partial t} + \frac{\partial}{\partial x_j}(wv_j) = 0, \tag{5}$$

where v_j is the wave velocity. By (2.47) we may to sufficient accuracy, neglecting the external forces X_i, write

$$\frac{\partial w}{\partial t} = \frac{\partial}{\partial x_j}\left(p_{ij} \frac{\partial u_i}{\partial t} \right). \tag{6}$$

By (5) and (6) we have

$$\frac{\partial}{\partial x_j}\left(wv_j + p_{ij} \frac{\partial u_i}{\partial t} \right) = 0. \tag{7}$$

For a plane wave advancing parallel to the x_1-axis, we find from (7) that at any particular instant

$$wv_1 = - p_{i1} \frac{\partial u_i}{\partial t}, \tag{8}$$

the integration constant being taken as zero.

We shall apply (8) to the particular case in which the waves are plane simple harmonic SH waves, the displacement being given by

$$u_i = (0, C\cos\{\kappa(x_1 - \beta t)\}, 0).$$

Thus $v_i = (\beta, 0, 0)$. By (2.32) and (2.13) we have

$$p_{12} = p_{21} = 2\mu e_{21} = \mu \frac{\partial u_2}{\partial x_1} = - \mu\kappa C \sin\{\kappa(x_1 - \beta t)\},$$

all the other p_{ij} being zero. Substituting into (8) we obtain

$$w = \mu\kappa^2 C^2 \sin^2\{\kappa(x_1 - \beta t)\}. \tag{9}$$

(We note that (9) can be independently derived using (2.54) and the fact (§3.3.6) that w is twice the strain energy per unit volume.) Integrating (9) over one wave-length, i.e. over the range

$$0 \leqslant x_1 \leqslant 2\pi/\kappa,$$

we find that the energy E, per unit area normal to the x_1-axis, in one wavelength is equal to $\pi\mu\kappa C^2$; i.e. by (3.56) and (4.5),

$$E = 2\pi^2 \rho \lambda C^2/\tau^2, \tag{10}$$

where λ is the wave-length, τ is the period of the wave, and ρ is the density of the medium.

It follows from (10) that the mean energy per unit length in a train of waves is proportional to the square of the amplitude, and inversely proportional to the square of the period.

Formulae similar to (10) may be likewise derived for the case of P and SV simple harmonic plane waves.

8.2 Movements of the surface due to an incident wave

We now consider the amplitudes of the vibrations of surface particles of the Earth model arising from various waves incident from below. It is sufficiently accurate here to neglect the curvature of the outer surface, and thus to make direct use of formulae in chapter 6 for plane waves incident against a plane boundary. We again assume that Poisson's relation holds for the material just below the outer surface, and so have e and f connected by (6.13), namely,

$$\cos^2 e = 3\cos^2 f. \tag{11}$$

We first consider an incident P wave emerging at the angle e. We may represent this near the outer surface (using the notation of §6.4) by

$$\phi = A_0 \exp\{i\kappa(x_3 \tan e + x_1 - ct)\}, \tag{12}$$

and the reflected P and S waves by

$$\phi = A \exp\{i\kappa(-x_3 \tan e + x_1 - ct)\}, \tag{13}$$

$$\psi = B \exp\{i\kappa(-x_3 \tan f + x_1 - ct)\}, \tag{14}$$

respectively. The ratios A/A_0, B/A_0 are then obtained from (6.14), f being determined from e using (11). By (12), (13), (14) and (5.1) we see that the components of the incident and the two reflected waves are given at the surface by

$$u_i = (i\kappa A_0, 0, i\kappa A_0 \tan e) \exp\{i\kappa(x_1 - ct)\},$$

$$u_i = (i\kappa A, 0, -i\kappa A \tan e) \exp\{i\kappa(x_1 - ct)\},$$

$$u_i = (-i\kappa B \tan f, 0, -i\kappa B) \exp\{i\kappa(x_1 - ct)\},$$

respectively. Hence the ratio of the horizontal component, A_h, of the amplitude of the motion of the outer surface to the amplitude, A_k say, of the incident wave alone is $(A_0 + A - B \tan f)/A_0 \sec e$. By (6.14) we then deduce that

$$\frac{A_h}{A_k} = \frac{12 \sin e \sec^2 e \tan f}{4 \tan e \tan f + (1 + 3 \tan^2 e)^2}. \tag{15}$$

Similarly, the corresponding ratio for the vertical component, A_v say, is

$$\frac{A_v}{A_k} = \frac{6 \sin e \sec^2 e (1 + 3 \tan^2 e)}{4 \tan e \tan f + (1 + 3 \tan^2 e)^2}. \tag{16}$$

If the ratio $A_v : A_h$ be denoted by $\tan \bar{e}$, we have by (15) and (16)

$$\tan \bar{e} = \frac{A_v}{A_h} = \frac{1 + 3 \tan^2 e}{2 \tan f}. \tag{17}$$

The angle $\bar{e}$ is called the *apparent angle of emergence*, and from (11) and (17) we find

$$2 \cos^2 e = 3(1 - \sin \bar{e}), \tag{18}$$

connecting the actual and apparent angles of emergence of P waves incident at the outer surface. In obtaining (18) we have, for ease of calculation, assumed Poisson's relation (4.7) to hold. It is not difficult to show that in the general case

$$2 \cos^2 e = \frac{\alpha^2}{\beta^2}(1 - \sin \bar{e}),$$

where α, β are the relevant P, S velocities; this relation was first obtained by Wiechert.

In a similar way, using formulae in §6.5, we may derive formulae corresponding to (15), (16), (17) and (18) for the case of SV waves incident at the outer surface.

In the case of incident SH waves, the amplitude of the surface ground movement is twice that of the incident waves. This follows immediately from (6.4) and (6.7).

8.3 Amplitude as a function of Δ

If, following an initial seismic disturbance, A_h and A_v are estimated as functions of Δ (for a particular pulse, assumed simple harmonic), we can by (17) and (18) compute $\bar{e}$, and then e, as functions of Δ. Since $\eta_0 \cos e = dT/d\Delta$ (i.e. equation (2) written in the present notation), we can then compute $(\eta_0)^{-1} dT/d\Delta$ an terms of Δ.

Moreover, from (4), (10), (15) and (16), we find

$$A_h^2 \propto \frac{4I \tan^2 e \sec^2 e \tan^2 f}{\eta_1 \sin \Delta (\eta_1^2 \tan^2 e - \eta_0^2 \sin^2 e)^{\frac{1}{2}} \{4 \tan e \tan f + (1 + 3 \tan^2 e)^2\}^2} \left| \frac{d^2 T}{d\Delta^2} \right|,$$

(19)

$$A_v^2 \propto \frac{I \tan^2 e \sec^2 e (1 + 3 \tan^2 e)^2}{\eta_1 \sin \Delta (\eta_1^2 \tan^2 e - \eta_0^2 \sin^2 e)^{\frac{1}{2}} \{4 \tan e \tan f + (1 + 3 \tan^2 e)^2\}^2} \left| \frac{d^2 T}{d\Delta^2} \right|.$$

(20)

In (19) and (20) the constant of proportionality depends on the properties of the material at the outer surface of the Earth model and on the wave-length λ and period τ of the particular pulse. With the use of (19) or (20), it would be theoretically possible, knowing A_h or A_v as a function of Δ, and knowing η_0, to estimate η_1, i.e. the value of r/v near the focus F.

Again, the proportionality of A_h^2 and A_v^2 to $|d^2 T/d\Delta^2|$ gives a means of using amplitude observations to check values of $|d^2 T/d\Delta^2|$ as estimated from the (T, Δ) relations.

It is evident, therefore, that observed amplitude variations in earthquakes have the possibility of providing information on P and S velocity variations within the Earth, to some extent independently of the method indicated in chapter 7. In practice, there are a number of complications, and because of these to date the amplitude method has given less precise results than the direct use of travel-time data. The chief complications arise from the heterogeneous character of rocks near the Earth's surface, and from the need for estimating energy losses associated with internal discontinuity surfaces (see § 8.4.2). For surface structures with layers or rapidly changing velocities, the results (19) and (20) need modification for frequency dependence. Nevertheless, amplitude considerations first drew attention to a discontinuity surface inside the Earth's central core (see § 13.1.3) and to the asthenosphere (Gutenberg, 1959).

As an application of amplitude theory we notice by (19) and (20) that in general A_h and A_v are relatively small when $d^2 T/d\Delta^2$ is small, and relatively large when $d^2 T/d\Delta^2$ is large. Smallness of $d^2 T/d\Delta^2$, i.e. smallness of curvature in the corresponding (T, Δ) curve, was seen in § 7.3.7 to be associated with a (not too great) decrease of velocity with increase of depth inside the Earth model; this circumstance was in turn seen to be associated with an increased spread of values of Δ against p (and therefore against e), and therefore diminished amplitude of the surface movements for a range of Δ. Also, largeness of $d^2 T/d\Delta^2$ was seen in § 7.3.4 to correspond to a fairly rapid increase of velocity with increase of depth. This entails a decreased spread of values of Δ against p, and hence entails increased surface amplitudes for a range of Δ. Thus, when markedly increased amplitudes are detected over a small range of Δ, there is a likelihood that the velocity increases markedly over a corresponding range of depth. Because of the complexities mentioned, it is, however, necessary to check from other evidence that this is in fact the cause of the increased amplitudes.

8.4 Loss of energy during transmission through the medium

So far we have ignored energy losses during wave transmission. There will be loss due to departures from perfect elasticity, but, as pointed out in §4.5, such losses are relatively slight in much of the Earth. The main energy losses arise from heterogeneity of the medium, including the effects of energy partitioning at discontinuity surfaces.

8.4.1 Gradual variation in properties

We consider first the case in which the three parameters specifying the elastic properties and density of a (perfectly elastic) medium vary gradually. More precisely, suppose that there is a large number n of thin homogeneous layers, the fractional changes in the parameters from layer to layer being of the order of $1/n$. By the use of formulae obtained in chapter 6, Jeffreys has shown that in general, for any particular wave (P, SV, or SH) encountering a boundary between a pair of the layers, all but the order of $1/n^2$ of the incident energy is transmitted in the same type.

For instance, in the case of SH waves incident at the angle $\frac{1}{2}\pi - f$ against a boundary separating a pair of layers M, M', whose properties are as defined in §5.1, we have by (6.8)

$$\frac{C}{C_0} = \left(1 - \frac{\mu' \tan f'}{\mu \tan f}\right)\left(1 + \frac{\mu' \tan f'}{\mu \tan f}\right)^{-1}. \tag{21}$$

If we put $\mu' = (1 + \delta)\mu$ and $\tan f' = (1 + \varepsilon)\tan f$, and take δ and ε to be of the order of $1/n$, we obtain

$$C/C_0 \approx \{1 - (1 + \delta + \varepsilon)\}\{1 + (1 + \delta + \varepsilon)\}^{-1}$$
$$\approx \tfrac{1}{2}(\delta + \varepsilon);$$

hence $(C/C_0)^2 = O(1/n^2)$. Thus the energy loss in the transmitted wave of the same type after passing through the n layers is in general only of the order of $1/n$ of the original energy, which is trifling when n is large.

An exception arises where the wave makes nearly grazing incidence with one of the boundaries. (In the case of SH waves, f is then nearly equal to $\frac{1}{2}\pi$

and it no longer follows that we can write $\tan f' = (1 + \varepsilon) \tan f$ with ε of the order of $1/n$.) A second exception occurs if the wave length is so great as to be comparable with the distance over which there is an accumulated finite change in the properties of the medium. There will then be an appreciable fraction of energy lost to derived waves. (The second exception approximates to a single surface of finite discontinuity.)

In this connection we should mention theory explaining seismic wave energy in the layered Earth not predicted by classical wave theory. For example, along a spherical interface such as the bottom of the crust (see §13.1.1) or core boundary (see §13.6.2) even a slight positive velocity gradient can produce a 'whispering gallery' effect and, by constructive interference of multiple reflections, set up a significant coherent wave pulse. Such pulses are called *interference head waves*. Another diffraction effect also occurs in which, at appropriate wavelengths, reflected energy is lost at a stratification by evanescent energy leaking through a ray inversion and refracting into the deeper structure (cf. Exercise 8.9.5). The effect is called *wave tunnelling* by analogy with a quantum mechanical argument.

8.4.2 Single discontinuity

We now consider the effect of a single finite discontinuity surface inside the Earth model, and take first the case of an incident SH wave. With our usual notation, it then follows that the transmitted energy is $\{1 - (C/C_0)^2\}$ times the incident energy, where C/C_0 is given by (21). A factor of the form $\{1 - (C/C_0)^2\}$ is called an energy *transmission factor*, and will need to be applied for each discontinuity surface that the wave traverses. Such factors need to be included on the right-hand sides of expressions such as (19) and (20) before these formulae may be used to interpret the amplitude variations observed at the outer surface.

The case of an incident P wave is more complicated than that of an incident SH wave. Consider a P wave incident at the angle $\frac{1}{2}\pi - e$ against the (plane) discontinuity surface separating two media M and M'. By an extension of the argument in §6.4, it is possible to derive an expression for the ratio A/A_0 in terms of e, where A_0 and A are relevant to the medium M and analogous to the A_0 and A in (12) and (13). (This involves the use of equations analogous to (13) and (14) for the medium M', in addition to the first two of the equations (6.12) of §6.2.2.) The precise result is algebraically complicated, but calculations such as those of Knott of the ratio A/A_0, for various values of e, indicate that when the media are not greatly different an incident P wave is reflected and refracted mainly into P waves. In this case, the reflected and refracted SV waves contain only a small fraction of the total energy involved. (An analogous result holds also for an incident *SV* wave.) Jeffreys accordingly assumed that a satisfactory first approximation to A/A_0 may be obtained by disregarding the reflected and refracted *SV* waves, and thus obtained the approximate formula

$$\frac{A}{A_0} \approx \left(\frac{\rho'}{\rho} - \frac{\tan e'}{\tan e}\right) \bigg/ \left(\frac{\rho'}{\rho} + \frac{\tan e'}{\tan e}\right). \tag{22}$$

The transmission factor in this case is thus approximately $\{1 - (A/A_0)\}^2$, where A/A_0 is given by (22).

The corresponding formula for an incident *SV* wave is less simple than (22), but numerical calculations indicate results broadly similar to the *P* case.

In practice, a further transmission factor is sometimes applied to the amplitude formulae in order to allow for energy losses through friction and scattering (see §4.5.3).

We assume that the amplitude frequency function can be factored in the frequency domain as

$$A(\omega, \Delta) = S(\omega)R(\omega)G(\Delta)\exp(-\omega t^*/2\pi), \tag{23}$$

where *S*, *R* are source and receiver response functions and *G* is the geometrical spreading factor. The exponential variable t^* can be expressed as

$$t^* = \pi \int \frac{ds}{Qv}, \tag{24}$$

where *Q*, here assumed to be independent of frequency, is defined in §4.5.3 and §5.7.2. By analogy with the results of §7.2.2, t^* is a measure of the average damping along the ray path. The slope of the logarithm of equation (23) may be used directly to find an average constant *Q* by forming spectral amplitude ratios (see §13.3.2). The variation of *Q* with depth in the Earth as a function of frequency is an inverse or tomographic problem that can be readily formulated in the same way as the velocity inverse problem described in §7.4.3 (cf. equation (7.52)).

There are special difficulties connected with the study of amplitudes of waves arising from 'near earthquakes' (see §7.6), for which the distance Δ does not exceed about $10°$. These difficulties are associated with present inadequate knowledge of features of the transition from layer to layer in the outer part of the Earth (see §13.6).

The effect of varying degrees of sharpness of discontinuities on the transmission and reflection of seismic waves in relation to the incident wave-lengths is discussed in some detail by Kennett (1983). Wave frequency must be taken into account.

8.5 Waves which change type

In the preceding sections, we investigated the energy conveyed only in waves which have been transmitted without change in type from a focus F to the outer surface of the Earth model, and which have not been reflected at any discontinuity surface. There will, of course, be other waves arriving at the outer surface which have been reflected at one or more surfaces of discontinuity (including internal discontinuity surfaces and the outer free surface itself), and which may have changed type when reflected or refracted.

The formulae (19) and (20) are still relevant to the case of a wave finally emerging in the P type, if I and η_1 be taken now to correspond to the energy and velocity (P or S) which the wave in question had on leaving F. It is necessary as before to work out transmission factors corresponding to each encounter with a discontinuity surface, and to apply these transmission factors to the right-hand sides of (19) and (20). The method follows the same principles as in the preceding sections, but the results are more complicated. Similar remarks apply to waves emerging in the SV type. In the case of a wave emerging as SH, the algebra is simpler, since an SH wave does not change in type on reflection or refraction in a spherically symmetrical Earth model.

8.6 Amplitudes corresponding to cusps in (T, Δ) curves

Rapid or sudden changes in the variation of the velocity v with the depth z can result in cusps in the corresponding (T, Δ) curves, examples being the curves shown in Figs. 7.5(c) and 7.8(c) of §§7.3.3, 7.3.8. It has commonly been assumed that such cusps are associated with abnormally large amplitudes, but Bullen has shown that this is by no means always the case.

For the case of §7.3.3, where v is a continuous function of z while dv/dz increases markedly in the vicinity of a certain depth, let C'', D'', as in Fig. 7.5(c), denote the cusps at the smaller and greater Δ, respectively. When the change in dv/dz takes the form of a simple discontinuous increase, there will be large amplitudes at C'' but not at D''. The reason is that while C'' corresponds to an ordinary minimum in p (i.e. $dT/d\Delta$) against Δ, D'' corresponds to a node-point maximum. Abnormally large amplitudes occur where $d\Delta/de$ is abnormally small, and therefore where $d\Delta/dp$ is abnormally small, since by (7.1) p is proportional to cos e, e being the angle of emergence at the surface. Since $d\Delta/dp$ vanishes at C'', but not at D'', the cusp C'' is associated with large amplitudes, while in general D'' is not.

On the other hand, it is possible to have large amplitudes at both cusps, if there is fairly sudden change in dv/dz spread over a small but finite range of depth. In this case, v and dv/dz are continuous, but d^2v/dz^2 may be discontinuous.

In the case of § 7.3.5, where v increases discontinuously with z, the (T, Δ) curve (Fig. 7.7(c)) takes the form shown in Fig. 7.5(c) if the branch $C''D''$ be added to include rays totally reflected at the discontinuity surface. For this case, it can be shown that in general the amplitudes are not abnormally large at either cusp.

These results show that what may appear to be minor changes in a velocity distribution can radically affect the associated variation of amplitude with Δ at the surface.

For the case of § 7.3.8, where there is a sudden diminution in v at a certain depth, the theory gives large amplitudes at the cusp D'' (Fig. 7.8(c)).

8.7 Amplitudes of surface seismic waves

We have seen that when a focus F is not too deep, sizeable surface waves are generated and spread over the outer surface. It is sufficiently accurate for problems on surface waves with periods less than about 50 s to neglect the curvature of the Earth's outer surface, and we have seen in § 3.5 that when there is no dispersion the amplitudes will then be inversely proportional to the square root of the distance from the source when the distance is appreciable. There is, however, the further effect of dispersion to be taken into account.

In § 3.3.5 we obtained the approximate formula (3.66) for the displacement, in the vicinity of a given place x and time t, of a wave subject to dispersion. In seismology, observations are made of the whole train of local earth motion at observing stations, and it is convenient, following Jeffreys, to use a related asymptotic approximation, namely,

$$y(x, t) = \frac{f(\gamma)}{\{\frac{1}{2}\pi|(d^2\kappa/d\gamma^2)|x|\}^{\frac{1}{2}}} \cos(\kappa x - \gamma t \pm \tfrac{1}{4}\pi), \tag{25}$$

where $f(\gamma)$ is analogous to $F(\kappa)$. The coefficient of the cosine factor in (25) indicates the amplitude to be expected.

The distance x appears only in the factor $x^{-\frac{1}{2}}$, which is common to all the wave groups. Thus the ratios of the amplitudes of waves of given periods would on this simplified theory be the same at all places on the Earth's outer surface. Apart from the factor $x^{-\frac{1}{2}}$, the amplitude would vary from wave group to wave group depending on the form of the initial disturbance and on the detailed character of the dispersion. In practice, departures from this ideal case are used to estimate source mechanisms and structural variations from the observed surface waves.

8.8 Reflectivity algorithms

Seismological investigations of the Earth's interior now rely not only on travel-times but also on amplitudes of seismic waves. The computation of synthetic seismograms aids significantly in the quantitative interpretation of observed seismograms. Two methods that allow fast computer calculation for many layers are now used, the ray-theoretical method and the reflectivity method. The ray-theoretical method, developed separately by Helmberger, Chapman, Müller and their associates, sums the elementary seismograms corresponding to the primary and multiple rays from the source to the point of observation. In the reflectivity method constructed by K. Fuchs in 1968 the numerical integration of the reflectivity (or plane wave reflection coefficient) of a layered medium is carried out in the horizontal wavenumber (or angle of emergence) domain. Multiplication with the source spectrum and inverse Fourier transformation yields the seismograms for the displacement components.

An extensive treatment of the complete response of a stratified elastic medium to propagating P and SV waves is given in the monograph of Kennett (1983) in which the two methods are synthesised. As an introduction we outline the essence of the reflectivity method in the special case treated by Fuchs and Müller. (See also Exercise 8.9.3.)

The model consists of $n-1$ plane, homogeneous and isotropic layers on top of a half-space called layer n (see Fig. 7.10). The jth layer is characterized by P and S velocities α_j and β_j, density ρ_j and thickness h_j. Assume an explosive point source in the free surface $z=0$ with the P wave in layer 1 spherically symmetric, and no shear and surface waves. Consider the compressional reflection from the reflecting zone which comprises the layers $m+1$ through n. The layers 1 to m are assumed to produce only transmission losses and time shifts, both for the P wave propagating downwards from the source to the reflecting zone and for the reflected P wave travelling upwards to the surface.

The compressional potential of the wave from the explosive point source is, from equation (3.82),

$$\phi_0(r, z, t) = \frac{1}{R} F\left(t - \frac{R}{\alpha_1}\right), \tag{26}$$

where $R^2 = r^2 + z^2$. It can be proved (see Morse and Feshbach, 1953, § 10.6) that its Fourier transform can be written in integral form as

$$\bar{\phi}(r, z, \omega) = \bar{F}(\omega) \int_0^\infty \frac{k}{iv_1} J_0(kr) \exp(-iv_1 z) dk, \tag{27}$$

where $\bar{F}(\omega)$ is the Fourier transform of the excitation function $F(t)$, $J_0(kr)$ the Bessel function of the first kind and order zero, i the imaginary unit, k the horizontal wave number, and

$$v_1 = (k_{\alpha_1}^2 - k^2)^{\frac{1}{2}}$$

the vertical wave number ($k_{\alpha_1} = \omega/\alpha_1$). This transform removes the time variation.

Next, the P wave moves across interface 2. The resulting Fourier transformed potential of the downgoing P wave in layer 2 has the same form as (27). The integrand includes additionally a transmission coefficient and a second term in the exponential function. This process is continued until layer m is reached. The compressional potential of the P wave in layer m, incident upon the reflecting zone, is

$$\bar{\phi}_1(r, z, \omega) = \bar{F}(\omega) \int_0^\infty \frac{k}{iv_1} J_0(kr) P_d(\omega, k)$$

$$\cdot \exp\left[-i\left(\sum_{j=1}^{m-1} h_j v_j + \left(z - \sum_{j=1}^{m-1} h_j\right) v_m\right)\right] dk, \tag{28}$$

where $P_d(\omega, k)$ is the product of the transmission coefficients of the interfaces $2, 3, \ldots, m$ for a downgoing wave, and

$$v_j = (k_{\alpha_j}^2 - k^2)^{\frac{1}{2}}, \quad v_j' = (k_{\beta_j}^2 - k^2)^{\frac{1}{2}} \tag{29}$$

are the vertical wave numbers of P waves and S waves, respectively, in layer j ($k_{\alpha_j} = \omega/\alpha_j, k_{\beta_j} = \omega/\beta_j$). Both types of wave numbers occur in $P_d(\omega, k)$. This function and related ones can be evaluated explicitly by the theory of § 8.4.

The reflection of the incident P wave (28) gives rise to a reflected P wave whose potential in layer m is

$$\bar{\phi}_2(r, z, \omega) = \bar{F}(\omega) \int_0^\infty \frac{k}{iv_1} J_0(kr) P_d(\omega, k) \tilde{R}_{pp}(\omega, k)$$

$$\cdot \exp\left[-i\left(\sum_{j=1}^m h_j v_j + \left(\sum_{j=1}^m h_j - z\right) v_m\right)\right] dk. \tag{30}$$

$\tilde{R}_{pp}(\omega, k)$ is the complex reflectivity (or plane wave reflection coefficient) of the reflecting zone.

The reflected P wave now moves upwards across the interfaces m, $m - 1, \ldots, 2$. Its potential in layer 1 is

$$\bar{\phi}_3(r, z, \omega) = \bar{F}(\omega) \int_0^\infty \frac{k}{iv_1} J_0(kr) P_d(\omega, k) \tilde{R}_{pp}(\omega, k) P_u(\omega, k)$$

$$\cdot \exp\left[-i\left(2 \sum_{j=1}^m h_j v_j - z v_1 \right) \right] dk, \qquad (31)$$

where $P_u(\omega, k)$ is the product of the transmission coefficients of the interfaces $2, 3, \ldots, m$ for an upgoing wave.

The aim is to find the vertical and the horizontal displacement at the free surface $z = 0$. Therefore, the final step is the reflection of the P wave at the free surface. The potential of the reflected P wave is

$$\bar{\phi}_4(r, z, \omega) = \bar{F}(\omega) \int_0^\infty \frac{k}{iv_1} J_0(kr) P_d(\omega, k) \tilde{R}_{pp}(\omega, k) P_u(\omega, k) r_{pp}(\omega, k)$$

$$\cdot \exp\left[-i\left(2 \sum_{j=1}^m h_j v_j + z v_1 \right) \right] dk. \qquad (32)$$

Similarly, the potential of the reflected S wave is

$$\bar{\psi}(r, z, \omega) = \bar{F}(\omega) \int_0^\infty \frac{k}{iv_1} J_0(kr) P_d(\omega, k) \tilde{R}_{pp}(\omega, k) P_u(\omega, k) r_{ps}(\omega, k)$$

$$\cdot \exp\left[-i\left(2 \sum_{j=1}^m h_j v_j + z v_1' \right) \right] dk. \qquad (33)$$

In (32) and (33), $r_{pp}(\omega, k)$ and $r_{ps}(\omega, k)$ are PP and PS reflection coefficients at the free surface. The horizontal and vertical displacements in layer 1 may then be determined from the potentials by substitution in the appropriate radial forms (see §5.1).

Integrals of the type (32) and (33) can be computed approximately by numerical integration. It turns out that it is sufficient for body wave studies to restrict the integration to real emergence angles e.

The most time-consuming step in the evaluation of (32) and (33) is the matrix computation of the reflectivity $R_{pp}(\omega, k)$ but this step needs to be done only once, because the reflectivity is the same for all r. The frequencies ω and the emergence angles e for which $R_{pp}(\omega, k)$ must be known are found by choosing the length T of the seismograms large enough to include all significant arrivals, i.e. the arrivals corresponding to the travel-time curve itself and multiply reflected phases from inside the reflecting zone, if their amplitudes are not negligible. Fourier and inverse Fourier transforms are performed with the aid of a fast Fourier transform algorithm.

The range of angles e is estimated from the apparent velocities in the time–distance range of interest. The horizontal wave number k is related to the apparent velocity c of arrivals at the surface by

$$k = \frac{\omega}{\alpha_m} \cos e = \frac{\omega}{c}.$$

Given the maximum and the minimum value of c, the corresponding angles $e_{1,2}$ come from

$$e_{1,2} = \cos^{-1} \frac{\alpha_m}{c_{1,2}} \tag{34}$$

With this choice of e_2, c_2 cannot be less than the maximum P velocity in the layers overlying the reflecting zone. As a consequence, all radicals occurring in the exponential functions of (32) and (33) are real throughout the range of integration.

For applications of synthetic seismograms to earthquake refraction studies, Bessel functions such as in (32) and (33), may be replaced by asymptotic approximations for large arguments given in §3.5. Synthetic seismograms computed by this method are plotted in Fuchs and Müller (1971) and elsewhere.

8.9 Exercises

1   By Fourier theory, a wave form $F(x - ct)$, corresponding to a plane wave of arbitrary shape advancing with speed c parallel to the x-axis, can be equated to a sum of harmonic plane wave forms

$$\sum_r a_r \cos\left\{ 2\pi\left(\frac{x}{\lambda_r} - \frac{t}{\tau_r}\right) + \varepsilon_r \right\},$$

where a_r is the amplitude, λ_r the wave-length, and τ_r the period in the rth mode and $\lambda_r = c\tau_r$. Prove that the average energy E per unit volume of the medium, being twice the kinetic energy (see §3.3), is given by

$$E = \lim_{b \to \infty} \left\{ b^{-1} \int_0^b \left[\sum_r \frac{2\pi a_r}{\tau_r} \sin\left\{ 2\pi\left(\frac{x}{\lambda_r} - \frac{t}{\tau_r}\right) + \varepsilon_r \right\} \right]^2 \rho \, dx \right\}$$

$$= \sum_r 2\pi^2 a_r^2 \tau_r^{-2} \rho.$$

2   Consider refracted P and S waves in a solid perfectly elastic half-space M_2 arising from an incident P wave in a fluid half-space M_1. (The problem is of interest concerning the question of excitation of S waves in the Earth's inner core – see §13.1.4.)

     Show that the energy in the incident P wave reaching unit area of the boundary in one period is (see Exercise 1)

$$2\pi^3 A_1^2 \tau^{-2} q^{-1} \rho_1 \sin 2e_1,$$

where A_1 is the incident P wave amplitude, $\tau = (2\pi/\alpha_1 q)\cos e_1$, $q = (2\pi/\lambda)\cos e_1$. Derive the corresponding expressions for energies leaving unit area of the boundary in refracted P and SV waves. Hence show that the ratio of energy in refracted P and SV waves is

$$R = \frac{B_2^2 \sin 2e_2}{C_2^2 \sin 2f_2} = \frac{(1 - \tan^2 f_2)^2}{4\tan e_2 \tan f_2},$$

where B_2 and C_2 are amplitudes of the two refracted waves, respectively.

3   Consider a surficial horizontal elastic layer, thickness H, physical constants α_1, ρ_1 over an elastic half-space, constants α_2, ρ_2. Use this model to study the effect of a layer of alluvium over basement rock on vertically propagating P waves with the form

$$u = A\exp\left(i\omega\left(-\frac{x_3}{\alpha} - t\right)\right).$$

Find the vertical displacement at the free surface and show that this reduces to $2A\exp(-i\omega t)$ when $H = 0$. Show that the amplification factor at the free surface is

$$\left|\left\{\cos\frac{\omega H}{\alpha_1} + iR\sin\frac{\omega H}{\alpha_1}\right\}^{-1}\right|,$$

where $R = \rho_1\alpha_1/\rho_2\alpha_2 < 1$. Hence prove that the maximum amplification occurs for wave periods $T = 4H/(2m + 1)\alpha_1$, where $m = 0, 1, 2,\ldots$. For $\alpha_1 = 570\,\text{m/s}$, $\alpha_2 = 2200\,\text{m/s}$, $\rho_1 = 1.7\,\text{g/cm}^3$, $\rho_2 = 2.2\,\text{g/cm}^3$ and $H = 100\,\text{m}$, show that a maximum amplification of about 5 occurs at $T = 0.7\,\text{s}$. (Hint. See §5.4.)

4   Show that a physically suitable time dependence of an explosive point source is the derivative of the excitation function $F(t)$ in §8.8. (The far-field displacement of the compressional wave from the source is proportional to this derivative.) Discuss the applicability of the following analytical form

$$F'(t) = \begin{cases} \sin\delta t - \dfrac{1}{m}\sin m\delta t & (0 \leqslant t \leqslant \tau) \\ 0 & (t \leqslant 0 \text{ and } t \geqslant \tau) \end{cases}$$

$$\delta = \frac{N\pi}{\tau}, \quad m = \frac{N+2}{N}, \quad N = 1, 2, 3,\ldots$$

5   The multiple core reflections P4KP (see §10.6) are observed for $p > p_d$ well beyond the cutoff distance defined by the ray parameter p_d corresponding to the P ray that grazes the core. Show, from the JB tables (cf. §10.9.1) that there is a caustic of $P4KP$ at the cusp B near $\Delta = 38°$ of the travel-time curve ABC (cf. Fig. 10.8). Further, give a physical argument why energy can tunnel down into the core for $p > p_d$.

9

Seismometry

The purpose of the seismograph is to record the ground movement at a particular point of the Earth's surface in an earthquake.

The most general type of local movement in an elastic body is represented by the form (2.11), the terms in which correspond to translation, rotation and strain, respectively. In the case of seismic waves, the translation movement usually receives most attention. (Local rotation effects are sometimes significant in regions close to the seismic source and are of interest in the study of earthquake effects on buildings.) Seismographs are usually constructed to record translational components of the local earth movement. As a rule, the components taken are the two horizontal components, north–south and east–west, and the vertical, or 'Z' component.

In §§9.1, 9.2, we describe ideal seismographs for measuring horizontal and vertical components, respectively, of a local Earth movement. Constructional details differ but the same type of differential equation representing the motion of the seismograph relative to the ground occurs in both. In later sections, we use this differential equation as a basis for discussing the connection between the relative motion of a seismograph and the actual motion of the ground. Finally, we offer brief comments on certain other types of instruments.

Because, in a book on the theory of seismology, we are interested in the broad principle of the seismograph rather than in practical instrumental details, we usually assume that the angular displacement of the seismograph is small, and ignore small quantities of the second order. A more complete discussion of seismographs requires consideration of higher-order terms (see §9.1.1). Finally, the usual modern seismological usage is to call the entire instrument that records the ground motion as a continuous function of time a seismograph, but the component that responds to the motion a *seismometer*. An instrument not recording time-dependence is a *seismoscope*.

9.1 The horizontal component seismograph

The seismometer consists in principle of a pendulum swinging about an axis which is at a small inclination i to the vertical and which is rigidly attached to the earth. The pendulum has the form of a bar BC with a mass at the end C which is connected to the axis AB at A by a light support AC (see Fig. 9.1); this pendulum, free to swing about AB, is the *boom*. Let M be the mass of the boom, and h the distance from AB of its centre of mass, G. The equilibrium position ABG_0 of the plane ABG is called the *neutral plane*.

During the passage of a seismic wave, let the component, denoted by u, of the displacement of the ground (and hence of AB) in the direction perpendicular to the neutral plane be given by $u = f(t)$, where t denotes time and f is any function. At any particular instant, let θ be the (assumed small) angle between the plane ABG and the neutral plane. Let X be the component, perpendicular to the neutral plane, of the resultant reaction force exerted by the axis on the pendulum, and Y the component in the neutral plane perpendicular to AB. Correct to the first order of small quantities, the horizontal displacement of G is $u + h\theta$. Hence we have

$$X = M(\ddot{u} + h\ddot{\theta}). \tag{1}$$

The weight of the boom is $Mg\cos i$ parallel to BA, together with $Mg\sin i$ in a direction perpendicular to BA and parallel to the neutral plane. The

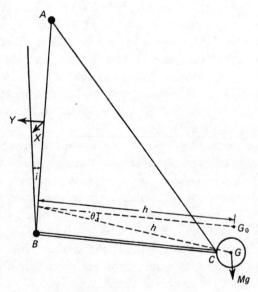

Fig. 9.1. Seismograph boom.

component of the acceleration of G in the direction of Y is of the second order, and so we have, correct to the first order,

$$Y = Mgi.\tag{2}$$

Considering the rate of change of angular momentum about an axis through G parallel to AB, we have, neglecting friction forces,

$$M\kappa^2\ddot{\theta} = -Xh - Yh\theta,\tag{3}$$

where κ is the radius of gyration of the boom about this axis. Eliminating X, Y we have

$$\kappa^2\ddot{\theta} + ghi\theta + h\ddot{u} + h^2\ddot{\theta} = 0,$$

i.e.

$$\ddot{\theta} + \omega^2\theta = -\ddot{u}/l,\tag{4}$$

where $l = (\kappa^2 + h^2)/h$, the *reduced pendulum length*, and $\omega^2 = gi/l$.

As will be seen in §9.4, it is usually desirable to arrange (by electromagnetic damping) that the pendulum of a seismograph be subject to an appreciable friction force. We take this force to be proportional to the first power of the speed of the pendulum. The equation (4) then needs to be modified by the inclusion of a corresponding damping term, so that the equation of relative motion of the seismograph is

$$\ddot{\theta} + 2\lambda\omega\dot{\theta} + \omega^2\theta = -\ddot{u}/l,\tag{5}$$

where λ is an index of the extent of damping.

9.1.1 Effect of tilt

The simplified analysis above does not take into account the sensitivity of the horizontal pendulum seismometer to tilting of the ground, linear accelerations along and normal to the boom, and variations in inclination i. The effect of these terms in the equation of motion has been considered for several seismic wave types by Byerly and P.W. Rodgers.

Fix the coordinate orientation in Fig. 9.1 so that the x, y axes are in the *horizontal* plane and let the neutral plane receive a tilt ψ about the y axis. Then (5) becomes, to the first order,

$$\ddot{\theta} + 2\lambda\omega\dot{\theta} + \omega^2\left[1 + \frac{g\Delta i - \ddot{u}_y}{l\omega^2}\right]\theta = -\ddot{u}_x/l + g\psi/l,\tag{5'}$$

when Δi is the variation in inclination. If all the kinetic parameters are assumed to vary like $\sin\Omega t$, it is easy to show that (5') takes the form of a Mathieu equation (see Exercise 9.11.2) for which there is an extensive literature. Solutions indicate that sizable corrections to recorded ground motions are required for Rayleigh waves at periods significantly longer than the free pendulum period. In particular, most horizontal seismometers respond primarily as tilt-meters to the graver eigen-vibrations of the Earth (see chapter 14). For these long periods, the tilt sensitivity is equal to $g/l\omega^2$.

9.2 The vertical component seismograph

The principle of the seismometer used for measuring the vertical component of seismic wave is indicated in the arrangement shown in Fig. 9.2. A pendulum ABG can rotate freely about a horizontal axis through A rigidly attached to the ground. The bulk of the mass of the pendulum is in the vicinity of the centre of mass G; AB, BG and GA are members of a light frame. Assume that adjustments have been made whereby G is level with A in the equilibrium position. The pendulum is supported at B by means of a spring BC, of natural length d say, which is rigidly attached to the ground at C, in the plane ABG. For ease of calculation we neglect the mass of the spring.

Let $AB = c$, $AC = b$, $CAB = \gamma$, $AG = h$, and let p be the perpendicular from A to BC. The tension of the spring is βq, where β is a constant of the spring and q is the extension. During the passage of the disturbance, let $u = f(t)$ be the upward vertical component of the ground displacement, let θ be the (assumed small) upward angular displacement of AG, and let Z be the

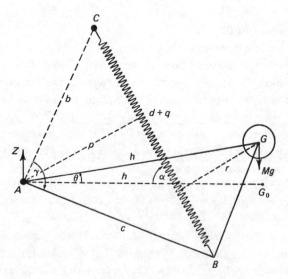

Fig. 9.2. Pendulum for vertical component of motion.

upward vertical component of the resultant force exerted by the axis on the pendulum at any particular instant.

Since the vertical displacement of G is then $u + h\theta$, correct to the first order, we have on resolving vertically

$$Z + \beta q \sin \alpha - Mq = M(\ddot{u} + h\ddot{\theta}),$$

where α is the angle BC makes with the horizontal. The angular momentum principle then gives

$$M\kappa^2\ddot{\theta} = -\beta qr - Zh,$$

approximately, where κ is the radius of gyration of the pendulum about an axis through G parallel to the axis of rotation, and r is the perpendicular from G to BC. Eliminating Z, we then obtain

$$M\kappa^2\ddot{\theta} = -Mh(\ddot{u} + h\ddot{\theta}) - Mgh + \beta q(h \sin \alpha - r)$$
$$= -Mh(\ddot{u} + h\ddot{\theta}) - Mgh + \beta pq. \tag{6}$$

Using the suffix 0 to correspond to the equilibrium position, from (6), $Mgh = \beta(pq)_0$. Also, correct to the first order in θ, we have

$$pq = (pq)_0 - \left(\frac{d(pq)}{d\gamma}\right)_0 \theta.$$

Hence (6) becomes

$$M(k^2 + h^2)\ddot{\theta} + \beta\left(\frac{d(pq)}{d\gamma}\right)_0 \theta = -Mh\ddot{u}.$$

If l be the reduced pendulum length, we rewrite this equation in the form

$$\ddot{\theta} + \omega^2\theta = -\ddot{u}/l,$$

which is the same as (4) obtained for the horizontal seismograph, except that ω is now a function of constants special to the vertical component seismograph. The inclusion of the frictional effect then gives the equation (5) again.

9.3 The indicator equation

In practice, the angle θ giving the relative motion of the seismographs in §§9.1, 9.2 is often so small that one could not measure its variation without amplification. In older seismographs levers attached to the pendulum magnify the relative rotation θ. The magnified motion is then recorded by a pen on smoked paper on a drum which rotates uniformly about an axis rigidly attached to the ground. In other seismographs use is made of a beam of light which is reflected from a mirror attached to the pendulum, and records on photographic paper on the rotating drum. The record containing the trace made by the indicator is called a *seismogram*.

The displacement, x say, of the indicator from its zero position will be approximately proportional to the angle θ, and in this case we introduce the *statical mechanical magnification* V_s, defined as the ratio $x/l\theta$. By (5) we then have

$$\ddot{x} + 2\lambda\omega\dot{x} + \omega^2 x = - V_s\ddot{u} = - V_s f''(t) \tag{7}$$

as the indicator equation.

The development of seismographs has now reached a stage in which any further increase in the sensitivity can, on account of microseisms and other noise (see §17.7), give no additional help in the recording of ordinary earthquakes.

9.4 Damping of seismographs

It would obviously be convenient if a seismograph could be designed so that its seismograms would give a direct measure of the actual earth movements, and we shall discuss the question as to how far this may be realised in practice.

We first note that the left-hand side of (7) is identical with that of equation (3.3). Therefore the complementary function of the solution of (7) takes one of the forms given in (3.8), and so depends on constants of the seismograph and not on parameters describing the earth motion (unless the latter is impulsive). This makes it desirable to arrange the instrumental constants so that the complementary function will be as insignificant as possible; this factor is taken into account in designing the amount of damping to be introduced. If the damping coefficient λ is less than unity, in which case the complementary function is

$$C\exp(-\lambda\omega t)\cos([1-\lambda^2]^{\frac{1}{2}}\omega t + \varepsilon), \tag{8}$$

then clearly the larger λ is, the less important is the complementary function; thus it is evident that unless there is some appreciable damping, the indicator could not give a close record of the actual earth movement. If $\lambda > 1$, the complementary function has by (3.7) the form

$$A\phi(t) + B\psi(t), \tag{9}$$

where A and B are constants, and

$$\phi(t) = \exp(-\alpha t), \quad \psi(t) = \exp(-\beta t),$$
$$\alpha, \beta = (\lambda \pm [\lambda^2 - 1]^{\frac{1}{2}})\omega. \tag{10}$$

Since $\beta < \alpha$, the term $B\psi$ of (9) becomes increasingly more important than the term $A\phi$ as time goes on, and this accordingly makes it desirable that β should be large, i.e. that any excess of λ over unity should be small. We conclude that a desirable feature of a seismograph is that the damping should be near the critical value, $\lambda = 1$. The corresponding form of the complementary function of the solution of (7) is then given by (3.8). Notice that the damping affects the complementary function and does not otherwise interfere with the response to the ground motion.

In other seismographs there are practical reasons for having λ somewhat less than unity (though still appreciable). In illustrations to follow we take $\lambda = 1/\sqrt{2}$ as a useful representative value. We have already pointed out (§ 3.1.2) that the ordinates of the graphs of the complementary functions in the cases $\lambda = 1/\sqrt{2}$, $\lambda = 1$ are not greatly different. Hence if λ is in the range

$$1/\sqrt{2} \leqslant \lambda \leqslant 1,$$

the damping will be reasonably satisfactory.

In the case of a seismograph for which $\lambda < 1$, the damping is frequently described in terms of the damping ratio ε, defined as the ratio of the amplitudes of x in two successive free swings of the seismograph (the Earth being at rest). By (3.6) it is easy to deduce that

$$\ln \varepsilon = \pi\lambda(1 - \lambda^2)^{-\frac{1}{2}}. \tag{11}$$

ε here is not to be confused with the phase constant ε appearing in (3.6). Knowledge of the damping ratio ε and of the free pendulum period $2\pi/\omega$ is sufficient to determine the constants in the indicator equation (7).

For a description in terms of the quality factor Q (see §4.5.3), we write $\lambda = 1/2Q$ and the logarithmic decrement is $\pi/2Q$, approximately ($Q \gg 1$).

9.5 Solution of the indicator equation

We now proceed to examine the nature of the solution of the indicator equation (7) corresponding to certain prescribed types of ground excitation. The solutions are by elementary methods but the reader can alternatively apply Laplace transformations for complete solutions or Fourier transformations for steady-state solutions.

9.5.1 Simple harmonic ground motion

Suppose that the relevant component of the ground motion takes the form $u = a\cos pt$, where a, p are constants. Then (7) becomes

$$\ddot{x} + 2\lambda\omega\dot{x} + \omega^2 x = V_s p^2 a\cos pt. \tag{12}$$

By (3.11) and (3.12) a particular solution of (12) is

$$x = aV_d \cos(pt - \delta), \tag{13}$$

where

$$\tan \delta = 2\lambda\omega p/(\omega^2 - p^2) \tag{14}$$

and

$$V_d = V_s p^2 \{(\omega^2 - p^2)^2 + 4\lambda^2\omega^2 p^2\}^{-\frac{1}{2}}. \tag{15}$$

The right-hand side of (13) is, apart from the phase difference δ, equal to V_d times the ground motion and therefore V_d is called the *dynamical mechanical magnification*.

When the earth movement is accurately simple harmonic with period $2\pi/p$, the dynamical magnification is constant and the seismogram will be an accurate record of the earth displacement (neglecting the effect of the complementary function). But the dependence of V_d on p shows that the magnification is different for different periods. Hence in general the seismogram will not be expected to be an accurate picture of the actual ground motion.

We notice from (15) that if there were no damping, there could be resonance if $p = \omega$. This gives a further reason for having seismographs damped; for if λ were very small, there would be abnormal exaggeration of those waves with periods near the natural period of the seismograph.

9.5.2 Impulsive ground motion

Suppose next that in the vicinity of the seismograph the earth is at rest prior to the instant $t = 0$, and that an impulsive disturbance then arrives which causes the relevant component $\dot{u}$ of the earth velocity to jump suddenly from zero to a fixed value, $\dot{u}_0$ say. The solution of (7) in these circumstances is found, for $t > 0$, to be

$$x = - V_s \dot{u}_0 t \exp(-\omega t) \tag{16}$$

if $\lambda = 1$, and

$$x = \frac{- V_s \dot{u}_0}{\omega(1 - \lambda^2)^{\frac{1}{2}}} \sin([1 - \lambda^2]^{\frac{1}{2}}\omega t) \exp(-\lambda\omega t) \tag{17}$$

if $\lambda < 1$.

We notice that the onset of the motion of the indicator is sharp in both cases. For the case $\lambda = 1$, we deduce from (16) that the maximum displacement on the seismogram occurs when $t = 1/\omega$. For the case $\lambda < 1$, we deduce from (17) that the maximum occurs when $\tan([1 - \lambda^2]^{\frac{1}{2}}\omega t) = (1 - \lambda^2)^{\frac{1}{2}}/\lambda$; taking the representative case $\lambda = 1/\sqrt{2}$, the last formula gives

$t = 1.1/\omega$. In the case $\lambda < 1$, x passes through its first zero after the time $t = 0$ when $\omega t = \pi/(1 - \lambda^2)^{\frac{1}{2}}$, for the case $\lambda = 1/\sqrt{2}$, this gives $t = 4.4/\omega$. After this time the displacement is always small compared with the first maximum for all cases where λ is in or near the range $1/\sqrt{2} \leqslant \lambda \leqslant 1$.

9.5.3 General ground motion response curves

We now derive a solution of the indicator equation (7) for a general earth motion $u = f(t)$, using the method of variation of parameters. We take as a trial solution of (7) the form

$$x = A(t)\phi(t) + B(t)\psi(t), \tag{18}$$

where ϕ and ψ are given by (10). In the trial solution (18) the right-hand side is the same as (9) except that A and B are now taken as functions of t and not as constants. By (18) we have

$$\dot{x} = A\dot{\phi} + B\dot{\psi}$$

and

$$\ddot{x} = A\ddot{\phi} + B\ddot{\psi} - V_s f''(t),$$

provided

$$\left.\begin{aligned}
\dot{A}\phi + \dot{B}\psi &= 0 \\[2mm]
\dot{A}\dot{\phi} + \dot{B}\dot{\psi} &= -V_s f''(t).
\end{aligned}\right\} \tag{19}$$

It follows (since $x = \phi$ and $x = \psi$ make the left-hand side of (7) zero) that (18) is a solution of (7) if $\dot{A}$ and $\dot{B}$ are given by (19), i.e. if

$$\dot{A} = \frac{V_s \psi f''}{\phi\dot{\psi} - \dot{\phi}\psi}, \qquad \dot{B} = \frac{-V_s \phi f''}{\phi\dot{\psi} - \dot{\phi}\psi}.$$

Hence a particular solution of (7) is seen to be (cf. (3.23))

$$x = V_s \phi \int \frac{\psi f''}{\phi\dot{\psi} - \dot{\phi}\psi} dt - V_s \psi \int \frac{\phi f''}{\phi\dot{\psi} - \dot{\phi}\psi} dt, \tag{20}$$

where ϕ, ψ satisfy (10).

Combining this solution with the relevant complementary function, we obtain an exact formula for the response of the seismograph to any given earth movement $u = f(t)$.

The usual way of specifying the overall response of a seismograph to seismic wave excitation is by means of a magnification or *response curve*, $R(\omega)$. For a sinusoidal excitation of a simple mechanical seismometer, its form is given by equation (15). Examples of representative response curves

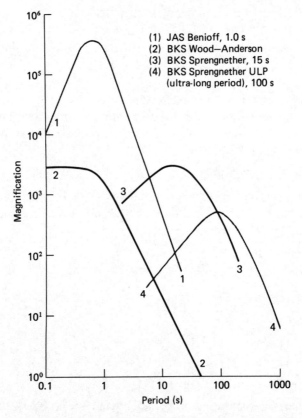

Fig. 9.3. Magnification curves for three types of seismograph operated at Jamestown (JAS) and Berkeley (BKS), California. Instruments (1) and (3) have magnifications like the standard seismographs of the WWSSN system. Times refer to the resonant periods of the pendulums.

are given in Fig. 9.3 where the ordinate is the ratio of seismogram amplitude to excitation displacement. Therefore, the magnification gives the sensitivity of the seismograph to displacement. Velocity and acceleration magnification curves are defined as $R(\omega)/\omega$ and $R(\omega)/\omega^2$, respectively.

Much seismological analysis is now carried out in the frequency rather than the time domain. Then, in complex variables, the overall response curve for the seismograph is

$$R^*(\omega) = R(\omega)\exp i\phi(\omega),$$

where the first factor is the amplitude response and $\phi(\omega)$ is the phase response spectrum. If the frequency spectrum of the recorded seismogram is

$S(\omega)$ then the actual ground motion spectrum $G(\omega)$ is computed from

$$G(\omega) = S(\omega)/R(\omega). \tag{21}$$

Equation (21) provides the link between synthetic and observed seismograms (see § 16.3.3). The time history of ground motion is obtained from the inverse Fourier transform of $G(\omega)$.

9.6 Computation of the ground motion from a seismogram

The important problem of inferring the earth movement from a seismogram is the inverse of that discussed in § 9.5, and we now look at the equation (7) again with this problem in view. Integrating (7) with respect to t, we obtain, using the suffix 0 at time $t = 0$,

$$\dot{u} = \dot{u}_0 - V_s^{-1}\left\{\dot{x} - \dot{x}_0 + 2\lambda\omega(x - x_0) + \omega^2 \int_0^t x\,\mathrm{d}t\right\}.$$

Integrating again, and using

$$\int_0^t \mathrm{d}t \int_0^t x\,\mathrm{d}t = t\int_0^t x\,\mathrm{d}t - \int_0^t xt\,\mathrm{d}t,$$

we obtain

$$u = u_0 + V_s^{-1}x_0 + (\dot{u}_0 + V_s^{-1}\dot{x}_0 + 2\lambda\omega V_s^{-1}x_0)t - V_s^{-1}x$$
$$- (2\lambda\omega + \omega^2 t)V_s^{-1}\int_0^t x\,\mathrm{d}t + \omega^2 V_s^{-1}\int_0^t xt\,\mathrm{d}t. \tag{22}$$

The equation (22) expresses the component of the seismic motion in terms of constants of the seismograph, the time t, and values of x which can be measured on the seismogram; and so gives a formal solution. In using equations like (22), recourse is had to numerical integration. Auxiliary experiments are made to test the reliability of the results; seismographs are placed on shaking tables to which known irregular motions are applied, and the results of the integration compared.

On account of numerical difficulties in treating equations such as (22), pains are taken to design seismographs so that certain important aspects of the ground motion may be rapidly read from seismograms. We proceed now to consider this matter.

9.7 Displacement and velocity meters and accelerometers

If the natural undamped period $2\pi/\omega$ of a seismograph is great compared with the predominant period of the incoming seismic wave, the indicator equation (7) approximates to

$$\ddot{x} = - V_s \ddot{u}. \tag{23}$$

Thus, if the seismograph were in equilibrium and at rest just prior to the onset of the disturbance, x from the seismogram would be approximately proportional to the actual component u of the earth displacement. A seismograph with this feature is a *displacement meter*. A drawback with such instruments is their instability when subjected to ground tilt, air pressure and temperature fluctuations, etc.

If, on the other hand, the natural period $2\pi/\omega$ were very short compared with the predominant period of the earth motion, the indicator equation approximates to

$$\omega^2 x = - V_s \ddot{u}. \tag{24}$$

A seismograph for which (24) holds is called an *acceleration meter* or *accelerometer*. In this case, a practical limitation arises in that, by (24), small $2\pi/\omega$ entails small x, for a given component $\ddot{u}$ of the ground acceleration (assuming that V_s is assigned). This small sensitivity in the accelerometer is, however, an advantage in recording strong ground motion (see §9.9.2).

In practice, the majority of modern seismometers are primarily velocity meters, because the seismometer motion x is converted to an output voltage proportional to $\dot{x}$ by means of a magnet and coil electro-magnetic transducer. Some examples are seismographs (1), (3), and (4) whose magnification curves are shown in Fig. 9.3.

9.7.1 Recording methods and timing

For many years, as mentioned in §9.3, the usual method of seismographic recording was a moving stylus which scratched smoked paper around a revolving and translating drum. Later other analogue devices, such as pen and ink recorders and a heated stylus marking special recording paper, were developed. Many installations, such as those at the World Wide Standard Stations, used a moving light spot on photographically sensitised paper or film. In practice, rather fine traces (about 0.1 mm wide) can be made by a needle on a smoked record but, in all cases, the highest measurable frequency depends upon the drum rotation speed, the sharpness of the stylus, focussing of the light spot, and so on.

It is usual to discuss the interval of variation of physical quantities in terms of the *dynamic range*, usually expressed in decibels. For ratios of power, this unit is defined as 10 times the logarithm to the base 10 of the ratio of the quantity to its equilibrium or reference level. Because power is proportional to squared amplitude, we have for two amplitudes A_1 and A_2, $dB = 20 \log_{10} A_1/A_0$. A tenfold order of magnitude change in frequency is 20 dB and a dynamic range of 10^7 is 140 dB.

At present more and more research depends on earthquake records stored on magnetic tape in either analogue or digital form. Suitable analogue playback facilities became widely available during the 1960s and much seismometry now incorporates this simple mode with playback either onto paper by pen and ink recorders or by some other analogue visual device. At the University of California, Berkeley, for example, earthquakes were recorded since 1964 at recording speeds as slow as 0.8 mm per second on magnetic tape in 16 channels with about 30 dB dynamic range for frequencies up to 20 Hz. With 14 data tracks on 1 inch tape at such slow speeds, one tape reel could store several weeks of data from three component seismographs.

The most recent development in seismological recording is the use of digital signals recorded on magnetic tape. These signals are usually in 12- or 16-bit words. The common 12-bit word gives a 72 dB (i.e. $20 \log_2 12$) dynamic range; the information is immediately accessible for processing in digital computers. Digital signals can be produced either discretely at the seismometer itself, or by analogue-to-digital conversion after initial storage in analogue form. A drawback with digital data, however, is the large amount of storage capacity required for normal observatory operations. As an example, consider digital recording at 20 samples per second, which is the minimum repetition necessary to represent seismic waves of 10 Hz frequency. At a tape density of 600 bits per centimetre, a tape speed of over 1 cm/s would be needed to give an order of magnitude less storage capacity than simple analogue recording.

Finally, it must be stressed that often seismometry requires not only adequate ground motion sensors, signal processors and recording apparatus, but also very precise timing devices. In modern practice, usually crystal clocks are used which are continuously checked against radio time signals. Commonly, the signals are obtained from commercial satellite transmitters and standard radio broadcasts such as provided by WWVB in the 15 to 60 KHz range and in the 2.5 to 25 MHz standard radio band. Such signals are received in most continental areas. Crystal oscillators that are contained in constant-temperature ovens are now commercially available. These provide time fluctuations as little as 0.1 s per year.

9.8 The dynamic ranges of seismic ground motion

9.8.1 Microseisms

In addition to ordinary earthquakes, seismograms record additional small earth movements called microseisms (see Fig. 3.5). These movements complicate accurate recording of earthquakes, but are of inherent interest because their form is related to the Earth's surface structure.

Some microseisms are purely, for example, those due to traffic or machinery, or to local wind effects and storms. Another class of microseisms shows features that are very similar on records traced at observatories distributed over a wide area. The features include approximately simultaneous occurrence of maximum amplitudes and similar noise energy spectra at all the observatories concerned. These microseisms may persist for many hours, and have more or less regular periods of from less than 5 to 8 or more seconds.

The largest amplitudes of these microseisms are of the order of 10^{-3} cm and occur in coastal regions; in central Asia, the amplitudes rarely exceed 10^{-4} cm. The amplitudes also depend to some extent on local geological structure. There is a fair correlation between the size of microseisms and the occurrence of stormy weather conditions in some adjacent region. At very quiet sites, microseisms may have less than 10^{-8} cm amplitudes at certain times.

Wiechert suggested that microseisms are generated by the action of rough surf against an extended steep coast, and follow-up studies have found a fair correlation between surf and microseisms recorded at nearby land stations. Gherzi noted the concurrence of large microseisms at Zi-ka-wei in China with cyclones far out over the ocean, and he put forward the hypothesis that microseisms are generated by atmospheric oscillations (or 'pumping') near the centre of the storm region. A.W. Lee, following detailed study of microseisms in Great Britain, gave evidence that they are essentially Rayleigh waves, and devised means of calculating their direction of advance.

Theoretical work of Longuet-Higgins and others has given strong support to the view that microseisms are generated when large standing waves are formed at sea, the microseismic periods being half the periods of the standing waves. However, not all microseisms appear to be generated in this way.

Longuet-Higgins derived the formula

$$P = p_0 + g\rho H + \frac{\rho}{\lambda}\int_0^\lambda \frac{\partial^2}{\partial t^2}(\tfrac{1}{2}\zeta^2)\mathrm{d}y \tag{25}$$

for the mean pressure P on a plane horizontal ocean floor at depth H below the undisturbed ocean surface. The formula (25) applies to one-dimensional wave propagation, where waves of wavelength λ advance in the direction of the y-axis (as in the analogous case of seismic surface waves treated in chapter 5); p_0 denotes the atmospheric pressure, ρ the density of the ocean (assumed incompressible), and t the time; the origin of z is taken on the undisturbed ocean surface, and the equation of the disturbed surface is $z = \zeta(y, t)$.

Consider a motion which in the first approximation consists of two sinusoidal waves of the same wavelength $\lambda = 2\pi/\kappa$, travelling in the positive and negative directions of y with speed c. Longuet-Higgins assumes for the actual motion of the ocean surface the form

$$\zeta = A_1 \cos\{\kappa(y - ct)\} + A_2 \cos\{\kappa(y + ct)\} + O(A^2\kappa), \tag{26}$$

where

$$\kappa^2 c^2 = g\kappa \tanh \kappa H, \tag{27}$$

and A is of the order of the larger of A_1, A_2. Apart from the second-order term $O(A^2\kappa)$, (26) and (27) correspond to the usual first-order solution for gravity waves on water. Substituting from (26) into (25), and simplifying, gives

$$P = p_0 + g\rho H - 2A_1 A_2 \kappa^2 c^2 \rho \cos 2\kappa ct, \tag{28}$$

neglecting a term of order A times the last term in (28). This equation gives the variation in the mean pressure P on the ocean floor to the second order of approximation.

When $A_2 = 0$, the last term in (28) vanishes. Thus in the particular case of a single progressive wave the mean pressure on the floor is constant and no microseisms are generated. But in general the last term of (28) is not zero. In the particular case of standing waves, for which $A_1 = A_2$, $= \tfrac{1}{2}a$ say, (26) and (28) become

$$\zeta = a \cos \kappa y \cos \kappa ct + O(a^2\kappa) \tag{29}$$

$$P = p_0 + g\rho H - \tfrac{1}{2}a^2 \kappa^2 c^2 \rho \cos 2\kappa ct. \tag{30}$$

The period of microseisms generated through the pressure fluctuations given by (30) would be $\pi/\kappa c$, or one-half the period $2\pi/\kappa c$ of the waves in (26). This matches the observed period relation and gives quantitative support to the theory that microseismic storms are generated when large standing waves are formed at sea. When the surface is flat, the centre of mass of the body of water in one wave-length is at its minimum height, and then rises as the crest forms, because of the net transfer of water from below the mean surface level to above. Since the weight of the water remains constant, the pressure at the ocean floor must supply the force which brings about this raising and lowering of the centre of mass, a fluctuation which takes place twice in each complete cycle. Thus, when ζ is given by (29), P fluctuates with one-half of the wave period $2\pi/\kappa c$, a result which is in agreement with (30).

Macelwane and Ramirez introduced in the 1940s 'tripartite' stations a few kilometres apart with coordinated timing, to study microseisms. In this way it was hoped to trace the movements of large storms at sea and to forecast hurricanes. In practice, there were difficulties at the time in securing the necessary precision but, nevertheless, until weather satellites, the method provided a valuable tool at some places, such as Hong Kong, in tracking cyclones at sea and constituted an early example of seismic arrays.

9.8.2 Frequency range

Types of seismic waves extend from P waves with frequencies in the audible range ($f > 20$ Hz) to free oscillations of the whole Earth with gravest period of 53 minutes (see § 14.2). Seismologists also are required to record especially high frequency waves in small rock specimens and in mining operations, and very low frequencies such as the 12 hour variation in earth tides. Consequently the frequency span of interest is from over 10 Hz to 10^{-4} Hz or about 100 db. Attenuation of waves in rock imposes the high frequency limits; as the size of the seismic sources increases the dominant frequencies in P and S waves measured at teleseismic distances fall to between 1 and 5 s. As mentioned in the previous section, the period of the microseisms corresponding to maximum power is between 5 and 8 s so that the frequency response of instruments is often separated by a cutoff at about 5 s period.

For small to moderate earthquakes, surface waves have frequencies extending from about 0.1 Hz out to 0.002 Hz for great earthquakes (magnitude > 7.5). At teleseismic distances, surface wave periods are limited at high frequencies to about 5 s because higher frequency surface waves are obscured by the dominant microseisms and scattered by the heterogeneity of the crust. In continental shields, surface waves with periods between 10 and 15 s are transmitted to great distances relatively efficiently, whereas across oceans Rayleigh waves with periods less than 15 s are attenuated more quickly. As discussed in § 14.3, since 1960 spectral measurements have been made on very-long-period seismographs and gravimeters of the free oscillations of the Earth, with eigen-periods ranging from hundreds of seconds to about one hour.

9.8.3 Amplitude range

There are also great demands on observatory seismograph design so that the full dynamic range of seismic wave amplitudes can be covered. Ground displacements in an earthquake magnitude range (see § 15.2) of -2 to over 8 range from 10^{-10} to 10^{-1} metres or about 180 dB. The background microseisms in the 5 to 8 s band often rise to ground amplitudes of 10 micrometres (10^{-3} cm or 10^4 nanometres). This value corresponds to an amplitude in short period P waves at $\Delta = 20°$ of about 10 nm and at $\Delta = 90°$ of 1 nm for an earthquake of magnitude about 4. For great earthquakes like the great Alaskan earthquake of 1964 (magnitude about 8.4), the ground amplitude of predominant P waves rises to about 1 cm at periods of 10 s at distant stations.

Surface waves also require that the seismographs have a large dynamic range. For example, at $\Delta = 80°$, a small earthquake with magnitude 3 would have a ground displacement in Rayleigh waves of about 10 nm and a great earthquake a ground displacement of order 1 cm at 80°. The signal dynamic range is about 120 dB in this case.

In the case of strong motions near to a seismic source, much larger wave amplitudes are measured (see § 17.4.1). Ground accelerations range up to twice gravity at high frequencies and ground displacements up to many centimetres at low frequencies. By contrast, a 1 μm displacement of a 1 Hz frequency seismic wave from a teleseism would correspond to a ground acceleration of only $4 \times 10^{-6} g$.

9.9 Modern seismographs

The theory of the seismograph as discussed in the preceding sections is basic to the simple pendulum type of design and is directly relevant to a number of seismographs used in practice, such as the Bosch–Omori seismograph. We now refer briefly to some other types.

Wiechert designed important early seismographs still in use in which the pendulum is vertical and inverted, maintained by small springs pressing against supports rigidly attached to the ground. The mass of the pendulum is large (up to several tons), and the seismographs record both horizontal and vertical components.

A key instrument for near earthquake magnitude purposes is the Wood–Anderson (see § 15.2). The principle of this instrument is the use of a vertical fibre which is under tension with a small attached mass and which suffers torsion during the passage of an earthquake wave.

The specifications of this seismograph are usually given as $T_0 = 0.8$ s, $\lambda = 0.8$, and static magnification 2800. Recent tests indicate, however, that many models in service have $V_s \approx 2100$.

9.9.1 The electromagnetic type

A major development took place when Galitzin introduced the idea of recording the seismogram by means of a ray of light from the mirror of a galvanometer through which passes an electric current generated by electromagnetic induction when the pendulum of the seismograph moves. The current is induced in a set of coils that are attached to and move with the pendulum in an independently set up magnetic field. If u is the relevant component of the ground displacement, the pendulum displacement θ and galvanometer displacement x (which is proportional to that shown on the seismogram) satisfy equations of the form

$$\ddot{\theta} + 2\lambda\omega\dot{\theta} + \omega^2\theta = -\ddot{u}/l + h\dot{x}, \tag{31}$$

$$\ddot{x} + 2\mu\Omega\dot{x} + \Omega^2 x = k\theta, \tag{32}$$

where λ, μ, ω, Ω, h, k and l are all instrumental constants.

In earlier instruments, the feed-back term $h\dot{x}$ in (31) was small and commonly neglected, in which case (31) and (32) are readily solvable for simple forms of the ground displacement u. We shall indicate solutions for the cases of (a) simple harmonic, (b) impulsive, ground motion taking $\lambda = \mu = 1$ (critical damping) and $\Omega = \omega$ (equal galvanometer and pendulum periods), in addition to neglecting the term in h.

For case (a), we take $u = a \cos pt$, for which (31) takes the form of (12) and so has a particular (the 'steady state') solution $\theta = aV' \cos(pt - \delta')$, where V' and δ' depend on ω, l and p. Substitution into (26) then gives

$$x = aV'' \cos(pt - \delta''), \tag{33}$$

where V'' and δ'' depend on ω, l, k and p, showing (cf. §9.5.1) that the galvanometer response is similar to that of the indicator in the simpler type of seismograph.

For case (b), taking an impulsive ground motion as in §9.5.2, the solution for x is found to be

$$x = -\frac{k}{l}\dot{u}_0(\tfrac{1}{2}t^2 - \tfrac{1}{6}\omega t^3)\exp(-\omega t). \tag{34}$$

It follows from (34) that the galvanometer does not here start with a finite velocity as does the indicator in the case of §9.5.2 (equation (16)).

Nevertheless, the initial acceleration of the galvanometer is appreciable, the first maximum of x being reached after time $1.3/\omega$, which is not greatly different from the corresponding time in §9.5.2.

When the term in h in (31) is not negligible, the solution is more complicated. There are, however, practical advantages to be derived from the seven adjustable constants in these equations, in contrast to the three constants in (5). In part for this reason, electromagnetic seismographs have superseded the simpler mechanical types. Various instrumental devices have been designed to realise optimum values of the seven constants for particular purposes. For example, instruments may be designed to produce seismograms in which the dynamical magnification is nearly constant over a wide range of periods, to reveal motions of unusually short or unusually long periods, or to enable specific parts of the seismic spectrum to be closely studied. Outstanding pioneering developments include those of Benioff, Willmore and Kirnos and an extensive modification of the Galitzin-type seismograph by Press and Ewing, resulting in the instruments adopted for the WWSSN (see §1.3.2); with these, the pendulum is operated at 15 s period, with a galvanometer period of 100 s.

Useful discussion of equations (31) and (32) and modern trends has been given by Willmore (1961) and McEvilly in Lomnitz and Rosenblueth (1976).

9.9.2 Signal enhancement. Digital processing

The electric voltages produced by motions of the pendulum seismometers or other sensors of ground motion are nowadays passed through considerable electronic circuitry to modify the motion.

Generally speaking, seismographs are divided into three types: short period; long (or intermediate) period; and ultra-long period or broadband instruments. Characteristic frequency response curves are drawn in Fig. 9.3. Short period instruments are used to record P and S body waves with high magnification of ground motion. For this purpose, the system response is shaped to peak at about 1 s or less. The long-period or intermediate-period instruments of WWSSN type have a response maximum at about 20 to 40 s. Again, in order to provide as much flexibility as possible for research work, the recent trend is towards the operation of very-broad-band seismographs often with digital representation of the signals. This is usually accomplished with very-long-period pendulums and electronic amplifiers which pass signals in the 0.05 to 50 Hz band. When the digital recording has 16 bits of information, broadband seismographs (such as those developed at Berkeley) have pass-band widths of 0.01 to 25 Hz and also a large dynamic range (96 dB) of amplitude in order to record small and large magnitude regional earthquakes and teleseisms.

Many seismographs now in operation, such as the WWSSN instruments, use the pendulum–galvanometer combination (§ 9.9.1). Electronic amplifiers were introduced in the 1960s after inherent noise qualities significantly improved. Modern amplifiers are used over frequency bandwidths in excess of 10 Hz with input noise levels superior to those of the galvanometer systems.

The representation of a seismic signal in digital form, either by direct recording or through an analogue-to-digital converter presents special problems. First, we note that any harmonic wave must have at least three samples per wavelength for a unique representation. If the sampling interval is Δt, the interval $2\Delta t$ exactly defines the harmonic with lowest frequency

$$f_N = \frac{1}{2\Delta t}. \tag{35}$$

This frequency is the Nyquist or *folding frequency*. Of course, higher frequency harmonics $f > f_N$ can be found that also pass through the three sample points defining f_N. The reader can show that these alias frequencies are $(2f_N \pm f)$, $(4f_N \pm f)$, and so on. Thus waves with these frequencies cannot be distinguished from harmonics with $f \leqslant f_N$ using only the digital sample available. The effect of aliasing is to reflect or fold back energy with $f > f_N$ into the wave spectrum and *it is essential* to remove harmonics with $f > f_N$ by filtering before spectra for $f < f_N$ are calculated.

Much attention is given in modern seismometry to filtering seismic signals in both continuous analogue or digital form. Although seismometer response provides initial filtering of the ground motion, the initial recorded response is readily modified further by electronic or numerical filters specially designed to remove aliasing, microseisms, long-period drift and other unwanted Fourier components. An example of a programmed filter to modify broad-band ground motion is shown in Fig. 4.2.

The subject of filter theory and design is large and beyond the present scope. We only remark that the main goals are usually to use linear filters that produce zero or linear phase shifts. The ideal low-pass filter has a response of unity in the given frequency pass-band and zero elsewhere. High-pass and band-pass filters may be constructed from a closely ideal low-pass filter such as the Butterworth filter. The frequency response function of such filters is often a fraction with numerator and denominator complex functions of frequency. The filter can then be specified by the number of zeros of the former and the poles of the latter. For example, the (Fourier) response spectrum of the nth order low-pass Butterworth filter (i.e. n poles) is, for $f < | f_N |$,

$$| B_L(\Omega) |^2 = \frac{1}{1 + \Omega^{2n}}, \tag{36}$$

where $\Omega = f/f_L$, $n = 1, 2, 3 \dots$

Note that $dB_L/d\Omega = 0$ for $\Omega \ll 1$ and $= -n$ for $\Omega \gg 1$. Further discussion will be found in Kanasewich (1981).

9.9.3 Strong-motion accelerometers and arrays

When seismic waves near to their source are to be recorded special design criteria are needed. First, instrument sensitivity must ensure that the largest ground movements stay on scale. From § 9.7 a seismometer capable of recording ground acceleration requires a natural frequency $\omega > \omega_g$ where ω_g is the highest frequency of ground motion of interest. For most seismological and engineering purposes this value is 20 Hz. Design principles for an accelerometer are thus easily met. The free frequency is high so the pendulum or its equivalent can be small and hence relatively free of tilt and drift. By comparison, displacement meters need a longer free period and pendulum with consequent instability. A further advantage of an accelerometer for strong-motion recording is that, because integration can be carried out more accurately than differentiation, ground velocity and displacement can be derived by summation (see § 17.4.1 and Fig. 17.2).

The accelerations to be recorded range up to twice gravity ($2g$). It turns out that this can be easily accomplished with short torsion suspensions, force-balance mass–spring systems, variable capacity transducers and other sensor types. A judicious choice of damping in electromagnetic accelerometers of about 0.7 of the critical value gives a response curve that is flat and directly proportional to ground acceleration from zero period to the natural frequency of the seismometer.

Because many strong-motion instruments need to be placed at un-attended sites in ordinary buildings for periods of months or years before a strong earthquake occurs, they usually record only when a trigger is started with the onset of motion. Short solid-state memories are now available, particularly with digital recording instruments that preserve the first few seconds before the trigger starts the permanent recording. In the past the usual recording has been on film strips for up to a few minutes duration. Now digitised signals are stored directly on cassette magnetic tape or on a memory chip. Commonly absolute timing has not been provided on strong-motion records but only accurate relative time marks; the present trend, however, is to provide Universal Time with special radio receivers or small crystal clocks. A detailed description of strong-motion instrumentation is given by D.E. Hudson (1982).

The prediction of strong ground motion and response of engineered structures in earthquakes depends critically upon measurements of the spatial variability of earthquake intensities near the seismic wave source. Towards this objective, installation of special arrays of strong-motion seismographs in highly seismic areas of the world was recommended at an international workshop in Hawaii in 1978. One of the first to become operational is located at the town of Lotung in the northeast corner of Taiwan, an area that suffers many earthquakes. This two-dimensional surface array, called SMART 1, consists of a centre element $C00$ and three concentric circles (inner I, middle M and outer O), each with 12 strong-motion seismographs having a common time base, and radii of 200 m, 1 km and 2 km, respectively (see Fig. 9.4). The inner ring controls the spatial aliasing since, for a wave velocity of, say, 1 km/s and element spacing of 100 m, the aliasing frequency is 5 Hz. Only one instrument in the I-ring is located more than 10 m from the position of perfect symmetry.

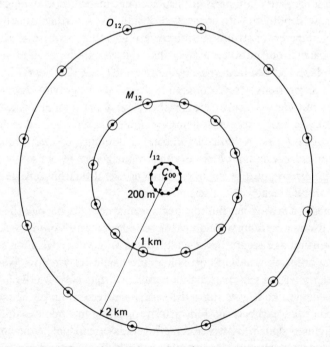

Fig. 9.4. The surface digital strong-motion array of accelerometers, called SMART 1, in Taiwan.

Each seismograph consists of a triaxial force-balance accelerometer, capable of recording $\pm 2g$, connected to a digital event recorder, that uses magnetic tape cassettes of conventional size. The accelerometers trigger on both vertical and horizontal components. Signals are digitised as 12-bit words at 100 samples per second. The Nyquist frequency is thus 50 Hz. Each instrument has a digital delay memory which stores the output for the force-balance accelerometers for approximately 2.5 s. The advantage over conventional design is that the first seismic waves are not missed, and complete seismograms of moderate-to-large earthquakes are obtained. Crystal clocks at each element are corrected every few days (and after each recorded earthquake) to Universal Time.

Installation began in September 1980 and, by January 1981, 27 instruments were in place. By December 1984, 29 earthquakes of local magnitude $3.8 \leqslant M_L \leqslant 6.9$ had triggered array elements. The largest, on 29 January 1981, was a strong earthquake centered 30 km from the centre of the array and felt over all of Taiwan. Its focal depth was approximately 11 km. The Richter magnitude was 6.9.

In this case, the seismic waves triggered all the 27 operational digital accelerometers, thereby providing comprehensive multi-dimensional recording of strong ground motion near to a significant fault rupture. In the 1981 Taiwan earthquake, the maximum horizontal and vertical accelerations recorded were $0.24g$ and $0.09g$, respectively. (A number of strong-motion accelerometers of conventional analogue type also triggered in the vicinity of SMART 1, allowing a crucial comparison between the two types of measurements.) Other important multi-site accelerogram sets have been obtained in Californian arrays in the 1979 Imperial Valley earthquake and the 1975 Oroville aftershocks.

Large aperture seismic arrays (linear dimension of order 100 km) have been used for over 15 years to improve discrimination between underground nuclear explosions and earthquakes by enhancing the signal-to-noise ratio (see §11.2.2).

In contrast, a strong-motion array of aperture 2 km is not designed to enhance small signals, but rather to determine the speed, direction of propagation, type of the evolutive wave components, and special spectral variations and phase relations. As in an array of radio telescopes, a seismic array, like SMART 1, allows wave correlations for consecutive time and frequency intervals. Thus, a computer algorithm can be used to insert appropriate time lags in the signals at each element, thereby steering the array response towards a known azimuth. One seismological aim is to follow the seismic dislocation as it moves along the rupturing fault. On the engineering side, the measurements address such key questions as the variability of the intensity of seismic waves over short distances, the contribution of seismic surface waves to the shaking at a given site and, in particular, the measurement of out-of-phase wave components over distances comparable to base dimensions of large engineered structures. Accelerograms from individual strong-motion seismographs cannot, in general, provide the resolution needed to resolve such questions.

9.9.4 Strain, tilt and other measurements

In 1935, Benioff made an important innovation by designing a strainmeter to measure a component of ground strain, instead of the usual ground displacement. During the passage of seismic waves, this instrument records variations in the distance between two points of the ground some 20 m apart. The variations are measured against a standard-length tube, originally of steel, but later replaced by a fused quartz extensiometer. The recording is electromagnetic, often with variable capacity transducers. The variable displacement of the two points gives one component of the strain tensor e_{ij} (§2.2 and equation (2.13). At the Earth's surface, three orthogonal strainmeters will determine the strain tensor completely. Routine measurement of pure rotation (equation (4.3)) has not been achieved although adequate rotation sensors are now available.

Instruments called *tiltmeters* have been developed to measure slow angular displacements of the ground extending over considerable periods of time (of the order of days and more). The design often incorporates a suspended pendulum or differential heights of fluid reservoirs connected by a tube. Records from tiltmeters are checked against local earthquake occurrence, with a view to finding correlations which might throw light on future earthquake occurrence.

Earthquake motions are sometimes detected on instruments designed for completely different purposes. A notable instance occurred when Ness, Harrison and Slichter recorded long-period earth motions on a LaCoste–Romberg tidal gravity meter following the main Chilean earthquakes of 1960 May 22, the periods ranging from 3 to 55 min (see §14.1).

Benioff also investigated properties of a seismograph designed to measure ground dilatations directly. The principle is simply the recording of the change in volume per unit volume of an underground (mechanical) cavity due to time changes in div u_i (equation (4.2)). Such *dilatometers* are routinely used in Japanese seismological observations, for example.

9.9.5 Portable seismographs and microprocessors. Telemetry

In addition to seismographs intended for recording inside earthquake observatories, other instruments, commonly called *geophones*, have been designed to record seismic waves in the field at short distances from small earthquake sources and explosions. The original purpose of the latter instruments was to make seismic surveys of subsurface rock structures (see § 13.2.1). These seismometers have short periods and usually work with electronic amplifiers.

New design principles were introduced in the 1950s that enabled very small sized seismometers with low internal noise to be produced. Because thermal noise due to Brownian motion is proportional to $(\lambda/m)^{\frac{1}{2}}$, very low damping permits the use of smaller masses, provided amplifiers and filters compensate for the reduced damping and seismometer signals. Rather than dissipative damping, the term in λ in (9.7) was provided by negative feedback forces to the mass itself.

Rapid modifications are now occurring in seismometry with the wide availability of inexpensive microprocessors based on the 16-bit silicon chip. Seismographic networks are converting from analogue recording on photographic paper, film or tape to broad-band digital recording. The output of the sensors is immediately digitised at various sample rates with a subsequent immediate processing by a built-in microprocessor. This high-speed processor is able to alter sample rates, change the gain ('gain-ranging'), filter and amplify signals, format the bit sequence, store data files, and trigger storage by detection of prescribed events. The seismograph is thus itself a microcomputer with programming capability.

It is now commonplace to site the seismometer some distance from the central observatory. Local cultural noise may be unacceptably high near the observatory but often also a network of stations in a broad seismically active region is needed. The seismic signals are telemetered back to the central recording system by telephone lines, ground radio systems or satellite radio transmission.

A commonly used telemetry system for seismological signals is based on frequency modulation (FM) in the audio range of 300–3000 Hz which is available on commercial voice-grade telephone lines. For example, seismometer output in the 0–25 Hz range is converted by a voltage-controlled oscillator (VCO) to tones in the audio band modulated ± 125 Hz by the signal. Several independent tones (various components and gains at one station, or many stations) can then be multiplexed onto one telephone line. At the receiver, the tones are demodulated by narrow-band filters which reconstruct the original analogue seismometer signal. A dynamic range of 50–60 dB can be obtained with relatively simple systems.

Telemetry links may now consist also of direct transmission of digital data by telephone line or radio link. Because of prescribed limits on transmission rates, care must be taken with sample rate and bit sequencing.

9.9.6 Ocean-bottom seismographs

The need for ocean-bottom seismometers to augment the global system of recording stations has been recently recognised, and comprehensive research efforts are under way to determine the feasibility of extensive long-term recording by instruments on the ocean bottom. Already, there is an operational OBS system in Japan for this purpose, where a semi-permanent seismograph system was placed on the sea floor off the Pacific coast of central Honshu using a cable. It has operated since 16 August 1978 and is expected to run for more than ten years. The cable extends a distance of 200 km across the Japan Trench, with a terminal OBS and three intermediate OBSs.

Because of the instrumental difficulties in maintaining permanent ocean-bottom instrumentation, a number of different systems have been considered. These include instruments which are placed in an ocean-bottom package, with signals transmitted to the ocean surface for retransmission or transmitted via cable to a shore-based station. Another system involves the recording portion of the device automatically releasing itself from the seismometers and floating to the surface for later recovery.

The use of ocean-bottom seismographs should ultimately secure a vastly improved global coverage of seismic waves and provide much new information on seismicity of oceanic regions. Stations on oceanic islands are useful but have the limitation that the island structure distorts waves when the wave-lengths are of the order of the dimensions of the island. Ocean-bottom seismographs will enable details of oceanic crustal structure to be determined, and because of the relative thinness of oceanic crust should enable clear seismic information to be gathered on the upper mantle. New data should be provided on focal mechanism, on the origin and propagation of microseisms, and on the character of ocean–continent margins.

9.10 Engineering response spectra

The Fourier spectrum of a recorded ground motion can be obtained, after correction if necessary for instrumental response, by the usual methods such as the Cooley–Tukey fast Fourier transform. Both amplitude and phase spectra (see § 14.3.2) are computed.

In engineering contexts, because of the interest in the motion of buildings and other engineered structures, the main attention is given to the response of a simple damped linear oscillator due to forcing accelerations. For mass m, damping parameter c and spring eigen-frequency ω, we have for relative displacement of the mass,

$$m\ddot{u} + c\dot{u} + \omega^2 u = -m\ddot{u}_g(t), \qquad (37)$$

where $\ddot{u}_g$ is the ground acceleration. This equation is, of course, identical with that of the simple seismometer (7) and the earlier discussion is immediately applicable. In the notation of (7), a suitable solution is given by the Duhamel integral (see Exercise 9.11.3).

$$u(t) = -\frac{1}{\omega_D} \int_0^t \ddot{u}_g(\tau) \exp[-\lambda\omega(t-\tau)] \sin[\omega_D(t-\tau)] d\tau, \qquad (38)$$

where $\omega_D = \omega(1 - \lambda^2)^{\frac{1}{2}}$ is the eigen-frequency of the *damped* system. For no damping or for practical purposes when $\lambda < 0.2$, $\omega = \omega_D$. We note that,

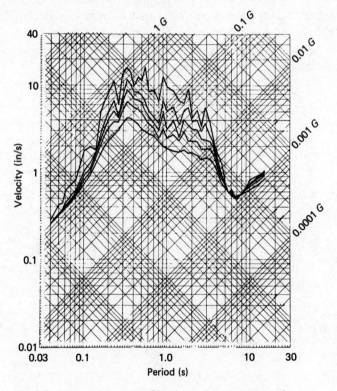

Fig. 9.5. Response spectra of horizontal components of a near earthquake, magnitude 6.9. Damping values 0, 2, 5, 10 and 20 per cent of critical. Here G denotes gravity.

from (38), $|u(t)|$ can be plotted as a function of ω to yield the Fourier amplitude spectrum. In strong motion work, however, the irregularity of the input motion $\ddot{u}_g(t)$ entails that careful numerical integration of (38) must be performed; sometimes a direct numerical integration of (37) is preferable.

For design purposes it is often sufficient to know only the maximum value of the response of the structure to seismic input. Thus, in terms of the simple oscillator model, Housner defined the response spectrum as the curve of maximum response of oscillators with different ω to the given input motion. The maximum value of the displacement $|u_{max}(t, \omega, \lambda)|$ is called S_d, the *displacement response spectrum*. In addition, an allied spectrum $S_v = \omega S_d$, called the *pseudo-velocity response spectrum*, is sometimes used. 'Pseudo' is attached to differentiate from the true velocity response spectrum $\dot{u}_{max}(t, \omega, \lambda)$. Similarly, maximum loads are investigated using $S_a = \omega^2 S_d$, or the *pseudo-acceleration response spectrum*. The S_d spectra for various levels of damping are plotted in Fig. 9.5 on the convenient tripartite log paper. Note that in these engineering developments the corresponding phase spectra are usually ignored although, for some structural types, failure is not independent of the time sequence of the wave arrivals.

9.11 Exercises

1   By scaling from seismograph response curves of Fig. 9.3, estimate the wave amplitude of a 5 s S wave as recorded on the BKS (Berkeley Station) Sprengnether (15 s) instrument if the BKS Wood–Anderson instrument showed an S wave amplitude of 2 mm at 5 s period.

2   Show that equation (5′) in §9.1.1 can be written in the form

$$\ddot{\theta} + 2\lambda\omega\dot{\theta} + \omega^2[1 - \mu\sin\Omega t]\theta = A\sin\omega t$$

for ground motion of the type $u = \bar{u}\sin\Omega t$, $\psi = \bar{\psi}\sin\Omega t$. Evaluate the terms μ and A and show that μ should be small for normal surface waves.

3   The equation for the linear displacement x of a pendulum under a time dependent force $f(t)$ is

$$\ddot{x} + 2\lambda\dot{x} + n^2 x = f(t) = (D - \alpha_1)(D - \alpha_2)x.$$

If $x = x_0$, $\dot{x} = u_0$ at $t = 0$ prove, by Laplace transforms, or use of an integrating factor, that a solution is

$$x(t) = \frac{1}{\alpha_1 - \alpha_2}[u_0(\exp\alpha_1 t - \exp\alpha_2 t) - x_0(\alpha_2\exp\alpha_1 t - \alpha_1\exp\alpha_2 t)$$

$$+ \int_0^t f(u)\{\exp\alpha_1(t-u) - \exp\alpha_2(t-u)\}\,du].$$

Show that the first terms are the eigen-response of the pendulum and the last is the response to the applied force.

4 Consider a simple seismograph with pendulum displacement x relative to equilibrium position given by

$$\ddot{x} + 2\lambda\dot{x} + \omega^2 x = k\ddot{\xi},$$

where ξ is ground displacement, and ω, λ, k are instrumental constants. Show that if the pendulum suddenly acquires a finite unit velocity $\dot{x} = k$, the recorded displacement when $\lambda < \omega$ is

$$x = \frac{k}{\gamma}\exp(-\lambda t)\sin\gamma t,$$

where

$$\gamma^2 = \omega^2 - \lambda^2.$$

5 An electromagnetic seismograph (see §9.9.1) is designed so that the two damping constants such as λ above are equal and the free periods are also equal so that $\omega^2 = 2\lambda^2$.

Prove that the instrumental response to unit impulsive velocity is proportional to

$$x = [(\lambda t \sin \lambda t - \sin \lambda t + \lambda t \cos \lambda t)\exp(-\lambda t)]/2\lambda^2,$$

where the reaction of the induced current on the pendulum is ignored. Show that the first recorded wave maximum for such an instrument occurs a little later than for the simple seismograph in 9.11.4.

6 The Alaskan earthquake of 27 March 1964 had an Ms magnitude of 8.4 and its epicentre was at a distance of 28° from Berkeley. The following trace amplitudes were observed on the Berkeley Wood–Anderson seismograms:

P wave   4 mm, $T = 4$ s, magnification $= 110$
S wave   10 mm, $T = 10$ s, magnification $= 20$
Rayleigh wave   75 mm, $T = 16$ s, magnification $= 10$

Assume a propagation velocity of 6.0 km/s for the P wave and 3.5 km/s for the S and Rayleigh waves.

The free oscillations of the earth were also observed following the Alaskan earthquake and the ground displacements observed at Berkeley for two of the modes were

$_0S_2$ 1.8 mm, $T = 54$ min

$_0S_6$ 1.2 mm, $T = 16$ min

Calculate the maximum ground strains and discuss whether the assumption of linear elasticity is valid in these cases. (See equations (2.11) and (2.13).)

7 Determine that an order $n = 4$ Butterworth filter is required to attenuate a seismic signal at $2f_L$ down by 24 dB relative to the value at $f = 0$.

Draw the response curve for the filter with gain normalised to unity. What is the attenuation at $2f_L$ for a three-pole filter?

8   Prove by an analysis of strain (see §2.2) that the differential movement in an earthquake of two points fixed on the ground surface will yield a measure of strain but not rotation.

Show further that three and not six surface strainmeters are needed to determine the strain tensor. How many would be needed at a point on the ocean floor?

9   Consider a Rayleigh wave with period 5 s, velocity 3.5 km/s and displacement amplitude 0.1 μm (1 micrometre $= 1\,\mu$m $= 10^{-4}$cm) from a moderate earthquake. Show that the ground acceleration, strain and rotation are 10^{-5} gal, 10^{-11}, and 10^{-6} arc sec respectively. What are the corresponding values for a similar wave from a great earthquake with period 20 s and amplitude 1 cm? (Cf. §15.2.)

10   Prove that an integral of the indicator equation (7) is the convolution

$$x(t) = \ddot{u}(t) * f(t) = \int_0^\infty \ddot{u}(t - \tau) f(\tau)\mathrm{d}\tau,$$

where $f(t)$ is the impulse response of the seismometer (see §3.1.5).

10

Construction of travel-time tables

There are more than a thousand seismological observatories distributed over the Earth's surface, each possessing one or more seismographs, and in the course of a single year hundreds of thousands of seismograms are obtained. The best treatment of this material requires the careful use of estimation theory, in conjunction with the theory discussed in the preceding chapters. The need for the use of statistics is more imperative than in some other branches of physics, and this is so whether use is made of extensive routine observations or whether a special study of original seismograms is made by an individual. In particular, it is crucial in observational seismology that some indication of the error should accompany every numerical estimate.

10.1 Parameters of earthquake location

The initial disturbance connected with an earthquake (see § 4.3) is confined to a limited region of the Earth's interior whose linear dimensions do not exceed the order of a few kilometres (see chapter 16). The centre of this confined region is called the *focus* (sometimes the *hypocentre*). The point of the Earth's outer surface vertically above the focus is the *epicentre*. Further quantitative description of an earthquake must be in terms of numerical parameters. For simplicity, it is convenient to take into account just a four-dimensional space with parameters time of origin, two epicentral coordinates (here taken as the colatitude θ and the east longitude ϕ), and depth of focus. To begin with, we shall assume the Earth to be spherically symmetrical about its centre.

10.2 Calculation of the epicentral distance and azimuth

As a preliminary to making use of seismograms, it is desirable that we should be able to connect the coordinates θ, ϕ of an epicentre Q (say),

231

with the corresponding coordinates θ', ϕ' of an observing station O (say). Let A, B, C, D, E, G, H, K be defined by

$$A = \sin\theta\cos\phi, \quad B = \sin\theta\sin\phi, \quad C = \cos\theta; \tag{1}$$

$$D = \sin\phi, \qquad E = -\cos\phi; \tag{2}$$

$$G = \cos\theta\cos\phi, \quad H = \cos\theta\sin\phi, \quad K = -\sin\theta; \tag{3}$$

and let A', B', ..., be corresponding constants for the station O. (A, B, C are direction-cosines of the line joining the Earth's centre to Q; etc.)

Tables giving A', B' and C' for many observatories were computed by the *International Seismological Summary*; these tables further give the heights of the observatories above mean sea-level, and above the surface of the standard (spherical) model Earth defined in §10.6.3.

We now introduce the *epicentral distance* Δ of O from Q, measured sometimes as the arc-length QO in kilometres, and sometimes (as will be the case in this book) as the angle subtended by QO at the Earth's centre; and also the *azimuth Z* which is the angle (measured from north through east) between the meridian line through Q and the arç QO. It is easily deduced from (1), (2) and (3) that

$$\cos\Delta = AA' + BB' + CC', \tag{4}$$

$$2(1 - \cos\Delta) = (A - A')^2 + (B - B')^2 + (C - C')^2, \tag{5}$$

$$2(1 + \cos\Delta) = (A + A')^2 + (B + B')^2 + (C + C')^2, \tag{6}$$

$$2 + 2\sin\Delta\sin Z = (A' - D)^2 + (B' - E)^2 + C'^2, \tag{7}$$

$$2 + 2\sin\Delta\cos Z = (A' - G)^2 + (B' - H)^2 + (C' - K)^2. \tag{8}$$

From (4), (5) or (6), Δ may be determined from knowledge of the coordinates of Q and O; and then from (7) or (8), Z.

Instead of (4) or (5), the approximate formula for Δ in radians

$$\Delta^2 \approx (\theta - \theta')^2 + (\phi - \phi')^2 \sin^2\tfrac{1}{2}(\theta + \theta') \tag{9}$$

is sometimes used when Δ is small. This gives a maximum error not exceeding about $0.0004\,\Delta^3$ degrees when Δ is not too great, provided Q and O are not within about $20°$ of the Earth's north or south pole. The formula (9) is sufficiently accurate for most work on near earthquakes, the error being ordinarily less than 1 km if $\Delta < 6°.5$.

10.3 Features of seismograms

Following the occurrence of an earthquake, the seismogram traced at an observing station is usually rather complicated in appearance.

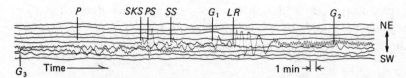

Fig. 10.1. Seismogram of a large earthquake, magnitude 6.8, in the Solomon Islands with $\Delta = 88°$ obtained on the Berkeley ultra-long-period NE component instrument (see Fig. 9.3). The Love pulses G1, G2 and G3 are marked. The paths are the direct and anticentre routes, respectively.

Fig. 10.1 is a copy of a seismogram traced at Berkeley by the north-east component of waves from a distant earthquake. The earlier part of the record shows the arrival of body elastic waves that have travelled by various routes from the earthquake source through the Earth's interior; the later part of the record is formed predominantly by surface waves which travel more slowly than the body waves. Except close to the epicentre, the surface waves usually have greater amplitudes than the body waves because of the slower rate of amplitude diminution with distance (see §§ 3.5, 5.6). (An exception occurs (see §§ 3.2.5, 15.3.2) when the focus is abnormally deep, for then the surface waves are abnormally small.) When the earthquake is a strong one, a seismogram at a distant station reveals the arrival of waves over a long period of time (sometimes many hours); the term *coda* is used to denote that part of the record which follows the earlier surface waves.

Seismograms show more or less prominent onsets of a number of specific pulses, including those due to the direct P and S waves (whose rays follow paths of least time from the focus to the observatory), and also possibly other pulses connected with waves that have suffered reflection or change of type at major discontinuity surfaces within the Earth, including the outer surface. Such displacements on the seismogram, associated with specific types of ray, mark the *phases* of the seismogram. In practice, the noisiness and complexity of a seismogram often make it difficult or even impossible to discern many of the phases that are expected to be present. Through long effort on countless seismograms, however, empirical travel-time tables have been constructed which give the times for particular families of rays in terms of an earthquake's focal depth and a station's epicentral distance. Derivation of these tables is a basic problem in seismology to which we shall now devote attention. Of special importance, the first P wave has travelled from an earthquake focus to a station by the path of least time and so is recorded by a seismograph that (apart from microseisms – see §9.8.1) is in an undisturbed state. This does not apply to any later pulse, and so it is usually possible to estimate the instant of onset of the first *P* pulse more precisely than with other pulses.

10.4 Estimation of *P* travel-time tables

The construction of the *P* travel-time tables is interlocked with the determination of values of the four main parameters (§10.1) for each earthquake used.

10.4.1 Equations of condition between hypocentre and table parameters. Geiger's and Inglada's methods

Suppose that, following successive trials from crude beginnings, a provisional *P* travel-time table has been constructed. Let *R* be the radius of the Earth stripped of its crust, and let *hR* denote the depth below the crust of an earthquake focus. Suppose that the table gives values $T(h, \Delta)$ of the *P* travel-times in terms of *h* and Δ, and let $\xi(h, \Delta)$ denote the errors in *T*, the correct times being thus $T + \xi$.

For an earthquake used in improving the table, take a tentative focus *F* and origin-time. Let *hR* now apply to this *F*, and take the tentative origin-time as zero. Let Δ apply to the angular distance from *F* to an observing station *O*, and let *Z* be the azimuth of *O* from *F*. (Uncertainty in the position of *O*, which will here be ignored, can be allowed for separately as in §10.2.) The suffix 0 will denote values of *T*, ξ, etc., which correspond to these particular values of *h* and Δ.

Let τ, *zR* and $\delta\Delta$ be the corrections needed to give the accurate origin-time, focal depth and epicentral distance from *O*. Thus

$$\delta\Delta \approx - x \sin Z - y \cos Z, \tag{10}$$

where *x* and *y* are the small east and north angular displacements which *F* needs.

Let *t* be the measured arrival-time of the first *P* onset at *O*, and ε the error in this measurement; *t* is also the 'observed' (O) travel-time of the pulse. The table value T_0 which corresponds to *h* and Δ will be called the 'calculated' (C) travel-time. Let

$$\mu = t - T_0 \tag{11}$$

denote the 'residual' (O − C).

On forming two expressions for the accurate travel-time from the true focus to *O*, we have, by a Taylor expansion, correct to the first order (provided $(\partial T/\partial \Delta)_0$ and $(\partial T/\partial h)_0$ exist),

$$t + \varepsilon - \tau = T_0 + \xi_0 - (x \sin Z + y \cos Z)\left(\frac{\partial T}{\partial \Delta}\right)_0 + z\left(\frac{\partial T}{\partial h}\right)_0, \quad (12)$$

and hence

$$\xi_0 - (x \sin Z + y \cos Z)\left(\frac{\partial T}{\partial \Delta}\right)_0 + z\left(\frac{\partial T}{\partial h}\right)_0 + \tau - \mu = \varepsilon. \quad (13)$$

(13) is an equation of condition between the five parameters x, y, z, τ and ξ_0, the quantities Z, $(\partial T/\partial \Delta)_0$, $(\partial T/\partial h)_0$ and μ being computable from explicit data. Similar equations can be formed for other observing stations and other earthquakes.

The above linearisation was essentially given first by Geiger in 1910 as a way to revise the location of an earthquake focus in the case of an assumed correct travel-time ($\xi_0 = 0$). It should be noted that equation (13) assumes that the angular displacements can be adequately treated as a Cartesian system on a plane tangent to the Earth near the true focus.

Because of the sphericity of the Earth, the algorithm is more efficient if a colatitude convergence factor is included by replacing x by $x \sin \varphi$ in (13). Note also that it may be advantageous in some circumstances to extend the Taylor expansion in (12) to include second-order terms $\partial^2 T/\partial \Delta^2$, $\partial^2 T/\partial h^2$. The first of these is proportional to the amplitude of the P or S wave being used (see (8.4)).

A simple hypocentre location method suitable for small seismographic networks or arrays and local earthquakes is based on the Pythagorean relation (Inglada, 1928) for the ith station,

$$(x_i - x_h)^2 + (y_i - y_h)^2 + (z_i - z_h)^2 = V^2(t_i - t_h)^2, \quad (14)$$

where x, y, and z are the Cartesian east, north and depth distances from a specific origin O and x_h, y_h, z_h are the corresponding hypocentral unknowns, $t_i - t_h$ is the travel-time for the seismic phase of velocity V, assumed constant. For V given, subtraction of successive station pairs yields as linear equations in the unknowns,

$$(x_{i+1} - x_i)x_h + (y_{i+1} - y_i)y_h + (z_{i+1} - z_i)z_h - V^2(t_{i+1} - t_i)t_h$$
$$= \tfrac{1}{2}\{V^2(t_i^2 - t_{i+1}^2) + x_{i+1}^2 + y_{i+1}^2 + z_{i+1}^2 - x_i^2 - y_i^2 - z_i^2\}. \quad (15)$$

The method with P or S readings at n stations yields $n - 1$ equations in 4 unknowns for that phase and up to $2(n - 1)$ overall. The problem may be extended to the case with side conditions when the mean velocity V also requires estimation. Often station elevations are approximately equal so that the coefficient of z_h is small; the focal depth z_h is then best found by putting $z_i = 0$ into (14) and solving directly for z_h^2 after the epicentral coordinates x_h, y_h are computed from (15).

10.4.2 Application of least-squares theory and inverse theory

When equations of the type (13) and (15) are formed for a set of earthquakes and observatories, the parameters of the earthquakes and the corrections for representative values of h and Δ can be estimated by least squares.

Let m be the total number of earthquakes used, n the average number of recording stations, and p the number of parameters needed to determine the P table to satisfactory accuracy. The total number of parameters would then be $4m + p$, and the number of equations of condition like (13) would be mn.

As a preliminary to discussing the use of the least squares method, we shall write the set of equations of condition like (13) in the form

$$\sum_{k=1}^{q} (a_{ik}x_k) - b_i = \varepsilon_i \quad (i = 1, 2, \ldots, r), \tag{16}$$

where $r = mn$, $q = 4m + p$, x_k is a typical one of the $4m + p$ parameters, and a_{ik}, b_i are known quantities. The normal equations corresponding to (14) are derived according to least squares theory,

$$\sum_{i=1}^{r} \left\{ \left(\sum_{k=1}^{q} a_{ik}x_k - b_i \right) a_{ij} \right\} = 0 \quad (j = 1, 2, \ldots, q).$$

In an analogous way to travel-time corrections, many seismological problems in this book reduce to the mathematical problem of solving a set of linear equations like (16) in the matrix form

$$\mathbf{Ax} = \mathbf{b}, \tag{17}$$

where $\mathbf{A}$ is $m \times n$; i.e. there are n parameters $\mathbf{x}$ and m observations $\mathbf{b}$. Examples occur with hypocentral estimation (§ 11.1.3), seismic prospecting (§ 7.6.4), finite elements (§ 12.3.2), density inversion (§ 13.5.3), and all linear least-squares regression (e.g. §§ 15.2.2 and 16.4.2).

Three situations arise. First, $m = n$ and solution is elementary. Secondly, $m > n$, and the principle of least-squares gives the solution

$$\mathbf{x} = (\mathbf{A}^{\mathsf{T}}\mathbf{A})^{-1}\mathbf{A}^{\mathsf{T}}\mathbf{b}, \tag{18}$$

as long as $\det \mathbf{A}^{\mathsf{T}}\mathbf{A} \neq 0$.

Thirdly, $m < n$, and solution is by the generalised or Penrose inverse

$$\mathbf{x} = \mathbf{A}^{\mathsf{T}}(\mathbf{A}\mathbf{A}^{\mathsf{T}})^{-1}\mathbf{b}, \tag{19}$$

as long as $\det \mathbf{A}\mathbf{A}^{\mathsf{T}} \neq 0$. This $\mathbf{x}$ is the solution with least-Euclidean length. Further discussion of the linear algebra involved may be found in texts such as Noble and Daniel (1975).

For data reduction and inverse problems, we need to take the analysis of the spaces defined by the matrices further. The key theorem, due to Eckart and Young, is discussed in Lanczos (1961).

First, an elementary theorem for any real symmetric $(n \times n)$ matrix $\mathbf{A}$ of rank $p \leqslant n$ is that it can be factored like

$$\mathbf{A} = \mathbf{U\Lambda U}^\mathrm{T} \tag{20}$$

where $\mathbf{\Lambda}$ is a diagonal matrix with the p non-vanishing eigenvalues λ_i of $\mathbf{A}$ as elements. The columns of $\mathbf{U}$ are the corresponding eigenvectors.

It is worth noting that from (20) we have

$$\mathbf{U}^\mathrm{T}\mathbf{AU} = \mathbf{\Lambda}. \tag{21}$$

Thus this multiplication transforms $\mathbf{A}$ into a new coordinate system that corresponds to the principal axes (see §2.1.4 and §2.2.2).

The more general and fundamental theorem is that an $m \times n$ matrix $\mathbf{A}$ of rank $p \leqslant n < m$ can be factored such that

$$\mathbf{A} = \mathbf{U\Lambda V}^\mathrm{T}, \tag{22}$$

where $\mathbf{\Lambda}$ is a $p \times p$ diagonal matrix whose diagonal elements are the non-zero *singular values* of $\mathbf{A}$. The p columns of $\mathbf{U}$ are the eigenvectors corresponding to the p non-zero eigenvalues of $\mathbf{AA}^\mathrm{T}$ (size $m \times m$) and the p columns of $\mathbf{V}$ are the eigenvectors corresponding to the p non-zero eigenvalues of $\mathbf{A}^\mathrm{T}\mathbf{A}$ (size $n \times n$) in order. We remark that a singular value of $\mathbf{A}$ is the positive square root of the real and non-negative eigenvalues of $\mathbf{A}^\mathrm{T}\mathbf{A}$ and $\mathbf{AA}^\mathrm{T}$. ($\mathbf{A}$ here is not square and, of course, has no eigenvalues itself.)

The significance of the factorisation (22) is both conceptual and computational. For example, consider the normal equations (18). By the theorem, the symmetric inverse term has the factor $\mathbf{\Lambda}^{-1}$ so that each equation has a multiplier $1/\lambda_i$. Therefore, if the eigenvalues are near zero, the value of $\mathbf{x}$ will become large. But in least-squares inverse problems (see, e.g., equation (13)), $\mathbf{x}$ is often supposed small which entails that convergence will not occur. A suggestion of Levenberg (1944) and Marquardt (1963) is of great help. Replace $\mathbf{\Lambda}$ by $\mathbf{\Lambda} + kI$ where k is a small (damping) constant, say about 0.01. The effect is not as severe as merely replacing the smallest eigenvalues by zero.

Seismological insights to guide experimental work also come from the analysis. A scheme to estimate deep Earth structure can now involve many tens of thousands of observed travel times leading to equations like (16). These can, in principle, be combined with a set of hundreds of linear equations from observations of the eigen-frequencies of the Earth. Because many observations may be close to repetitions, there is much redundancy in the data and many equations may define almost parallel spaces. The singular values of **A** may range greatly in size and the *effective* rank is low. Such analysis, through (22), allows observations to be selected or *winnowed* to improve computational stability, precision, and parameter resolution. A simple indicator of the spread of eigenvalues is the *condition number*, defined usually as the ratio of the greatest to the least singular value (or its logarithm).

The parameter *resolution* in the least-squares case can be more formally assessed by computing the $n \times n$ matrix that spans the n-dimensional parameter space

$$\mathbf{R} = \mathbf{HA}, \tag{23}$$

where

$$\mathbf{H} = (\mathbf{A}^T\mathbf{A} + k\mathbf{I})^{-1}\mathbf{A}^T. \tag{24}$$

Note that when all eigenvalues are present and $k = 0$, $\mathbf{R} = \mathbf{I}$, but otherwise $\mathbf{R} \neq \mathbf{I}$ and the relative size of the elements of **R** indicates the relative resolution of the parameters **x**. Numerical examples for epicentre location are left as computer exercises.

The observation space can be similarly examined by defining the *information density* matrix, size $m \times m$,

$$\mathbf{Q} = \mathbf{AH}. \tag{25}$$

As before when $k \sim 0.01$, $\mathbf{Q} \neq \mathbf{I}$ and the relative size of the elements tells which data supply most information. The expression has been used, for example, by Uhrhammer (1980) to analyse the optimal design of a seismographic network.

Finally, it is convenient here to define in this notation the covariance matrix

$$\mathbf{C} = \mathbf{HH}^T; \tag{26}$$

in the usual terminology, the statistical variances and covariances of the solution **x** are the elements of $\sigma^2\mathbf{C}$, where σ^2 is the variance of one observation (see §11.2.2).

10.4.3 Jeffreys' method of successive approximation.
Summary values

In practice, the number n of recording stations may be as large as about 100, the number p of parameters may be as great as 50, so that the number m of earthquakes must be considerable to achieve robust results.

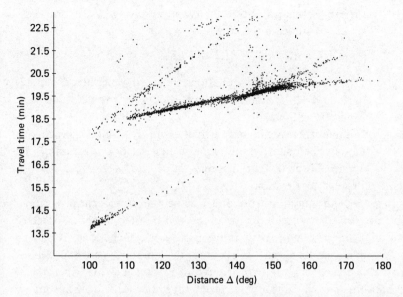

Fig. 10.2. Computer plot of 3275 travel-times (shallow-focus sources) reported to the ISC seismographic stations for a seven-month period with $100° < \Delta < 180°$.

The direct process of reducing the formal solution (16) to a useful numerical result therefore requires the aid of digital computers, which were not available in earlier days when much of the basic work was done. Jeffreys derived an effective iteration process which is still applicable.

As a guide to the scope of the problem, observed travel-times of seismic waves for $100° < \Delta < 180°$ are plotted in Fig. 10.2 as reported from one year (1980) of global earthquakes listed in the *International Seismological Catalogue*. Concentrations of travel-times, corresponding to particular seismic wave types and paths through the Earth, and also the background scatter, are evident. Such statistical distributions are the raw material for the theoretical analysis described below.

For the remainder of this subsection, we drop the term in z in (13) and follow the earlier stages of the procedure as described in §10.4.1. Provided the table corrections ξ_0 are not too great, it can then be shown that the following will be a convergent process:

(i)   Find by least squares from equation (13) first approximations to the x, y, τ for each separate earthquake considered, taking all the other parameters as zero; this is equivalent to ignoring all errors in the provisional travel-time table, and so solving by least squares a set of equations of the type

$$(x \sin Z + y \cos Z)(\partial T/\partial\Delta)_0 - \tau + \mu = 0. \tag{27}$$

(ii)   Using these first approximations to x, y, τ, determine from (13) (putting $\varepsilon = 0$) preliminary estimates of ξ_0 for various values of Δ for each earthquake.

(iii)   Combine these preliminary estimates of ξ_0 by a suitable process (to be described shortly), obtaining a first set of corrections to be applied to the travel-time table.

(iv)   Proceed by iteration.

In practice, the question of deciding whether the errors in the provisional table are sufficiently small for the process to be valid is not a source of serious trouble, because the iterative process is self-checking. It can be shown, moreover, that even when some of the table errors are appreciable, there will be convergence with the use of earthquakes for which the P arrival-times are available at three groups of stations which are at roughly the same epicentral distance and are in widely different azimuths. At an early stage of the approximation, it may be necessary to censor or group the readings of earthquakes; lack of rapid convergence being the criterion.

In treating the preliminary estimates of ξ_0 obtained during the process (ii), it is convenient to arrange these estimates, which we now assume are random variables and denote as ξ, in groups or *bins* corresponding to suitable ranges of values of Δ; the size of a bin depends on the number and consistency of the observations and on the curvature of the travel-time curve in the vicinity.

The process (iii) of finding the best values of the corrections to be applied to the travel-time table is complicated by the distribution of the ξ for a given range of Δ departing from the normal law. Jeffreys has taken this into account by expressing the distribution as a probability function

$$f(\xi) = (1 - s)\pi^{-\frac{1}{2}}h \exp\{-h^2(\xi - \xi_0)^2\} + sg(\xi); \tag{28}$$

in this expression, h and s are constants, the latter being small, and $g(\xi)$ is a function which varies slowly with ξ within a range up to a few times $1/h$ on either side of ξ_0, and whose integral over ξ is unity. The law (28) thus corresponds to an abnormal error occurring with probability s in addition to a normal error distributed about the true value with precision constant h.

The related processes of smoothing, trend elimination, and interpolation still cause difficulties in the treatment of observed distributions such as in

Fig. 10.2. Standard smoothing methods only apply precisely to equally-spaced samples and unequally-spaced data produce unknown fluctuating correlations between the estimated parameters. Thus, in the usual least-squares curve fitting, such as cubic splines, fluctuations in the data density along the time axis introduce varying weights of undetermined magnitude. Further, the smoothed points have, in general, covariances of significant size.

In seismology the most common requirement has been the smoothing of observed travel-times T of seismic waves as a function of distance Δ. Because derivation of the corresponding seismic velocities along the path uses an inverse functional that contains derivatives $dI/d\Delta$, subjective selection of a smoothing process often determines the form of the velocity variation. (Recently, physical parameters like velocity associated with earthquake occurrence have been plotted against time in an attempt to detect trends which might be forerunners of impending major earthquakes (see §15.5). Here, subjectivity in the smoothing procedures can introduce fallacious inferences.)

The method explained below is based on the method of summary values developed by Jeffreys. The method applies to both dense and scarce data in the presence of random errors when no definite functional form is known. In such cases it must be suspected *a priori* that the form of the curve changes radically from one end of the range to the other. In other words, a smoothed value must be obtained from the observations in its immediate neighbourhood rather than over the whole range as is often done with disastrous consequences in primitive least-squares curve fitting. The method of summary values selects a convenient subrange as standard interval. Linear and quadratic curves are then fitted by least squares in each interval. The summary points are the intersections of the two curves which, of course, may always be computed. At each such point it may be shown that the uncertainties of the estimated values are independent. The summary points, therefore, not only give the linear trend in each range, but also take account of the local parabolic curvature.

Consider $n \geqslant 4$ points (x_i, y_i) of equal weight in an interval of the argument. Let the pair of summary points in the interval be (X_1, Y_1) and (X_2, Y_2). The linear least-squares fit to the data points, passing through the summary points, is

$$y = \frac{Y_2(x - X_1)}{(X_2 - X_1)} + \frac{Y_1(X_2 - x)}{(X_2 - X_1)}. \tag{29}$$

The normal equations for the ordinates are

$$
\begin{bmatrix}
\sum \dfrac{(x_i - X_2)^2}{(X_2 - X_1)^2} & -\sum \dfrac{(x_i - X_1)(x_i - X_2)}{(X_2 - X_1)^2} \\[2ex]
-\sum \dfrac{(x_i - X_2)(x_i - X_1)}{(X_2 - X_1)^2} & \sum \dfrac{(x_i - X_1)^2}{(X_2 - X_1)^2}
\end{bmatrix}
\begin{bmatrix} Y_1 \\ Y_2 \end{bmatrix}
$$

$$
=
\begin{bmatrix}
-\sum \dfrac{y_i(x_i - X_2)}{(X_2 - X_1)} \\[2ex]
\sum \dfrac{y_i(x_i - X_1)}{(X_2 - X_1)}
\end{bmatrix}.
\tag{30}
$$

The quadratic least-squares fit to the data points, also passing through the summary points, is

$$
y = \frac{Y_2(x - X_1)}{(X_2 - X_1)} - \frac{Y_1(x - X_2)}{(X_2 - X_1)} + A(x - X_1)(x - X_2).
\tag{31}
$$

The normal equations for the ordinates in this case are

$$
\begin{bmatrix}
\sum \dfrac{(x_i - X_2)^2}{(X_2 - X_1)^2} & -\sum \dfrac{(x_i - X_1)(x_i - X_2)}{(X_2 - X_1)^2} & -\sum \dfrac{(x_i - X_1)(x_i - X_2)^2}{(X_2 - X_1)} \\[2ex]
-\sum \dfrac{(x_i - X_1)(x_i - X_2)}{(X_2 - X_1)^2} & \sum \dfrac{(x_i - X_1)^2}{(X_2 - X_1)^2} & \sum \dfrac{(x_i - X_1)^2(x_i - X_2)}{(X_2 - X_1)} \\[2ex]
-\sum \dfrac{(x_i - X_1)(x_i - X_2)^2}{(X_2 - X_1)} & \sum \dfrac{(x_i - X_1)^2(x_i - X_2)}{(X_2 - X_1)} & \sum (x_i - X_1)^2(x_i - X_2)^2
\end{bmatrix}
$$

$$
\begin{bmatrix} Y_1 \\ Y_2 \\ A \end{bmatrix}
=
\begin{bmatrix}
-\sum \dfrac{y_i(x_i - X_2)}{(X_2 - X_1)} \\[2ex]
\sum \dfrac{y_i(x_i - X_1)}{(X_2 - X_1)} \\[2ex]
\sum y_i(x_i - X_1)(x_i - X_2)
\end{bmatrix}
\tag{32}
$$

It follows immediately, when (30) are substituted into the first two equations of (32), that

$$
A\sum (x_i - X_1)(x_i - X_2)^2 = 0
\tag{33}
$$

and

$$
A\sum (x_i - X_1)^2(x_i - X_2) = 0,
\tag{34}
$$

from which X_1 and X_2 can be determined.

In particular, subtracting (33) and (34) yields

$$
\sum (x_i - X_1)(x_i - X_2) = 0.
\tag{35}
$$

Let $\sum x_i = n\bar{x}$, $\xi = x - \bar{x}$, $\xi_1 = X_1 - \bar{x}$, $\xi_2 = X_2 - \bar{x}$; and put $\sum \xi^2 = n\mu_2$, $\sum \xi^3 = n\mu_3$, summed over n data points.

It then follows by substitution that ξ_1, ξ_2 are the roots of the quadratic

$$t^2 - (\mu_3/\mu_2)t - \mu_2 = 0. \tag{36}$$

Hence we can compute X_1 and X_2 and the corresponding ordinates follow by inverting the matrix in (30).

More generally, each point is associated with a weight $w_i = \sigma^2/\sigma_i^2$, where σ^2 is the variance of a y_i sample value of unit weight. The weights can be introduced in appropriate places in the above equations. In the usual way, if we operate on (30) with the variance operator, we find

$$\text{var } Y_1 = \frac{(X_1 - X_2)^2 \sigma^2}{\sum w_i(x_i - X_2)^2} \tag{37}$$

and

$$\text{var } Y_2 = \frac{(X_1 - X_2)^2 \sigma^2}{\sum w_i(x_i - X_1)^2}. \tag{38}$$

Also, calculation yields $\text{cov}(Y_1, Y_2) = 0$, so that the summary points have the special property that the uncertainties of their ordinates are uncorrelated.

The summary gradient for the interval in question is, from (29),

$$dy/dx = (Y_2 - Y_1)/(X_2 - X_1) \tag{39}$$

and the corresponding abscissa is $x = (X_1 + X_2)/2$.

Because the covariance vanishes, the variance of the gradient is simply $(\text{var } Y_1 + \text{var } Y_2)/(X_2 - X_1)^2$.

Summary value smoothing is also a valuable tool in estimation of the smooth gradients (or slowness curve) of the time series. The argument, apparently previously overlooked, is as follows: the summary linear trend in an interval is the slope of the linear form. However, this slope should be taken not at the midpoint of the sample interval, but midway between the summary points. At this abscissa, the slopes of the linear and the quadratic forms coincide so that at the point the slopes of both the linear trend and the curvature are summarised.

10.4.4 Uniform reduction and robust estimation

Jeffreys has derived a useful method for determining the parameters in (28). When the number of observations is large enough, it is sufficiently accurate to subtract, from each frequency total in the distribution of travel-time table corrections ξ, a constant number just sufficient to isolate by zeros a central frequency group, to which normal error theory may be applied. This method is called *the method of uniform reduction*, and has been successfully applied in seismology.

For example, the following table shows values of ξ obtained from a set of Japanese earthquakes for $0° < \Delta < 20°$; the first row gives ξ in seconds, and the second row the frequencies. By inspection, a uniform reduction of 2 may be made to each entry in the second row, giving the third row,

-9	-8	-7	-6	-5	-4	-3	-2	-1	0	1	2	3	4	5	6	7	8	9
0	2	2	1	3	4	7	6	13	22	14	12	11	4	3	2	0	2	1
0	0	0	0	1	2	5	4	11	20	12	10	9	2	1	0	0	0	0

Jeffreys evolved a method of ascertaining weights to be attached to the preliminary ξ, giving modified ξ to which normal error theory may be applied to determine the best estimates of the ξ_0 and the precision constant h.

When the best estimates of table corrections ξ_0 are obtained in this way for the centres of the bins, it is finally necessary to apply suitable smoothing. This may be carried out by the method of summary values, but with special attention to places where the slope of the (T, Δ) curve changes abruptly.

For more complete details of the various methods introduced by Jeffreys in this section of seismology, the reader is referred to *Theory of probability* by Jeffreys (Oxford, Clarendon Press, 1961).

The more general problem, common in seismology when dealing with travel-times, surface-wave dispersion, spectrograms, and so on, is to find a background of scattered large residuals on the flanks of the central cluster (from multiple branches, higher modes, erratic time control, errors in picking coherent onsets, etc.). We require 'safe' estimates of the coefficients of the summary curves, i.e. parameter values that are not seriously dependent on the abnormal values. This basic statistical problem has led to the development of *robust estimates*. These aim at a high performance even with very diverse sampling. (The mode and median, for example, are often more robust estimates of central tendency than the mean.)

Robust techniques can be illustrated in polynomial regression. Consider the set of n linear equations for the nth degree polynomial in x,

$$y = x\beta + \varepsilon. \tag{40}$$

In general, the estimation of β is done by minimising a function ρ of the residuals ε_i

$$\sum_{i=1}^{m} \rho\left(\frac{y_i - x_i\beta}{S}\right), \tag{41}$$

where S is the measure of scale.

Some proposed forms of ρ are

(i) $\rho(\varepsilon) = \varepsilon^2/2$ (Gauss)

(ii) $\rho(\varepsilon) = |\varepsilon|$ (Least absolute residuals)

(iii) $\rho(\varepsilon) = \begin{cases} \frac{1}{2}\varepsilon^2, & \text{for } |\varepsilon| < c \\ c|\varepsilon| - \frac{1}{2}c^2, & |\varepsilon| > c \end{cases}$

(iv) $\rho(\varepsilon) = \begin{cases} |\varepsilon|\left(1 - \dfrac{\varepsilon^2}{c^2}\right)^2, & |\varepsilon| < c \\ 0, & \text{else.} \end{cases}$

The forms (iii) and (iv) have an outer and an inner part and hence may be thought of as the result of multiplication by a weight factor in two parts – the *biweight*. In a numerical estimation, the scale S may be an assumed standard error or, for safety using a computer in robust estimation from scattered data, the interquartile range. The biweights $\omega_i(\varepsilon)$ are then defined at each stage of the iteration from the residuals $\varepsilon_i = y_i - x_i\beta$ from the previous fit. In each iteration a multiplicative weight factor such as

$$w_i = \left(1 - \frac{\varepsilon_i^2}{c^2}\right)^2, \tag{42}$$

is used in the least-squares inversion where the cut-off c is selected by trial, say. In a less formalised way, the form (iii) above is close to Jeffreys' method of uniform reduction, given earlier to deal with outlying residuals in constructing travel-time curves. Further, in the automatic iterative method for locating earthquake hypocentres (still used in essentials by the International Seismological Centre) the scheme outlined above was found valuable (see Bolt, 1960, pp. 434–5).

10.4.5 Regional variations and focal depths

An important question is how far the assumption of spherical symmetry in the Earth is reliable. The two main sources of deviation are the Earth's ellipticity of figure (to be considered separately in §10.7), and differences in subsurface structure in different geographic regions. The latter make it desirable first to group together earthquakes with epicentres in particular tectonic regions and later combine the results for the whole Earth.

When allowance is made for ellipticity, for epicentral distances exceeding 20° the travel-times are still not quite independent of the locations of epicentres and observing stations. The greatest systematic differences are between the times for paths under the Pacific and under continental regions for distances of the order of 50°. In continental regions, there are some differences between shield and mountainous regions. With P waves for $\Delta > 20°$, the time differences do not appear to exceed 2s or so in travel-times up to 10 min, although the effects may be more marked in observed slownesses $dT/d\Delta$. The lateral variations in structure responsible for the differences are likely (see § 13.1) to be largely confined to the outermost 700 km of the Earth. For all these variations, apart from its ellipticity, the Earth is, broadly speaking, close to spherical symmetry, and the theory of chapters 7 and 8 is relevant.

Further refinements include the investigation of systematic errors in arrival-time data from particular observing stations. These errors may be due to local geological structure, the height of a station above mean sea level, poor time service, microseisms, or other cause. The application of statistical weights to the data from different stations, taken in conjunction with knowledge of the types of seismographs, leads to a higher precision.

In order to take account of layering in the crust, recourse is had to theory of § 7.3.5, which treats the case of a velocity jump downwards across a single discontinuity surface concentric with the outer surface. For two such surfaces the (T, Δ) curve takes the form shown in Fig. 10.3, and various travel-time tables have assumed this model. Geographic regional variations are, however, specially marked in the crust and contribute substantially to uncertainties in the tables.

Towards examining effects of focal depth, let E and F be the epicentre and focus of an earthquake, and P an observing station at distance Δ, where Δ is sufficiently great for the ray EP to penetrate below the crust. Let v denote the P velocity at depth z below the Earth's surface, let $z = f$ at F, and let H be the crustal thickness. Let T be the travel-time along the ray FP, and $T + \delta T$ along EP. (cf. Fig. 15.1).

When $f < H$, i.e. when the focus is inside the crust, the dependence of T on f is best indicated by formulae developed in § 7.6.1 (where h is equal to

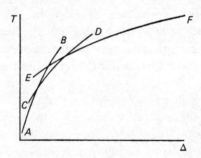

Fig. 10.3. Travel-time curve for a three-layered crust.

the present f). For comparison with the case $f > H$, it is useful to note that, to a first approximation in which squares of f/R are neglected, it can be shown that

$$\delta T \approx \int_0^f \phi(v, \Delta)dz, \tag{43}$$

where

$$\phi(v, \Delta) = \{v^{-2} - (dT/Rd\Delta)^2\}^{\frac{1}{2}}.$$

When F is below the crust, let $f = H + hR$ (h being here defined as in § 10.4.1). For this case, it can be shown that

$$\delta T \approx \int_0^H \phi(v, \Delta)dz + hR\phi(v_0, \Delta), \tag{44}$$

where v_0 is the P velocity just below the crust. (Similar theory of course also applies to S.) The integral in (44) (like that in (43)), ranges only inside the crust, and its total fluctuation for varying Δ is found to be less than 2 s. On the other hand, the term in h, through its dependence on $dT/d\Delta$, can vary markedly with Δ. For example, for $h = 0.05$ (see Table 10.1), $\delta T = 9, 36, 41$ s when $\Delta = 10°, 50°, 100°$, respectively; for $h = 0.10$, the corresponding δT are 8, 60, 71 s.

Thus significant depth of an earthquake focus can be readily detected from the fact that the residuals μ (of § 10.4.1), taken against table-times for a surface focus, show systematically increasing negative values as Δ increases. This result provides one of the principal methods of estimating focal depths.

Another complication in setting up the tables arises from the fact that in many earthquakes the principal P movement at the focus is rather small compared with the principal S movement. Thus a wave may ascend from the focus to the outer surface in the S type, and after the reflection there (near the epicentre) the reflected P wave may be recorded at a station some distance away as the first P movement. The significance of this phenomenon was first noticed by Stechschulte and Scrase, who introduced the notation sP to denote the corresponding phase on the seismogram. The notations sS, pP, pS have analogous meanings; thus, pP corresponds to a wave which ascends from the focus to the outer surface in the P type, and then after reflection travels in the P type to the recording station.

When the phases P and pP (or S and sS, etc.) are both recorded, the differences between their arrival-times for various Δ provide a further important means of estimating focal depths.

We finally mention the possibility of using amplitude theory to check various features of the tables. For instance, we have pointed out (§ 8.3) that the occurrence of abnormal amplitúdes for a range of values of Δ may assist in assessing corresponding values of $d^2T/d\Delta^2$. Amplitude data need to be used, however, with caution, since amplitudes are very dependent on conditions near an observatory site, and, further, are sensitive to small changes in d^2v/dz^2 and even d^3v/dz^3 (see §8.6).

10.5 Use of digital computers. Tomography

The iterative character of the foregoing procedures makes them eminently suited to the use of electronic computers.

Bolt, for example, in 1960 devised a program based on (13)(with the term in ξ_0 omitted) whereby successive approximations to trial origin-times, epicentres and focal depths are derived at high speed. The program was adapted in 1961 for use in preparing the *International Seismological Summary* (§11.2.1) and later was incorporated into the algorithm used by the *International Seismological Centre*.

A recent related seismological development dependent upon high-speed computation is *tomography*. Long in use in nuclear medicine and radio astronomy, it refers to the numerical reconstruction of an image of internal structure from data expressed as a functional (line integrals or Radon transform) of the function (parameter) to be imaged. It is a case if inverse theory applied, say, to the use of travel-time data to define Earth structure. In its simplest form it was used (not by this name) by R.D. Oldham to infer the Earth's core and generalised recently for computer algorithms, to the block perturbation method of Aki, Lee and others. (See also §7.4).

As an illustration, consider two-dimensional seismic cross-hole exploration with seismic sources S_i (e.g. explosions) at equal intervals down one vertical hole and receivers R_i (geophones) at similar intervals down an offset parallel hole. Suppose that the area between the holes is mathematically mapped onto a square area, subdivided into a grid of square elements e_{ij}, within which the (constant) velocity is α_{ij}. Now consider a seismic ray from S_i to R_j, say. (As a first iteration, for small velocity changes between elements, Snell's law is often ignored.) Then the travel time along ray S_3R_2, for example, where S_3 is on the side of e_{31} and R_2 on e_{23} is

$$T = d_{31}/\alpha_{31} + d_{21}/\alpha_{21} + d_{22}/\alpha_{22} + d_{23}/\alpha_{23}, \tag{45}$$

where d_{ij} are the ray segments in corresponding elements.

For many rays, (45) becomes

$$\mathbf{Dx} = \mathbf{T},$$

where $x_{ij} = 1/\alpha_{ij}$ is the slowness. $\mathbf{D}$ is an $m \times n$ matrix where m is the number of rays and n is the number of square elements. Because T is observed, x_{ij} can be mapped by inverting D. Interesting problems of resolution and convergence arise in practice because matrices may be large and ill-conditioned and some elements poorly sampled by rays. Refraction at the boundaries should be included using ray tracing (see §7.1.1).

Nevertheless, in the form outlined above, tomographic algebraic reconstruction beautifully illustrates the frustrations and dilemmas of seismological exploration of the Earth's deep interior. Although nowadays inversion with m of order 10^4 for P and S rays is available from earthquake catalogues, the above problems and lack of uniqueness theorems indicate caution before transferring tomographic solutions to the actual Earth. The Radon transform is discussed by S.R. Deans (1983).

10.6 Travel-time tables other than P

As pointed out in §10.3, the times of onset on seismograms are in general more precisely determined for P than for other phases. In consequence, the P travel-time tables are the best determined and are ordinarily used in origin-time and epicentral determinations. The process of constructing travel-times for other phases is therefore not usually complicated by having to take account of uncertainties in origin-times and epicentres.

Nevertheless, there is the possible complication that the main S wave may leave a source at a different instant from the main P wave, so that the origin-time found for P may not necessarily be the same as that for S. This complication has been referred to as the 'Z phenomenon'. (Actually the Z phenomenon is less important than it was at one stage thought to be.)

The greater uncertainties in the tables other than the P are accentuated by difficulties in identifying onsets after the first P. An apparently prominent onset is sometimes the result of a fortuitous reinforcement of a number of relatively unimportant ground movements. Also it is often difficult to decide which phase is associated with a particular onset among the genuine ones to be expected. Thus in practice there is inevitable misidentification of a proportion of onsets after the first. Efforts are made to reduce misidentifications both by statistical treatment of data and by instrumental procedures.

There are, however, other factors which help to reduce the table uncertainties. For instance, among the different significant earthquake phases observed, the results for one phase can often be used to check and improve the results for another phase; instances of the interdependence of different travel-time tables are given in §10.6.2. Again, readings of deep-focus earthquakes can be used with advantage to improve certain of the tables.

10.6.1 Notation used for phases read on seismograms

We have already referred to the phases P, S, and the group pP, pS, sP, sS. Reflection at the ocean bottom is pwP, etc.

Phases corresponding to waves that have suffered reflection at the Earth's outer surface (or possibly at one of the crustal discontinuity surfaces), the wave initially leaving the focus in a direction away from the outer surface (in contrast to pP, etc.), are denoted as PP, PPP, SS, SSS, PS, SP, PPS, etc. For instance, PS corresponds to a wave which is of P type before the reflection and of S type afterwards. In addition, there are phases such as pPP, sPP, sPS, etc. A few cases are illustrated in Fig. 10.4.

Further important phases are associated with the presence of a discontinuity surface which occurs at a depth of about 2900 km below the outer surface, separating the 'central core' of the Earth from the 'mantle' (see chapter 13). The symbol c is used to indicate an upward reflection at this discontinuity. Thus if a P wave travels down from a focus to the discontinuity surface in question, the upward reflection into the S type is

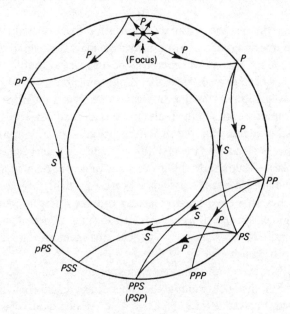

Fig. 10.4. Notation for mantle rays.

recorded at an observing station as the phase *PcS*; and similarly with *PcP*, *ScS*, *ScP*. A phase such as *ScSP* would correspond to the reflection into the *P* type, at (or near) the outer surface, of a wave that had before reflection approached the outer surface as *ScS*. Phases such as *pScS* are also significant, the symbol *p* corresponding as before to an initial ascent to the outer surface in the *P* type. Dif *P* and dif *S* are *P* and *S* waves diffracted around the Earth's core.

The symbol *K* is used to denote the part (of *P* type) of the path of a wave which may lie inside the central core. Thus the phase *PKS* corresponds to a wave that starts in the *P* type, is refracted into the central core into the *P* type, and refracted back into the mantle in the *S* type in which it finally emerges. Phases such as *PKKP*, etc., correspond to waves that have suffered an internal reflection at the boundary of the central core. Some cases of phases corresponding to waves that have penetrated into the central core are indicated in Figs. 10.5, 10.6. In cases where the travel-time is a two-valued function of the distance (see Fig. 7.8(c)) the suffix 2 is sometimes used for the upper branch; this is the case, for instance, with *PKP*, the phase corresponding to the upper branch being denoted as *PKP*$_2$. (For more precise details concerning *PKP*, see § 10.9.2).

Lehmann's discovery of the existence of a further discontinuity surface inside the central core (see chapter 13) made it necessary to introduce further basic symbols. For paths of waves inside the central core, the symbols *i* and *I* are used analogously to *c* and *K* for the whole Earth; thus *i* indicates reflection upwards at the boundary between the outer and inner portions of the central core, and *I* corresponds to the part (of *P* type) of the path of a wave which lies inside the inner portion. Thus, for instance, discrimination now needs to be made between the phases *PKP*, *PKiKP* and

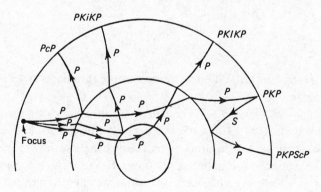

Fig. 10.5. Notation for core rays.

PKIKP; the first of these corresponds to a wave that has entered the outer and not reached the inner portion of the central core, the second to one that has been reflected upwards at the boundary between the two portions, and the third to one that has penetrated into the inner portion.

Similar lower-case notation has been introduced by Bolt to define waves reflected from secondary discontinuities. Thus *PdP* is the *P* wave reflected from a discontinuity depth *d* km in the upper part of the Earth, and *PKhKP* is a *PKP* wave reflected from a discontinuity around the inner core.

By combining the symbols *p, s, P, S, c, K, i, I, d, h* in various ways, we can set down notation for all the main phases associated with body earthquake waves. The symbol *J* has been introduced to correspond to *S* waves in the inner core, following evidence (§ 13.3) on the solidity of the inner core. (The symbol *Z* was once proposed for *S* waves in the outer core, but there is now strong evidence (§ 13.3) that the outer core is fluid.)

10.6.2 Relations between different travel-time tables

We now show some connections between travel-times for different phases.

A very simple instance is the connection between the travel-times of *P*, *PP*, *PPP*, etc. Thus, if we consider foci at the outer surface and assume that reflections take place at the outer surface, the travel-time T_{PPP} for *PPP* for a given Δ is given by

$$T_{PPP}(\Delta) = 3T_P(\tfrac{1}{3}\Delta),\tag{46}$$

where T_P is taken from the *P* travel-time table.

A case such as *PS* is dealt with by using the fact that the parameter *p* of a seismic ray is by (6.1) and (7.1) the same before and after a reflection. Hence by (7.2) the travel-time T_{PS} for a given distance Δ is given by

$$\left.\begin{array}{c} T_{PS}(\Delta) = T_P(\Delta_1) + T_S(\Delta_2), \\[2mm] \Delta_1 + \Delta_2 = \Delta \quad \text{and} \quad dT_P/d\Delta_1 = dT_S/d\Delta_2. \end{array}\right\}\tag{47}$$

where

Use of the last of these equations involves matching gradients in the *P* and *S* travel-time tables.

The equations (46) and (47) are actually used in preparing theoretical travel-time tables for *PPP* and *PS*; and similarly with *PP*, *SS*, *PPS*, etc. We note, however, that there are discrepancies between these theoretical tables and the indications of seismogram readings; the latter show considerable scatter. This is to be expected, since reflections are liable to occur at any of the discontinuities in the Earth's upper layers, and such reflections will

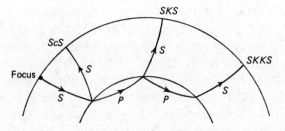

Fig. 10.6. The S, ScS, and SKS family of rays.

(depending on the angles of incidence, etc.) sometimes carry more energy than a wave reflected right from the outer surface.

Another instance of a relation between travel-times is that for the phases ScS, SKS and $SKKS$, which incidentally has proved useful in evolving tables for them. Thus, if $T_{ScS}(\Delta)$ be the travel-time for ScS for the distance Δ, etc., we have, neglecting focal depth,

$$T_{ScS}(\Delta_1) + T_{SKKS}(\Delta_3) = 2T_{SKS}(\Delta_2),$$

where

$$\Delta_1 + \Delta_3 = 2\Delta_2$$

and

$$dT_{ScS}/d\Delta_1 = dT_{SKKS}/d\Delta_3 = dT_{SKS}/d\Delta_2. \tag{48}$$

The derivation of (48) is easily seen from a study of Fig. 10.6.
It is clear that many other relations similar to those exhibited can be set down.

10.6.3 Types of travel-time tables for body waves

Useful instrumental records of earthquakes began to accumulate towards the end of the nineteenth century, and the beginning of the present century saw the introduction of travel-time tables by Oldham. The first widely used tables were constructed by Zöppritz in 1907. Oldham in 1906 drew attention to the existence of the central core, and in 1914 Gutenberg prepared a comprehensive set of tables, including tables for a number of phases corresponding to waves that penetrate into the central core.

The Zöppritz tables were later adapted by Turner for use in the *International Seismological Summary* for earthquakes occurring between 1918 and 1929.

In 1930, Jeffreys, starting from the Zöppritz–Turner tables, inaugurated a series of successive approximations towards improved tables, and was joined by K.E. Bullen in 1931. In 1935 the first Jeffreys–Bullen tables were produced. Substantial refinements were incorporated in a new set of 'JB' tables first published in 1940. These tables give travel-times for the phases *P*, *S*, *pP*, *sS*, *PP*, *PPP*, *PS*, *SP*, *PPS*, *SPP*, *PSP*, *SS*, *SSP*, *PSS*, *SPS*, *SSS*, *PcP*, *PcS*, *ScP*, *ScS*, *ScSP*, *PKP* (sometimes denoted *P'*), *PKS*, *SKP*, *SKS*, *PKKP*, *PKKS*, *SKKP*, *SKKS*, *SKKKS*, *PcPPKP*, *PcSPKP*, *ScSPKP*, *SKSP*, *PKPPKP*, *PKPPKS*. Auxiliary tables are given for *sP*, *sPKP*, *sPcP*, *sPKS*, *sSKP* and *pPKP*. Tables are included for *K*, *KIK* and I, corresponding to paths within the central core. Compatible tables for near-earthquake phases are also included. The tables are in a form which enables focal depth to be readily taken into account.

The JB tables relate to a standard model Earth in which each surface of equal *P* (or *S*) velocity in the interior is spherical and encloses the same volume as the corresponding surface of the actual Earth. The aim of the tables is to serve as a standard for the 'average' global earthquake.

The next step is to allow for regional geographic deviations from the average. Auxiliary tables allowing for the Earth's ellipticity of figure are already available (see §§ 10.7, 10.9.3), and progress, as yet far from final, has been made in assembling allowances for further regional differences.

Travel-time tables and charts have been independently constructed by a number of other seismologists (see § 10.9.2). Many tables are special to intensive studies of the seismograms of relatively small sets of earthquakes. Such tables have, however, directed attention to special points of interest; for example, Byerly's tables for the Montana earthquake of 1925 June 28 gave the first indication of peculiarities in the *P* travel-times near 20° (see chapter 13). The contemporary tables of Gutenberg and Richter were in the main in good accord with the JB tables.

With the large increase of first-class seismological observatories since World War II, there has been a big increase in reliable arrival-time data. The application of nuclear explosions to seismic problems has further provided a quantity of valuable data directly on travel-times (see chapter 1). The newer data showed the JB tables needed some correction.

An important recent set of comparative tables was constructed in 1968 for *P* and associated phases by Herrin and others and for *PKP* by Bolt. More precise revisions are a major part of contemporary seismological research.

10.7 Effect of the Earth's ellipticity

In 1933, Comrie, and Gutenberg and Richter, pointed out that the use of geocentric instead of geographic latitudes (which had always been used previously) would reduce errors in travel-time tables due to neglect of the Earth's ellipticity. (The *geographic latitude* of a place is the angle between the normal to the level surface at the place and the plane of the equator; the *geocentric latitude* is the angle between the radius vector from the Earth's centre to the place and the plane of the equator.)

We introduce spherical polar coordinates (r, θ, ϕ), the origin being at the Earth's centre, and take $\theta = 0$ at the focus of a particular earthquake. To begin with we take the focus to be at the outer surface, and we consider the travel-time along any ray. For the model Earth defined in §10.6.3, we have, using notation in chapter 7,

$$T = \int ds/v,$$

$$\left(\frac{ds}{d\theta}\right)^2 = \left(\frac{dr}{d\theta}\right)^2 + r^2 + r^2 \sin^2 \theta \left(\frac{d\phi}{d\theta}\right)^2.$$

For the actual Earth, we write correspondingly

$$T' = \int ds'/v,$$

$$\left(\frac{ds'}{d\theta}\right)^2 = \left(\frac{dr'}{d\theta}\right)^2 + r'^2 + r'^2 \sin^2 \theta \left(\frac{d\phi'}{d\theta}\right)^2,$$

and we assume v to be known as a function of r. (This will be the case, to sufficient accuracy, using travel-times as determined neglecting ellipticity and using the theory of §7.4.)

Since for the model we have $d\phi/d\theta = 0$ along a ray, it follows that for the actual Earth $d\phi'/d\theta$ is small. Hence, writing

$$\delta r = r' - r, \quad \delta T = T' - T,$$

we have, correct to the first order of small quantities,

$$\delta T = \int_0^\Delta v^{-1} \left\{ \left(\frac{dr'}{d\theta}\right)^2 + r'^2 \right\}^{\frac{1}{2}} d\theta - \int_0^\Delta v^{-1} \left\{ \left(\frac{dr}{d\theta}\right)^2 + r^2 \right\}^{\frac{1}{2}} d\theta$$

$$= \int_0^\Delta \tfrac{1}{2} v^{-1} \left\{ \left(\frac{dr}{d\theta}\right)^2 + r^2 \right\}^{-\frac{1}{2}} \left\{ 2 \frac{dr}{d\theta} \frac{d\delta r}{d\theta} + 2r\delta r \right\} d\theta$$

$$= p \int_0^\Delta r^{-2} \left(\frac{dr}{d\theta} \frac{d\delta r}{d\theta} + r\delta r \right) d\theta,$$

where in obtaining the last line, use has been made of the relation (7.15). This gives the formula obtained by Jeffreys, namely,

$$\delta T = p \left[r^{-2} \frac{dr}{d\theta} \delta r \right]_0^\Delta + p \int_0^\Delta \left(\frac{d^2 r^{-1}}{d\theta^2} + r^{-1} \right) \delta r \, d\theta. \tag{49}$$

The first term in (49) is readily seen (using (7.1) and the relation $|rd\theta/dr| = \tan i$, see Fig. 7.3) to be equal to $[v^{-1}\delta r \cos i]_{0,\Delta}$; this represents the time along the extra lengths at the two ends of the actual ray projecting outside the outer boundary of the model (such an extra length and time will of course be negative if δr is negative). The second term involves an integration along the whole length of the ray. On expressing the second term in the form

$$-p \int_0^\Delta (\rho \sin^3 i)^{-1} \delta r d\theta$$

(see (7.71)), where ρ is the radius of curvature of the ray, we see that if for a particular ray δr everywhere has the same sign, this second term will be opposite in sign to the first (assuming, which is sufficiently near to the actual circumstances, that ρ is always positive). This result is to be expected from the fact that where δr is positive the internal layers of higher velocities are displaced outwards slightly, and so cause an increase in velocity along the corresponding part of the actual ray, above that pertaining to the Earth model.

The equation of a surface of equal velocity within the actual Earth is given to sufficient accuracy by

$$r' = r\{1 + \varepsilon(\tfrac{1}{3} - \cos^2 \theta')\}, \tag{50}$$

where ε is the ellipticity of the surface, and θ' denotes the geocentric colatitude of a point. Hence

$$\delta r = \varepsilon r(\tfrac{1}{3} - \cos^2 \theta'). \tag{51}$$

By (49) and (51), it is possible, using formulae in chapter 7, to express the ellipticity correction in the form

$$\delta T = \varepsilon_0 [(\tfrac{1}{3} - \cos^2 \theta')(\eta^2 - p^2)^{\frac{1}{2}}]_{0,\Delta} + p^{-1} \int_0^\Delta \varepsilon(\tfrac{1}{3} - \cos^2 \theta')\eta^3 \frac{dv}{dr} d\theta, \tag{52}$$

where ε_0 is the ellipticity of the Earth's outer surface.

In chapter 13 we shall indicate a method of estimating values of ε throughout the Earth. Hence, using (52), it is possible to construct tables giving values of the ellipticity correction δT to be applied for any particular ray.

The above theory assumes the focus to be at the outer surface of the Earth. The ellipticity corrections are small, however, and it has been shown that the tables for surface-focus earthquakes give the corrections to sufficient accuracy for all deep-focus earthquakes as well.

10.8 Travel-times of surface waves

Seismic surface waves with periods up to the order of a minute, unlike body waves, are largely confined to the vicinity of the Earth's outer surface. The records of surface waves are thus much influenced by crustal structures and may vary with the azimuth from the epicentre. In addition they show marked dispersion and several modes may be present (see chapter 5). Hence in most respects they do not lend themselves to the construction of global travel-time tables as in the case of body waves, and require special surface ray tracing.

In routine observatory work, it has been the custom to denote the onset on a seismogram of the first surface waves (recognised by their longer periods) by the symbol *L*.

Stoneley showed that the frequency distributions of travel-times of *L* for given ranges of Δ are strongly double-humped, and pointed out that the two peaks would correspond to the onsets of Love and Rayleigh waves, respectively. The onsets are denoted as *LQ* and *LR* (*Q* for *Querwellen*, an alternative name for Love waves) and Stoneley found mean travel-rates to be about 4.43 and 3.97 km/s, respectively.

10.9 Numerical results

We now give a number of typical travel-times from the Jeffreys–Bullen tables. For more complete details, the reader is referred to Jeffreys and Bullen, *Seismological Tables* (Brit. Assoc. for Adv. of Science, Gray-Milne Trust, 1958). In § 10.9.3 an indication will be given of the ellipticity corrections; and in Fig. 10.7, of the times for *LQ* and *LR*.

10.9.1 The Jeffreys–Bullen seismological tables

The JB tables apply to the model Earth as defined in § 10.6.3. They are given in graphical form in Fig. 10.7.

The *P* tables give the travel-time *T* in terms of the epicentral distance Δ for $0° \leqslant \Delta \leqslant 105°$ for fourteen different focal depths, corresponding respectively to foci at the level of the outer surface and at the levels $h = 0.00, 0.01, \ldots, 0.12$, where *h* denotes fractions of the distance between the base of the crustal layers and the centre of the model. Typical results are shown in Table 10.1.

There is a complication in the vicinity of $\Delta = 20°$, due to the (T, Δ) relation being of the form indicated in Fig. 7.5(*c*). The main *P* tables apply to the branch giving the smallest value of *T* for a given Δ, separate tables being

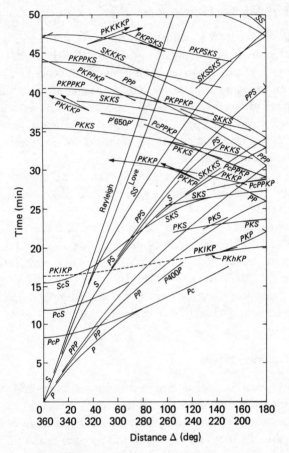

Fig. 10.7. Jeffreys–Bullen travel-time curves for surface focus (1940), with additions.

included for the other branches; the notation P_d, P_u, P_r is used for phases corresponding to the three branches $A''B''$, $C''B''$, $C''D''$, respectively (the letters d, u, r being initial letters of the words *direct, upper* and *refracted*). Auxiliary tables are given for special use with short epicentral distances, and also for foci at specific levels within the crustal layers.

Similar remarks apply to the S tables (including the same complication near $\Delta = 20°$).

The P and S tables are supplemented by tables giving the excesses of the travel-times of pP over P and of sS over S. For the case $pP - P$, typical results are shown in Table 10.2.

Table 10.1 *P travel-times*

Δ	Surface focus min s	$h = 0.00$ min s	$h = 0.05$ min s	$h = 0.10$ min s
10°	2 28.0	2 24.4	2 18.8	2 19.9
50°	8 58.0	8 53.1	8 21.6	7 58.1
100°	13 48.4	13 43.1	13 7.1	12 37.3

Table 10.2 *pP − P travel-times*

Δ	$h = 0.00$ s	$h = 0.05$ min s	$h = 0.10$ min s
30°	10	1 6	—
50°	10	1 12	1 58
100°	11	1 23	2 22

These tables are useful in the study of deep-focus earthquakes.

Tables for other phases are given in detail for a surface focus, and there are auxiliary tables giving sufficiently accurate allowances for focal depth.

With phases such as the group *PSS, SSP, SPS*, the travel-times for a surface focus are, of course, the same in each case, but separate tables for the focal depth allowances have to be used for *PSS* on the one hand, and *SSP, SPS* on the other.

Table 10.3 gives extracts from the tables for certain other important phases, corresponding to a surface focus (focal depth allowances are needed as with *P*).

We note incidentally from the *PKP* table that a disturbance will travel from a surface focus *F* to the *anticentre* (the point on the opposite side of the Earth cut by the diameter through *F*) in a little more than 20 min.

The two values in the above table for *SKS* at Δ = 100° correspond to two branches in the travel-time curve. These two branches are due to the excess of the *P* velocity near the outer boundary of the central core over the *S* velocity near the base of the mantle (see chapter 13), and the circumstances are as in Fig. 7.7(c). One branch runs from about Δ = 62° to Δ = 133°, and the other from about Δ = 99° to Δ = 180°. The two branches meet near Δ = 133°, so that there is only one *SKS* phase for 133° ≤ Δ ≤ 180°. The

Table 10.3 *Additional travel-times*

Δ	PcP min s	ScS min s	PKP min s	PKP$_2$ min s	SKS min s
0°	8 34.3	15 35.7	—	—	—
10°	8 39.0	15 44.6	—	—	—
50°	10 18.3	18 47.8	—	—	—
100°	13 48.5	25 20.7	—	—	⎰24 27.0
					⎱25 24.9
143°	—	—	19 33.5	19 33.5	26 43.9
180°	—	—	20 12.2	22 10.6	27 13.5

first branch crosses the main S curve at $\Delta = 83°$; when $\Delta > 83°$, SKS precedes S.

10.9.2 Tables for *PKP*

For reference, the basic form of the PKP travel-time curve is drawn in Fig. 10.8. The travel times from the DF branch $(PKIKP)$ are now known within 1 s and for AB within 2 s, but the crucial location of the cusps B, D and C are still not very exact. For the caustic B, Jeffreys (1939) gives $\Delta_B = 143.0°$; Qamar (1973) has $\Delta_B = 144°$. For D, Δ_B ranges from $110°$ (Jeffreys) to $121°$ (Choy and Cormier, 1983). The cusp C is, like B, in a portion of the time–distance diagram where branches overlap. Diffraction (head waves), not accounted for by ray theory, also sets in near these cusps.

The travel-time curve for the type PKP is particularly complicated in the vicinity of $\Delta = 143°$ where there is a cusp of the type indicated in the curve of Fig. 7.8(c). This cusp is a consequence of a sudden diminution in the P velocity across the boundary of the central core. The upper branch, AB, corresponding to refraction, runs from $\Delta = 143°$ to $\Delta \geqslant 180°$, and is the one commonly denoted as PKP_2. The lower branch, BC, runs continuously only from $\Delta = 143°$ to about $\Delta = 155°$; this is because of the velocity jump between the outer and inner parts of the central core (see chapter 13), which results in there being another branch, DF, corresponding to refraction through the inner core which runs to $\Delta = 180°$.

The concave upward branch CD (running back to even shorter distances) corresponds to reflection from the inner core. Additional discontinuities just above the inner core boundary would give rise to additional branches between DB and Y in Fig. 10.8, but most precursor observations are now believed to arise in another way (see § 13.6.4).

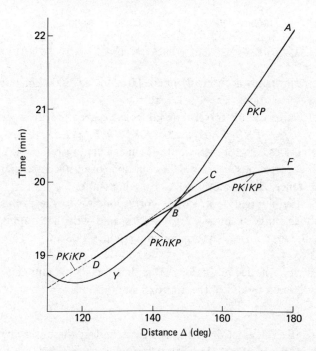

Fig. 10.8. The *PKP* travel-time curve with branches *AB* and *BC* (outer-core transit only), and *DF* (*PKIKP* rays). Waves reflected from the inner core *PKiKP*, corresponding to the dashed receding branch *CD* with a continuation back to $\Delta = 0$ beyond the cusp *D*. The light line *Y* is the locus of theoretical minimum times of arrival of PKP rays scattered near the mantle-core boundary (Haddon and Cleary 1974).

There are similar complications with a number of other phases. But the above cases illustrate the main features.

10.9.3 Ellipticity tables

The following table gives values of the ellipticities of internal strata of equal (*P* or *S*) velocity within the Earth as calculated by the method to be indicated in § 13.7. In this table, *d* denotes the depth in kilometres below the Earth's outer surface and ε the ellipticity; values are also given for η, where

$$\eta = \frac{d(\ln \varepsilon)}{d(\ln r)}, \tag{53}$$

where *r* denotes the distance from the Earth's centre (η as here defined is, of course, distinct from the η we have used in previous sections). This parameter η is useful in a number of problems in geophysics.

d	0	1000	2000	3000	4000	5000	6000
η	0.56	0.48	0.46	0.16	0.21	?	0.00
ε	0.00337	0.00309	0.00280	0.00257	0.00242	0.0021	0.0021

The values of ε are probably accurate within 0.000 02 for $d \leqslant 3000$ km, but may have errors up to about 0.0004 below this level.

Ellipticity corrections to JB travel-time tables have been prepared for the phases P, S, PKP, PKS, SKP, SKS, $SKKS$, PcP, PcS, ScP, ScS, corresponding to the use of geocentric latitudes in calculating values of Δ. The tables give the corrections in terms of the colatitude of an epicentre, and the epicentral distance and azimuth of an observing station.

For P waves, the ellipticity corrections corresponding to the use of geocentric latitudes range from -0.9 to $+1.0$ s, and with geographic latitudes from -1.7 to $+2.7$ s. The corresponding ranges for S are -1.7 to $+1.7$ and -3.3 to $+5.1$ s.

When geocentric latitudes are used, the application of the ellipticity corrections is facilitated by use of the approximate formula

$$\delta T = f(\Delta)(h_0 + h_1), \tag{54}$$

where h_0, h_1 are the values of δr at the epicentre and at the observing station, and $f(\Delta)$ is a function of Δ alone given by auxiliary tables (δT and δr are as defined in §10.7). The values of δr range from -14 km at the Earth's poles to $+7$ km at the equator. In the case of P, $f(\Delta)$ ranges from zero at $\Delta = 0°$ to 0.07 s/km at $\Delta = 105°$; in the case of S, from zero to 0.13 s/km.

Formulae of the form (54) may also be used in the case of the following phases, the numbers in brackets giving the range of variation of $f(\Delta)$ in s/km: PKP (0.07 to 0.10); SKS and $SKKS$ (0.10 to 0.14); PcP (0.05 to 0.07); PcS and ScP (0.08); ScS (0.09 to 0.13).

For PSK and SKP, the forms

$$\delta T = f(\Delta)(h_0 + h_1 + \tfrac{1}{4}h_1), \tag{55}$$

$$\delta T = f(\Delta)(h_0 + h_1 + \tfrac{1}{4}h_0), \tag{56}$$

respectively, are necessary, with $f(\Delta)$ varying from 0.08 to 0.11 s/km.

The maximum errors in the ellipticity corrections obtained using the approximate formulae (54), (55) and (56) are within 0.3 s for P, and within 0.5 s for most of the other phases. More accurate formulae, suitable for computer programs, can be found in Bullen (1939) and Dziewonski and Gilbert (1976).

In earlier work in seismology, geographic latitudes were always used in calculations of epicentral distances, and earlier tables giving the constants A', B', C' (§10.2) for the world's observatories were computed using geographic latitudes. Geocentric latitudes are now regularly used in seismic calculations.

A point of interest, noted by Jeffreys, is that seismological data alone enable the ellipticity of the Earth's outer surface to be estimated within an uncertainty of one-sixth.

10.9.4 Statistical treatment of velocity and travel-time table estimation

Optimal experimental design and estimation of uncertainties for observations on a sphere are not simple. We give only a few sample procedures (see Uhrhammer, 1979).

The inversion integral (7.40) can be expressed as a sum,

$$r_n = \exp\left\{-\frac{1}{\pi}\sum_{i=1}^{n} w_i \cosh^{-1}\left(\frac{p_i}{p_n}\right)\right\}, \tag{57}$$

where w_i is a weighting function which depends upon the numerical integration procedure, $p_i \equiv p(\Delta_i)$ and $p_n \equiv p(\Delta_n)$. Given the variance of $p(\Delta)$, the variance of r_n can be estimated using the standard method for estimating the approximate variance of an arbitrary function

$$\text{var}(r_n) = \left(\frac{\partial r_n}{\partial p_1}\right)^2 \text{var}(p_1) + \cdots + \left(\frac{\partial r_n}{\partial p_n}\right)^2 \text{var}(p_n). \tag{58}$$

In equation (58) the covariance terms were neglected and the formulation assumes that the first derivative terms in the Taylor expansion about the population means are dominant. Computing the partial derivative terms and substituting v_n for r_n/p_n, we obtain

$$\text{var}(r_n) = \frac{v_n^2}{\pi^2}\sum_{i=1}^{n-1}\frac{w_i^2}{p_i^2 - p_n^2}\{p_n^2 \text{ var}(p_i) + p_i^2 \text{ var}(p_n)\}. \tag{59}$$

The variance of v_n is

$$\text{var}(v_n) = \left(\frac{1}{p_n}\right)^2 \text{var}(r_n) + \left(\frac{-r_n}{p_n^2}\right)^2 \text{var}(p_n) \tag{60}$$

and the assumptions are the same as for equation (58). The second term in the brackets on the right side of equation (59) is the dominant one for most depth ranges.

In order to check the selection of earthquake sources and stations in terms of geographical location, a test of randomness of a distribution of points on a sphere has been developed. Suppose that the distribution of unit vectors on a sphere follows the probability distribution of Fisher,

$$C\exp(\kappa \cos\theta), \tag{61}$$

where θ is the angle between a unit vector and a reference direction (to the north pole, say) and κ is a precision constant.

We wish to test the hypothesis that $\kappa = 0$ which implies that (61) is constant; this is the case when the geographical distribution is random (with respect to the reference point).

Let R be the length of the resultant of N randomly directed unit vectors

$$R^2 = (\sum l_i)^2 + (\sum m_i)^2 + (\sum n_i)^2, \tag{62}$$

where l, m and n are direction cosines. R is approximately distributed as

$$(N\chi^2_{\alpha,3}/3)^{\frac{1}{2}}, \tag{63}$$

where $\chi^2_{\alpha,3}$ is a chi-squared variate, α is the percentage significance point, and there are three degrees of freedom.

In the usual way, to define a significance test, for each N we find a value of $R(= R_0)$ that will be exceeded, on the hypothesis of randomness, with a specified probability α (i.e. $P(R \geqslant R_0) = \alpha$). If the observed value of R exceeds R_0, the hypothesis of randomness is rejected. On the other hand, if R is less than R_0, the hypothesis of randomness stands.

For a sphere, the significance point of R is given by

$$R_0^2 = N\chi^2_{\alpha,3}/3 = NF_{\alpha,3,\infty}, \tag{64}$$

where $F_{\alpha,3,\infty}$ is Fisher's distribution with three degrees of freedom. Suppose that the distribution of unit vectors, with respect to the reference point, does not cover the entire sphere, but a spherical cap of angle Δ; then we can express R_0 approximately in terms of Δ.

$$R_0^2(\Delta) = N \cdot [N + \tfrac{1}{2}(1 - \cos \Delta)(F - N)]. \tag{65}$$

The two extremes are $R_0(\pi) = (NF)^{\frac{1}{2}}$ and $R_0(0) = N$. Then R_0 can be easily computed from (65) with $\alpha = 0.05$ and $F_{\alpha,3,\infty} = 2.60$, for any values of N and Δ.

There are two remaining vital questions, azimuthal and radial density of the observing stations about the epicentres. Theoretically, the azimuthal station density D_i would be constant if the stations were randomly distributed in azimuth with respect to the epicentres. A convenient test for randomness, in this case, would be to group the observations into $30°$ wide azimuth cells, say, and use the χ^2 test statistic to test the hypothesis that the stations are randomly distributed in azimuth, where

$$\chi^2 = \sum_{i=1}^{k} (O_i - E_i)^2/E_i; \tag{66}$$

O_i is the observed number in cell i and E_i is the expected number (total number (n) divided by the number of cells (k)). The hypothesis is rejected if

$$\chi^2 > \chi_0^2 = \chi_{\alpha,(k-1)}^2 \tag{67}$$

for $(k-1)$ degrees of freedom at the α per cent confidence level.

Weighting by surface density can be used to reduce the effect of non-random surface distribution of the stations or epicentres (which would lead to non-random sampling of the interior). Suppose there are N stations located at coordinates (ϕ_i, λ_i) on the Earth's surface. Define the density at a station i on the surface as

$$D_i = \sum_{j=1}^{N} \cos\frac{\theta_{ij}}{2}, \tag{68}$$

where θ_{ij} is the angular distance between stations i and j.

The normalized weight w_i for the station i is defined as

$$w_i = \frac{D_{\min}}{D_i}, \tag{69}$$

where $D_{\min}$ is the minimum density (of all the D_is). The range of w_i is $0 < w_i \leqslant 1$.

The above method of weighting may be applied to the azimuthal distribution of observing stations with respect to the epicentres and vice versa. The weight of each individual observation is the normalised product of the weights. A problem is to determine if the weights w_i are significantly correlated from the correlation coefficient matrix given by equation (26).

10.10 Exercises

1 A deep-focus earthquake occurred near the Tonga Islands and good readings of pP as well as P were observed at the following stations:

Station	Distance Away ($\Delta°$)	pP minus P times
Berkeley	65°	65 s
Tucson	80°	67 s
St Louis	90°	70 s

Using ray diagrams and the velocity distribution Table A.1, estimate the depth of focus of the earthquake from the $pP - P$ interval.

2 At an epicentral distance of 153°, the travel time of the surface focus PKP wave is 20 min 10.0 s and the gradient $dt/d\Delta$ is 4.2 s/deg. For the same ray parameter p, the corresponding travel-time of PcP is 11 min 30 s to a distance of 68°.

Sketch these two rays from a surface focus in the Earth and also the ray $PKKP$ with the same ray parameter. Calculate, using the constancy property for p, that the travel-time of this $PKKP$ wave is 28 min 50 s at $\Delta = 238°$.

3   At $\Delta = 60°$, the surface focus ray PcP has a travel-time of 10 min 56.6 s and a ray parameter $dT/d\Delta = 4.0$ s/deg. The ScS ray with the same parameter travels to a distance of $\Delta = 25°$ with a time of 16 min 28.9 s.

Calculate the travel-times and distances of emergence at the surface of the corresponding PcS and $PcPPcP$ rays (i.e. with the same $dT/d\Delta$).

4   For shallow-focus earthquakes, t_{pP-P} is a function of the P wave velocity (α) in the crust, the distance to the station (Δ) and the depth (h) of the hypocentre. Show that

$$t_{pP-P} = \left(\frac{\partial t}{\partial h}\right)_\Delta h = \frac{2Rh}{\alpha} \bigg/ [R^2 - \alpha^2 p_\alpha^2(\Delta)]^{\frac{1}{2}},$$

where $p_\alpha(\Delta)$ is the slowness of the ray ($p_\alpha(\Delta) = dt(\Delta)/d\Delta$) and R is the Earth's radius. Solve for h, and, assuming $\alpha = 6.6$ km/s, find h using the slowness values from the JB P tables for $50 < \Delta < 55°$.

5   To first order in $\delta r/r$, prove that the difference between travel-times on an oblate spheroid and a sphere is

$$\delta t_\varepsilon = \sum_{i=1}^{N} \varepsilon_i(\tfrac{1}{3} - \cos^2 \theta_i) \cdot (\eta_i^2 - p^2)^{\frac{1}{2}} |_{\eta_i^-}^{\eta_i^+}$$

$$+ \frac{1}{p} \int_0^\Delta \varepsilon(\tfrac{1}{3} - \cos^2{}^h)\eta^3 \frac{dv}{dr} d\theta$$

where the summation term is due to the ellipticity of first-order discontinuities and the integral term is due to the ellipticity of internal regions of equal velocity: ε is the layer ellipticity, η is a model parameter ($r/v(r)$), dv/dr is the velocity gradient, p is the ray slowness ($dT/d\Delta$) and θ is the colatitude.

For surface ellipticity $\varepsilon_0 = 0.00337$, compute the ellipticity adjustments for S and P individually and then prove that the range of the ellipticity adjustment for t_{S-P} is $-1.0 + 0.8$ s.

11

The seismological observatory

Because the greater part of the data used by the seismologist comes from analysis of seismograms obtained at observatories, we include here a short account of the organisation of observatory work.

A typical earthquake observatory has three main functions, (i) the recording of local ground motion due to distant and near earthquakes, (ii) storage and interpretation of seismograms, and (iii) data dissemination. In the following sections we comment on the equipment and routine procedure in observatories, and on the process of compilation of data on local and world-wide earthquakes.

11.1 Inside the observatory

A well-equipped observatory has seismographs that record the vertical and two horizontal (usually the NS and EW) components of the local ground motion, and has several instruments for each component.

The ground motions which can be measured cover a wide dynamic range of both amplitude and frequency, with periods ranging from fractions of a second to the order of an hour. For simple ground motion of the form $u = a \cos pt$, equations such as (9.15) show that the response of a particular seismograph varies with the period $2\pi/p$ of the motion. Hence it is desirable to have a variety of seismographs each concerned primarily with a different section of the spectrum. Seismographs are classified into 'short', 'intermediate' and 'long', according as the effective instrumental periods $2\pi/\omega$ are between about 0.1 and 2 s, 2 and 15 s, and 15 s and 1 min, respectively. Recently 'ultra-long' period instruments capable of measuring ground-motion periods in excess of a minute have been developed. A seismogram of a large teleseism ($\Delta = 88°$) recorded at the University of California, Berkeley, Seismographic Station on a very long-period seismograph is shown in Fig. 10.1. Short-period seismographs are mainly used for recording near earthquakes, where the predominant periods are normally less than a second. Long-period seismographs are particularly valuable for more distant earthquakes which generally contain surface waves with periods in excess of 10 s. Many observatories, such as World Wide Standard Stations (WWSSN), have only short- and long-period seismographs; but more and more of the better observatories operate instruments with at least three widely separated periods.

When the observatory is in a seismically active region, it usually has in addition a strong-motion seismograph (§ 9.9.3), of low sensitivity, which will not be put out of action by a strong local earthquake.

A limited number of observatories contain special facilities which are used to investigate less routine aspects of earthquakes sometimes associated with earthquake prediction programmes. Special instruments such as tilt and strain seismographs may have their signals telemetered from sites removed from the observatory proper (see § 9.9.4).

The careful selection of the site for a seismological observatory is important. It is desirable that seismographs should be set up on a hard rock foundation in an area remote from cultural activities since extraneous local 'noise' such as tilt is more readily introduced through soft ground. Some observatories are seriously troubled with microseisms (see § 9.8.1), which may mask important parts of the record of an earthquake. When the microseisms have a prevailing period, it is possible to reduce their effects by appropriate filter systems.

It is evident (see § 9.7.1) that a highly accurate time service is necessary in conjunction with the running of the seismographs. Methods of indicating the time on a seismogram are to have the pen raised from the paper, or the light shut off or displaced in the case of photographic recording, at the sixtieth second of each minute. Specially formatted time codes are now used on many seismograms, particularly digital recordings. The error in the clock giving these signals is recorded at regular intervals of a day or less. Accuracy to 0.01 s in recording seismograms is now commonly sought, and accuracy to 0.1 s is required in the reading of most body wave phases.

Since the times of occurrence of natural earthquakes are not known in advance, most seismographs are in operation continuously or fitted with trigger devices, perhaps with 'pre-event' memories. In visual recording the rate of rotation of the drum is generally between 0.8 and 6 cm/min; the faster rates are desirable in near-earthquake recording.

Seismograms and magnetic tapes are removed from the transports when complete, developed in the case of photographic recording, and read directly or by analogue playback devices. Various measuring devices are used in the process of reading, which involves noting the amplitude and time of each onset on the seismogram that is judged to be significant. The development of this judgment involves long experience. After the readings have been made, travel-time tables, and charts and diagrams derived from the tables, are used to identify phases (see Fig. 10.7). Usually there remain unidentified phases which should be reported because they may later prove to be important in seismological research.

It is imperative that travel-time tables, or devices derived from them, should not be decisive until after the seismograms have been read. Prior to the production of the JB tables, this was not a serious matter because the earlier tables contained so many large errors that they could be used only as a general guide; in fact the unidentified readings of earlier days, when subjected to appropriate statistics, provided some of the most valuable data used in evolving the JB tables. The reliability of modern tables, however, sometimes tempts a reader of seismograms to select onsets near the expected arrival-times of particular pulses and to disregard equally prominent onsets arriving at other times.

The result is lack of objectivity which can impede further improvements to the tables. A clear example of this type of data censorship can be seen in Fig. 10.2 where most readings on the DF branch of PKP begin at $\Delta = 110°$. This distance has no theoretical justification but happens to define the initial value given in the JB tables for this phase.

The readings when made are transmitted to various centres (see §11.2) which correlate the readings of particular earthquakes at different stations, and make estimates of epicentres, origin-times, focal depths, magnitudes and other features of the earthquakes. In the transmitted details, the symbols i and e (corresponding to *impetus* and *emersio*) are commonly prefixed before the times for individual phases, indicating respectively a sharp and a gradual onset.

11.1.1 Interpretation of seismograms

An experienced reader of seismograms will usually be able to decide quickly the epicentral distance and size of a particular earthquake and whether it is deep-focus. A deep-focus earthquake is detected by the (usually) relatively small amplitudes of the surface waves, by observations of phases such as pP, sS, $sScS$ (see §15.3.2), and by marked deviations in the arrival-times of particular pulses from those for a shallow earthquake (see tables in §10.9.1).

We have already mentioned that the identification of particular phases on a seismogram requires some skill. Moreover, the appearance of a seismogram depends to a considerable extent on characteristics of the recording seismograph; for example, the phases PcP, ScS are best detected at shorter distances by short-period vertical component seismographs incapable of recording long surface waves.

There are a number of guiding principles which help in reading seismograms of shallow earthquakes. The following selection of results derived from travel-time and amplitude data and based on experience is of help in this connection:

(i)   Except close to the epicentre, the first movement for distances up to 105° corresponds to the main P phase; movements corresponding to the main S phase are fairly large up to about 100°.

(ii)   If $\Delta > 83°$, the phase SKS precedes S; but if $\Delta < 100°$, SKS is weaker than S.

(iii) PS, SP, PPS, PSP, SPP cannot appear unless $\Delta > 40°$; but if $\Delta > 80°$, PS and PSP are stronger than P and S on long-period instruments.

(iv)   For $105° < \Delta < 142°$, the first prominent phase is PP, and the second is PPP or PKS (depending on the particular value of Δ); PS, PSP are large; SKS is weaker than $SKKS$.

(v)   For $130° < \Delta < 140°$, PKS is usually the largest phase in the earlier part of the seismogram; after PKS, a fairly large number of phases arrive at close intervals and close comparison with travel-time tables is usually necessary. SKP is stronger for deeper earthquakes and arrives first.

(vi)   Near $\Delta = 142°$, PKP appears strongly; for $143° < \Delta < 180°$, the first phases correspond to the branches of PKP.

(vii) Near $\Delta = 160°$, it is frequently difficult to pick out distinct phases, the motion often appearing very complicated but without clear signs of new onsets; SS and sometimes SSS may be large.

(viii) Near $\Delta = 180°$, there is an appreciable interval between movements due to body waves and those due to surface waves, and the seismogram may have the appearance of being a record of two separate disturbances.

Errors that are frequently made in the reading of seismograms are:

(i)   the identification of PP and PS as P and S when Δ is about 115–120°, leading to a false estimate of about 80° for Δ;

(ii)   the identification of SKS as S when $\Delta > 83°$;

(iii) the pP phase is sometimes mistaken for the first arrival (P) of a very-deep-focus earthquake when there is a weak P phase;

(iv)   the PcP phase is sometimes mistaken for the first arrival (P) for earthquakes of 20–50° distance;

(v)   multiple core phases are sometimes mistaken for first arrivals of a second earthquake;

(vi)   care must be taken to identify correctly *Pn*, *P**, *Pg*, *Sn*, *S**, and *Sg* for local earthquakes. The identification is very dependent on the type, magnitude, and location of the earthquake as well as the type of recording instrument. In some cases *Pn* and *Sn* are not discernible (see § 17.4).

(vii)   Failure to identify small precursors to certain reflected phases (e.g. *PKPdPKP*);

(viii)   Use of Love and Rayleigh wave arrivals in determining distance when the surface waves have travelled a mixed oceanic–continental path. The distance will be too great assuming an oceanic path.

(ix)   Unidentified phases are often not listed. They may be valuable for research.

Recording and interpretation of surface waves is also an important part of observatory work. Empirical average dispersion curves for fundamental Love and Rayleigh waves and their overtones (see §§ 5.4, 5.5) have been constructed for various wave paths (see Fig. 11.1) and these are used,

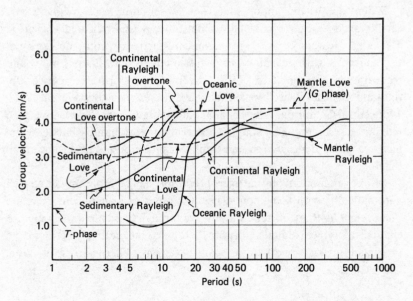

Fig. 11.1. Group velocity dispersion curves as a function of period for Love and Rayleigh waves. (After curves prepared by J. Oliver and colleagues.)

together with the theoretical results of chapter 5, to identify recorded surface waves (see Exercise 11.3.1).

11.1.2 Determination of hypocentres and earthquake size

At some observatories it is the custom to make provisional estimates of the epicentres, etc., of the more important earthquakes. These estimates give early information locally about particular earthquakes, and they serve as first approximations in the calculations subsequently made by the large coordinating centres.

In the case of a single observatory, an earthquake's epicentre can often be estimated from the readings of three perpendicular component seismograms. For example, for a shallow earthquake the epicentral distance, if less than 105°, is indicated by the interval between the arrival-times of P and S; the azimuth and angle of emergence are indicated by a comparison of the sizes and directions of the first movements shown in the seismograms and by the relative sizes of later phases, particularly surface waves. It should be noted, however, that in certain regions the first movement at a station arrives from a direction differing from the azimuth towards the epicentre. The explanation is usually in terms of anomalous structures under the station (see chapter 12).

When data from more than one observatory are available, an earthquake's epicentre may be estimated from the indicated epicentral distances. Nowadays, in many seismically active regions, networks of seismographs with telemetry transmission and centralised timing and recording are common (see §9.9.5). The first major permanent network of this kind was established by the University of California, Berkeley, in 1960–1 for central California. Whether analogue or digital recording is used such integrated systems greatly simplify observatory work: multi-channel signal displays make identification and timing of phase onsets easier and more reliable.

Moreover, modern on-line microprocessors can be programmed to pick automatically, with some confidence, the onset of a significant common phase, such as P, by correlation of wave forms from the parallel network channels. A typical digital algorithm for a phase picker first generates from the signal time-series, $y(i)$, from separate stations a characteristic function, CF, such as

$$CF(i) = |y(i)| \tag{1}$$

The program then compares the current value, CV(i), of $y(i)$ to the maximum expected value of the CF(i) in the absence of a seismic onset.

This predicted value, PV, is a running time average of the signal (such as the standard deviation of $y(i)$) averaged over a suitable time window. If CV/PV becomes greater than a given threshold at a given number of network stations the arrival of a significant onset is registered.

Such an automatic system is often best operated jointly with an experienced seismogram reader to minimise false alarms and to include phases later than the first arrival. The system can be further programmed in the microprocessor to provide automatically magnitudes and hypocentre locations for special purposes (see § 15.3.4). Earthquakes below magnitude 3, often referred to as microearthquakes, may now be mapped by such means with the use of dense telemetry networks of short-period seismographs.

Details on the applications of microearthquake networks are given by Lee and Stewart (1981).

11.1.3 Group estimation of earthquake parameters

Observatories with access to signals from a regional network can now greatly improve the precision of hypocentres, magnitude, and source parameters by the use of simultaneous group estimation. The method is a special application in analysis of variance in which arrival-time or amplitude measurements at network stations for a group of earthquakes in a region are pooled. The statistical advantage of such group reduction is that a sample of measurements is accumulated for each station. The mean and variance allow detection of incorrect readings and provide stable estimates of the station arrival times.

The major application has been to group or *joint* hypocentre location. Results have been shown to be more reliable estimates of the relative adjustments between individual epicentres in the group of earthquakes selected. Systematic displacements of the group as a whole are minimised by the imposition of side conditions. The simplest proposal to obtain realistic side constraints is to include in the simultaneous group reduction the observed times from an earthquake with a prior well-determined location. For convenience, the adopted parameters of this *master earthquake* may be held fixed.

Appropriate computer algorithms have been tested in a number of circumstances. Theoretically they involve an extension of the matrices given in § 10.4.2 to include all the equations of condition. For m earthquakes and n stations in the network, the problem is one of determining up to $4m + n + l$ unknowns, where l is the number of crustal parameters to be adjusted (e.g. seismic velocities). There are $m \times 2n$

observations if both *P* and *S* are read at each network station. Such problems are soluble in the least-squares sense.

For some tectonically active regions, such as parts of Japan, New Zealand, California, Nevada, and Alaska, suitable well-determined hypocentres from earthquakes or explosions are readily available as master events. In other regions, such as the South Pacific and South America, tradeoffs between precision and detectability require calibration with temporary installations of mobile seismographs.

11.1.4 Abnormal observations. The *T*-phase. Precursors

Not all detection and identification of seismic onsets at seismographic stations are routine. Occasionally, more unusual phases are recorded which, if recognised, provide valuable clues on source location and properties or structure of the Earth.

An illustration that occurs at observatories near ocean margins is a high-frequency wave, called the *T*-phase, which propagates mainly in a low-velocity layer, called the SOFAR channel, in the ocean, where the wave speed is about 1.5 km/s. The *T*-phase was first noticed by D. Linehan in 1940 and has been recorded, for example, in California from Hawaiian earthquakes (see Fig. 11.2).

Other phases often propagating strongly across continental paths are called *Lg* with velocity about 3.50 km/s, *Rg* with velocity 3.05 km/s, and *Sa* with velocity 4.4 km/s. All appear to be associated with Airy phases of higher modes of surface waves (see § 5.5.1). The prominent *Lg* phase in the eastern United States is the basis for an important magnitude scale (see § 15.2.1). Another illustration is the long-period *PL* wave that sometimes develops after the *P* onset due to coupling in the crust of *P* and locally dispersed *S* waves.

A few examples of unusual phases that provide important constraints on deep interior structure are the high-angle reflection *PKiKP* from the

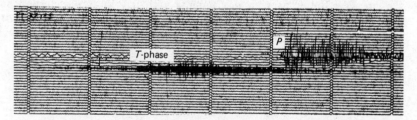

Fig. 11.2. Vertical-component seismic wave train followed by *T*-phase from Fiji earthquake recorded at Pt Reyes Station, California, in 1963.

Earth's inner core (see § 13.5.4), the precursors $PKhKP$ (see § 13.6.4), the multiple reflection $PmKP$ on the inside of the boundary between core and mantle (see § 13.6.3) and the precursor phase $P'640P'$ reflected from the bottom of an upper mantle discontinuity at 640 km depth (see Fig. 11.2 and § 13.6.1). It is even likely that a few observatories have recorded $PKIIKP$, the phase singly reflected *inside* the inner core (see § 13.6.5). Some detail on these phases is provided in the descriptive book by Bolt (1982).

11.2 International seismological organisations

From 1841 onwards, reports on earthquake investigations appeared intermittently in the General Reports of the British Association, and from 1881 were prepared regularly by Milne. Milne subsequently (at Shide in the Isle of Wight) produced the 'Shide circulars' which dealt with earthquakes from 1899. The earliest known list of instrumentally recorded earthquakes with computed origin-times and epicentres is that for the period 1899–1903.

The International Association of Seismology decided in 1906 to establish an international station at Strasbourg. After difficulties caused through World War I, a Central Bureau was re-established at Strasbourg (now in France) under the direction of E. Rothé, and arrangements were made for the preparation and publication at Oxford of the *International Seismological Summary (ISS)* in succession to the Shide circulars and British Association Reports. The first *ISS*, that for earthquakes of 1918, appeared in 1923.

In the 1970s, the activity at Strasbourg was extended to a European and Mediterranean Seismological Centre with support from European countries. The objectives were the rapid determination of regional hypocentres, rapid data exchange and publication of a regional bulletin.

International seismological services were also provided at an early stage by other organisations. Specially valuable were the services of the United States Coast and Geodetic Survey which from readings by telegram from a world-wide selection of stations issued, within a month of the occurrence, *Preliminary Determination of Epicenter* (PDE) lists, giving rapidly estimated origin-times and focal locations of all larger earthquakes and published global earthquake bulletins. Since 1973 this service has been carried on by the United States Geological Survey through its National Earthquake Information Service (*NEIS*) in Colorado.

In addition, various national and regional centres control networks of stations and act as intermediaries between individual stations and the international organisations. Examples of longstanding national centres are the Japan Meteorological Agency, and the Canadian Seismograph Network operated by the Department of Energy, Mines and Resources, Ottawa. These centres normally undertake the estimation of magnitude, epicentres, origin-times and focal depths of local earthquakes. By long-standing tradition, individual stations and regional centres make their readings, and copies of their seismograms where desired, available to stations and research centres throughout the world. This is the raw material from which much of the knowledge of the Earth's interior ultimately emerges.

11.2.1 International seismological catalogues

From 1918 to 1963, the *ISS* was the most comprehensive publication on earthquake occurrence. By use of methods described in chapter 10, the *ISS* in this period determined origin-times and epicentres for all earthquakes sufficiently well read to make determinations possible. For earthquakes after 1953, the growing quantity of data, however, forced a restriction in the determinations to earthquakes of magnitudes 6 and more.

For earthquakes from 1918 to 1929, the Zöppritz–Turner travel-time tables were used in preparing the ISS; from 1930 to 1936, the preliminary JB tables of 1935; and from 1937 onward, the JB tables of 1940. For each earthquake treated, the *P*, *S*, *PKP*, *SKS* and *SKKS* arrival-times together with the residuals (O – C) were given. Various other features have been included in the past. Examples are the very valuable 'additional readings' which were given from 1918 to 1951.

From 1964, the ISS was replaced by the *International Seismological Centre*, now at Newbury, England. It receives about 80 000 readings each month from about 1200 stations worldwide and preliminary estimates of position are supplied by national and regional agencies and observatories for about 1600 earthquakes. The readings are reanalysed and correlation of previously unassociated phases results in the detection of about 200 new earthquakes a month.

The *ISC* publishes monthly, with about a two-year delay, a *Bulletin* giving all available information on each of about 1500 to 2000 earthquakes. Over 600 stations report phases for the largest of these earthquakes. A *Regional Catalogue of Earthquakes* and a *Bibliography of Seismology* are also published.

Continuous surveillance of global seismicity is maintained in the United States by the *NEIS* which receives daily from many U.S. and overseas observatories readings of earthquake phases by telex or telephone. As well, seismic signals are telemetered to *NEIS* in Golden, Colorado from over 60 U.S. stations providing a real-time capability for earthquake location. The *NEIS* routinely calculates and publishes, in a relatively short time, extensive lists of global earthquake locations and other parameters. Each year these earthquake reports include as many as 7000 events. In general, all hypocentres are computed using the 1940 Jeffreys–Bullen tables for P times and the 1968 *PKP* travel-time tables by Bolt. Fault-plane solutions (see chapter 16) are determined where possible for earthquakes with magnitude above 6.5.

11.2.2 Global digital networks

In 1982 the *Directory of World Digital Seismic Stations* listed over 100 digital stations (see Fig. 1.2). The main types are mentioned in § 1.3.2. About thirty of these stations constitute the *Global Digital Seismograph Network* (*GDSN*) which has become a major source of earthquake data for research.

Digitised seismic signals are recorded on tapes at the *GDSN* stations. At weekly or fortnightly intervals, the tapes are mailed to the Albuquerque Seismological Laboratory of the US Geological Survey where the recordings are reviewed and placed on temporary disk files. They are then edited into network-day tapes that contain data from all network stations for an overlapping 26 hour interval. These tapes are copied and distributed through several data centres.

An illustration of the information available in a relatively short time, for at least moderate-sized earthquakes, is given in Fig. 11.3. Wave forms associated with the P and *PKP* phases recorded at 9 *GDSN* stations are computer plotted as a function of azimuth around the earthquake focus. Wave-form analysis can then be made visually or by computer-aided correlation algorithms to estimate properties of the earthquake source and the propagation paths. Such experimental capability opens the way for higher resolution theoretical studies than possible with only information on, say, arrival-times. For further discussion of Fig. 11.3 the reader is referred to Exercise 11.3.2. Such plots, taken from a 1983 Preliminary Determination of Epicenter listing, are produced now routinely (see Sipkin, 1982).

Finally, we remark that the basis of the design of new regional and global digital seismic networks is the concept of *synthesis mapping*, developed in astronomy for very long arrays of radio telescopes. Seismic networks or arrays may be regarded as ensembles of interferometers which provide a

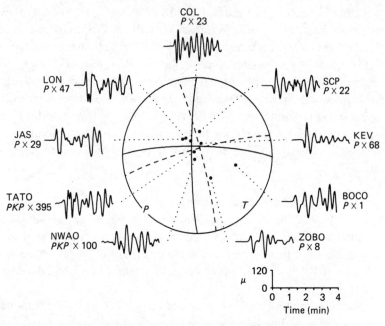

Fig. 11.3. *P* and *PKP* waves recorded at worldwide digital
seismographic stations from an earthquake south of Panama of 19
August 1982. The fault-plane solution with the tension (*T*) and
pressure (*P*) axes is shown. (Courtesy National Earthquake
Information Service.)

synthesis of special response and produce maps of the 'seismic brightness' of
the Earth. Optimal correlator algorithms must be constructed for this
synthesis mapping that yield clean representations of seismic sources and
paths (see § 10.5).

There are two types of networks: *fixed redundancy* and *variable
redundancy* networks. At present, regional stations are at fixed locations
often with a considerable amount of redundancy. The second type of
configuration combines a basic fixed network with movable stations that
are located to maximise resolution and information.

Optimal design of such networks is assisted by the theory of § 10.4.2. For
example, consider the effectiveness of estimation of earthquake location
parameters, given in § 10.4.1, by a specified network of stations. The least-
squares analysis leads to the covariance equation (10.26) and we may define
the real square matrix

$$\mathbf{D} = \sigma^2 \mathbf{C}. \tag{2}$$

The elements d_{ii} of the principal diagonal of $\mathbf{D}$ are estimates of the uncertainty in the unknown earthquake parameters; the off-diagonal elements of $\mathbf{D}$ are used to estimate the linear correlation coefficients ρ_{ij}, where $\rho_{ij} = d_{ij}^2/d_{ii}d_{jj}$, among the unknown parameters. The α per cent confidence interval for a linear correlation coefficient ρ is estimated from

$$\rho^2 \leqslant t^2/(t^2 + n - 2), \tag{3}$$

where $t = t_{(1-\alpha/2,n-2)}$ is the value of Student's distribution at the α per cent probability level with $(n-2)$ degrees of freedom.

We may also consider the length a of the principal axis of the error ellipsoid $\varepsilon^T \mathbf{D} \varepsilon$ (see (2.6)). For data variances equal and independent, from (2), using (10.20), we obtain

$$a^2 = \sigma^2 \sum_{i=1}^{n} \Lambda_{ii}^{-2}. \tag{4}$$

Knowledge of the range of values spanned by the principal axis of the error ellipsoid for the parameters of earthquakes located in the vicinity of the station network is a most useful piece of information for analysing the performance of the network. A quantitative measure is efficiency η in per cent, defined by

$$\eta(\%) = 100a(\text{min})/a(\text{net}) \tag{5}$$

where $a(\text{min})$ is the theoretical minimum length that the principal axis in (4) can attain, and $a(\text{net})$ is the computed length of this axis for the network under consideration. The efficiency η can be contour-plotted, as a function of the coordinates of the epicentre, for a useful representation of how the network under study compares with a theoretically optimal network.

11.3 Exercises

1   Identify the body wave phases on the vertical-component seismogram of the shallow earthquake shown in Fig. 10.1 from the JB travel-time tables (Fig. 10.7). Then use the curves in Fig. 11.1 to identify the various recorded branches of the fundamental Rayleigh waves. Is there evidence of higher modes (overtones) present?

2   The polar wave diagram in Fig. 11.3 shows vertical component wave forms recorded by long-period seismographs at 9 seismographic stations around the world. Computer analysis of these and other data led to publication by the US National Earthquake Information Service of the earthquake parameters listed in Table 11.1. After reading chapters 10, 15, and 16 of this book, interpret the tabulated output and list the algorithms used in the computations.

Table 11.1

Earthquake : South of Panama $M_s = 7.0$ *(BRK)* 19 *Aug* 1982
OT : 15 59 01.5, *Lat* : 6.718°N, *Long* : 82.680°W, *Depth* : 10 *km*
No. of sta. : 199, *Standard error* : 1.2 *s.*

Fault plane solution (deg): *P* Waves
Fault planes   Principal axes

	Strike	Dip	Slip		PLG	AZM
NP1	255	83	355	(T)	1	120
NP2	346	85	187	(P)	9	210

Moderately well-controlled strike-slip fault type mechanism. NP2
designated as the fault plane.

Moment tensor solution (deg): No. of sta. 8, Scale 10^{26} dyne cm
Best double couple   Principal axes

	Strike	Dip	Slip		VAL	PLG	AZM
NP1	271	80	6	(T)	1.80	11	136
NP2	181	84	170	(N)	-0.23	79	331
				(P)	-1.58	3	226

$M_0 = 1.7 \times 10^{26}$ dyne cm.

12

Seismic waves in anomalous structures

Most of the theory described to this stage deals with highly idealised media and boundary conditions. For obvious reasons in an introductory text, we have concentrated upon perfectly elastic, homogeneous and isotropic continua with planar or spherical geometry. We have, however, pointed out to the reader at various places in the text that real media and geophysical applications deviate in various ways from these mathematical simplifications; in particular, we considered (§ 5.7) in some depth the propagation of damped seismic waves in linear visco-elastic media.

In this chapter we explore the modifications to the theory required when restrictions on isotropy, homogeneity, and simple geometry are relaxed. A large body of results is now available on each of these problems in electromagnetic, acoustic, hydrodynamical, as well as seismological contexts. Much of the present research in seismology is aimed at the determination of not only precision measurements of major structure, but also more highly resolved structural detail and variations in physical properties. In the search for these 'second-order' properties, more attention must be given to the effects of physical and structural anomalies on seismic waves. As usual, the direct problems must be solved before the inverse ones can be attacked.

12.1 Anisotropic media

12.1.1 Equation of motion and determinantal conditions

Considerable emphasis has been given recently to the propagation of seismic waves in anisotropic media (see § 2.5). For homogeneous perfectly elastic media, the equations of motion without body force are, from (2.9) and (2.27),

$$\rho \frac{\partial^2 u_i}{\partial t^2} = A_{ijkl} \frac{\partial^2 u_k}{\partial x_j \partial x_l}. \tag{1}$$

The fourth-order tensor of 21 independent elastic moduli greatly complicates the analysis carried out in chapters 4 and 5 for wave propagation in isotropic media so that matrix notation is essential. In (1), we can always first rotate the Cartesian system to bring the direction of phase propagation along the x_1 axis and the x_3 axis vertical.

It is often convenient in anisotropic analysis to write the stress–strain relations (2.27) as

$$p_i = c_{ij}e_j \quad (i, j = 1, \ldots, 6). \tag{2}$$

Then, for example, it is straightforward to prove, by the symmetry transformations developed for derivation of the isotropic case, that for *transverse* isotropy there are five elastic constants $A = c_{11}$, $F = c_{12}$, $L = c_{44}$ and two others C and N (see §2.5).

By the usual approach, we substitute the equation for plane waves

$$u_i = a_i \exp\left[i\omega(q_k x_k - t)\right], \tag{3}$$

into (1). Here a_i is the amplitude vector defining the polarisation of particle motion and $q_k = (1/c, 0, 0)^T$ is the slowness vector and c is the phase velocity. Substitution gives three linear equations which may be solved if

$$\det M_{ik} = 0 \tag{4}$$

where

$$M_{ik} = A_{ijkl}q_j q_l - \rho\delta_{ik}.$$

The expression for $\det M$ is a homogeneous sextic polynomial in q. As an eigenvalue problem, we omit the time factor and write

$$(\mathbf{M} - \rho c^2 \mathbf{I})\mathbf{a} = 0, \tag{5}$$

where $\mathbf{M}$ is now the 3×3 matrix with elements A_{i1k1}.

$\mathbf{M}$ is real, symmetric and positive definite so that its three eigenvalues are all real and eigenvectors orthogonal. The roots can be identified with a quasi-P wave (qP) and two quasi-S waves ($qS1$ and $qS2$). These waves are, however, not, in general, purely dilatational and rotational (§4.1). For isotropic media, $\mathbf{M}$ is diagonal ($\lambda + 2\mu, \mu, \mu$) with roots α, β, β, the latter corresponding to SH and SV waves, say. Physically, in an anisotropic medium, there are *three* body waves with orthogonal polarisations that are fixed by the direction of phase propagation in the given anisotropic symmetry system, i.e. cubic, monoclinic, orthorhombic, etc.

If the q_k are real, then equation (4) represents the three *slowness surfaces*. The geometry of the wave fronts is defined by reciprocals of these surfaces. The determinantal equation is usually too complicated for closed solutions to be practicable. Numerical values have been computed for transversely isotropic (see §2.5) and various mineral symmetries. Generally the qP slowness surface is everywhere convex and *inside the* $qS1$ and $qS2$ surfaces. The topology of wave surfaces can, however, be quite complicated with variations in concavity and convexity, cusps, and singularities where the shear wave surfaces touch or intersect. Illustrations are given in the review by Crampin (1981).

Another consequence of the dependence of slowness q_j on direction is that wave number κ is a vector rather than the scalar quantity of isotropic media. Body waves, surface waves and group velocities are all affected. Equation (3.67) becomes

$$C = (\partial \omega / \partial \kappa_1, \ \partial \omega / \partial \kappa_2, \ \partial \omega / \partial \kappa_3)^{\mathsf{T}} \tag{6}$$

and the wave energy may now not be transported parallel to the phase velocity. As a result, treatment of reflection and refraction of plane body waves in anisotropic media is less tractable than for an isotropic medium. Although the three plane waves travelling in the same direction are orthogonally polarised, they are not generally radial and transverse. Therefore, anomalous conversions occur between pairs of qP and qS waves at the boundaries.

12.1.2 Surface waves in anisotropic media

For surface waves, a number of special cases have been examined. For example, Stoneley demonstrated for cubic symmetry (i.e. coefficients $A = B = C$, $F = G = H$, $L = M = N$ of the equation (2.56) non-zero) that Love-type waves exist only when certain planes of crystal symmetry are the free surface. In such a case, the general surface motion degenerates into a superposition of Love and Rayleigh type waves.

The cubic case gives, from (2.54),

$$W = \tfrac{1}{2}A(e_{11}^2 + e_{22}^2 + e_{33}^2) + F(e_{22}e_{33} + e_{33}e_{11} + e_{11}e_{22})$$
$$+ \tfrac{1}{2}L(e_{23}^2 + e_{31}^2 + e_{12}^2).$$

Then the stress components p_{ij} follow from (2.53), and the three equations of motion, from (2.9), are

$$\rho \frac{\partial^2 u_1}{\partial t^2} = \frac{\partial}{\partial x_1}\left(A\frac{\partial u_1}{\partial x_1} + F\frac{\partial u_2}{\partial x_2} + F\frac{\partial u_3}{\partial x_3}\right) + L\frac{\partial}{\partial x_2}\left(\frac{\partial u_1}{\partial x_2} + \frac{\partial u_2}{\partial x_1}\right)$$
$$+ L\frac{\partial}{\partial x_3}\left(\frac{\partial u_1}{\partial x_3} + \frac{\partial u_3}{\partial x_1}\right), \tag{7}$$

and two similar.

Let solutions associated with the free surface $x_3 = 0$ be of the type

$$u_j = U_j \exp\{ -\kappa q x_3 + i\kappa(lx_1 + mx_2 - ct)\}, \tag{8}$$

where l and m are direction cosines of the wave front in the plane $x_3 = 0$ and displacements decrease with depth. Substitution from (8) in (7) yields three equations in U_j, so that for a solution

$$
\det \mathbf{M} = \begin{vmatrix} Al^2 + Lm^2 - \rho c^2 - Lq^2 & lm(F+L) & lq(F+L) \\ lm(F+L) & Ll^2 + Am^2 - \rho c^2 - Lq^2 & mq(F+L) \\ lq(F+L) & mq(F+L) & \rho c^2 + Aq^2 - L \end{vmatrix}
$$

$$(9)$$

There are up to three non-zero roots q_1^2, q_2^2, q_3^2. The boundary conditions at $x_3 = 0$ (see § 5.2) lead to a further 3×3 determinantal equation for q^2 of the form

$$\det \mathbf{N} = 0. \tag{10}$$

The equations (9) and (10) together define the allowable solutions for c^2 and q^2 and back-substitution gives the displacement ratios in the usual way. Elimination of q^2 among the six equations would, in principle, provide the dispersion equation (see Exercise 12.5.9) but the formal algebraic solution does not ensure in any case that a real wave exists. Stoneley has shown that waves with amplitude falling off exponentially with depth from the free face of a cubic crystal do not usually exist; for rock salt (NaCl), however, Rayleigh-type waves do exist with particle motions elliptical in a plane inclined to the direction of propagation.

For more general symmetries, studies show that solutions for surface waves involve complex roots and the determinantal equations contain both κ and c. Directions can be found in which Rayleigh-type waves propagate, but then they are usually damped. It has been suggested that if characteristic polarisations of qS and surface waves can be isolated observationally, inferences could be drawn on the extent and type of aligned anisotropy in the Earth's crust and lithosphere. Models incorporating plausible anisotropy show, however, that dispersion curves are little affected by the anisotropy, with group velocity variations in the first few generalised modes not usually exceeding a few per cent. Unfortunately, dipping layers and other structural heterogeneities also produce a mixing of wave types, modal conversions, and rotations of particle orbits. Experimentally, it is thus extremely difficult to disentangle the mix of causes.

12.2 Heterogeneous media. WKBJ approximation

In many seismological problems, assumptions of a homogeneous medium or homogeneous layers are not adequate (see § 4.3). Aside from questions of anisotropy, discontinuities and strong velocity gradients in

Earth structure lead to serious inadequacies in the use of simple ray theory. Frequency dependence of the amplitudes of body waves becomes observationally significant in such cases and an extension of the classical ray theory (see §7.1.1) is needed when rays are near such turning points as cusps (see §7.3.3), caustics (see §3.7), and shadow zones (see §7.3.8).

Considerable improvement in ray theory for heterogeneous media is obtained by using the **WKBJ** approximation (named after Wenzel, Kramers, Brillouin and Jeffreys).

We again consider equations with Sturm–Liouville form (e.g. the *SH* wave equation (5.28))

$$\frac{d^2 y}{dz^2} + r(z)y = 0. \tag{11}$$

If $r(z)$ is constant in a region, the solution is purely sinusoidal

$$y = \begin{cases} A\cos r^{\frac{1}{2}}z + B\sin r^{\frac{1}{2}}z, & r > 0 \tag{12} \\ A\exp\left[(-r)^{\frac{1}{2}}z\right] + B\exp\left[-(-r)^{\frac{1}{2}}z\right], & r < 0. \tag{13} \end{cases}$$

For slow variation in $r(z)$, (12) and (13) still approximately describe the essentially oscillatory motion for $r > 0$ and exponential motion for $r < 0$, but $r^{\frac{1}{2}}$ varies with z so that A and B vary with z. At $r = 0$, there is a turning point with transition between the two types of motion.

Put $y = \exp\eta(z)$ so that (11) becomes

$$\zeta' + \zeta^2 + r(z) = 0, \tag{14}$$

where $\zeta = \eta' = d\eta/dz$. We can iterate to a solution of this non-linear first-order differential equation as follows.

First, let $r(z) < 0$. Then

$$\zeta^2 = -r(z) - \zeta', \tag{15}$$

so that, neglecting ζ', $\zeta = \pm(-r)^{\frac{1}{2}}$, and thus (for the positive sign)

$$\zeta' = -r'/2(-r)^{\frac{1}{2}}. \tag{16}$$

Substitute (16) in (15), to obtain, approximately

$$\zeta = (-r)^{\frac{1}{2}} + r'/4(-r).$$

Therefore,

$$\eta = -\tfrac{1}{4}\ln(-r) + \int^z (-r)^{\frac{1}{2}}dz$$

and

$$y = (-r)^{-\frac{1}{4}}\exp\left\{\int^z (-r)^{\frac{1}{2}}dz\right\}. \tag{17}$$

The complete approximate WKBJ solution, corresponding to $\pm \zeta$, is

$$y = (-r)^{-\frac{1}{4}} \left[A \exp \left\{ \int^z (-r)^{\frac{1}{2}} dz \right\} + B \exp \left\{ -\int^z (-r)^{\frac{1}{2}} dz \right\} \right],$$
(18)

where A and B are constants.

In the case $r(z) > 0$, we obtain a similar expression to (18) with trigonometric functions in place of exponentials. (In the integrals, the lower limits can be chosen for convenience.)

It should be noted that in the neighbourhood of the turning point $r(z) = 0$, the WKBJ method is invalid. For the range of ray parameters corresponding to this case, other methods such as the Airy approximation (see § 3.7) or the Langer–Olver approximation must be used. More detail on the latter may be found, for example, in Aki and Richards (1980).

The theory outlined here is of importance in computing synthetic seismograms for waves that have nearly grazed the outer core, or the Mohorovičić discontinuity, or passed near low velocity zones, and for the *PKP* and associated caustics. Total internal reflections in a continuously varying layer can take place with change of phase, blunting and oscillation of pulses. The behaviour is analogous to that of a mirage in optics.

12.3 Topographic and structural variations

We have now reached the point in geophysics where we would like to consider solutions for non-spherically-symmetric Earth models. Moreover, in predicting strong ground motion for engineering design, two- and three-dimensional crustal models are required. Solutions to such problems usually require numerical methods.

The finite element method is one such scheme. In seismology, both finite difference and finite element methods have been shown to be applicable to the computation of surface wave excitation of geologically complicated structures; a number of problems have been solved to demonstrate the effect of surface topography and buried structures on planar and cylindrically propagating P and S waves; dynamic finite difference and finite element models of fault rupture have been used to compute fault ruptures. In this section, some numerical methods are summarised with emphasis on general techniques. No effort is made to compare the efficacy of the competing methods. The great usefulness of both formulations is evident from their wide application in physical and engineering contexts.

Seismological problems span a wide range of scale. In exploration geophysics, there is the problem of detecting a buried anomalous body by examining variations in seismic waves that have passed through it. At the same scale, there is the problem of interpreting the intensity of seismic shaking in terms of surface topography. On a larger scale, determination of crustal structure depends upon the calculation of the modal conversion and dispersive effects that mountain ranges and oceanic trenches have on propagating seismic surface waves. At the largest scale, there is the problem of determining the effects that the irregular distribution of continents and oceans has on the periods of free oscillations of the Earth.

Although differing in scale by the order of 10^3, these sample problems have in common the elastic wave equation.

12.3.1 Finite difference methods

The wave equation is a hyperbolic partial differential equation (3.80) for which a number of important analytic methods of solution have been discussed in previous sections for certain important seismological applications. In many cases involving arbitrarily varying media and boundaries, however, recourse must be made to numerical integration to find a solution. For the second-order wave equation, two supplementary conditions are normally required to specify a solution. In one case, numerical values of the solution and its derivative may be given at one boundary leading to a *marching* problem, or alternatively numerical values of the solution are given at both boundaries, defining a *jury* problem. Sometimes mixed boundary conditions apply. When conditions are specified at the source boundary, it is convenient to solve numerically by a step-by-step algorithm, with numerical values calculated at successive pivotal points of a prescribed grid. Most, but not all, step-by-step methods use finite difference formulae which we discuss here. It should be noted, however, that hyperbolic equations have real and distinct *characteristic* curves (see §7.1.1) and their use often leads to the most satisfactory numerical analysis.

We seek to approximate the derivatives in (3.80) by finite differences. If $y(x)$ is tabulated at equal intervals h, i.e. for arguments $x_n = x_0 + nh$, denote $y(x_n)$ by y_n. Three schemes of differences are then available called *forward* ($\triangle$), *backward* (∇), or *central* (δ), where

$$\triangle y_n = \nabla y_{n+1} = \delta y_{n+\frac{1}{2}} = y_{n+1} - y_n$$

and

$$\delta^2 y_n = y_{n+1} - 2y_n + y_{n-1}.$$

For wave propagation marching problems, it is preferable to use the forward difference formulae so that the differences follow the motion into the solution space. The simplest problem is for the Laplace operator $(\partial^2/\partial x^2 + \partial^2/\partial y^2) f$ with f specified at points on a rectangular boundary. Starting at a particular point (x_0, y_0) of the grid dividing up the rectangle, in central difference notation,

$$h^2 \partial^2 f_0/\partial x^2 = (\delta_x^2 - \tfrac{1}{12}\delta_x^4 + \tfrac{1}{90}\delta_x^6 - \cdots) f_0,$$

the suffix x denoting differencing in the x direction and the index the order of the finite difference. A similar expression holds for the y direction. Another accurate formula for the quantity $\nabla^2 f$, in terms of the function values, is

$$6h^2 \nabla^2 f_0 = 4(f_1 + f_3 + f_5 + f_7) + (f_2 + f_4 + f_6 + f_8) - 20 f_0,$$

where here ∇^2 is the Laplacian (not the backward difference) operator and the subscripts define alternating grid points anticlockwise around a square grid, size h, surrounding f_0.

For the wave equation, we need to equate to the spatial forms above the similar appropriate finite difference formula for $\partial^2 f/\partial t^2$ and step time forward recursively. Detail on these schemes is given in books on numerical analysis; a discussion with seismic wave application is given by Boore (1972). Boore suggests, for the two-dimensional S wave equation, the recursive form that retains only second differences,

$$v_{m,n}^{t+1} = 2v_{m,n}^t - v_{m,n}^{t-1} \tag{19}$$
$$+ \beta^2 h_t^2 \left\{ \frac{v_{m+1,n}^t - 2v_{m,n}^t + v_{m-1,n}^t}{h_x^2} + \frac{v_{m,n+1}^t - 2v_{m,n}^t + v_{m,n-1}^t}{h_y^2} \right\},$$

where grid intervals are h_t, h_x, h_y and the subscripts locate the space grid points and the superscripts the time step.

12.3.2 Finite element methods

In the finite element construction, displacements throughout the structure are written as functions of the displacements at a finite number of nodes by means of interpolation functions. In the most general case,

$$\mathbf{U}(x, y, z, t) = \mathbf{N}(x, y, z)\mathbf{u}(t). \tag{20}$$

$\mathbf{U}(x, y, z, t)$ are the three components of displacement at a general point in space and time, $\mathbf{u}(t)$ is the vector of the three components of displacement at all the n nodes at time t and $\mathbf{N}(x, y, z)$ is the $3 \times 3n$ interpolation matrix. It is highly banded because it interrelates only those nodes which are part of the same element. The interpolations performed by $\mathbf{N}$ can be of any order of splines; however most computations use linear elements.

The relation (20) allows the separation of space and time in the formulation of the field variables. This separation of variables simplifies the calculations of potential strain energy because taking the derivatives of the field variable requires taking the derivatives of only the linearly varying functions of $\mathbf{N}$. Then, the energy terms are integrated in space over the whole finite element model. Finally, the Euler–Lagrange equation is applied to Hamilton's principle giving

$$\mathbf{M\ddot{u}} + 2\mathbf{L\dot{u}} + \mathbf{Ku} = \mathbf{F}(t). \tag{21}$$

In seismological terminology, $\mathbf{M}$ is the density matrix, $\mathbf{L}$ the damping matrix, $\mathbf{K}$ the elastic modulus matrix, and $\mathbf{F}(t)$ is the source amplitude vector used to incite the appropriate wave motion within the model.

For simplicity, $\mathbf{L}$ is usually taken to be a linear combination of the density and modulus matrices. This way of forming $\mathbf{L}$ does not result in a Q value that is constant for all frequencies, but the desired value of Q may be chosen for a particular frequency.

The simultaneous set of ordinary differential equations (21) replaces the relevant partial differential equations for wave motion in an elastic solid. The field variable $\mathbf{u}$ can have a single scalar value at each point, for example, the horizontal displacement of an SH wave or Love wave motion. Or $\mathbf{u}$ might have two values at each point, corresponding to vertical and horizontal components of P and SV waves or Rayleigh wave motion. To date, most of the finite element modelling of Earth structures has been restricted to less than three dimensions, because of the large storage and computational demand required in the latter case. However, a limited number of axisymmetric problems have been treated.

Equation (21) has been solved three different ways. One approach is to recast it into its eigenvalue form (without damping),

$$(\omega^2\mathbf{M} - \mathbf{K})\mathbf{u} = 0. \tag{22}$$

Here $\mathbf{u}$ is the displacement for a given frequency ω. Equation (22) has been used to solve successfully a number of problems involving propagation of Love and Rayleigh waves through two-dimensional complex structures.

A second approach is to integrate (21) numerically in the time domain. Rewrite as

$$\mathbf{\ddot{u}} = \mathbf{M}^{-1}(\mathbf{F}(t) - \mathbf{Ku}). \tag{23}$$

$\mathbf{M}$ is here approximated by its diagonalised or 'lumped mass' form. This approximation serves to simplify the inversion of $\mathbf{M}$ as well as to eliminate the problem of spurious precursors that sometimes appear in such time domain solutions. Then starting with the initial (source) conditions of the mesh, integration of (23), for example by a Runge–Kutta scheme, yields $\mathbf{u}(t)$, the seismograms at specified nodal points of the mesh.

A third approach involves the use of Rayleigh's principle (see § 3.2.2). In engineering and physics applications, it has been long realized that (22) could be put into the form of the Rayleigh quotient,

$$\omega^2 = \frac{\mathbf{u}^T\mathbf{Ku}}{\mathbf{u}^T\mathbf{Mu}}. \tag{24}$$

The full implications of linking Rayleigh's principle with the finite element method for seismic wave motion in complicated geological structures was not explored until 1976. It is not necessary to attempt an analytic solution for the complex structure if a known eigenfunction for a similar modified regular structure is substituted in (24). Then, if $\mathbf{M}$ and $\mathbf{K}$ are the mass and stiffness matrices for the complex structure, (24) allows ω to be calculated for the latter, correct to the second order in the eigenfunction. The effectiveness of (24) in the case of Rayleigh waves in a layered Earth model has been demonstrated. Although the application of a variational principle to the finite element form has been explored in one dimension, its greatest effectiveness will probably be found in solving three-dimensional vibration problems. In the case of the estimation of the eigen-periods of the free oscillations of a non-radially-symmetric model of the Earth, the known eigenvectors for a radially symmetric Earth model can be used in (24). Since the largest departures of the non-radially-symmetric Earth from the radially symmetric Earth are on the order of 10 per cent, the Rayleigh quotient yields eigen-frequencies correctly to within a few per cent.

A serious drawback in dynamic finite element modelling of wave propagation is the unwanted reflections from the boundaries of the model that contaminate the direct signal.

It is desirable to be able to eliminate these reflections and thus simulate an infinite medium. The problem can be overcome by constructing a model of sufficient size that the solution is complete before the reflections arrive. But this is not always feasible, as the model size is limited by computer storage and cost.

The following technique for completely eliminating reflections is due to W.D. Smith (1974). The analytical formulation is exact, independent of both frequency and incidence angle. It involves the superposition of solutions and is thus more costly than a single solution, but it allows solutions uncontaminated by reflections without the construction of a very large model.

The scalar wave equation for *SH* waves is almost trivial. The equation of motion is (5.4b) where u is the displacement and β the velocity. For a boundary at $x_1 = 0$ and incidence at angle i an incident plane wave of angular frequency ω and unit amplitude may be expressed as

$$u = \exp(i\omega/\beta)(x_1 \cos i + x_2 \sin i - \beta t). \tag{25}$$

The plane is that of the incident and reflected waves, and the normal. The reflected wave is then

$$u = A \exp(i\omega/\beta)(- x_1 \cos i + x_2 \sin i - \beta t). \tag{26}$$

For a free boundary at $x_1 = 0$, the Neumann condition $\partial u/\partial x_1 = 0$ gives $A = 1$. For a fixed boundary, the Dirichlet condition $u = 0$ gives $A = -1$. Addition of the two solutions entirely eliminates the reflection. The same relationship holds if propagation is damped, since the imaginary exponents are replaced by complex ones, and the mathematics is otherwise unchanged. For curved wave fronts, one may (see § 5.7) take an image source which gives rise to the reflected wave. The Neumann condition requires the opposite sign. Addition eliminates the reflection.

The vector wave equation is a special case. The two solutions which are added are the following:

Boundary value problem 1. The x_1 displacement and the tangential stress are set to zero on the boundary. That is,

$u_1 = 0$ (Dirichlet for u_1),

$$p_{12} = \mu[(\partial u_1/\partial x_2) + (\partial u_2/\partial x_1)] = 0. \tag{27}$$

Hence (27) becomes

$$\partial u_2/\partial x_1 = 0 \quad \text{at} \quad x_1 = 0 \quad \text{(Neumann for } u_2\text{)}. \tag{28}$$

Boundary value problem 2. The x_2 displacement and the normal stress are set to zero on the boundary. i.e.

$u_2 = 0$ (Dirichlet for u_2),

$$p_{11} = (\lambda + 2\mu)(\partial u_1/\partial x_1) + \lambda(\partial u_2/\partial x_2) = 0, \tag{29}$$

hence

$$\partial u_1/\partial x_1 = 0 \quad \text{at} \quad x_1 = 0 \quad \text{(Neumann for } u_1\text{)}. \tag{30}$$

Displacements may be expressed in terms of two potentials (cf. equation (5.1)).

$$u_1 = (\partial\varphi/\partial x_1) + (\partial\psi/\partial x_2), \tag{31}$$

$$u_2 = (\partial\varphi/\partial x_2) - (\partial\psi/\partial x_1). \tag{32}$$

The potential function φ describes longitudinal (P) motion, and ψ describes transverse (S) motion.

The potentials for P incidence are

$$\varphi = \exp\left[(i\omega/\alpha)(x_1 \cos i + x_2 \sin i - \alpha t)\right] \qquad \text{(incident } P)$$
$$+ A \exp\left[(i\omega/\alpha)(-x_1 \cos i + x_2 \sin i - \alpha t)\right] \quad \text{(reflected } P\text{), (33)}$$
$$\psi = B \exp\left[(i\omega/\beta)(-x_1 \cos j + x_2 \sin j - \beta t)\right] \quad \text{(reflected } SV\text{),} \quad \text{(34)}$$

where $\alpha = P$ velocity, $\beta = S$ velocity, $i = $ incidence angle, and $j = SV$ reflection angle.

Solution 1
From (27) and (31)

$$(1 - A)(i\omega \cos i/\alpha) \exp\left[(i\omega/\alpha)(x_2 \sin i - \alpha t)\right]$$
$$= - B(i\omega \sin j/\beta) \exp\left[(i\omega/\beta)(x_2 \sin j - \beta t)\right].$$

So

$$\sin i/\alpha = \sin j/\beta \quad \text{(Snell's Law)} \tag{35}$$

and

$$(1 - A) \cos i/\alpha = - B \sin j/\beta \tag{36}$$

From (29) and (32)

$$0 = (\partial^2 \varphi/\partial x_1 \partial x_2) + (\partial^2 \psi/\partial x_1^2)$$
$$= - (1 - A)(\omega^2 \cos i \sin i/\alpha^2) \exp\left[(i\omega/\alpha)(x_2 \sin i - \alpha t)\right]$$
$$+ B(\omega^2 \cos^2 j/\beta^2) \exp\left[(i\omega/\beta)(x_2 \sin j - \beta t)\right];$$

So

$$(1 - A)(\cos i \sin i/\alpha^2) = B(\cos^2 j/\beta^2). \tag{37}$$

From (36) and (37), $A = 1$ and $B = 0$. This corresponds to a P reflection in phase with the incident wave, and no SV reflection.

Solution 2

$$u_2 = 0 \text{ gives } (1 + A)(\sin i/\alpha) = - B(\cos j/\beta), \tag{38}$$
$$\partial u_1/\partial x_1 = 0 \text{ gives } (1 + A)(\cos^2 i/\alpha^2) = B(\cos j \sin j/\beta^2). \tag{39}$$

So $A = -1$ and $B = 0$. This implies a P reflection out of phase with the incident wave, and no SV reflection.

Therefore addition of the two solutions will exactly cancel the reflections. The case of incident SV waves is exactly parallel. It is simple to show that there is no reflected P wave in either solution, and that the two reflected SV waves are of opposite sign. Addition eliminates all reflections.

As for the scalar wave equation, the presence of a damping term does not affect the formulation.

In both the solutions, mode conversion is suppressed, so in fact the reflections are those for the vector wave equation.

The two solutions can also be formulated in terms of image sources. *P* and *SV* reflections at a *free* boundary cannot be treated by image techniques, because of the mode conversion problem. The requirement that one displacement be zero, however, suppresses the conversion and allows the use of images. The image sources are of opposite sign in the two solutions, and they therefore disappear on addition.

Reflections of surface waves are also eliminated by this formulation. For *Love waves*, the particle motion is horizontal, and transverse to the direction of propagation. At any particular depth, therefore, the elastic conditions to be satisfied at a boundary are exactly those for *SH* reflection in two dimensions, i.e. transverse particle motion in the plane of the incident wave and the normal. As was indicated above, the addition of the two solutions cancels the reflection.

For *Rayleigh waves*, the particle motion is in the vertical plane containing the propagation direction. At a free boundary there will be a reflected Rayleigh wave, together with body waves whose generation is necessary to satisfy the stress conditions at the boundary. Smith has shown that if only the reflected Rayleigh wave is considered, reflections of opposite sign correspond to the Dirichlet conditions for normal and parallel displacements, respectively.

When more than one face of the model is required to be nonreflecting, more solutions must be added to eliminate multiple reflections. In general, if reflections are required to be eliminated on n surfaces, 2^n solutions must be added. So eight solutions would be necessary at a three-dimensional corner.

In practice it may not be necessary to add all the solutions to cancel the highest order reflections. Because these arrive later, it may be possible to obtain the required synthetic record before they arrive. High order reflections which cannot be cancelled occur when a ray path encounters the same face more than once. These reflections are of such high order, and their travel time so long, that they are of little practical importance.

Fig. 12.1 shows a model which demonstrates the technique. It is an assemblage of 576 square elements, and was excited at the upper left-hand corner by a square pulse. Four solutions were added to eliminate reflections on the right-hand and lower faces. The displacements observed at the points marked in Fig. 12.1 are shown in the subsequent figures. The frequency characteristics of the model are determined by the size of its

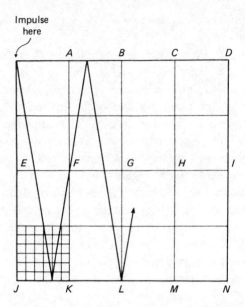

Fig. 12.1. Finite element test model. All elements of the same size but fine detail has only been partly indicated. The points A to N are those at which the displacements of Fig. 12.2 are shown. The ray path corresponds to one high order of reflection which is not cancelled. The model is 1 km square, P and S velocities are 5.5 km/s and 3.3 km/s, respectively, and there is no damping. (Courtesy W.D. Smith, 1974.)

elements. Wave-lengths shorter than the element length are severely attenuated, so the square input pulse becomes smoothed. The finite element solution also contains a resonance effect at a frequency characteristic of the size of the elements. If the time series is low-pass filtered, the ringing can be almost entirely removed.

Fig. 12.2 illustrates a solution of the scalar wave equation. No damping or forcing terms were used; the excitation was applied as a displacement pulse. The solution for free boundaries is shown in Fig. 12.2(a), and in Fig. 12.2(b) can be seen the mean of the required four solutions. The only reflection occurring in Fig. 12.2(b), visible on traces D and I to N, corresponds to the ray path shown in Fig. 12.1 and to the corresponding path nearly parallel to the x axis. This reflection is not cancelled since it involves two encounters with a non-reflecting face, and so the reflected wave always has a positive sign. If the model incorporates damping, the high order reflections may be damped out completely, and this would of course allow excitation by a long time series, without contamination from reflections.

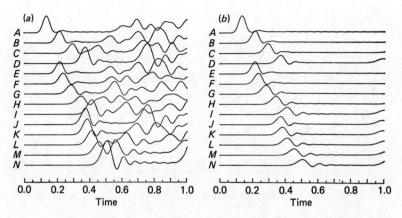

Fig. 12.2. Calculated displacements for the scalar wave equation for SH motion in two dimensions. In (a) reflections are not cancelled. In b) the sum of the four solutions necessary to eliminate reflections on the lower and right hand faces of the model is shown. Time is in seconds. (Courtesy W.D. Smith, 1974.)

12.3.3 Numerical results. A mountain and oceanic–continental transition

Following the San Fernando, California, earthquake of February 1971, there was much interest in the question of amplification of ground motion by surface topography. D. Boore showed that *SH* motion can be significantly amplified by the presence of a mountain or a ridge, which he modelled in two dimensions by finite differences. Field observations tend to support the conjecture that topography can play an important role.

The finite element method may also be usefully applied to this type of problem. An analysis by W.D. Smith was restricted to two dimensions. There is no difficulty in programming the three-dimensional case, but the computer storage required makes it impracticable with currently available facilities. The finite element model had the sides of the mountain sloping at 20° to the horizontal; the material was homogeneous. A displacement pulse was applied along the entire base, and allowed to propagate up to the surface as a plane wave.

In order to determine the amplification due to the mountain, the source function may be removed by taking the Fourier spectra of the traces computed from (23), and dividing by the spectrum that is obtained in the absence of the mountain. This is shown in Fig. 12.3 for three locations. Dimensionless frequency is given by the ratio of the half-width of the

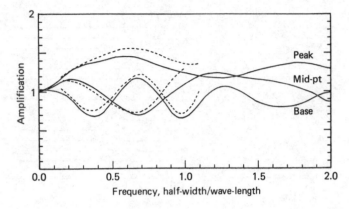

Fig. 12.3. Fourier spectra of SH waves at the base, mid-point, and peak of the mountain. The broken curves at each location are the finite difference results of Boore (1972). (Courtesy W.D. Smith, 1979.)

mountain to the wave-length. The analytical value of 1.0 at zero frequency is reproduced to within a few per cent.

The spectra for vertical P and S wave incidence were also computed. The amplitude observed on the horizontal component at the peak of the mountain for SV incidence reached a maximum of 1.8. Amplification due to P incidence reached a maximum of 1.2. The base of the mountain experienced attenuation at most frequencies, and the peak experienced mostly amplification.

Now that ocean bottom seismometers with broadband recording are becoming available (see § 9.9.6) there is a greater need for numerical results on wave propagation through realistic models of continental boundaries. There is also a critical requirement to allow for the effect of continental margins in inversions from surface wave dispersion to oceanic structure. The propagation of fundamental-mode Love waves for both directions of normal incidence was analysed by Drake and Bolt across a finite element model of a representative (nonsubducting) continental boundary.

An ocean bottom seismometer, 235 km WNW of Berkeley and beneath 3.9 km of water, was operated from 1966 to 1972 by Lamont Geological Observatory. Because the present finite element study is two-dimensional, consider Love waves that propagate normally to the continental boundary and adopt the finite element model shown in Fig. 12.4 to represent a region extending 200 km WSW of the Berkeley seismographic station to a possible ocean bottom seismometer site 120 km from the continental shelf and beneath 4.5 km of water. The thicknesses of the layers, compressional

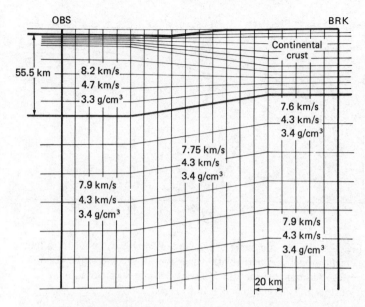

Fig. 12.4. Finite element model from an ocean-bottom seismometer site in the Pacific Ocean to Berkeley, California. (Courtesy L.A. Drake and B.A. Bolt, 1980.)

velocities, shear velocities, and densities of the horizontally layered structures at the ends of the finite element model are shown in the figure.

Key results are now summarised. Between periods of 60 and 17.5 s, the phase velocity of fundamental-mode Love waves across the continental boundary model is slightly greater (less than 1 per cent) than the mean of the phase velocities of fundamental-mode Love waves in the horizontally layered structures at the ends of the model (see Fig 12.5). By contrast, between periods of 15 and 4.3 s, the phase velocity of the fundamental-mode Love waves across the finite element model is, in general, slightly less (about 2 per cent) than the mean of the phase velocities of fundamental-mode Love waves in the horizontally layered structures at the ends of the model. At a period of 60 s, over 90 per cent of the energy of incident fundamental-mode Love waves is transmitted in waves of the fundamental Love mode. At a period of 20 s, only about 10 per cent of the energy of incident fundamental-mode Love waves is transmitted in waves of the fundamental Love mode. At a period of 15 s, approximately 2 per cent of the energy of incident fundamental-mode Love waves is transmitted in waves of the fundamental Love mode.

This loss of energy from oceanic fundamental-mode Love waves mainly

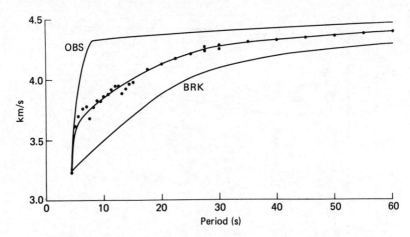

Fig. 12.5. Dispersion curve of the fundamental Love mode for a finite element model, compared with dispersion for the site of the ocean-bottom seismometer and the Berkeley Seismographic Station. (Courtesy L.A. Drake and B.A. Bolt, 1980.)

to waves of the continental first higher Love mode at a continental boundary has two important consequences. First, it is an additional cause of contamination by waves of the first higher Love mode when phase velocity measurements are made at continental stations of fundamental-mode Love waves travelling from an ocean. Earthquakes with sources at a depth of 100 km under an ocean can generate considerable energy in waves of the first higher Love mode, and fundamental-mode Love waves and Love waves of the first higher mode have the same group velocity at periods of about 20 and 60 s. Near these periods, Love waves of the first higher mode from an oceanic earthquake arrive at a continental boundary at approximately the same time as the fundamental-mode Love waves. Thus, measurements of the phase velocity of oceanic fundamental-mode Love waves suffer interference from contamination from waves of the oceanic first higher Love mode. In the same way, near the continental boundary, measurements of the phase velocity of continental fundamental-mode Love waves are affected by contamination from waves of the continental first higher mode.

The calculations show that this can be especially true at periods between 9 and 20 s, because for these periods, approximately 90 per cent of the energy of oceanic fundamental-mode Love waves is transferred to waves of the continental first higher mode at the continental boundary.

The second consequence of this transfer of energy from oceanic fundamental-mode Love waves to waves of the continental first higher mode at a continental boundary is that it seriously affects estimates of attenuation of oceanic fundamental-mode Love waves.

Fig. 12.6(*a*) shows the variation of displacement with depth of the first three Love modes at a period of 30 s at the Berkeley seismographic station. The depth of the finite element model is 1070 km, and the modes are normalised. The fundamental mode has a large displacement in the lithosphere, or uppermost 45 km of the model, while the first higher mode has a large displacement below a depth of 45 km in the low-velocity zone. When oceanic fundamental-mode Love waves at a period of 30 s reach the continental region, more of their energy propagates in waves of the first higher mode (53 per cent) than in waves of the fundamental mode (46 per cent).

Fig. 12.6(*b*) shows the variation of displacement with depth of the first three Love modes at a period of 10 s at the oceanic site. The base of the finite element model is fixed at a depth of 370 km. The fundamental mode has a comparatively small displacement at the top of the oceanic sediments and a much larger displacement in the low-velocity zone below a depth of 60 km from the surface of the ocean. The first higher mode has one zero crossing,

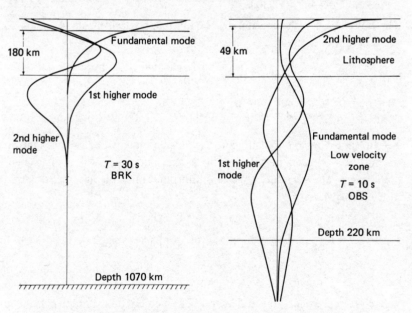

Fig. 12.6. Radial eigenfunctions for Love modes at Berkeley (*a*) and at the ocean-bottom seismometer (*b*). (Courtesy L.A. Drake and B.A. Bolt, 1980.)

and the second higher mode has two zero crossings in the low-velocity zone.

We now examine as a further illustration of the method the estimation of changes in eigen-frequency for mantle Rayleigh waves worked by Smith and Bolt.

Let the displacements be $\mathbf{u} = (-\mathrm{i}u_x, u_z)\exp(\mathrm{i}(\omega t - \kappa x))$, where u_x, u_z are real. κ is the wave number and ω the angular frequency. The engineering strains are

$$
\left.
\begin{aligned}
e_{xx} &= -\mathrm{i}(-\mathrm{i}\kappa)u_x = -\kappa u_x \\
e_{zz} &= \frac{\partial u_z}{\partial z} \\
2e_{xz} &= -\mathrm{i}\frac{\partial u_x}{\partial z} - \mathrm{i}\kappa u_z.
\end{aligned}
\right\}
\tag{40}
$$

The potential strain energy averaged over x and t is

$$
2V = \tfrac{1}{2}(\lambda + 2\mu)\left[\kappa^2 u_x^2 + \left(\frac{\partial u_z}{\partial z}\right)^2\right] - \lambda\kappa u_x\frac{\partial u_z}{\partial z} + \tfrac{1}{2}\mu\left(\frac{\partial u_x}{\partial z} + \kappa u_z\right)^2
\tag{41}
$$

where $\lambda(z)$ and $\mu(z)$ are the Lamé parameters. The mean kinetic energy is

$$
2T = \tfrac{1}{2}\rho\omega^2(u_x^2 + u_z^2).
\tag{42}
$$

The Euler–Lagrange equation may be invoked to derive from (41) and (42) the differential equations

$$
\left.
\begin{aligned}
-\rho\omega^2 u_x &= -(\lambda + 2\mu)\kappa^2 u_x + (\lambda + \mu)\kappa\frac{\partial u_z}{\partial z} + \mu\frac{\partial^2 u_x}{\partial z^2} \\
-\rho\omega^2 u_z &= (\lambda + 2\mu)\frac{\partial^2 u_z}{\partial z^2} - (\lambda + \mu)\kappa\frac{\partial u_x}{\partial z} - \mu\kappa^2 u_z
\end{aligned}
\right\}
\tag{43}
$$

For the finite element modelling we use the one-dimensional element shown in Fig. 12.7 aligned parallel to the z-axis and with two displacement components at each node. For linear interpolation of displacements within an element we have

$$
\mathbf{u} = \begin{bmatrix} u_x \\ u_z \end{bmatrix} = \begin{bmatrix} 1 - \dfrac{z}{h} & 0 & \dfrac{z}{h} & 0 \\[2mm] 0 & 1 - \dfrac{z}{h} & 0 & \dfrac{z}{h} \end{bmatrix} \begin{bmatrix} u_{1x} \\ u_{1z} \\ u_{2x} \\ u_{2z} \end{bmatrix}
$$

$$
= \mathbf{N}\mathbf{u}_\mathrm{e}, \text{ say}
$$

where the suffix refers to the elements. The element density matrix is given by

$$
\mathbf{M}_\mathrm{e} = \int_0^h \mathbf{N}^\mathrm{T}\rho\mathbf{N}\,\mathrm{d}z.
\tag{44}
$$

Fig. 12.7. Finite element used for the Rayleigh wave problem. There are two displacement components at each mode, propagation is in the X direction and Z is vertically upwards.

The average strains are $\mathbf{e} = \mathbf{B}\mathbf{u}_e$ and the stresses are $\mathbf{p} = \mathbf{H}\mathbf{e}$, where

$$\mathbf{B} = \begin{bmatrix} -\kappa\left(1 - \dfrac{z}{h}\right) & 0 & -\kappa\dfrac{z}{h} & 0 \\[2mm] 0 & -\dfrac{1}{h} & 0 & \dfrac{1}{h} \\[2mm] -\dfrac{1}{h} & \kappa\left(1 - \dfrac{z}{h}\right) & \dfrac{1}{h} & \kappa\dfrac{z}{h} \end{bmatrix}$$

and

$$\mathbf{H} = \begin{bmatrix} \lambda + 2\mu & \lambda & 0 \\ \lambda & \lambda + 2\mu & 0 \\ 0 & 0 & \mu \end{bmatrix}. \tag{45}$$

The element elastic moduli matrix is then

$$\mathbf{K}_e = \int_0^h \mathbf{B}^{\mathrm{T}}\mathbf{H}\mathbf{B}\,\mathrm{d}z. \tag{46}$$

The global elastic moduli and density matrices $\mathbf{K}$ and $\mathbf{M}$ are accumulated in the usual way. For n elements in the model, each representing one layer, they will be $(2n + 2)$ square, but with a bandwidth of only 6. Construction of the vector $\mathbf{u}$, containing all the $(2n + 2)$ displacement components, enables the evaluation of the energy terms $\mathbf{u}^{\mathrm{T}}\mathbf{K}\mathbf{u}$ and $\mathbf{u}^{\mathrm{T}}\mathbf{M}\mathbf{u}$ to be made. Their ratio is ω^2. With the wave number specified at the outset and the frequency calculated from (24), phase velocity may be determined.

Table 12.1 *Angular frequency, period and phase velocity for a wave number of* 0.009 909 *km*$^{-1}$

	Angular frequency ω(rad/s)	Period T(s)	Phase velocity c(km/s)
Dorman & Prentiss	0.041 92	149.9	4.23
Finite element	0.042 03	149.5	4.24

Dorman and Prentiss (1960) published mode shapes for mantle Rayleigh waves for a set of discrete frequencies in a horizontally layered Earth with a P and S velocity distribution derived by Gutenberg and the density distribution known as Bullen Model A (see § 13.5). The period and phase velocities for one of these frequencies are reprinted in Table 12.1. For a period of about 150 s the horizontal and vertical wave displacement in the fundamental Rayleigh mode for their Earth model are perceptible to depths of between 500 and 750 km. The mode shape in our comparison computation is approximated by 19 linear elements, corresponding to layers of thickness from 20 to 100 km, to represent the Earth to a depth of 800 km. Modelling with a variety of element sizes presents no computational problem, because the global matrices $\mathbf{K}$ and $\mathbf{M}$ (size 40 × 40) are built up element by element.

As shown in Table 12.1, for the wave number, 0.009 909 km^{-1}, the finite element expression (24) is used with the published mode shape for the same Gutenberg model. Table 12.1 lists the resulting values for angular frequency, period and phase velocity. Agreement is very close; the small discrepancies of 0.25 per cent between these estimates and those of Dorman and Prentiss probably arise from the linear finite element modelling.

The effect of changes in the model may be estimated with ease by simply changing elements in the elastic moduli matrix. The original mode shape may be retained as an approximate eigenfunction for the new model.

12.3.4 Variational methods

Variational methods such as Rayleigh's principle are commonly invoked (see § 3.2.2) in the determination of periods of free oscillation. To recapitulate, it states that when the mean potential and kinetic energies of the system in free oscillation are equated, the solution for the angular frequency takes a stationary value. Meissner in 1926 observed that it may be used to determine group velocity for Love waves without resorting to numerical differentiation. In 1961 Jeffreys extended this work to Rayleigh waves, and further noted that perturbations in the eigen-frequency due to changes in the structure can be determined without repeating the entire calculation. The original form of the mode shape is retained, on the assumption that it approximates that for the modified structure. Because of the stationarity, the new eigen-frequency is obtained with an error of only second order. This observation led to the publication of universal dispersion tables for Love and Rayleigh waves and for terrestrial eigen-vibrations by D.L. Anderson, J. Derr and others from which the effects of changing the structure on the period of surface waves, or torsional and spheroidal oscillations, with given wave number could be determined.

Consider Love waves propagating in an elastic half-space with vertical variation of properties $\rho(z)$, $\mu(z)$. Put

$$u_2 = U(z)\cos(\kappa x - \gamma t). \tag{47}$$

Then the kinetic and potential energy T and V, averaged over time and area, are

$$\bar{T} = \frac{1}{4}\int_0^\infty \gamma^2 \rho U^2 \, dz, \tag{48}$$

and, from (2.55),

$$\bar{V} = \frac{1}{4}\int_0^\infty \kappa^2 \mu U^2 \, dz + \frac{1}{4}\int_0^\infty \mu \left(\frac{dU}{dz}\right)^2 \, dz. \tag{49}$$

Then, by conservation of energy,

$$\gamma^2 I_0 = \kappa^2 I_1 + I_2, \tag{50}$$

where

$$I_0 = \int_0^\infty \rho U^2 \, dz, \quad I_1 = \int_0^\infty \mu U^2 \, dz, \quad I_2 = \int_0^\infty \mu \left(\frac{dU}{dz}\right)^2 \, dz. \tag{51}$$

If an approximation to the unknown $U(z)$ is substituted in (51), by Rayleigh's principle, (50) yields the phase velocity $c = \gamma/\kappa$, correct to the second order.

It is worthy of note that the method obviates the use of the boundary condition, and, of value from a numerical point of view, computations involve only integration. Indeed, by applying first variations to (50), we can obtain robust formulae for group velocity and effects of structural variation. First, vary only γ, κ in (50) leaving I_0, I_1, I_2 unchanged. Therefore,

$$\frac{\gamma \, d\gamma}{\kappa \, d\kappa} = \frac{I_1}{I_0}$$

or

$$cC = I_1/I_0, \tag{52}$$

where C is the group velocity.

Meissner's formula (52) should be compared with the numerical differentiation formula (3.68) for group velocity. Now, with κ fixed, vary γ, ρ, μ. Then (50) gives

$$\kappa^2(c + \delta c)^2(I_0 + \delta I_0) = \kappa^2(I_1 + \delta I_1) + I_2 + \delta I_2.$$

A similar, but slightly more complicated result can be derived for Rayleigh waves in a vertically inhomogeneous half-space. The reader can show that

$$\gamma^2 I_0 = \kappa^2 I_1 + 2\kappa I_2 + I_3 \tag{53}$$

where

$$I_0 = \int \rho(U_1^2 + U_3^2)\mathrm{d}z$$

$$I_1 = \int \{(\lambda + 2\mu)U_1^2 + \mu U_3^2\}\mathrm{d}z$$

$$I_2 = \int \left\{-\lambda\frac{\mathrm{d}U_3}{\mathrm{d}z}U_1 + \mu U_3\frac{\mathrm{d}U_1}{\mathrm{d}z}\right\}\mathrm{d}z$$

$$I_3 = \int \left\{(\lambda + 2\mu)\left(\frac{\mathrm{d}U_3}{\mathrm{d}z}\right)^2 + \mu\left(\frac{\mathrm{d}U_1}{\mathrm{d}z}\right)^2\right\}\mathrm{d}z.$$

The dispersion equation is, by Rayleigh's principle and varying κ and γ only.

$$cCI_0 = I_1 + I_2/\kappa. \tag{54}$$

The inverse problem of identifying small changes in ρ, λ and μ that produce small changes in phase velocity is greatly simplified by the formula, from (53),

$$(c + \delta c)^2(I_0 + \delta I_0) = I_1 + \delta I_1 + \frac{2}{\kappa}(I_2 + \delta I_2) + \frac{1}{\kappa^2}(I_3 + \delta I_3). \tag{55}$$

Two related theoretical problems of geophysical importance are the determination of transmission coefficients appropriate to an anomalous region whether the incident wave is travelling vertically or, as in the oceanic–continental transition in § 12.3.3, horizontally. Unfortunately, the variation technique used above with Rayleigh's method is invalid because phase shifts occur within each wave cycle. The appropriate variational method, developed by Schwinger and Levine for wave barriers in quantum physics, has been applied to seismological problems by Lapwood, J.A. Hudson and others. The algorithm is described well in Moiseiwitsch (1966).

A typical result is for the reflection coefficient f for a plane P wave incident in the z direction on an anomalous (e.g. visco-elastic) layer with properties $M(z)$ and bounded width. Correct to the second order,

$$f = 1 + \frac{I_1}{I_2 - I_3}, \tag{56}$$

where

$$I_1 = \frac{1}{2i\kappa} \int_{-\infty}^{\infty} \tilde{Z}(z)M(z)\exp(i\kappa z)dz \int_{-\infty}^{\infty} \exp(-i\kappa\zeta)M(\zeta)Z(\zeta)d\zeta,$$

$$I_2 = \int_{-\infty}^{\infty} \tilde{Z}(z)M(z)Z(z)dz,$$

$$I_3 = \frac{1}{2i\kappa} \int_{-\infty}^{\infty} \tilde{Z}(z)M(z) \int_{-\infty}^{\infty} \exp(i\kappa|z - \zeta|)M(\zeta)Z(\zeta)d\zeta dz.$$

Here $Z(z)$ and $\tilde{Z}(z)$ describe the incident and reflected (or adjoint) waves. We can determine f with a second-order error from (56) if we use in I_1, I_2, I_3 functions approximating to $Z(z)$ and $\tilde{Z}(z)$ with first-order error.

As a first approximation, insertion into the integrals I_1, I_2, I_3 of the values $Z(z) = A\exp i\kappa z$, $\tilde{Z}(z) = A\exp(-i\kappa z)$ yields a form which is satisfactory when the perturbation of the incident wave is small.

12.4 Laboratory model seismology

One approach to the study of elastic wave propagation through geologically realistic models is to use scaled physical laboratory models. Commonly pulses of elastic energy are applied by piezoelectric crystals. The source is typically driven by a DC pulse generator up to 1000 volts amplitude and 10 μs duration, thus exciting waves at frequencies of about 50 kHz. The crystals are designed to excite and receive both P and SV or SH waves. The receiving crystal connects the displacement at a point in the models to a measurable voltage that can be amplified, filtered, and displayed on an oscilloscope. The oscilloscope sweep is triggered simultaneously with each source pulse. Pulse repetition (at, say, 15 pulses per second) refreshes the output trace before the last has faded, creating a single complete 'seismogram'.

The models are of varied construction. In two-dimensional studies, thin sheets of plastic or metal are glued edge-on together to form layered media or cut in the shape of wedges or lenses. The dimensional studies of Earth structure may be explored by moulding spherical plastic balls with radial or angular variations in properties, a liquid outer core and so on. Individual modes of free oscillations of the Earth can be detected by sweeping the receiver filter frequency across the free oscillation spectrum appropriate to the model scale.

In its simplest form, model seismology can be described as an analogue computer which simulates seismic wave fields for structural models either to check analytical results or to predict motions in complicated structures. Computed seismograms are highly reproducible and the experimental cost is relatively low. There are, of course, problems of realistic scaling of elastic moduli, damping effect, and body forces. Further details are given by Northwood and Anderson (1953).

12.5 Exercises

1   Consider a transverse isotropic elastic material with the following stress–strain relations (1 axis is vertical):

$$p_{11} = Ce_{11} + F(e_{22} + e_{33})$$
$$p_{22} = Fe_{11} + Ae_{22} + (A - 2N)e_{33}$$
$$p_{33} = Fe_{11} + (A - 2N)e_{22} + Ae_{33}$$
$$p_{12} = Le_{12}$$
$$p_{13} = Le_{13}$$
$$p_{23} = Ne_{23}.$$

Show that the velocities of P, SV and SH waves in the x_3 direction are $(A/\rho)^{\frac{1}{2}}$, $(L/\rho)^{\frac{1}{2}}$ and $(N/\rho)^{\frac{1}{2}}$. In the x_2 direction?

2   Show that the displacement of SH waves in a medium continuously varying in the vertical direction is given by

$$\mu V'' + \mu' V' - \kappa^2 \rho(\beta^2 - c^2)V = 0,$$

where $\beta^2 = \mu/\rho$, prime denotes d/dz, and $v = V(z) \exp i\kappa(ct - x)$.

Transform the equation to normal form using $V = \mu^{-\frac{1}{2}} Z$ and use the WKBJ method (with Z measured from the depth where $c = \beta$) to show that, for downward travelling SH waves,

$$v = r^{-\frac{1}{4}}A[\sin\{\kappa(ct - x + L) + \pi/4\} - \sin\{\kappa(ct - x - L) - \pi/4\}],$$

where

$$L = \left| \int_0^z \left(\frac{c^2}{\beta^2} - 1\right)^{\frac{1}{2}} dz \right|.$$

Hence show that total reflection occurs as if at depth $\kappa L = \pi/4$ with a change of phase.

3   The P potential is given by

$$\frac{d^2\Phi}{dr^2} + \frac{2}{r}\frac{d\Phi}{dr} + \left\{k^2(r) - \frac{l(l+1)}{r^2}\right\}\Phi = 0. \tag{1}$$

Show from the WKBJ method that the approximate ray solution is

$$\Phi \sim \frac{1}{r(q_\alpha)^{\frac{1}{4}}}\exp\left[\pm i \int^r q_\alpha(\zeta)d\zeta\right], \tag{2}$$

where

$$q_\alpha^2 = k^2 - (v^2/r^2)$$

and $l(l+1)$ is replaced by $v = l + \frac{1}{2}$.

Why does (2) not hold when q_α varies rapidly or near a turning point of a ray $(q_\alpha(r) = 0)$? How is diffraction dealt with in the latter case? What is the physical meaning to the first factor on the right-hand side of (2) in terms of energy? Indicate briefly how to solve (1) numerically using (i) finite differences, and (ii) Runge–Kutta methods.

4   Consider the case of Love waves travelling in a horizontal layer, thickness $-h$, and elastic parameters ρ, μ, β over a welded elastic half-space with constant density ρ_1, and variable rigidity $\mu_1(1 + z/d)^2$ where d is constant and z is the depth coordinate.

Show that suitable forms for the displacement u_2 (i.e. that satisfy equation (5.4b)) are

$$u_2 = \begin{cases} A \sec \kappa\sigma h \cos \kappa\sigma(h + z), & -h < z < 0 \\ A \exp(-\kappa\sigma_1 z), & z > 0, \end{cases}$$

where $\sigma^2 = c^2/\beta^2 - 1$, $\sigma_1 = 1 - c^2/\beta_1^2$.

From the boundary conditions, u_2, $\mu du_2/dz$ must be continuous on $z = 0$. As an alternative, obtain the dispersion equation by equating the mean kinetic and potential energy over a cycle,

$$c^2\left\{\rho \sec^2 \kappa\sigma h\left(h + \frac{1}{2\kappa\sigma}\sin 2\kappa\sigma h\right) + \rho_1/\kappa\sigma_1\right\}$$

$$= \mu \sec^2 \kappa\sigma h\left\{(1 + \sigma^2)h + \frac{1 - \sigma^2}{2\kappa\sigma}\sin 2\kappa\sigma h\right\}$$

$$+ \mu_1(1 + \sigma_1^2)\left\{\frac{1}{\kappa\sigma_1} + \frac{1}{d\kappa^2\sigma_1^2} + \frac{1}{4d^2\kappa^3\sigma_1^3}\right\}.$$

Show that this dispersion relation between wave number κ and phase velocity c is correct to the second order in κ.

5   Consider a seismic P wave travelling vertically downwards through an anomalous layer, width $2a$, and attenuation constant $Q_\alpha(z)$. At small wavelengths, show that the transmission coefficient predicted by ray theory is

$$f = \exp\left(i\omega \int_{-a}^{a} \frac{dz}{\alpha(z)} - \frac{\omega}{2}\int_{-a}^{a} \frac{dz}{Q_\alpha(z)\alpha(z)}\right).$$

6   The displacement for Love waves in a heterogeneous half-space is

$$v = V(z)\cos(\kappa x - \gamma t).$$

Derive the average kinetic and potential energy over a cycle, then write down the quotient for γ^2 in terms of three integrals. Suppose μ is constant, and $\rho = \rho_0 (1 + \exp(-z))$.

Assuming that the displacements are approximately $V = V_0 \exp(-z)$, evaluate the Rayleigh quotient. What is the precision of your estimate of γ even though the eigenfunctions are only approximate?

7   The modified form of Bessel's equation for torsional oscillations of a radially varying Earth is

$$\mu\left\{\frac{d^2 W}{dr^2} + \frac{2dW}{r\,dr}\right\} + \frac{d\mu}{dr}\left\{\frac{dW}{dr} - \frac{W}{r}\right\} + \left\{\omega^2\rho - \frac{l(l+1)}{r^2}\mu\right\}W = 0. \qquad (1)$$

Derive from this the integral expression $I(W)$ which when stationary, yields the solution $W(r)$ of the differential equation. How would you use the integral to obtain the fundamental torsional eigen-period? [Hint: (i)

If $$I = \int_{x_0}^{x_1} \{p(x)y'^2 + q(x)y^2 + 2r(x)y\}\,dx$$

is stationary, y satisfies

$$p(x)y'' + p(x)'y' - q(x)y = r(x) \text{ in } (x_0, x_1).$$

(ii) The integrating factor to put (1) into standard form is r^2.]

8   Consider the finite element formulation for Love waves in which the sth propagating mode has the form

$$\mathbf{u}_s = \alpha_s \mathbf{U}_s \exp \mathrm{i}(\omega t - \kappa_s x),$$

where α_s is the mode participation factor (see §12.3.2).

Prove that the rate of energy transmission per unit width of wave front is

$$E_s = \tfrac{1}{2}\omega\kappa_s|\alpha_s|^2,$$

and that the group velocity is

$$C_s = d\omega/dk = \frac{\kappa_s/\omega}{\mathbf{u}_s^{\mathrm{T}}[\mathbf{M}]\mathbf{u}_s},$$

where $\mathbf{M}$ is the mass matrix for the finite element mass.

9   Show for cubic symmetry in the notation of §12.1.2, with $l = 1$, that the quadratic for the phase velocity surface is

$$(A - \rho c^2 - Lq^2)(L - \rho c^2 - Aq^2) + q^2 (L + F)^2 = 0.$$

From the boundary conditions prove that in this case the wave velocity can be found from

$$\left(1 - \frac{A}{L}R\right)\left(1 - \frac{F^2}{A^2} - R\right)^2 = R^2 (1 - R),$$

where $R = \rho c^2/A$. Note that there is no dispersion in this case.

10 Show that for cubic symmetry there are only three elastic constants, A, F and L.

Consider whether a Love type motion can exist (i.e. $u_3 = 0$). Try

$$u_1 = V_i \exp\{-q_i \kappa x_3 + i\kappa(lx_1 + mx_2 - ct)\}$$

for $l = 1$, $m = 0$, and show the resulting equations are satisfied by $u_1 = 0$, $u_2 \neq 0$ if

$$\rho c^2 - L(1 - q^2) = 0.$$

By examining the boundary conditions, can you say whether a Love wave can ever exist in this case?

13

Seismic waves and planetary interiors

The extraction from observed seismograms of seismic wave velocities, density, and elastic and inelastic parameters in the Earth is a mathematical inverse problem subject to statistical variability, incomplete resolution, and lack of uniqueness. Yet, from the theory discussed in earlier chapters, it is possible to derive much information concerning the Earth's interior.

It was once thought that the Earth was composed of a comparatively thin crust resting on material in a fluid or near-fluid state. Seismology has shown that this is not the case, but the terms *crust* and *crustal layers* continue to be used in connection with the outer part of the Earth.

Seismological data on even deeper structure come from several sources, including body waves in earthquakes and explosions (see § 10.4), the dispersion of surface waves from distant earthquakes (§ 5.5), and eigenspectra from large earthquakes (§ 5.7). We proceed to discuss these sources of evidence in this and the next chapter.

13.1 Major discontinuities within the Earth

13.1.1 Existence of a crust. Oceanic and continental structures

In 1909, A. Mohorovičić detected two distinct pairs of P and S phases on seismograms within 10° of the epicentre of the Kulpa Valley (Croatia) earthquake of 1909 October 8, and inferred the presence of a marked structural discontinuity some distance below the surface of the Earth.

For phases corresponding to rays in the 'upper layer' (above the discontinuity), Mohorovičić used the notation $\bar{P}$ and $\bar{S}$. He used a form equivalent to (7.81) to arrive at a $\bar{P}$ speed ranging from 5.53 to 5.68 km/s in this layer, and initially gave 54 km as the depth to the discontinuity. Similar pairs of phases were found by S. Mohorovičić and Gutenberg with two German earthquakes. Subsequent studies in Europe, and later over the whole globe, showed that the *Mohorovičić discontinuity* (sometimes called the 'M discontinuity') is world-wide, though its average depth is considerably less than 54 km and it is not always a sharp transition.

In the Tauern (Austria) earthquake of 1923 November 28, Conrad found evidence of a third distinct *P* phase, which he called *P**, with speed about 6.3 km/s. This was supported in studies of the Jersey and Hereford earthquakes of 1926 by Jeffreys, who also identified a companion phase *S**. Jeffreys re-named the phases $\bar{P}$ and $\bar{S}$ as *Pg* and *Sg*. The two layers associated with *Pg* and *P** have been called the *granitic* and *intermediate* layers, respectively, and the boundary between them the *Conrad discontinuity*. The notation *Pn* and *Sn* is used in near-earthquake studies for phases corresponding to rays which penetrate below the Mohorovičić discontinuity. Most early near-earthquake studies gave *Pn* speeds of the order of 7.8 km/s, though Conrad had inferred 8.1 km/s with the Schwadorf earthquake of 1927 October 8.

Sometimes provision is also made for phases P_s and S_s, corresponding to speeds of order 4.7–5.0 and 2.8–3.0 km/s, in a thin sedimentary layer above the granitic layer.

In 1937, Jeffreys combined a number of European studies, and, on the assumption of homogeneous upper layers with horizontal boundaries, gave the following speeds:

Pg	5.57 ± 0.02 km/s	*Sg*	3.36 ± 0.01 km/s
P*	6.50 ± 0.03	*S**	3.74 ± 0.03
Pn	7.76 ± 0.03	*Sn*	4.36 ± 0.02

This model is formally connected with the JB tables, but needs some corrections, especially for *Pn* and *Sn*. (See the discussion in §13.2.1 and Fig. 4.2.)

For comparison, in central California, in 1939 Byerly inferred a two-layer crustal model with $\alpha = 5.6$, $\beta = 3.3$ km/s in the upper layer; $\alpha = 6.7$, $\beta = 3.9$ km/s in the second layer; and $\alpha_n = 8.0$, $\beta_n = 4.7$ km/s.

Observations of surface waves also played an important part in establishing the reality of a crustal layer (see §5.4). Early work of Tams, Angenheister and Gutenberg indicated that surface seismic waves over paths in the Pacific region travel faster than and are differently dispersed from those in continental regions, and suggested that there are significant differences in the crustal structures.

The first quantitative investigations with surface waves were made in 1928 by Stoneley, taking a model crust with a single layer of thickness *H* (with velocities derived for the European upper layer) resting on an olivine subcrust. He found that observations of the dispersion of Love waves (see § 5.4.2) yielded $H = 19$ km for Eurasia as against $H = 10$ km for the Pacific region, and therefore that the crustal structures of the two regions differ significantly. A similar conclusion emerged from early Rayleigh wave investigations.

Other assumed crustal models led to similar differences, and all pointed to the sub-Pacific crust being significantly thinner (5 to 6 km) than the Eurasian. Investigations were extended by Wilson, Ewing, Press and others to the Atlantic and Indian Ocean regions, for which oceanic crusts about 6 km thick were also found.

With the development of the seismic survey method, it became possible to investigate oceanic crustal thicknesses more directly. Work of Hill, Raitt, Ewing and collaborators yielded values ranging from 9 to 14 km for the depth of the Mohorovičić discontinuity below the ocean surface in parts of the Pacific and Atlantic regions.

The corresponding P speeds in an assumed single crustal layer below the ocean sediments range from 6.5 to 6.8 km/s, and the Pn speeds from 7.9 to 8.2 km/sec. Ewing and collaborators constructed theoretical dispersion curves based on these results and found satisfactory agreement with Love and Rayleigh wave observations in a number of cases.

Since this early definitive work, the availability of multiple surface wave paths between the globally distributed WWSSN stations (see § 1.3.2) has provided much more detailed information on structural variability from surface wave analysis. Higher-mode wave trains have been recorded and used in specially developed inversion schemes (see § 7.4.3) and the measured wave spectrum extended out from periods of less than 10 s to hundreds of seconds (mantle modes) and finally to the terrestrial eigen-vibrations (see § 5.6.6). Among the main results has been the probable need to introduce some transverse anisotropy in the upper mantle in order to explain simultaneously dispersion curves of Love and Rayleigh waves (see § 12.1.2) and the mapping of large global structural variations.

13.1.2 Existence of a central core

The existence of a central core within the Earth appreciably different from the outer shell had been suggested by Wiechert in 1897. Seismological evidence for such a core was put forward by Oldham in 1906, and Gutenberg in 1913 estimated from seismic waves the depth of its boundary as 2900 km below the Earth's outer surface. Work of Jeffreys in 1939 gave the depth as 2898 ± 3 km.

Direct evidence of the existence of a central core is twofold. First, there is the occurrence of a 'shadow zone' for P waves emerging between epicentral distances of about 105° and 142°; in this shadow zone the amplitudes of P waves are much reduced (but diffracted waves dimly illuminate the zone (cf. Fig. 10.2)). Such a shadow zone would correspond to the presence of a discontinuity surface in which the P velocity sharply diminishes from above to below. Waves beginning at $\Delta = 142°$ correspond to the phase PKP. The main observations correspond to the case described in §7.3.8, and as Fig. 10.2 shows very clearly, there are two branches in the travel-time curve of PKP for $\Delta > 142°$.

Secondly, evidence comes from reflected waves. The depth determination of Jeffreys for the core boundary was made using observations of travel-times of ScS and PcP (see Fig. 13.1). The prominence of these reflected phases from the core surface on many short-period seismograms not very distant from the epicentres is evidence of the sharpness of the discontinuity. Repeated core reflections on the inside of the core boundary such as $P7KP$ confirm the inference (see Fig. 13.4).

The region of the Earth outside the central core and below the crust is called the *mantle*. Both P and S waves are transmitted throughout most of the mantle, but only P waves have been detected passing through the central core. (The reader should ponder, however, why weak, slow core S waves would be hard to detect on seismograms.)

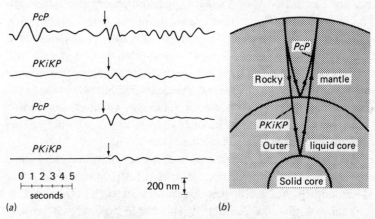

(a)   (b)

Fig. 13.1. (a) Seismograms from LASA from an underground nuclear explosion in Nevada on 19 January 1968. The upper pair are the signal sum of 25 seismometers; the lower pair are velocity-filtered from the complete array. Time proceeds from left to right; the vertical scale shows the magnitude of ground movement involved. (b) PcP and $PKiKP$ reflections ($\Delta = 10.9°$). (Courtesy B.A. Bolt and A. Qamar, 1970.)

13.1.3 Discontinuities in the mantle

Let α and β denote the P and S velocities at distance r from the Earth's centre, or depth z below the surface.

Below the Mohorovičić discontinuity, the mantle is characterised by much smoother variation of α and β with z than in the crust above. But there are some significant abnormalities, especially in the outermost 1000 km, or 'upper mantle'.

Early work of Byerly and Lehmann indicated a sharp change of gradient in the P travel-time (T, Δ) curve near 20°. Examination by Jeffreys and Bullen of well-recorded earthquakes confirmed this result for both P and S waves, and the phenomenon came to be called the '20° discontinuity'. On data available up to 1939, Jeffreys derived P and S velocity distributions in which α and β increase steadily as r decreases to $z = 413$ km, while the gradients $d\alpha/dz$ and $d\beta/dz$ increase discontinuously at 413 km and then steadily diminish until at $z \approx 1000$ km the gradients resume normal values; α and β are continuous in the solution throughout this whole range of depth.

This model corresponds to features described in § 7.3.3, including triplication of the (T, Δ) curves.

In order to derive any velocity distribution precisely, full details are required of the corresponding (T, Δ) curve, including details of upper branches and cusps. In practice, when triplication occurs, it is difficult to detect on seismograms all the needed arrival-times of onsets after the first so that there is appreciable uncertainty in velocity values corresponding to the 20° discontinuity estimated from P arrivals alone.

Over a period of years starting from 1939, Gutenberg used seismic amplitude data to infer negative P and S velocity gradients at depths of 100 to 150 km. When these negative velocity gradients violate the condition $dv/dr < v/r$ (see § 7.2.3), the problem of estimating values of α and β in the upper mantle becomes more complicated (see § 7.4.4).

The next major revision occurred when greater resolution of the seismic gradient $p = dT/d\Delta$ was provided by the large aperture seismic arrays built in the 1960s and 1970s. Careful array measurements, such as those of L.R. Johnson, showed that the Jeffreys P (and hence probably S) velocity gradients at depths from 200 to 800 km required significant changes (although the P and S *velocities* needed smaller adjustments). The major changes required were sharp *jumps in velocity* near 400 km and 650 km depth with concave downward P velocity gradients above and below the latter depth. A comparison of the Jeffreys velocity model with one that incorporates these revised curvatures is given in Fig. 13.2.

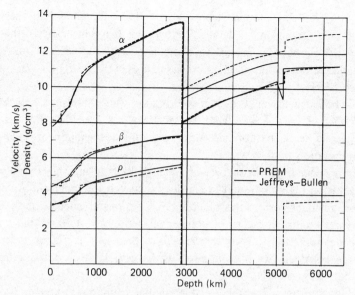

Fig. 13.2. Comparison of seismic velocities and density in the Earth models of Jeffreys and Bullen A, and PREM. (See text.)

Other evidence on the structure of the upper mantle comes from analyses of surface waves of periods over 1 minute. Most inversions are compatible with Gutenberg's negative velocity gradient, but some points of consistency and uniqueness require further critical examination before definite conclusions can be made. Inferences are complicated by the possibility of elastic anisotropy in the upper mantle (see § 13.2.3).

In the lower mantle, i.e. where $1000 \text{ km} < z < 2900 \text{ km}$, the variation of the P and S velocities appears to be less complicated with some regional differences. Most travel-time studies yield fairly uniform and normal P and S velocity gradients down to at least 2700 km depth, but there is evidence of a reduction in gradients taking place between 2700 and 2900 km (see § 13.6.2).

13.1.4 Discontinuities in the central core

The shadow zone already referred to for the range $105° < \Delta < 142°$ is not devoid of observations of P phases. For a number of years these observations had been attributed to diffraction associated with the boundary of the central core and the 142° cusp of PKP waves, but Lehmann in 1936 pointed out that the amplitudes were sufficiently great to suggest reflection from an *inner core* in which the P velocity is significantly higher than that in the surrounding central core. In 1938 Gutenberg and Richter, using *PKP* data, found strong support for Lehmann's hypothesis, and supplied quantitative details. Jeffreys later applied Airy's theory of diffraction near a caustic to show explicitly that diffraction was an inadequate explanation of observed features of the *PKP* pulse for $\Delta < 142°$ (see §3.7).

The existence of the inner core is now well-established and reflected *PKiKP* waves are seen from its boundary (see Fig. 13.1). For want of a better name, the part of the 'central core' outside the inner core is called the 'outer core'.

Evidence from *PKP* core waves led Jeffreys to infer the existence of a transition region, about 150 km thick, between the inner and outer core, and an inner core radius of 1250 km. The formal calculations yielded a negative $d\alpha/dz$ in this transition region. Jeffreys postulated that α is here proportional to r (this corresponds to $d\alpha/dr = \alpha/r$; cf. §7.2.3), and showed that this postulate is compatible with the JB tables but not uniquely required by them.

In contrast, Gutenberg's velocity determinations gave a transition region in which there is a rapid increase in $d\alpha/dz$ at the bottom of the outer core, then $d\alpha/dz$ diminishes steadily and fairly rapidly with increasing depth until the inner core proper is reached. The net effect was an increase in α of about 10 per cent inside the transition region.

The more recent evidence, obtained from seismic arrays, of *PKiKP* recorded with $\Delta < 20°$, is crucial evidence favouring a velocity distribution of the Jeffreys type against the Gutenberg type.

Throughout the outer core proper, on the Jeffreys solution, $d\alpha/dz$ is fairly steady and normal, while inside the inner core it is steady but small and positive.

Fig. 13.2 compares the *P* and *S* velocity distributions derived by Jeffreys in 1939 from the JB tables with a more recent solution, called Preliminary Reference Earth Model PREM (see §13.2.3 and §13.5.3). The JB velocities relate to the Earth model defined in §10.6.3 and were in broad agreement with values derived at the time by other seismologists, notably Gutenberg and Richter, who also used extensive travel-time data.

The figure includes no values of *S* velocities for the Jeffreys solution in the central core. The reason for this will be discussed in §13.3.

As already mentioned, the JB velocity values are in need of corrections immediately below the crust. Moreover, starting from a few tens of kilometres below the crust, for radially symmetric average Earth models, the principal revisions in the upper mantle are in the velocity gradients rather than in the velocities themselves. There are other appreciable revisions needed, notably at the bottom of the outer core.

In spite of such modifications, based on the accumulation of much data since 1939, there are reasons why the JB tables are still used for many purposes. First, these tables have the advantage of being consistent with a large quantity of travel-time data analysed according to explicit statistical procedures, and also with subsequent computations relating to the physical structure of the Earth's interior. Secondly, evidence on the corrections is still conflicting in a number of respects. As with all mathematical models, it is important to bear in mind the uncertainties and simplifications when applications of the velocities are being made to geophysical problems.

13.1.5 Division of the Earth's interior into shells

In 1940–2, in work on the Earth's density variation (see §§ 13.4 and 13.5), Bullen found it convenient to divide the Earth into regions according to depth, and used the nomenclature A, B,...,G for this purpose. The division, which was based on the velocity distributions entailed by the JB travel-time tables, is shown in the Table 13.1. For convenience of com-

Table 13.1 *Specification of internal shells of the Earth* (1942)

Region	Level	Depth (km)	Features of region
—	Outer surface	—	—
A			Crustal layers
—	Base of crustal layers (distant R from the Earth's centre)	33	—
B			Steady positive P and S velocity gradients
—	$0.94R$	413	—
C			Transition region
—	$0.85R$	984	—
D			Steady positive P and S velocity gradients
—	$0.548R = R_1$	2898	—
E			Steady positive P velocity gradient
—	$0.40R_1$	4982	—
F			Negative P velocity gradient
—	$0.36R_1$	5121	—
G			Small positive P velocity gradient
—	Earth's centre	6371	—

putation, the depths in the third column are given to the nearest kilometre but, as the preceding discussions have indicated, are not determined to this order of accuracy. The gradients mentioned in the last column are with respect to increase of depth.

In 1950, the region D was subdivided into D' and D", extending from 984 to 2700 km, and 2700 to 2900 km, respectively, following evidence (see §13.6.2) that the diminution in velocity gradients inside D" is physically significant.

The nomenclature in the table continues to serve as a useful basis in discussing the Earth's interior but requires significant modifications of the regions B and C (see Bolt, 1982, Table 4.1).

13.2 *P* and *S* velocity distributions in the Earth and Moon

We now turn our attention to a brief overview of more recent estimates of *P* and *S* velocities in the Earth's interior. A thorough analysis of the large literature would be out of place in a book on seismology theory. We emphasise that the evolution of seismological research is in the direction of the resolution of heterogeneity and structural variations. It remains crucial, however, to compare such perturbations with simpler radially averaged Earth models.

13.2.1 The crust

In recent years there have been many special studies and publications reviewing the structure and values for seismic velocities in the crust. The interested reader is referred, for example, to the monograph by the American Geophysical Union (see Mueller, 1977).

The basic theory of seismological refraction and reflection methods has been given in §7.6. We only indicate here a few representative results for velocities in the crust.

In Europe, the crustal structure has been found to be complex with the normal thickness of 30–32 km disturbed by the central rift system and the Alps. The rift structure, with smaller thickness, traverses Europe from the North Sea through the Rheingraben to the Rhône Depression; the Alpine collision region has crustal thicknesses reaching 55 km. Similar conclusions on complexity of structure and thickness hold for other continental areas. In particular, there is now much evidence that a two-layered continental crust with a Conrad discontinuity is too simple. A mid-crustal discontinuity is not found in many seismic surveys while other discontinuities are regional in extent. Reflection surveys reveal variations on a scale smaller than 10 to 100 km that typifies seismic refraction methods (see Fig. 7.11).

Usually, in normal continental regions there is a superficial low-velocity sedimentary layer underlain by a zone with positive $d\alpha/dz$. This may be followed by a layer in which P velocities in some places fall from 6.0 km/s to 5.6 km/s. The middle part of the crust is characterized by a heterogeneous zone with $\alpha \approx 6.0$ to 6.3 km/s. The lowest layer of the crust (about 10 km thick) has significantly higher P velocities ranging up to $\alpha \approx 7$ km/s.

The composition of the crustal rocks is now believed to be complex on a regional scale. The best evidence points to an upper crust consisting of metamorphic rocks such as gneisses and schists underlain by granite intrusions. The middle crust is composed of granitic gneiss and migmatites and the major constituents of the lower crust are differentiated metasedimentary igneous and volcanic rocks. This description is consistent with an exposed crustal section visible on a geological profile through the Ivrea zone of the southern Alps.

Over the period 1930 to 1935, Gutenberg (for the Alps) and Byerly (for the Sierra Nevada in California) produced the first seismic evidence on 'mountain roots', that is, a marked increase in the depth to the Mohorovičić discontinuity under certain mountain ranges. This seismological work confirmed earlier arguments based on gravity anomalies and the isostatic model. Similar work has been done in other mountainous regions and, in many cases, it appears that the depth extends to approximately an additional 30 km. Apart from major mountain systems, large variations in crustal structures occur along the margins between continents and oceans, and at tectonic plate boundaries such as mid-ocean ridges, and 'arcuate structures' i.e. island arcs with associated deep trenches.

Early key seismic surveys of continental margins included the work of Worzel and Shurbet in 1955 near the New Jersey coast, and of Tuve and Tatel in 1958 in the Andes region. Changes were found in the depth of the Mohorovičić discontinuity from 30 to 15 km, spread over a horizontal distance of about 200 km approximately inversely with water depth, with more gradual changes of depth on either side of this range of distance. Tuve and Tatel found values of crustal thickness ranging from 46 to 56 km under the Andes to 10–15 km just offshore in the Pacific Ocean. Oceanic crustal velocities were estimated to be about $\alpha = 6.6$ and $\beta = 3.4$ km/s.

In the deep ocean the Mohovoričić discontinuity is at a depth of about 11 km. Under a sedimentary layer of about 1 km thickness, the lower layer of the oceanic crust is inferred to consist of tholeiitic basalt that formed where extrusions of hot basalt at mid-ocean ridges were being added to the upper part of lithospheric plates as they were spreading away from the ridge crests. This crustal layer cools as it moves away from the ridge crest and its seismic velocities increase correspondingly, reaching values of $\alpha \approx 6.8$ and $\beta \approx 3.8$ km/s at distances beyond the flanks of the ridges.

In summary, it is now well established that the crustal thickness lies between 30 and 40 km in normal continental regions. (There is therefore no reason for Earth models to deviate from the value 33 km that is used in the JB tables.) The reduced depth to the Mohorovičić discontinuity under the oceans is well confirmed. Most studies show that Pn and Sn speeds are about 8.2 and 4.7 km/s in many regions, though speeds below 7.8 and 4.4 km/s and above 8.5 and 4.9 km/s are appropriate in others. Pulses arriving at certain epicentral distances have been interpreted as reflections from the Mohorovičić discontinuity, indicating that the early assumption of a sharp discontinuity may be reasonable at least in some regions (see Fig. 7.11).

The intense effort spent on crustal investigations has led to the application of full wave theory to many problems with a view to extracting information from detailed analysis of wave forms (cf. § 8.8). The mathematical approach to these problems has been supplemented by optimal filtering techniques (§ 9.9.4) in order to provide finer identification of crustal structure.

13.2.2 The lithosphere

The lithosphere is defined as the outer, more-or-less rigid, shell of the Earth above the asthenosphere. It contains the crust and tectonic plates and extends to depths of about 100 km into the mantle. Below this depth the *asthenosphere* is defined as extending a further 200 km or so. The reader should be aware that, unlike the shells of Table 13.1, the distinction between the lithosphere and asthenosphere is not based primarily on seismic velocities but rather mainly from dynamical considerations.

The theory of surface waves and its application to crustal and upper mantle structure has been extended in order to study the variations in this geophysically important shell. Dispersion curves for many different assumed structures have now been constructed with the aid of digital computers. Multiple comparisons make it possible to eliminate many types of lithosphere structure from consideration in particular regions, and narrow the available variations. Questions of uniqueness of many structural changes defined by inversion methods still, however, need examination.

A basic method of inferring regional structures from phase velocity measurements of surface waves was devised by Press in 1956. Individual phases are correlated from record to record taken at a network of stations in the region. When there are sufficient seismographs in the region, even when the paths are mixed, the method can resolve the average structure in a region whose linear dimensions are of the order of several wave lengths. By a major application of computer algorithms, D.L. Anderson, J.H. Jordan and others have lately explored the method for mapping global structural variations in the lithosphere.

Immediately below the crust, the seismic speeds are so high as to fit laboratory data on only a small number of ordinary rocks. Peridotite, which consists largely of the mineral olivine (magnesium–iron orthosilicate) has values of k/ρ and μ/ρ of the right order, and olivine composition was first suggested by Adams and Williamson in 1923. The ultrabasic silicate interpretation has led to the adoption of a mean value of about 3.32 g/cm^3 for the density just below the crust. Available velocity–density correlations make it fairly unlikely that this value needs serious change. Eclogites also give reasonable agreement with Pn and Sn velocities, and Birch considered an eclogite consisting of an assemblage of olivines, pyroxenes and garnets as being a possible constituent just below the crust. In that case ρ might be greater than 3.32, but an eclogite composition becomes somewhat improbable if it requires ρ to exceed 3.4.

It should be noted that when the values of the densities at the various depths are determined, values of the elastic parameters k and μ are simply derived using (4.4) and (4.5).

13.2.3 The deep interior. Recent solutions

Many recent studies use classical ray-theory procedures (see chapter 7) with travel times from short-period body phases. Inversion from times to velocities is usually by the Herglotz–Wiechert integral or linear perturbation theory. As well as the main phases P, S, PKP and SKS, direct measurements for other phases have now been used, such as PdP, ScP, dif PcP, dif ScP, $SKKS$, $PKhKP$ (precursors to $PKIKP$ for $120° < \Delta < 142°$), $PKiKP$ and $PmKP$, $m = 2, \ldots, 13$ (multiple reflections inside the mantle–core boundary).

Because present controversies on fine interior structure usually entail travel-time differences of less than 2 s, alternative research tools with higher resolving power must be found. For example, measured amplitudes of seismic waves can test models using appropriate wave theory, as in a recent application where observed amplitudes of the $P'640P'$ precursors were found to be consistent with the superposition of waves scattered from a thin boundary layer at about 640 km depth (see § 13.6.1).

Theory now also permits the computation of synthetic seismograms of ground displacement with frequencies as low as 1 Hz (see § 16.3.3). Account may be taken of plausible source models and diffraction effects along the path of the propagating body waves. At least for simple sources and seismograms from calibrated digital instruments, it is then possible to make direct overlays between the observed and predicted pulse pattern for models of interest. Although no completely quantitative inversion procedures are at present available, trial and error permit a rejection of competing models and a narrowing of possibilities.

In applications of the above method to both mantle and core studies, the effect of damping on the pulse shapes can be significant and there is often a tradeoff between Q values and gradients in the elastic parameters. The effect may be important when the method is applied to achieve superior resolution in transition zones, where damping may also be anomalous.

An illustrative core structure study of this type is by Choy and Cormier (1983) who compute theoretical wave forms appropriate for a deep-focus earthquake in Brazil. Comparison was made with PKP waves of various branches (see Fig. 10.8) digitally recorded at standard stations over a distance range of $127.2°-165.7°$. A representative comparison from the work is given in Fig. 13.3.

Precursors to the main seismic phases have now received theoretical attention. One interesting case involves two explanatory hypotheses, not necessarily exclusive, that were advanced to explain the precursors to the $PKIKP$ core waves (see Fig. 10.8). First, in 1962, Bolt suggested that they are waves called $PKhKP$, reflected from a discontinuity near the base of the outer core. Travel-times of the first scattered arrivals could be explained by an additional branch GH of the PKP curve. Further analysis showed that a very small velocity jump (less than 0.02 km/s) would suffice to explain the amplitudes, with a small velocity gradient at the bottom of the outer core and a large velocity gradient at the top of the inner core (shell G).

The alternative hypothesis, proposed by Cleary and Haddon in 1972, was that the PKP precursors are due to scattering from irregularities near the mantle–core boundary. Travel-times could not discriminate between hypotheses but wave slowness $(dT/d\Delta)$ measured across arrays could. Most such slowness measurements are clearly in agreement with the scattering hypothesis. The possibility of ambiguities and large biases from wave interference and near array structure has been noted, however, and some experimenters find some $dT/d\Delta$ values consistent with reflecting interfaces near the base of the outer core.

From the better travel-time inversions, the P velocities in most of the lower mantle and outer core, averaged over 50 km or so, must be

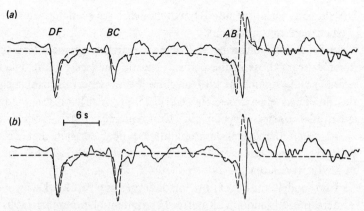

Fig. 13.3. Comparison of synthetic wave forms (broken lines) and observed ground displacements (solid lines) of core waves from the deep-focus earthquake in Brazil on 11 July 1978 at SHIO ($\Delta = 156.3°$). In (b), the comparison velocity model is CAL8 with a weak P velocity gradient in shell F. In (a), the comparison is with PREM with a higher velocity gradient at the bottom of the outer core (see Fig. 13.2). The pulse BC is significantly affected by a structure near the inner core boundary. (Courtesy G.L. Choy and V.F. Cormier, 1983).

considered to be now resolved to within 0.03 km/s (see Fig 13.2). Although further refinements are marginally possible, inherent sources of scatter in travel-time and slowness studies (such as source-location error and structural variations) preclude much greater precision in average velocity estimates by using the Herglotz–Wiechert method (see §7.4.1). Further application is likely to use differential travel-times (e.g. $PKP(AB) - PKIKP$) from individual sources and seismograms as constraints to *regional* models.

Finally, we must emphasise that most current *average* seismic velocities and densities inferred for the terrestrial interior depend heavily on eigen-spectra measurements.

The free oscillations of the Earth are described by ordinary differential equations with elastic parameters (and hence seismic velocities) as coefficients (see §14.2). The problem is, given the measured eigen-frequencies of the system, to determine these coefficients. Mathematical formulations show that, without infinite spectra, unique solutions for the coefficients cannot be determined by inverse procedures (e.g. Sabatier 1978; Hald 1980). Nevertheless, from more than 1000 eigen-frequencies of normal modes that have already been measured, Earth models have been derived that closely predict these measurements.

In 1965, Jeffreys warned that eigen-spectra for an anelastic Earth were different from those of the perfectly elastic one (see § 5.7.3). The effect is geophysically significant, with eigenfrequency observations differing from the predicted elastic case by 8 s (0.25 per cent) for the fundamental spheroidal eigen-frequency $_0S_2$. Consequently, Earth models that are solutions to the eigen-vibration inverse problem but that do not incorporate anelasticity are suspect and should be neglected in making quantitative inferences.

Two models that do allow for damping are PREM (Dziewonski and Anderson, 1981) and CAL8 (derived by Bolt and Uhrhammer, 1981). The P and S velocities and densities for PREM are plotted in Fig 13.2 and numerical values for PREM and CAL8 are tabulated in the Appendix; a description of their construction methods is given in § 13.5.3.

13.2.4 The lunar interior

The Moon is the only other planetary body at present that has been subject to seismological exploration of structure. The complete seismic data set (from 1969 to 1977) is not large (see § 1.3.4) with few seismological stations and a limited number of usable sources. Nevertheless, inversions of travel-time data have led to a reasonably well resolved interior structure, with some significant differences between published solutions, and variable uncertainties in estimated parameters.

In brief, the Moon has a well-defined crust, a solid radially differentiated mantle and, with lower resolution, a possible partially molten central core. A likely layered crustal model has a 4 to 12 m surface layer (regolith) with $\alpha \approx 100$ m/s grading to 960 m/s at 1.5 km depth. At 4 km, the P velocity jumps to $\alpha = 4.7$ km/s. Two crustal layers below this depth have representative velocities of $\alpha = 5.1$ km/s and $\beta = 3.0$ km/s to 20 km depth, and 6.8 and 3.9 km/s, respectively, from 20 to 60 km depth.

Average velocities in the upper mantle appear to decrease from $\alpha = 7.74$, $\beta = 4.49$ km/s to $\alpha = 7.46$, $\beta = 4.25$ km/s between 270 and 500 km depth. In the middle mantle (depths between 500 and 1000 km) $\alpha \approx 8.26$, $\beta \approx 4.65$ km/s with standard errors of 5 per cent. Attenuation parameter values in the upper and middle mantle are of order $Q_P = 5000$, $Q_S = 3000$ and $Q_P = 1500$ and $Q_S = 1000$, respectively. Few rays that penetrated below 1000 km depth were recorded so that very tentatively, there may be a lower mantle of higher attenuation below this depth and, less convincingly, a small separate core of radius 200 to 400 km.

13.3 The states of the Earth's mantle and core

13.3.1 Solidity and fluidity

Because S as well as P waves are transmitted through essentially all parts of the Earth's mantle, this portion of the Earth is solid, i.e. has rigidity (μ), down to a depth of about 2900 km.

This result is in contrast to the evidence from seismograms that only P and not S waves are transmitted through the outer core. It indicates (see equation (4.5)), that rigidity in the central core is inappreciable, i.e. that the central core is in a fluid state. Support of this conclusion comes from other studies such as on the amplitudes of SKS waves. These amplitudes are much greater than would be the case if the rigidity μ in the region E of the core were comparable with its incompressibility k. (There are, however, reasons for believing that the *inner* core may be solid; see §13.5.2.)

It is well to appreciate that the evidence that the rigidity in the core is small or zero does not come entirely from seismology. Astronomical observations on the movements of the Earth's poles, together with observations of the Earth's tidal movements, give information on the rigidity of the Earth as a whole. Because the rigidity throughout the Earth's mantle is known to good accuracy (see §13.5) from observations of S seismic waves and knowledge of the density distribution, the global data yield an estimate of the mean rigidity of the central core. Calculations made by Takeuchi in 1951 and Molodenski in 1955 show directly in this way that the rigidity in the outer core is at most a small fraction of the incompressibility. (This applies strictly to behaviour under stresses with periods up to 12 h.) In addition, measurements of certain modes of the eigenfrequencies of the Earth (see §14.3) provide evidence on the average elastic properties of the core. Various comparisons between observations and theoretical values have demonstrated that a plausible fit can be made only if the outer core has essentially zero rigidity.

A key observational result for the core is the sharp increase in α from outer to inner core. From (4.4), we have $k + 4\mu/3 = \rho\alpha^2$, so that this increase in α must be due to a sharp increase in k or μ (or both), because ρ is an increasing function of depth. If the inner core were fluid, there would be no increase in μ, in which case the increase in k would have to be so large as to violate known physical constraints (see §13.5.2). For this reason, Bullen inferred in 1946 that the inner core is solid. From physical data on the variation of k with atomic number, he estimated that any increase in k near the inner core boundary is unlikely to exceed one-fifth of the increase needed to keep the inner core fluid.

A solid inner core would transmit *S* seismic waves, so that there is the possibility of confirming its solidity by detecting *PKJKP* readings on seismograms, where *J* corresponds to an *S* wave passage through the inner core. For research purposes, theoretical travel-times of the phase *PKJKP* have been prepared by Bolt and O'Neill in 1964 but energy calculations show that the phase is only on the border of observability. Several investigators have claimed that they have identified *PKJKP* on occasions, but more evidence is required before a positive conclusion can be drawn. At the same time, failure to detect *PKJKP* does not disprove the solidity of the inner core, because the nature of its boundary and a low value for the damping parameter Q_μ in it could inhibit the transmission of detectable *S* waves.

Once seismic velocity and density distributions are estimated and pressure and solid/liquid state conditions determined, the major state parameter required is temperature. It is unfortunate that, although thermodynamic conditions do affect the seismic velocities to some extent (see § 2.4.5), the temperature distribution in the Earth cannot be calculated directly from seismic wave measurements. Much effort has been put into the indirect estimation of the internal temperature variation, usually using solid state theory (see, e.g., equation (13.12)). The essential arguments are set out by Verhoogen (1980). In summary, the temperature *T* in the mantle is likely to rise from about 1000 K at the bottom of the lithosphere to 4500 K at the mantle–core boundary. Assuming adiabatic conditions in the liquid outer core (from (11*b*) below) and putting $(\partial T/\partial p)_a = \gamma_G T/k$, where γ_G is the dimensionless Grueneisen's ratio (see § 2.4.5) we have

$$\frac{dT}{dz} = \frac{g\gamma_G T}{\alpha^2},$$ (1)

which gives a temperature at the inner core boundary of about 5750 K for plausible $\gamma_G (\approx 1.2)$. At the centre of the Earth the temperature may reach 6000 K.

The average chemical compositions of the mantle and core have been inferred from the estimated depth distributions of seismic velocities, densities, anelastic properties, and temperatures. The most direct method is to make correlations between these values and corresponding laboratory measurements on likely rock and mineral constituents. On this basis, probable constituents of the mantle are mixed oxides of magnesium, aluminium, calcium and iron with silicates. Eclogite, dunite, and peridotite consisting of olivines (Mg_2SiO_4–Fe_2SiO_4), pyroxenes, and garnet are the most often suggested. In the upper part of the mantle, water and other ubiquitous compounds and elements may be present in various proportions.

In the core, the commonly held view is that iron is the major constituent. In 1963, Birch noticed that the relation between $\alpha = (k/\rho)^{\frac{1}{2}}$ and density in a wide range of elements and compounds depends systematically on their mean atomic weight. On this basis, the seismological Earth models suggest that the outer core (estimated mean atomic number 24) is mainly iron (mean atomic weight 26) with 10 to 15 per cent lighter elements. The correlation also suggests that the inner core is almost pure iron. The inference on iron and its alloys is supported by shock wave experiments which can simulate core pressures. The suggested candidates for alloying elements include oxygen, silicon, sulphur, and magnesium. Cosmic abundances of elements and meteorite compositions are also used to support various Fe mixes, including a few per cent nickel, particularly in the inner core.

13.3.2 Anelastic properties

Amplitudes of seismic waves depend upon geometrical spreading, energy partition at reflecting boundaries, and damping. Estimates of damping come from calculated ratios of amplitude spectra of wave pulses recorded from the same event on the same seismogram.

For a damping factor $\exp(-\pi ft/Q)$, where t is the travel-time along the ray, average Q values may be determined from the spectral ratios (see §8.4.1). For example, let $R(f)$ be the spectral ratio of $P7KP$ to $P4KP$ (see Fig. 13.4). Then, to the first order

$$\ln R(f) = a - \{(3\pi T/Q) + b\} f, \tag{2}$$

where T is the time for one core leg and a, b are constant. Regression of (2) against log (spectral ratio) yields Q. Application of (2) has given $Q \approx 10\,000$ for P waves through the outer core. A similar procedure in the inner core gives Q values two orders of magnitude less. In 1977, Bolt estimated $Q = 450 \pm 100$ from the spectral ratio for $PKIIKP/PKiKP$ as recorded by the Large Aperture Seismic Array (LASA) in Montana. If the doubly reflected phase is identified correctly, the last estimate should be firm.

The measured decay of recorded eigen-vibrations of the Earth can also be used to infer average anelastic properties of the interior (see §14.3.3 for mathematical details). The rates of decay of the oscillations of order l are usually given as $\beta_l = \gamma_l/2Q_l$ (cf. §5.7.3). Because each Q value is related to a particular wave-length and radial displacement in the Earth, the estimated values can themselves be inverted to determine corresponding damping properties for specified depths in the Earth. Those eigen-vibrations with

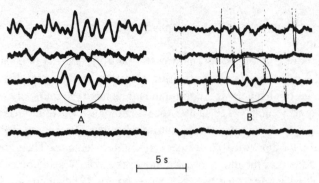

Fig. 13.4. Seismograms from the Jamestown station in California, showing *P4KP*, labelled A, and *P7KP*, labelled B, phases from an underground nuclear explosion in Novaya Zemlya ($\Delta = 68.9°$). The event occurred on 12 September 1973, and the records are from the vertical component short-period seismograph. Ground motion of *P7KP* is about 10^{-9}m.

significant particle motions in a particular shell thus provide a measure of damping within that shell. The inferred Q_P value for the lithosphere is about 200, for the lower mantle about 1000; in the outer core it has a very high value, of order 10^4, in agreement with the *PmKP* measurements mentioned above.

13.4 The Earth's density variation

Since the famous experiment of Cavendish in 1799, who used a modified form of apparatus devised by Michell, it has been known that the mean density of the Earth is about 5.5 g/cm^3. This implies that in the Earth's deep interior the material is considerably denser than typical surface rocks. A recent estimate of the Earth's mean density is 5.517 g/cm^3, using the value of 6.670×10^{-8} c.g.s. units for the constant of gravitation G. The corresponding value of the mass of the Earth is 5.977×10^{27} g. The uncertainty in each of these figures is about 1 part in 1300.

A second observational result which must be fitted by any representation of density in the Earth is that the moment of inertia I about the polar axis is $0.3308Ma^2$, where M and a are the mass and mean radius of the Earth. This value, which is accurate to one part in 300, has been derived from studies of the figure of the Earth and related data on the motion of artificial satellites (see Cook, 1980).

13.4.1 Early models of density variation

Let ρ, p, k and μ denote the density, pressure, adiabatic incompressibility and rigidity at depth z below the Earth's surface (or distance r from the centre).

Early representations of density variation in the Earth were mathematical relations, plausible physically but arbitrarily set down to study large-scale terrestrial and planetary phenomena. Before seismological evidence on interior structure was available only very limited resolution of density could be obtained. Laplace, for example, took in 1825 a homogeneous Earth model in which dk/dp equals 2. This equation of state yields the density law

$$\rho = Ar^{-1} \sin Br. \tag{3}$$

In 1848 Roche took the simpler density law

$$\rho = A - Br^2, \tag{4}$$

which has a density gradient zero at $r = 0$. Knowledge of M and I enables A and B in either (3) or (4) to be determined, and thence ρ as a numerical function of r. Roche's law gives 2.18 and 10.53 g/cm^3 for the densities at the surface and centre of the model. Wiechert allowed for a central core, but had to take the densities ρ_0, ρ_1 in the mantle and core as constants in order to get a determinate result.

13.4.2 Equations for density gradient from seismology

An outstanding importance of the P and S velocity distributions is that through (4.4) and (4.5) they lead to well-determined values of k/ρ and μ/ρ throughout much of the Earth's interior. If independent information were available on some third explicit function of any of ρ, k, and μ, the values of all three quantities could be computed. In practice, the procedure is more involved, and rests on a variety of evidence and the application of inverse theory.

By (4.4) and (4.5)

$$k/\rho = \alpha^2 - 4\beta^2/3 = \phi, \quad \text{say}, \tag{5}$$

and

$$\mu/\rho = \beta^2, \tag{6}$$

where α and β denote the P and S velocities. It is sufficiently accurate for most purposes to take μ as zero in the outer core (see §13.3), so that k/ρ, μ/ρ and the important quantity ϕ are observationally determined within fairly close limits all the way from the Mohorovičić discontinuity to the base of the region E.

As a close approximation, let us represent the stress at a point of the interior in terms of the mean $(-p)$ of the principal stresses (see §4.7). Strictly, ρ is a function of the pressure p, the temperature T, and an indefinite number of parameters specifying the chemical composition. It is convenient temporarily to take the entropy S as a variable instead of T. For the variation of ρ with z in a chemically *homogeneous* region of the Earth, we can then write

$$\frac{d\rho}{dz} = \left(\frac{\partial \rho}{\partial p}\right)_S \frac{dp}{dz} + \left(\frac{\partial \rho}{\partial S}\right)_p \frac{dS}{dz}. \tag{7}$$

By (2.36), we have for adiabatic changes of a chemically homogeneous material

$$k\,d\rho = \rho\,dp. \tag{8}$$

Let g denote the gravitational force per unit mass at distance r from the centre, and m the mass inside the sphere of radius r. By hydrostatic theory and the theory of attraction,

$$dp = g\rho\,dz, \tag{9}$$

where

$$g = Gm/r^2. \tag{10}$$

By (5), (9) and (10), the first term on the right-hand side of (7) is equal to $Gm\rho/r^2\phi$.

Next, let α_p denote the coefficient of thermal expansion at constant pressure, and τ the excess of the temperature gradient over the adiabatic gradient, at depth z. By standard thermodynamical theory,

$$\alpha_p = -\rho^{-1}(\partial\rho/\partial T)_p, \tag{11a}$$

and

$$\tau = \frac{dT}{dz} - \left(\frac{\partial T}{\partial p}\right)_S \frac{dp}{dz} \tag{11b}$$

$$= \left(\frac{\partial T}{\partial S}\right)_p \frac{dS}{dz}.$$

Thus

$$\tau = -\frac{1}{\rho\alpha_p}\left(\frac{\partial \rho}{\partial S}\right)_p \frac{dS}{dz}. \tag{12}$$

By (12), the second term on the right-hand side of (7) is equal to $-\alpha_p\tau\rho$. Hence (7) becomes

$$\frac{d\rho}{dz} = \frac{Gm\rho}{r^2\phi} - \alpha_p\tau\rho. \tag{13}$$

The form (13), neglecting the term in τ, was used in 1923 by Adams and

Williamson. In 1936, Bullen applied it to determine the Earth's density distribution. The term in τ in (13) is due to Birch, though derived by him in another way. Birch estimated that a departure of 1 K/km from an adiabatic temperature gradient could affect the right-hand side of (13) by the order of 10 per cent. Work of Verhoogen (1980) on temperature variation makes it likely that the actual departure is appreciably less than 1 K/km at most depths. Bullen showed that when the density distribution is derived by the method to be outlined in § 13.5.1, a 10 per cent error arising from neglect of the term in τ in (13) would lead to errors in the computed mantle densities nowhere exceeding 0.07 g/cm^3.

It follows that the simplified form

$$\frac{\mathrm{d}\rho}{\mathrm{d}z} = \frac{g\rho}{\phi}, \tag{14}$$

is a satisfactory approximation in parts of the Earth that are chemically homogeneous, adiabatic and devoid of phase changes.

13.4.3 Extension to inhomogeneous layers. The index η

We can now generalise (13) to allow for inhomogeneity. Simply, let us write

$$\theta(\eta, \tau) = \frac{\phi}{g\rho}\frac{\mathrm{d}\rho}{\mathrm{d}z} \tag{15}$$

$$= \eta - \alpha_p \tau \phi/g, \quad \text{say.} \tag{16}$$

The state index θ was first used by Bolt (1957) in density studies. Comparison of (14) and (15) shows that the coefficient θ is an index of the state of a region, being equal to unity in an adiabatic homogeneous region and greater than unity elsewhere. If we neglect temperature effects, because $k = \phi\rho$, we have

$$\frac{\mathrm{d}k}{\mathrm{d}p}\frac{\mathrm{d}p}{\mathrm{d}z} = \rho\frac{\mathrm{d}\phi}{\mathrm{d}z} + \phi\frac{\mathrm{d}\rho}{\mathrm{d}z},$$

and, by (9),

$$\frac{\mathrm{d}\rho}{\mathrm{d}z} = \frac{g\rho}{\phi}\frac{\mathrm{d}k}{\mathrm{d}p} - \frac{\rho}{\phi}\frac{\mathrm{d}\phi}{\mathrm{d}z}$$

$$= \frac{\eta g\rho}{\phi} \quad \text{from (16).} \tag{17}$$

In this case,

$$\theta = \eta = \frac{\mathrm{d}k}{\mathrm{d}p} - g^{-1}\frac{\mathrm{d}\phi}{\mathrm{d}z}. \tag{18}$$

The form for η was derived by Bullen in 1963 and is related to the Brunt–Vaisäla frequency in convection theory. The differential equation (15) is an important generalisation of the Adams–Williamson equation which it has replaced for Earth modelling purposes. (See also §13.6.2.)

Representative values of θ can be easily computed from the values in the Appendix. In the lower mantle $\theta \approx 1$ for both models. A feature of the PREM model is that $\theta = 1.00$ to the first decimal throughout the entire core. A strong constraint is implied because any linear perturbation of basic parameters in the inversion algorithm will produce fluctuations in θ. By contrast, the CAL8 inversions do not start with an Adams–Williamson core and do not converge exactly to unity; θ varies between 1 and 1.3 in E and, in G, θ decreases like $1/r$ with depth.

13.5 The inverse problem of density determination

13.5.1 Bullen's procedure
Integration of (14) and

$$\mathrm{d}m = 4\pi r^2 \rho \mathrm{d}r, \tag{19}$$

yields a density distribution for any range of r for which ϕ is known, provided starting values of m and ρ are available at some point.

Let m' and ρ' be the values of m and ρ just below the Mohorovičić discontinuity. In 1936, Bullen used (14) and (19) to compute a trial density distribution for the mantle between 33 and 2900 km depth, taking ρ' equal to 3.32 g/cm^3 and m' equal to the mass of the Earth minus a conventional allowance for the crust (see §13.2.1).

To test the trial distribution, the corresponding moment of inertia of the mantle was computed and subtracted from the known I for the Earth, with the result $I_1 = yM_1 R_1^2$, where $y = 0.57$ and I_1, M_1 and R_1 are the moment of inertia, mass and radius of the central core. This value of y exceeds 0.40 for a uniform sphere and so entails a substantial decrease of density with depth inside the central core – a conclusion that Bullen rejected for stability reasons.

The essential assumptions are: (a) the value for ρ'; (b) neglect of the term in τ; (c) neglect of possible changes of chemical composition and phase. In regard to (a), the calculations showed that ρ' would need to be at least 3.7 g/cm^3 to reduce y from 0.57 to 0.40, while all available evidence is against so high a value of ρ'. In regard to (b), it can be shown that any temperature effect could only increase y. It therefore followed, with strong probability, that the faulty assumption was (c) and hence that there are marked in homogeneities in the mantle between crust and core. Seismograms indicated that at least a large part of the changes occurs high up in the mantle, most probably in the top 600 km.

It followed also that the simplified equation (14) is inapplicable throughout all of the regions B, C and D. The trial procedure was then modified by abandoning the use of (14) in C where the abnormal velocity gradients occur. Indeterminacy immediately arose through lack of an equation for $d\rho/dz$ in C, and consequently also of starting values for ρ and m at some point of D. However, the simple dependence of α and β on ρ, according to (4.4) and (4.5), indicates that discontinuities in α, β, $d\alpha/dz$ and $d\beta/dz$ should be accompanied by corresponding discontinuities in ρ and $d\rho/dz$. Thus, on the velocity distributions of §13.2, Bullen assumed continuous variation of ρ from B through C to D, a discontinuous change in $d\rho/dz$ at the boundary between B and C, and continuity in $d\rho/dz$ through C to D. By assuming in C a quadratic law of density variation with depth, with constants chosen to fit the requirements just stated, it is then possible to derive a unique density distribution for the mantle if sufficient is known about the density in the central core.

In 1936–42, Bullen applied (14) also to the central core. Here, starting values of m and ρ are limited to the datum that $m = 0$ at $r = 0$, and there is indeterminacy through lack of evidence on the value of the density ρ'' at the centre of the Earth. But it transpired that powerful controls on the permissible density values throughout the regions B, C, D and E were supplied by moment of inertia criteria. The procedure gave $12.3 \, \text{g/cm}^3$ as the lower bound to ρ'', and showed that increasing this value of ρ'' by $5 \, \text{g/cm}^3$ affected the formally computed densities elsewhere by maximum amounts of only $0.03 \, \text{g/cm}^3$ in the mantle and $0.4 \, \text{g/cm}^3$ in the outer core.

It thus became possible to compute the Earth's density distribution to fairly good precision throughout all but the regions F and G, i.e. through about 99 per cent of the Earth's volume. The question of the density in the inner core will be considered in §13.5.4.

In 1940–2, the calculations were used with the JB velocities to derive density distributions on two fairly extreme hypotheses: (i) $\rho'' = 12.3$; (ii) $\rho'' = 22.3 \, \text{g/cm}^3$ (the latter value being taken arbitrarily).

A model with density values midway between those of the hypotheses (i) and (ii) was called Model A. Strictly, the non-linearity innate in (14) and (19) forbids the adoption of simple means but Bolt verified that the effect of non-linearity is trivial. Fig. 13.2 shows the density values for Model A from the crust to the base of the region E.

By (4.4) and (4.5), the density determination carries with it the determination of the elastic parameters k and μ, and hence, also, of the Lamé parameter λ, Young's modulus E and Poisson's ratio σ.

From (10), the variation of g is also derived from the density distribution. Calculations showed that g keeps within 1 per cent of $990 \, \text{cm/s}^2$ down to a depth of $2400 \, \text{km}$ (see Table A.2). It is a useful simplification for many purposes to treat g as constant down to this depth. The maximum value of g occurs near the mantle–core boundary. Inside the whole central core, g diminishes monotonely from this maximum to zero at the centre. Finally, the pressure distribution is obtained by (9) by numerical integration (see values in Table A.3). Further details are to be found in Bullen's book on density.

13.5.2 Bullen's compressibility–pressure hypothesis

The incompressibility k, the density ρ and the rigidity μ are three parameters whose values indicate primary physical characteristics of the material at any point of the Earth's interior. A striking feature of the Model A results was that, whereas the changes in ρ and μ at the mantle-core boundary are very large, the formally indicated change in k is merely (a reduction of) 5 per cent. The uncertainties in the postulates underlying Model A, moreover, permitted the change in k to be zero. (See also Table A.3.)

A second feature emerged following an examination by Bullen in 1949 of dk/dp. For a chemically homogeneous region in which the term in τ is neglected, we have by (5), (8) and (9)

$$dp/d\rho = \phi,$$

and hence

$$\frac{dk}{dp} = \frac{d(\phi\rho)}{\phi d\rho}$$

$$= 1 + \frac{\rho d\phi}{\phi d\rho}$$

$$= 1 + g^{-1}\frac{d\phi}{dz}. \tag{20}$$

Values of $d\phi/dz$ come directly from the seismic velocity data, and are about $2.2 \times 10^3 \, \text{cm/s}^2$ between 2500 and 2700 km depth (i.e. in the lowest 200 km of D'), and $2.1 \times 10^3 \, \text{cm/s}^2$ in the outermost 200 km of E. Also g is approximately $1000 \, \text{cm/s}^2$ at these depths. Because D' and E are likely to be nearly uniform in chemical composition, from (20), $dk/dp = 3.2, 3.1$ near the base of D' and top of E, respectively. Thus the suggestion arises that dk/dp, in addition to k, is nearly continuous between mantle and core.

It is necessary to note that $d\phi/dz$ falls continuously to about zero inside D″, so that application of (20) would give values of dk/dp falling to unity inside D″. However, this fall implies a marked variation of composition or temperature and so invalidates the use of (20) inside D″. Thus the above method fails to determine a value for dk/dp in D″, and provides no evidence against the near-continuity of dk/dp between mantle and core. The question of the heterogeneity of D″ will be considered further in § 13.6.2.

The contrast near the mantle–core boundary between the apparently slight changes in k and dk/dp on the one hand, and the very large changes in ρ and μ on the other leads to a hypothesis on the variation of compressibility with pressure.

Bullen in 1949 stated the hypothesis in the form that, to good approximation, k is a smoothly varying function of p for the materials present in the Earth below a depth of 1000 km. In 1950, Bullen set up a second Earth Model, B, in which a central postulate is that k and dk/dp vary smoothly with p at all depths below 1000 km.

In conjunction with the known mass and moment of inertia of the Earth, the procedure was found to entail severe restrictions on density in the upper mantle. In particular, between depths of about 200 and 1000 km, the procedure precluded the marked reduction in ρ, with decreasing depth, that occurs in Model A.

These results are now mainly of historical interest. On finite strain theory, dk/dp has the same value, at a given strain ε, for all elastic materials, regardless of composition, if expansion of strain energy in powers of ε is limited to the quadratic term (see § 2.7). There is now much experimental evidence, however, that a satisfactory equation of state requires inclusion of higher powers of ε.

13.5.3 Linear inversion. Tradeoff curves

We now give an introductory account of the current method of construction of Earth models for density and other elastic parameters. The description is best given by means of a comparison in assumptions used in computing the Earth models CAL8 and PREM (cf. § 13.2.3).

In the development of the model CAL8 (see Appendix) sets of observations of body-wave travel-times, surface-wave velocities and eigen-periods were combined simultaneously. These included differences in travel times between pairs of seismic phases that have similar paths in the upper mantle (e.g. $PP–P$, $PcP–P$, $PKiKP–PcP$, $ScS–S$, $SKKKS–SKKS$). Love and Rayleigh wave dispersion values for predominantly oceanic paths were used for periods between 60 and 200 s.

The P and S velocities adopted for the first iterations were close to recent empirical travel-time tables. For $25° < \Delta < 97°$, calculated P times fit the 1968 P tables within ± 0.3 s (see § 10.9.1). For the core, the 1968 PKP times of Bolt give similar agreement for the DF branch and lie within ± 1 s for the AB branch. For $25° < \Delta < 80°$, the S residuals lie between ± 0.8 s and for SKS ($85° < \Delta < 120°$), the CAL8 residuals range from -0.3 to $+1.8$ s against standard times.

At the outset, the class of Earth models must be chosen in terms of the incorporation of structural details previously determined using high resolution body waves such as $P'dP'$, PcP, $PKKKKP$, and $PKiKP$ (see § 13.5.4). Long period eigen-vibrations cannot provide sharp resolution of such first-order discontinuities.

A general assumption must also be made that variations in the P velocity, α, are accompanied by concomitant variations in the S velocity, β, and density, ρ. This plausible assumption has not always been followed in certain Earth models.

In CAL8 and PREM, first-order discontinuities were incorporated in the upper mantle at depths of 400 and 640 km (670 km in PREM). This assumption follows results that are now rather standard, particularly the consequences of reflections of the type $P400P$ and $P'640P'$ reported near these depths (see § 13.6.1). In CAL8, low-velocity layers were introduced in α and β centred at a depth of 120 km while in PREM there was the additional assumption of minor transverse isotropy located in the upper mantle. At the base of the mantle there is evidence of a mean reduction of about 2 per cent in β and perhaps α before the core is reached. Arguments such as those in § 13.5.2 then indicate that there would be a simultaneous increase in density. These properties in shell D'' were incorporated in CAL8 but not in PREM.

In the core, minor structure in any transition shell F was ignored in both models with the P velocity increasing slowly with depth to the inner core boundary, now known definitely to be a first-order discontinuity in structure. In the inner core itself the P velocity curve increases monotonely slowly.

In earlier iterations of CAL8 the radii of the inner core boundary and the boundary between the mantle and outer core usually converged to within a few kilometres of 1216 km and 3486 km; these values were fixed in the final iteration of CAL8.

Similarly, the density distribution in the zeroth iterate was developed by means of numerous previous trials. The Adams–Williamson equation (13) was not used at any depth in CAL8 but was assumed in shells D and E in

PREM. In the usual way, the crust of both radially symmetric models is meant to simulate, to the first order, some average of oceanic and continental properties.

All inversions were assumed to be linear. Travel-time and eigenfrequency residuals provide n linear equations of condition of the form (see §10.4.2)

$$\mathbf{Ax} = \mathbf{b}, \tag{21}$$

for n observations, and m unknown model adjustments x_i. Each equation can be weighted to allow for relative uncertainties in the observations. The elements of $\mathbf{A}$ correspond to the first partial derivative term in the Taylor expansion of functions $T(\rho_1, \rho_2, ...)$ where T is expressed in a common time dimension, and the subscripts on ρ denote some discrete parametrisation of the density (or other unknown parameter) as a function of radius $\rho(r)$. Typically, for example, $x_i = \delta\rho_i$.

In a number of inversion procedures, difficulty is encountered by the discrete parameterisation of the functions. Subsequent smoothing and additional integrations are needed. If polynomials are fitted by divided differences or Lagrangian interpolation, discontinuities may result in the derivatives at each tabulated point. A more suitable method is to fit cubic spline functions between specified points, called *knots*. The splines are piecewise continuous with continuous first and second derivatives at the internal knots.

The placing of knots is critical and in Earth models location should be guided by the structural boundaries. For density, function values at specified radii are available. For n cubics there are $n - 1$ internal knots, but $n + 3$ equations to be determined from the observations. Thus, in general, additional end conditions must be provided to prevent spurious curvature.

Let $z = r/R$ be the dimensionless radius. Suppose for the density in the Earth there are n knots $K_l(z_l)$, $l = 1, ..., n$. Each spline can be written

$$\rho = a_l z^3 + b_l z^2 + c_l z + d_l \quad (z_l < z < z_{l+1} \quad (l = 1, ..., n-1)). \tag{22}$$

The model function is now expressed in terms of the coefficients. That is,

$$T(\rho_1, \rho_2, ...) = T(a_l, b_l, c_l, d_l) \quad (l = 1, ..., n - 1). \tag{23}$$

If the new variables are assumed independent (small variations only),

$$O - C = \delta T = \sum \frac{\partial T}{\partial a_i} \delta a_i + \sum \frac{\partial T}{\partial b_i} \delta b_i + \sum \frac{\partial T}{\partial c_i} \delta c_i + \sum \frac{\partial T}{\partial d_i} \delta d_i. \tag{24}$$

Continuity at knots provides three linear side-conditions, in general, per knot. The system reduces to the matrix form (21) and its inversion follows the procedures of §10.4.2.

Next, we wish to determine at any radius the resolution of density (or other computed parameter) given by the particular parameterisation and set of seismic data with their experimental uncertainties. In other words, at a particular radius, we wish to know the estimated statistical error in a density value, given the amount of spread in the calculated density function there. Clearly, a desirable aim in providing this information is to minimise both the spread (i.e. maximise the resolution) and the statistical uncertainty, simultaneously. As Backus and Gilbert (1970) have shown, such a joint minimisation can be accomplished in a number of ways with mild assumptions.

Development is straightforward if the observational residuals y_i are assumed normal: $y_i \cap N(0, \sigma^{*2})$, say. (For different distributions the argument below is not changed in essentials.)

Consider the radius z_0. The functional relation between the observations and the density variation is

$$y_i = \int_0^1 G_i(z)\delta\rho(z)\,dz \quad (i = 1,...,n) \tag{25}$$

where $G_i(z)$ is the data kernel appropriate to the observation.

We seek a weighted average of the density variation about a given z_0 so that the spread is small. First, form the linear estimate

$$\overline{\delta\rho(z_0)} = \sum a_i y_i \tag{26}$$

$$= \int_0^1 \sum a_i G_i(z)\delta\rho(z)\,dz$$

$$= \int_0^1 H(z, z_0)\delta\rho(z)\,dz. \tag{27}$$

In (27), $H(z, z_0)$ is analogous to the probability frequency function. Because it has a convenient form to control the spread, we thus put

$$H(z, z_0) = \frac{1}{s(2\pi)^{\frac{1}{2}}} \exp\{ -(z - z_0)^2/2s^2\}, \tag{28}$$

where s is the 'spread' and $\int_0^1 H\,dz = 1$.

The spread function H may be now represented by a series of discrete values ε_i for an assumed spread s. Thus, fix s and let z take the discrete values $z_0 \pm j\delta z, j = 0,..., (m-1)/2$.

Now the weights a_i may be calculated from the m linear equations formed from (27). That is

$$\mathbf{aG} = \mathbf{H}(z, z_0), \tag{29}$$

where $\mathbf{G}$ is $n \times m$, so that, for a generalised inverse (§ 10.4.2),

$$\mathbf{a} = \mathbf{G}^{-1} \frac{1}{s(2\pi)^{\frac{1}{2}}} \begin{bmatrix} \varepsilon_1 \\ \vdots \\ \varepsilon_m \end{bmatrix}. \tag{30}$$

Here we can take $m < n$ so that ordinary least squares may be used to determine $\mathbf{a}$.

The second problem is to incorporate the uncertainties of the observations y_i. From (26), by the addition theorem for the normal variate,

$$\overline{\delta\rho}(z_0) \cap N(0, \sigma^{*2} \sum a_i^2). \tag{31}$$

Now the uncertainty in the estimates of $\delta\rho(z_0)$ is measured by its standard deviation. That is

$$\text{standard deviation of } \overline{\delta\rho}(z_0) = \sigma = \sigma^*(\sum a_i^2)^{\frac{1}{2}}. \tag{32}$$

It follows from (30) and (32) that the uncertainty is proportional to the reciprocal of the spread,

$$\sigma \propto 1/s. \tag{33}$$

Equation (33) defines the *tradeoff curve*. Of particular interest, the form of (33) is a rectangular hyperbola, a result in line with other approaches to this analysis.

The uncertainty can be minimised by minimising $\sum a_i^2$, but, from (33), a small statistical uncertainty is accompanied by large spread, i.e. poor resolution.

For CAL8, calculation of this kind indicates that, in the lower mantle D, and outer core E, an uncertainty in density of 0.2 g/cm^3 results from averaging over a shell 400 km thick. Such estimates should be provided with all Earth models for comparison purposes.

13.5.4 Direct use of seismic waves

We have seen in previous sections how estimates of seismic P and S velocities can provide a method of estimating the state of the deep interior. It is perhaps extraordinary that, in addition, measurements of travel-times, amplitudes and spectra of seismic waves at the Earth's surface can also provide rather direct estimates of jumps in density, elastic moduli and pressure at first-order discontinuities in the Earth.

An early example was the algorithm for computing the density jump at the boundary between mantle and core derived by Båth. He suggested measuring the polarity of the PcP reflection from the core as a function of Δ. The location of zeros in the reflection or scattering function (see § 6.5) depends upon the required ratio of densities above and below the core–mantle boundary.

Another illustration comes from Bullen's initial attempts to find stronger evidence that the inner core was solid. He considered, from the ray theory of §6.1, the partition of energy at the inner core boundary (assumed sharp) from an incident PKP wave into a refracted $PKIKP$ wave and an (assumed) $PKJKP$ wave (see §10.6.1). In fact, because the absolute energy in each phase from a given source is uncertain due to many unknown properties along a ray path (transmission coefficients, attenuation, source mechanism, etc.) he actually compared the ratio of each phase with that of the (observed) $PKIKP$ phase for corresponding rays, i.e. having the same ray parameter p. It turned out that Bullen's estimates of partition coefficients based on Jeffreys' velocities and Model A densities (see Fig. 13.2) are numerically divergent from the most likely values but the theory of §6.5 provides a method of calculating more realistic amplitude ratios.

The wave $PKiKP$, reflected from the inner core at a small angle of incidence, also provides a way, with due regard for experimental difficulties, to estimate a bound to the density jump at the boundary of the inner core (ICB). A demonstration of the method was carried through by Bolt and Qamar (1970) using short period (~ 1 s) $PKiKP$ and PcP waves recorded by the Large Aperature Seismic Array (LASA) in Montana (see §9.9.3) from an underground nuclear explosion in Nevada ($\Delta \sim 10°.9$).

The directly measured amplitude ratio, $A = \text{amp } PKiKP / \text{amp } PcP$, should be a reliable estimator for two reasons. First, the phases shown in Fig. 13.1 are recorded by the same seismographs with identical processing; no instrumental or crustal adjustments are needed. Secondly, the downgoing $PKiKP$ and PcP rays leave the explosive source at almost the same angles of emergence ($e \approx 89°$, $87°$, respectively). Consequently, no nodal surfaces of source radiation complicate the energy partition and we may assume that the wave amplitudes in the downgoing $PKiKP$ and PcP rays are initially essentially equal.

The identified phases in Fig. 13.1 have residuals of only a few seconds against standard travel-time tables. Further, processing with the LASA array indicates slownesses of about 0.5 and 1.2 s/deg for the recorded $PKiKP$ and PcP waves. These raw values are close to the expected values of $p = 0.3$ and $p = 1.0$ s/deg, respectively. Measurements of A from four independently processed groups of signals from different subclusters of the array yielded a mean of $A = 0.38$ with a variation of about 5 per cent.

This raw value for A must, however, be adjusted for the relative geometrical spreading η_g, of the two waves, as well as the relative frictional attenuation η_q. Given the P velocity distribution down to the inner core boundary, the value of η_g can, in principle, be determined exactly. By theory, for the same source and receiver, apart from a factor $1/4$ involving the angles of incidence,

$$\eta_g^2 = \left| \left(\frac{dp}{d\Delta} \right)_{PKiKP} \middle/ \left(\frac{dp}{d\Delta} \right)_{PcP} \right|. \tag{34}$$

For steep reflections, the seismic rays are approximately straight lines. In this case, we easily derive

$$\frac{dp}{d\Delta} = \tfrac{1}{2} p \cot \frac{\Delta}{2} - \frac{1}{t} p^2, \tag{35}$$

where Δ is the epicentral distance and t is the travel-time of the phase in question. From equation (34), $\eta_g = 0.47$ for $\Delta = 10°.9$. In principle, η_g can be checked from the curvature of the empirical travel-time curves; in practice, uncertainties in measuring curvature are high. Present standard tables suggest $\eta_g \approx 0.5$–0.6.

The $PKiKP$ waves reflected from the inner core have additional frictional attenuation compared with the PcP waves. The former travel a further distance of more than $3500 \, \text{km}$ through the liquid outer core. (Special effects due to the outer boundary and physical state of any transition shell around the inner core are ignored throughout this argument.) Suppose the attenuation of the wave amplitude varies like (see (8.24))

$$\eta_q = \exp(-\pi t / Q\tau), \tag{36}$$

where τ is the period, and Q the attenuation parameter.

In the liquid outer core, the Q value for P waves of period $1 \, \text{s}$ is probably in excess of $10\,000$. Seismic measurements indicate that Q values for P waves increase to about 1500 near the base of the solid mantle. From equation (36), these values given $\eta_q \geqslant 0.85$.

The required density ratio d_1 at the inner-core boundary arises as follows. Let subscripts $0, 1, 2, 3, 4$ refer to the terrestrial surface, the top and bottom of the mantle–core and inner-core boundaries, respectively. Let e and f be the angles at which P and S waves emerge at a boundary.

Consider the ray system generated by an incident P wave at the inner core in terms of the P and S potential functions ϕ and ψ; incident and reflected waves in the liquid core may be represented by (see §6.2),

$$\begin{cases} \varphi_3 = A_3 \exp \{ i\kappa (z \tan e_3 + x - ct) \} \\ \quad\quad + C_3 \exp \{ i\kappa (-z \tan e_3 + x - ct) \\ \psi_3 = 0, \end{cases} \tag{37}$$

where $2\pi/\kappa$ is the wavelength, and z is along the radius towards the centre. Similarly, the potentials for refracted P and SV waves in the inner core (assumed solid) are

$$\begin{cases} \varphi_4 = D_4 \exp\{i\kappa(z\tan e_4 + x - ct)\} \\ \psi_4 = B_4 \exp\{i\kappa(z\tan f_4 + x - ct)\} \end{cases}. \tag{38}$$

Three boundary conditions apply: zero tangential stress (to a close approximation) and continuity of normal displacement and normal stress. The resulting three equations in the unknown amplitudes A_3, C_3, D_4, B_4, are

$$\begin{bmatrix} \tan^2 f_4 - 1 & 0 & 2\tan e_4 \\ 1 & -\tan e_3 & -\tan e_4 \\ 2\beta_4^2\tan f_4 & d_1\alpha_3^2\sec^2 e_3 & -\beta_4^2(\tan^2 f_4 - 1) \end{bmatrix} \begin{bmatrix} B_4/A_3 \\ C_3/A_3 \\ D_4/A_3 \end{bmatrix}$$

$$= \begin{bmatrix} 0 \\ -\tan e_3 \\ -d_1\alpha_3^2\sec^2 e_3 \end{bmatrix} \tag{39}$$

The third equation contains the required $d_1 (= \rho_3/\rho_4)$. Equation (39) together with appropriate triads from the mantle–core boundary continuity conditions may be solved simultaneously for A as a function of adopted values for d_1 and β_4, the (unknown) shear velocity in the inner core.

From recent seismological studies, the values adopted as likely, in usual units, for density ρ, P velocity α, S velocity β are: $\rho_1 = 5.51, \alpha_1 = 13.20$, $\beta_1 = 6.99$; $\rho_2 = 9.70$, $\alpha_2 = 7.99$; $\rho_3 = 12.35$, $\alpha_3 = 10.31$, $\alpha_4 = 11.23$. The reflection coefficient for steep PcP rays is roughly proportional to a term $|\rho_2\alpha_2 - \rho_1\alpha_1|$ which is quite sensitive to the uncertainties of the particular combination of densities and velocities. The adopted values give a high (but not maximum) value to this term. The choice is consistent with seeking a lower bound to d_1 (and hence a high reflection coefficient for $PKiKP$ there).

A result of the computation is that a fairly extreme minimum value of d_1 is 0.86; a more likely minimum is perhaps 0.875. For the plausible value of $\rho_3 = 12.17$, these ratios suggest a maximum density jump at the inner core boundary of less than $1.8\,\text{g/cm}^3$.

It is remarkable that seismological evidence alone can show that, 1200 km from the Earth's centre, the density does not exceed $14.0\,\text{g/cm}^3$. (The pressure at this radius is $3.3 \times 10^{12}\,\text{dyne/cm}^2$.) Only a few years ago, central densities in the Earth as high as $22\,\text{g/cm}^3$ were seriously considered (cf. Model A). Uncertainties in the above bound need, of course, to be checked from additional observations of recorded PcP and $PKiKP$ amplitudes. Nevertheless, the result $\rho_4 < 14.0$ is in close agreement with recent core models (see Table A.2) and also with the results of shock wave experiments on iron, that the Earth's central density is about $13 \, \text{g/cm}^3$.

13.6 Stratification of the shells

We now give a brief outline of some recent inferences on fine Earth structure based on the application of seismological theory to observations of seismic waves. It must be stressed that all the structural details described are subject to various uncertainties and caveats. We repeat for emphasis the propositions of § 7.4.3, § 13.5.3 and elsewhere concerning the non-uniqueness of inverse processes and the essential probabilistic nature of geophysical inferences. Another consideration of importance is that the description in this chapter is predominantly in terms of a radially symmetric Earth model. We have indicated that this description turns out to explain a major part of available seismological measurements and theory, but we repeat (see, e.g., chapter 12) that the actual Earth structure deviates to various degrees from the radially symmetric model. Much present research is aimed at an elucidation of lateral deviations from full symmetry, but many problems are in a state of flux and the detail is too extensive to summarise here.

13.6.1 The upper mantle

In terms of a globally characteristic structure, there are three major structural changes in most seismically based Earth models of the upper mantle (see Fig. 13.2). First, a low-velocity layer coinciding roughly with the asthenosphere is defined by both P and S wave velocities but is more pronounced for β and under oceanic regions. The decrease in β is about 5 per cent and is spread out over depths from 80 to about 200 km. Because the effect of increasing pressure on α and β can be reversed by sufficiently high temperature gradients (see (16)), a gradient of $dT/dz \approx 10 \, \text{K/km}$ (perhaps with partial melting) could explain the decreased velocities.

It should be noted that the model PREM (but not CAL8 – see Appendix) has a slightly decreasing density gradient throughout the low-velocity layer. Resolution of $d\rho/dz$ is low in Region B (see § 13.5.3). The problem is further complicated by the introduction in PREM of about 2 to 4 per cent transverse anisotropy in the upper 200 km of the mantle for both P and S waves.

Inversions of short-period P wave reflections such as $P400P$ and $P'650P'$ show sharp discontinuities at about 400 km and 650 km that are widespread in the upper mantle. Such changes are incompatible with a

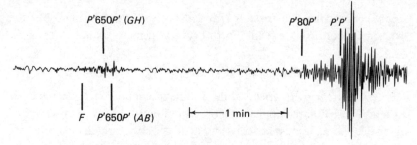

Fig. 13.5. Seismogram at Jamestown, California, from the underground nuclear test on 14 October 1970 on the Russian island of Novaya Zemlya. The large phase $P'P'$ was produced by a P wave reflected under Antarctica. It was preceded by $P'80P'$ reflected from a structure 80 km below the surface of Antarctica. Two minutes earlier the record shows a doublet $P'650P'$ evidently reflected from a layer 650 km below the surface. The origin of the wave train starting at F is unknown.

homogeneous Region C, as recognized by Bullen (see §13.1.1). The sharpness of the boundaries (< 5 km thick) is apparent from the impetus nature of the reflected waves (see Fig. 13.5). Calculations of reflection coefficients (see §6.2) indicate jumps in α and β of about 3 per cent at the 400 km discontinuity and about 5 per cent at the 650 km discontinuity. It has been suggested that changes in chemical composition or phase state (or both) occur at these discontinuities. Inversion of olivine to a spinel structure, first proposed by Jeffreys and Bernal as an explanation of the 20° discontinuity (see §13.2.1) has been further verified experimentally as plausible.

13.6.2 The shell D″

Direct use of the Herglotz–Weichert inversion algorithm near the base of the mantle is unsatisfactory (see §7.4.1) because for $\Delta > 95°$, P and S wave travel-time observations become more scattered. There is no discontinuity in the (T, Δ) curve near 100° for either wave so that the radius of the mantle–core discontinuity and the velocity distribution just above it must be estimated from other wave measurements. Usually times of PcP and ScS are combined with P and S times to solve for $\alpha(r)$, $\beta(r)$ and r_c. On the Jeffreys 1939 solution, the gradients of α and β both fall to near zero in D″ the lowest 200 km of the mantle (see §13.5.2).

In 1980, Bolt demonstrated a direct method to find r_c and the smooth velocity variation $\alpha = ar^b$ (see (7.33)) in D″. If the velocity α_1 is assumed known at a radius $r_1 > r_c$, say, the travel-time τ of a normal PcP ray through D'' from r_1 to r_c can be computed (see Exercise 13.8.7). The time τ is known from observed PcP times and a given velocity structure for $v > r_1$. The value of the ray parameter p_1 is known from the P time curve but p_c is required. This key value comes from determining the slope of the diffracted P wave for $\Delta > 105°$. In the core shadow, dif P is observed out to considerable Δ (see, e.g., Fig. 10.2) generally with wave-length increasing with Δ in accordance with theory (see equation (3.96)). (Somewhat surprisingly, short-period P waves are also observed in the core shadow, providing evidence for scattering from lateral inhomogeneities in D'' (see also § 13.2.3).) Linear regression fits the dif P arrival-times closely for $105° < \Delta < 115°$ yielding $dT/d\Delta = p_c$. Substitution of this value and τ in the equation of Exercise 13.8.7 gives b and hence a from Mohorovičić's law. Computation of r_c and v_c then follows (see Exercise 13.8.7).

Experimental results suggest a value $p_c = 4.55 \pm 0.05\,\text{s/deg}$ for the slowness of P waves around the core boundary. Some recent solutions using this datum indicate that $d\alpha/dr \approx 0$ in D'' for $r_c < r < 3600\,\text{km}$ with $r_c \approx 3486\,\text{km}$ (see Appendix). A slight decrease in P velocity in the lowest few tens of kilometres of D'' is not precluded.

Diffracted SH waves are also well observed in the core shadow for $\Delta < 105°$. Because SH motion is not refracted into the liquid core (unlike incident SV motion), the dif SH pulse, often with *impetus* onsets, is well observed on transverse wave component seismograms. A characteristic dif SH wave slowness is $p_c \approx 8.40 \pm 0.07\,\text{s/deg}$. For $r_c = 3486\,\text{km}$, this value yields (see Exercise 13.8.7) $\beta_c = 7.24\,\text{km/s}$.

The discussion in § 13.5.2 showed that the generalisations (16) and (18) in conjunction with the k–p hypothesis are applicable in D'' to determine dp/dz. The hypothesis requires that, inside D'', $dk/dp \approx 3$. Because $d\phi/dz \approx 0$ inside D'', we therefore obtain $\eta \approx 3$ in D''. Thus D'' is characterised by an anomalous downward density increase or an abnormally high temperature gradient, or both. Perhaps there is an admixture of core material in a thin transition shell or D'' is a thermal boundary layer in which heat is transferred only by conduction, thus requiring a steeper gradient than in convecting regions above and below.

13.6.3 The outer core (shell E)

Present seismic evidence does not rule out a transition shell E' (called also E_1 by Bullen) up to about 200 km thick at the top of the liquid core. The cubic splines used throughout E in PREM and CAL8 prevent the eigen-spectra inversion from discriminating for or against E'. Resolution depends upon high-frequency body-wave analysis. The Herglotz–Wiechert inversion of Jeffreys in 1939 used observed travel-times of *SKS* and *PKP*. Because only the former bottom in a shell between $r = 3480$ and $r = 3000$ km, the velocity in this region is entirely dependent on the *SKS* waves. However, before 83°, *SKS* is obscured by the direct *S* waves on seismograms. For this reason, Jeffreys was forced to interpolate the *K* travel-time curve from 35° back to 0°, using a cubic spline. He obtained a velocity at the core top of 8.10 km/s. The assumed extrapolation of *K* was reduced to only 14° by Hales and Roberts (1970) by using differences in arrival time of *SKKS* and *SKS*. They found $\alpha = 7.91$ km/s at the core. Caution must be used with *SKKS* because of phase shifts relative to *SKS* (Choy 1977). It is of interest that the smoothed model PREM has a value $\alpha = 8.06$ km/s, close to the Jeffreys value. There is no definitive work precluding $7.8 < \alpha < 8.2$ km/s, or 5 per cent variation at the top of the outer core. Below the speculative boundary layer E' there is little evidence for first- or second-order discontinuities in a monotonic velocity distribution in E.

13.6.4 The shell F

The peculiarity of Jeffreys' decrease in velocity just outside the inner core (see Fig. 13.2) was shown by Bolt in 1968 to be unnecessary on the *PKP* data. It could be replaced by a constant velocity shell about 450 km wide with $\Delta\alpha \approx 0.4$ km/s at its upper boundary and, like the Jeffreys solution, a sharp inner core boundary. At first, numerous independent studies tended to confirm the existence of an F shell with smaller $\Delta\alpha$ and a weak velocity increase with depth. Even with the discontinuities removed (see § 13.2.3), shell F persisted because of the change in velocity gradient obtained from Herglotz–Wiechert inversions. The solution used in CAL8 has α increasing by only 0.15 km/s from $r = 1800$ to $r = 1250$ km compared with 0.34 km/s in the PREM model.

For a small velocity gradient, the second term in (18) is negligible so that, assuming normal compression the value of η is 3 to 4, entailing moderate changes in constitution in F. (The Jeffreys velocity decrease gave the unlikely value $\eta \approx 30$.)

There is now some independent evidence from core wave shapes on seismograms that $d\alpha/dr$ remains almost constant with depth, at least up to about 100 km from the inner-core boundary. The inference follows from comparison of synthetic with broadband seismograms of *PKP* (see Fig. 13.3) with the use of a range of core velocity models. The same results imply also strong gradients in the *P* and *S* velocities in the upper 200–300 km of the inner core. Definite conclusions are premature because few earthquakes have as yet been analysed in this way and the effect of anomalous conditions near the source and along the wave paths must be checked.

13.6.5 The inner core (shell G)

Among other less direct evidence, observed travel-time differences *PKiKP* minus *PcP* and *PKIIKP* minus *PKiKP* at short distances are compatible with a radius of 1216 ± 2 km and an average *P* velocity in G of 11.14 ± 0.02 km/s.

Evidence for significant rigidity at the inner core is fourfold. First, travel-times of *PKP* and amplitudes of *PKiKP* at nearly normal incidence fix closely the *P* velocity and density on each side of the boundary. On Bullen's argument, for the jump in compressibility to remain plausible (less than 10 per cent, say), β must jump from 0 to about 3.1 km/s. Secondly, one observation of the seismic body wave phase *PKJKP* with a shear wave leg in the core has been reported by using the LASA array but subsequently not confirmed; the average velocity was $\beta = 2.95 \pm 0.1$ km/s. Thirdly, observed eigen-vibrations with significant particle motions in the inner core provide information. For example, core modes $_6S_2$ and $_7S_3$ indicate $\bar{\beta} = 3.52 \pm 0.03$ km/s (Masters and Gilbert 1981). Fourthly, smooth solutions for PREM and CAL8, both with data sets incorporating a number of core-sensitive modes, have $3.50 < \beta < 3.67$ and $3.49 < \beta < 3.60$ km/s, respectively. Pending a full analysis of resolution and uniqueness, an average rigidity $\mu \approx 1760$ kbar is about all that can be adopted. Suggestions have also been made that there is a relatively rapid increase in β with depth at the

Table 13.2 *Average core parameters (i) PREM (ii) CAL8*

Location	α/(km/s)		β/(km/s)		ρ/(g/cm³)		θ		$Q(1\ Hz)$
	(i)	(ii)	(i)	(ii)	(i)	(ii)	(i)	(ii)	
Top of E	8.06	8.09	0	0	9.90	9.82	1	1.1	—
Middle of E									
($r = 2500$ km)	9.34	9.37	0	0	11.19	11.20	1	1	*ca.* 10 000
Bottom of F	10.36	10.19	0	0	12.17	12.17	1	0.2	—
Top of G	11.03	10.89	3.50†	3.49†	12.76	13.34	1	1.3	—
Middle of G									
($r = 500$ km)	11.22	11.29	3.64	3.60	13.03	13.58	1	0.3	450

†$\beta \approx 3.0$ km/s is perhaps more likely.

top of G. The inversion solution of CAL8 yields an abnormally small θ towards the Earth's centre. (In PREM, the value $\theta = 1$ is assumed.)

Comparisons at key points are given in Table 13.2. The present situation is that more statistically precise and rigorous evidence from seismology is needed because details of constitution, such as the presence of partially molten transition shells, have crucial implications for core dynamics and equations of state. The condition of part of the inner core may be close to melting; its average Poisson ratio is about 0.44 compared with 0.5 for a fluid and 0.35 in the solid mantle.

13.7 Ellipticities of surfaces of equal density within the Earth

Knowledge of the density distribution also enables the ellipticity ε involved in § 10.9.3 to be computed. If the moment of inertia (about a diameter) of the material enclosed within a sphere of radius r of the model is equal to ymr^2, where m denotes the mass of this material, the coefficient y is readily calculated from knowledge of the density distribution. The value of η, as defined by (10.53), for the corresponding level surface within the actual Earth is then found by using the approximate equation

$$1 - \tfrac{3}{2}y = \tfrac{2}{5}(1 + \eta)^{\frac{1}{2}} \tag{40}$$

obtained by Radau and Darwin in investigations of the figure of the Earth. The validity of this equation depends on the form of the Earth's density variation.

The fact that the values of η and ε given in § 10.9.2 have of necessity to be calculated first from density values determined using P and S velocities deduced from travel-time tables set up prior to the use of ellipticity corrections is of small consequence, since the ellipticity corrections are (see § 10.9.2) so small.

13.8 Exercises

1 In a continental area, assume the crust consists of an upper layer, thickness 15 km and $\alpha = 6.0$ km/s, overlying a lower layer thickness 10 km and $\alpha = 7.0$ km/s. In the upper mantle, let $\alpha = 8.0$ km/s. Calculate arrival-time curves for refracted arrivals from a surface explosion and determine whether the critically refracted P wave along the boundary between the crustal layers is ever a first arrival. What are the best epicentral distances for seismographs to observe critical-angle reflections from the interfaces? Determine the length of the line of seismographs required to estimate the upper-mantle velocity to within 0.05 km/s using first arrivals. (Assume times are read to 0.01 seconds and distances have negligible error.)

2   With the geometrical properties of types of seismic waves in mind, explain
    (a) why the horizontally polarised S wave (SH) is observed at distances
    much greater than 105°, i.e. in the shadow of the Earth's core, yet the
    companion vertically polarised (SV) wave is not seen there.
       (b) why multiple reflections of core waves like PKKKKP do not soon
    lose most of their energy into refracted waves at the point of reflection.
       (c) in what way the above observations strengthen the inference that
    the outer core is liquid.

3   For paths across the oceans, distinctive long-period Love waves ($T > 20$ s)
    concentrated in a short time interval, corresponding to a velocity of about
    4.4 km/s, are observed on seismograms from horizontal component
    seismographs. Explain this in terms of dispersion curves. Would the Love
    wave train look the same for continental paths?

4   Given that in the Earth there is a discontinuous decrease in the P velocity
    at the mantle–core boundary and a discontinuous increase in the P
    velocity at the outer-core–inner-core boundary. Would you expect large,
    normal, or small amplitudes to be observed on seismograms for the
    following waves and distances: (a) P at 130°, (b) PKP at 142°, (c) PKIKP at
    110°, (d) PKIKP at 140°, and (e) PKIKP at 180°?

5   A way of inferring which shells in the Earth's interior have changes in
    physical properties is to calculate the homogeneity index.
       In region D″ at the bottom of the mantle, the seismic velocities change
    little with depth. Show that this suggests either rapid temperature
    increases there or mixing in of heavier minerals or both. Why does a
    significant decrease in P velocity with depth in region F in the core seem
    unlikely? [Hint: Assume $dk/dp \approx 3.0$ in D″ and F.]

6   The velocity of P waves in the inner core is known to jump to a value 10
    per cent greater than the velocity of P waves in the liquid outer core.
       Because $\alpha = [(k + 4\mu/3)/\rho]^{\frac{1}{2}}$ and $\mu = 0$ in the outer core, give an
    argument why a plausible way to account for the jump in α is to give μ a
    finite value in the inner core. Hence, estimate the upper value for average μ
    and shear velocity in this solid inner core.

7   Prove that in the situation of § 13.6.2 the travel-time τ of a normal PcP ray
    through layer D″ for $r_c < r < r_1$, is

$$\tau = \{(\eta_1^2 - p^2)^{\frac{1}{2}} - (\eta_c^2 - p^2)^{\frac{1}{2}}\}/(1 - b),$$

    where $\eta = r/\alpha$ and p is the ray parameter (see § 7.1). Then, show that, given
    τ, p_1 and p_c,

$$\ln r_c = (\ln a\eta_c)/(1 - b),$$

    and

$$\alpha_c = \pi r_c/180 p_c.$$

14

Long-period oscillations and the Earth's interior

There has been a remarkable extension in the range of periods of ground movements that can be recorded, enabling the whole inversion spectral interval to be filled in from ordinary surface-wave periods of a minute to periods of the order of hours such as occur in Earth tidal movements. In particular, it has become possible to measure free vibrations of the whole Earth and thus to provide important additional evidence on the Earth's interior.

In the theory of seismic waves developed in earlier chapters, the source energy has been usually looked upon as being transmitted outward from the focus in the form of P and S and ordinary surface waves. The emphasis has been on the wave motions of travelling disturbances which affect only part of the Earth at any one time, rather than on what is happening to the globe. All such wave motions can, however, be considered from a more general standpoint as belonging to some mode of vibration of the whole Earth, although for periods of a few minutes or less the modes would be of high order.

14.1 Historical background

The vibrations of a perfectly elastic solid sphere were first considered by Poisson in 1829. Kelvin and G.H. Darwin later developed important theory on the straining of an elastic sphere, with applications to Earth tidal problems.

In 1882, Lamb discussed in some detail the simpler modes of vibration of a uniform sphere and showed that two distinct classes of vibration are possible (see §5.6). This classification continues to be relevant for more complicated spheres such as the commonly used Earth models.

In his notable work of 1911, Love investigated the statical deformation and small oscillations of a uniform gravitating compressible sphere. He obtained a period of 60 min for the slowest mode in his model.

350

Alterman, Jarosch and Pekeris in 1959 simplified Love's analysis by working entirely in spherical polar coordinates. They reproduced a number of Love's results, and extended them to various non-homogeneous spherical Earth models. The more difficult cases required at least six differential equations of the first order to be solved subject to boundary conditions, and the degree to which electronic computers are needed makes it clear that the calculations would have been mostly impossible in Love's day. In the course of the computations, Pekeris used both direct numerical integration procedures and variational methods. The recent extension to heterogeneous models makes possible realistic comparisons between theory and observation.

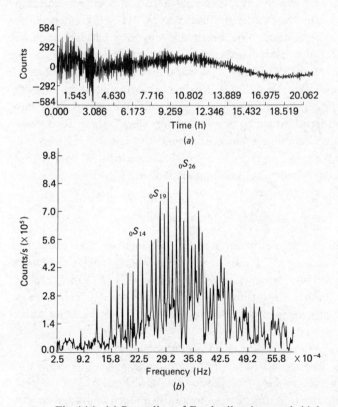

Fig. 14.1. (*a*) Recording of Earth vibrations and tidal oscillations after the Indonesia earthquake of 19 August 1977 ($M_s = 8.0$). The vertical ground motion was measured at Whiskeytown station of the University of California at Berkeley. (*b*) Frequency spectrum calculated from the seismogram in (*a*). Three spheroidal S modes are identified. (Courtesy R.A. Hansen, 1982.)

Other important calculations of the eigen-vibrations of realistic Earth models are referred to in Bullen (1975) and Lapwood and Usami (1981).

Following the Kamchatka earthquake of 1952 November 4, Benioff found indications on his strain seismograph (§ 9.9.2) of two unexpected long wave periods of about 57 and 100 min. He suggested that these periods related to fundamental oscillations of the Earth excited by the earthquake, thereby stimulating others to carry out the theoretical calculations already mentioned.

During the following seven years, there were no reports of similar long-period waves from earthquakes, and many seismologists attributed Benioff's readings to some purely instrumental effect. Then in 1960 at the Helsinki meeting of the IASPEI, there was a dramatic announcement by several independent observers of the first unequivocal measurements of the graver modes of oscillations of the Earth following the Chilean earthquakes of May 1960. It is now established conclusively that these earthquakes, the largest of which had a magnitude of $8\frac{1}{2}$, set the whole globe into vibration for many days. The vibrations were recorded in North and South America, Europe and Japan. Since that time instrumentation and analysis have improved, leading to estimates of eigen-frequencies of many hundreds of modes from diverse large earthquakes around the world. In Fig. 14.1, we reproduce the seismogram and Fourier eigen-spectrum of over 20 h of Earth vibrations following an earthquake with $M_s = 8.0$.

14.2 Numerical results for Earth models

A radially symmetric Earth model is defined for present require-ments when ρ, α and β, or ρ, κ and μ, are given as functions of r. Calculations have been carried out for many models including PREM (§ 13.5). Density values are combined with various sets of values of α and β, and models are defined with mass-average properties corresponding to the known mass and moment of inertia.

In some early studies, values of ρ were combined with values of α and β which did not have matching discontinuities, and caution is needed for interpreting the results for such cases.

We now consider the numerical computation of eigenvalues and -vectors for, at least to the first order, reasonably realistic Earth models.

14.2.1 Torsional oscillations

First, the torsional oscillations (see § 5.6.2) can be effectively treated numerically using transfer matrix theory (cf. § 5.5.2).

In the free oscillation problem it is convenient to use a vector $\tilde{\mathbf{y}}$ defined by

$$\tilde{\mathbf{y}} = r\mathbf{y}, \tag{1}$$

and write the fundamental equation (5.35) as,

$$d\tilde{\mathbf{y}}/dr = \tilde{\mathbf{u}}(r)\tilde{\mathbf{y}}. \tag{2}$$

The modified coefficient matrix $\tilde{\mathbf{u}}$ for the torsional oscillation of a transversely isotropic sphere (cf. §2.5) is

$$\tilde{u}_{11} = -\tilde{u}_{22} = 2/r,$$

$$\tilde{u}_{12} = 1/L, \quad \tilde{u}_{21} = \frac{(l-1)(l+2)N}{r^2} - \omega^2\rho, \tag{3}$$

where r is the radius of the mid-point of a layer, and the eigenvalues are given by

$$\tilde{v}^2 = \frac{1}{L}\left[\frac{(l-1)(l+2)N}{r^2} - \omega^2\rho\right] + \frac{4}{r^2}. \tag{4}$$

From (4) and (5.36) we obtain for the layer matrices (§5.5.2)

$$E_{11}(h) = 1 + 2\sinh^2\frac{\tilde{v}h}{2} + \frac{2}{r}\frac{\sinh\tilde{v}h}{\tilde{v}},$$

$$E_{12}(h) = \frac{1}{L}\frac{\sinh\tilde{v}h}{\tilde{v}}, \tag{5}$$

$$E_{21}(h) = \frac{1}{L}\left[\frac{(l-1)(l+2)N}{r^2} - \omega^2\rho\right]\frac{\sinh\tilde{v}h}{\tilde{v}},$$

$$E_{22}(h) = 1 + 2\sinh^2\frac{\tilde{v}h}{2} - \frac{2}{r}\frac{\sinh\tilde{v}h}{\tilde{v}}.$$

14.2.2 Spheroidal oscillations

The calculation for the full spheroidal oscillation case by the matrix method is almost impossible. In 1972, Takeuchi and Saito derived an expression for the eigenvalues of the coefficient matrix in the special case of isotropy and no gravity,

$$2\tilde{v}^2 = -\left[\frac{\omega^2\rho}{\lambda + 2\mu} + \frac{\omega^2\rho}{\mu} - \frac{2l(l+1)}{r^2} - \frac{3\lambda + 10\mu}{(\lambda + 2\mu)r^2}\right]$$

$$\pm\left\{\left[\frac{\omega^2\rho}{\lambda + 2\mu} - \frac{\omega^2\rho}{\mu} + \frac{\lambda - 2\mu}{(\lambda + 2\mu)r^2}\right]^2 + \frac{12l(l+1)(\lambda + 4\mu)}{(\lambda + 2\mu)r^4}\right\}^{1/2}. \tag{6}$$

The layer matrix may be generated by the use of a computer, but according to experience the speed of calculation is significantly less than the direct integration of the equations of motions given below.

Suppressing the detail which is available in Alterman, Jarosch and Pekeris (1959) we reduce the equations (5.55) to the following set of six first-order ones.

$$\dot{y}_1 = -2(1 - 2\gamma^2)\frac{y_1}{r} + \frac{y_2}{\rho\alpha^2} + l(l+1)(1 - 2\gamma^2)\frac{y_3}{r} \tag{7}$$

$$\dot{y}_2 = -\rho\{\omega^2 r^2 + 4gr - 4(3 - 4\gamma^2)\beta^2\}\frac{y_1}{r^2} - 4\gamma^2\frac{y_2}{r}$$

$$+ l(l+1)\rho\{gr - 2(3 - 4\gamma^2)\beta^2\}\frac{y_3}{r^2} + l(l+1)\frac{y_4}{r} - \rho y_6 \tag{8}$$

$$\dot{y}_3 = -\frac{y_1}{r} + \frac{y_3}{r} + \frac{y_4}{\rho\beta^2} \tag{9}$$

$$\dot{y}_4 = \rho\{gr - 2(3 - 4\gamma^2)\beta^2\}\frac{y_1}{r^2} - (1 - 2\gamma^2)\frac{y_2}{r}$$

$$- \rho\{\omega^2 r^2 - 2\beta^2[2l(l+1)(1 - \gamma^2) - 1]\}\frac{y_3}{r^2}$$

$$- 3\frac{y_4}{r} - \rho\frac{y_5}{r} \tag{10}$$

$$\dot{y}_5 = 4\pi G\rho y_1 + y_6 \tag{11}$$

$$\dot{y}_6 = -4\pi G\rho l(l+1)\frac{y_3}{r} + l(l+1)\frac{y_5}{r^2} - \frac{2y_6}{r}, \tag{12}$$

where $\gamma^2 = \beta^2/\alpha^2 < 1$ and ρ and g are the unperturbed values for density and gravity at radius r. The coefficients of the equations are functions of α, β, ρ, g, which are the empirical parameters most directly available. The new variables y_i are defined to be

$$y_1 = U, \quad y_2 = p_{rr}, \quad y_3 = V,$$

$$y_4 = p_{r\theta}, \quad y_5 = P, \quad y_6 = \dot{P} - 4\pi G\rho U,$$

where p_{rr} and $p_{r\theta}$ are stress components and U, V, and P are related to displacement as in §5.6.2. Here displacements are restricted to the solid mantle but modification for motion in the liquid can readily be made by applying appropriate boundary conditions. Equations (7) to (12) are then integrated by the Runge–Kutta method through homogeneous shells of thickness equal to the integration step. The shells are chosen sufficiently thin to obtain the required accuracy of integration and to represent adequately the variation of α, β, ρ and g with r. Across each shell boundary all the y_i are taken continuous. At the free surface, $y_2 = y_4 = 0$; and the equality of the internal and external gravitational potentials yields $y_5 = -Ry_6/(l+1)$.

The crux of the method used to isolate the eigen-vibrations is the property of linearity of the equations. For a fixed l, three integrations to the surface are performed commencing from a prespecified depth $d = \varepsilon\lambda$ (λ is wave-length, $\varepsilon > 1$), with initial values $y_1 = y_2 = y_5 = 0$ and three linearly independent sets of values of y_2, y_4, y_6. In general, these integrations yield three independent sets of values of the dependent variables at the free surface, namely, $y_i^{(j)}$, $j = 1, 2, 3$. Since the actual solution of (7) to (12) which satisfies the surface boundary conditions is a linear combination of the solutions now available, we have at the free surface

$$
\begin{bmatrix}
y_1^{(1)} & y_1^{(2)} & y_1^{(3)} \\
y_2^{(1)} & ---- & \\
y_3^{(1)} & ---- & \\
y_4^{(1)} & ---- & \\
y_5^{(1)} & ---- & \\
y_6^{(1)} & ---- &
\end{bmatrix}
\cdot
\begin{bmatrix} a \\ b \\ c \end{bmatrix}
=
\begin{bmatrix}
U \\ 0 \\ V \\ 0 \\ P \\ -(l+1)P/R
\end{bmatrix}
\tag{13}
$$

where a, b, c are undetermined multipliers, *provided* that ω is the eigen-frequency $\bar{\omega}$. In practice we calculate

$$
\det(\omega) =
\begin{vmatrix}
y_2^{(1)} & y_2^{(2)} & y_2^{(3)} \\
y_4^{(1)} & y_4^{(2)} & y_4^{(3)} \\
y_6^{(1)} + \dfrac{l+1}{R}y_5^{(1)} & y_6^{(2)} + \dfrac{l+1}{R}y_5^{(2)} & y_6^{(3)} + \dfrac{l+1}{R}y_5^{(3)}
\end{vmatrix}
\tag{14}
$$

for a series of frequencies ω, $\omega + d\omega$...starting below the frequency of the expected eigen-frequency. The search is stopped when $\det(\omega)$ changes sign, and the eigen-frequency $\bar{\omega}$ corresponding to $\det(\omega) = 0$ is then estimated by inverse interpolation (using Lagrange's four-point formula) from the computed values of $\det(\omega)$. The smallest root of $\det(\omega) = 0$ corresponds to the fundamental mode of vibration while larger roots correspond to the higher modes (overtones) for a particular l. This method of searching for the correct eigen-frequency is based on the assumption that for the particular l there exists a solution of (7) to (12) in which the magnitudes of U, V, and P converge to zero with depth.

The variation of U, V, and P with depth, which is also of interest, may be determined by integration of (7) to (12) downward after the eigen-frequency has been found. For this computation, upward integration of (7) to (12) using the known eigen-frequency provides a set of surface values of U, V, and P from which a, b, and c are determined by (13) under the normalising condition that U is unity at the surface.

For example, the surface value of V is $ay_3^{(1)} + by_3^{(2)} + cy_3^{(3)}$ where

$$a = \begin{vmatrix} y_2^{(2)} & y_2^{(3)} \\ y_4^{(2)} & y_4^{(3)} \end{vmatrix} \bigg/ \begin{vmatrix} y_1^{(1)} & y_1^{(2)} & y_1^{(3)} \\ y_2^{(1)} & y_2^{(2)} & y_2^{(3)} \\ y_4^{(1)} & y_4^{(2)} & y_4^{(3)} \end{vmatrix}$$

and similarly for b and c.

Downward integration of (7) to (12) is then performed, and the desired amplitude distribution of U, V and P is obtained. Results of this procedure are shown in Fig. 14.2 for $l = 30, 80$, and 110. The characteristic amplitude distributions of these solutions throughout the mantle identify them as waves of the Rayleigh type.

Since U, V and P in the above discussion are radial factors, the vertical and horizontal displacements, u_r and u_θ, and the gravitational disturbance p are given by (5.60), corresponding to the displacements in ordinary Rayleigh waves.

The ultimate justification of a particular choice of ε to define the starting depth $\varepsilon\lambda$ for a vibration with wave-length $\lambda = 2\pi R/(l + \frac{1}{2})$ is that, after

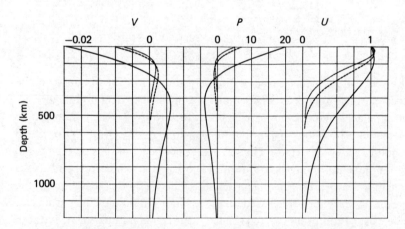

Fig. 14.2. Relative amplitudes of eigenvectors as functions of depth. $l = 30$, bottom curves; $l = 80$, dashed curves; $l = 110$, top curves. (Courtesy B.A. Bolt and J. Dorman, 1961.)

following the above procedure, the computed variations with depth of U, V and P all show close convergence to zero for a depth less than $\varepsilon\lambda$. This is exemplified in Fig. 14.2.

14.2.3 Model splitting. The solotone effect

Most early studies of free oscillations restricted the model structure to spherical symmetry, although even then coupling between modes due to the rotation and ellipticity was considered (see § 5.6.5). Recently, quantitative studies have incorporated deviations from spherical symmetry due to heterogeneity in the normal mode analysis. Theoretical calculations have been done using perturbation theory for torsional and spheroidal modes. Anelasticity has now also been incorporated and Woodhouse and Dahlen (1978) give relevant formulae for the first-order perturbation theory.

Direct variational methods have also been used including the finite element method. The analogous model of a vibrating inhomogeneous string shows that the effect of vibrational coupling is sometimes significant, especially when two modes are very close in the frequency domain.

Although methods for using normal modes to invert for the heterogeneity of the Earth have provided useful results, up till 1982 they all neglect coupling of modes. Direct calculations suggest that this omission can sometimes lead to significant discrepancies. A study by Tanimoto and Bolt (1983) illustrates the main points for torsional mode coupling for a realistic Earth model structure. Anelasticity was included in this work; the large matrix triangulation necessitated that only couplings between the closest multiplets in the frequency domain be considered.

The equation of motion for normal oscillations is given by (2.9) and the rotational splitting parameter by (5.65). In visco-elastic problems, the stress p_{ij} becomes a convolution of the relaxation function and strain instead of a multiplication of elastic constants and strain (cf. § 4.5.1). In both elastic and visco-elastic problems, in the s-domain of the Laplace transform, equation (2.9) becomes

$$\rho s^2 \bar{u}_i = \frac{\partial}{\partial x_j}\left(A_{ijpq}(s)\frac{\partial \bar{u}_p}{\partial x_q} \right), \tag{15}$$

the difference being that A_{ijpq} is independent of s in the elastic case. (This follows from the correspondence principle of § 5.7.1.) The bar denotes a Laplace transformed quantity. Because we treat only transformed quantities in the following, we now drop the bar.

The eigenvalue problem in the form (15) for the visco-elastic case differs considerably from the elastic case in that the eigenvalue s appears on both sides of the equation. Rigorous evaluation of s requires numerical iteration, but fortunately in most seismological situations, s dependence on the right-hand side of (15) can be ignored.

Construction of 'true' eigenfunctions can be accomplished with high accuracy by using only those eigenfunctions which have close eigen-frequencies in the average Earth. In this case, because the frequency range of the basis functions (see equation (16)) is small, dispersion is negligible. It follows that the real part of A_{ijpq} is essentially independent of s. In addition, Earth models are obtained without taking into account disper-sion and are assumed to represent closely elastic parameters for the range of normal mode frequencies. Also the assumption is made of constant Q (see §4.5.3) within the frequency range of the basis functions. With these assumptions, the ratio of the real part to the imaginary part of A_{ijpq} is constant, and it follows that the imaginary part of the right-hand side of (15) is also independent of s.

Under these two conditions (neglect of dispersion and constant Q within the frequency band of basis functions), the problem reduces to the ordinary form of the eigenvalue problem, although s and $\bar{u}_i$ are complex quantities in general.

Let n, l, m be the overtone, angular order and azimuthal order number, respectively. Specifically, we expand the 'true' eigenfunction as

$$u_i = \sum_{q=1}^{N} a_q v_{iq} \tag{16}$$

where a_q is a complex coefficient, v_{iq} is an eigenfunction of an average Earth with i denoting a component in a given coordinate system and q a certain pair of (n, l, m). For example, in the torsional case, the eigenfunctions are given by (see (5.59))

$$(v_{rq}, v_{\theta q}, v_{\phi q}) = W_{nl}(r) \cdot \left(0, \frac{1}{\sin \theta} \frac{\partial Y_l^m}{\partial \phi}, -\frac{\partial Y_l^m}{\partial \theta} \right). \tag{17}$$

The upper limit N of summation is the number of basis functions included.

On substituting (16) in (15), multiplying by $v_{ip}^{T}(p = 1, 2, \dots N)$, and integrating over the volume of the Earth, we obtain the following matrix equation

$$s^2 \mathbf{Ta} = \mathbf{Ua}, \tag{18}$$

where $\mathbf{a}$ is an eigenvector whose elements are the coefficients of the expansion (16) and $\mathbf{T}$ and $\mathbf{U}$ are matrices whose elements are

Table 14.1 *Calculated eigen-periods (s) and Q of a pair of modes in an average Earth model. The case has very close eigen-periods but the average Q values are not as close.*

	Eigen-periods	Q
$_1T_{11}$	359.32	249.78
$_0T_{20}$	361.62	158.92

$$T_{pq} = \int \rho v_{ip}^{\mathrm{T}} v_{iq} \mathrm{d}v$$

$$U_{pq} = - \int A_{ijkl} \frac{\partial v_{ip}^{\mathrm{T}}}{\partial x_j} \frac{\partial v_{kq}}{\partial x_l} \mathrm{d}v.$$

In the above formulations, the problem reduces to the generalised eigenvalue–eigenvector problem for general (non-Hermitian), complex matrices. T and U are non-Hermitian owing to the incorporation of anelasticity. In order to solve this problem, algorithms are available, which solve the generalised eigenvalue problem for complex matrices such as the ones in (18).

In Earth model calculations the sizes of the matrices are moderately large. Only slight perturbation from spherical symmetry and small attenuation (large Q values) are required with realistic cases, so that both T and U are diagonally dominant and the solution algorithm behaves well.

For preliminary discussion a simple Earth model is usually adopted which has only oceanic and continental structures. (Continental shelves are treated as continent.) The continental parameters are from Precambrian Shield structure and shear wave velocities and Q values can be assigned from plausible correlations between many published estimates. The difference between such structures persists down approximately to a depth of 1000 km with the main differences in the upper 450 km.

Tanimoto and Bolt examined coupling effects in several cases that were chosen because they have very close eigen-frequencies in the average Earth model. Sample periods and Q in a representative average model are shown in Table 14.1.

Results of eigen-frequency spectra, calculated from (18), are shown in Fig. 14.3. The spectrum plotted as A on the left-hand side is the result of

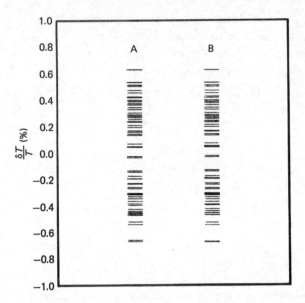

Fig. 14.3. Splitting of eigen-frequencies for the coupled torsional oscillations $_1T_{11}$ and $_0T_{20}$. (Courtesy T. Tanimoto and B.A. Bolt, 1983.)

applying first-order perturbation theory to each angular order separately and then plotting the results together. The B spectrum on the right-hand side results when the modal coupling effect is allowed for in the basis functions. The spectra look remarkably similar except for fine structures represented by a change in line density. At first sight, the implication appears to be that there are insignificant coupling effects in these cases. This conclusion is incorrect, however, as can be seen in Fig. 14.4 from coefficients of the eigenfunction expansion, the vector **a** in (18). In each multiplet (see § 5.6.5), m decreases from left to right; thus, the line furthest left corresponds to $m = 1$ and the line furthest right to $m = -1$. Because the coefficients are complex, only the moduli are plotted here. Although each mode has dominant energy in one angular order, it can also have some energy associated with the other angular order of up to 22 per cent as a ratio of the largest eigenfunction.

Fig. 14.5 shows the results for Q values. These values were calculated as the ratio of the real part over the imaginary part of an eigenvalue s. In this way, Q is simultaneously found with the corresponding eigen-frequency. Fig. 14.5 indicates significant differences between cases A and B, and demonstrates the coupling effect of non-radial Earth structure on Q values.

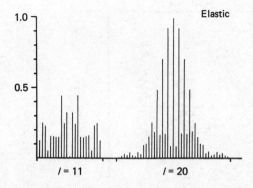

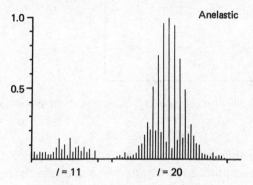

Fig. 14.4. Comparisons of coupling between elastic and anelastic cases for $_1T_{11}$ and $_0T_{20}$. Note the complete symmetry around the centre of each multiplet ($m = 0$) in the elastic case. The stronger coupling effects in the elastic case are clearly seen. (Courtesy T. Tanimoto and B.A. Bolt, 1983.)

In most observational studies of Q values for normal modes, Q is assumed to be constant for all singlets in a multiplet. The numerical results referred to here support this assumption within close limits because the maximum scatter of Q within a multiplet does not exceed 5 per cent for all cases considered. At the same time, the largest coupling of eigenfunctions is, at most, 30 per cent in the cases considered. For other pairs of possible interfering modes with closer eigen-frequencies, the scatter can be larger and comparable to the standard deviations reported in the literature (Hansen and Bolt, 1980). The analysis illustrated in Fig. 14.5 emphasises another aspect of an intrinsic difficulty associated with accurate Q determinations from eigen-vibrations. In the extreme case, when the eigenfunction has comparable energy in two multiplets, the Q values of the corresponding mode can be anywhere between the two Q values in the

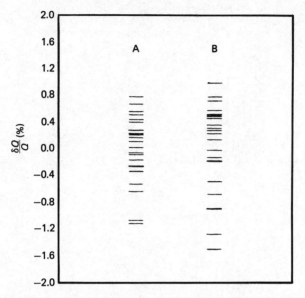

Fig. 14.5. Splitting of Q for the case $_1T_{11}$. (Courtesy T. Tanimoto and B.A. Bolt, 1983.)

average model. Thus, Q estimates for possible interfering modes should be viewed with caution. This caution especially applies to modes in the high-frequency range, where many candidates for coupling can be found in both torsional and spheroidal spectra.

In summary, significant mechanical couplings occur between eigen-vibration modes with angular order as low as 10 (period about a few hundred seconds). Although the effect is minor in the eigen-spectrum, eigenfunctions may have comparable components in the neighbouring multiplet. This property requires that inversion methods using observed spectral lines (see §13.5.3) be extended to include the effect of mode coupling.

Determination of normal oscillation Qs for interfering modes should be viewed with care, because a first-order change in the eigenfunctions can change the values of Q considerably. Differences in Q decrease the couplings. A similar effect arises in torsional-spheroidal couplings due to ellipticity and rotation of the Earth.

An unrelated but curious property of the spectra of terrestrial eigen-vibrations, not predicted by the theory for a homogeneous elastic sphere (§5.6), has been found by Anderssen, Cleary, Lapwood and others. According to Sturm–Liouville theory (see §3.8) for equations like (5.58) the eigen-frequencies of radial overtones, say, in torsional oscillations should, as n becomes large, be asymptotic to $n\pi/t$, where t is time for the diametrical S wave. For Earth models with discontinuous radial functions of the elastic parameters, computation shows eigen-frequencies that deviate about the asymptote. The variation has been called the *solotone effect* and evidently arises from wave resonances between the discontinuities. Lapwood and Usami discuss what is known on the matter in detail.

14.3 Estimation of observed eigen-spectra

A basic need in the measurement of terrestrial eigen-spectra is a general algorithm for simultaneously estimating eigen-frequencies, amplitudes, phases and damping coefficients. This section outlines such a method, formulated in a statistical context so that variances of each estimate can also be obtained. Further detail is given in Bolt and Brillinger (1979).

From the beginning of work on the Earth's free vibrations, the emphasis has been on estimation of the spectral eigen-frequencies, but few estimates have been accompanied by statistical uncertainties. This requirement is important because independent frequency estimates have been seen to differ by up to 0.5 per cent on occasion (e.g. 2 s for $_0T_{14}, {}_0T_{17}$) and it is difficult to know how to combine the separate estimates.

Fewer measurements are available of the actual ground displacements in each eigen-vibration, partly because some key recording instruments were not calibrated for impulse response, but also because some methods of spectrum estimation used could not provide the true amplitudes. Current work on terrestrial eigen-vibrations stresses not only measurements of the ground amplitudes but also the damping of amplitudes. Recent estimates of the damping constant (see § 5.7.3) show considerable scatter and indicate the difficulty of precise measurements of the amplitude decay rate. Further, there are questions on the amount that Q depends on frequency. Progress clearly depends upon more systematic use of statistical analysis of the time series.

14.3.1 Fourier analysis

The impulse response, $s(t)$, of a wide variety of stable geophysical, mechanical and electromagnetic linear systems with finite dissipation is a linear combination of decaying cosine waves,

$$s(t;\theta) = \sum_{k=1}^{K} \alpha_k \exp\{-\beta_k t\} \cos\{\gamma_k t + \delta_k\} \quad (t \geq 0), \tag{19}$$

where $\theta = \{\alpha_k, \beta_k, \gamma_k, \delta_k, k = 1,..., K\}$ with $\alpha_k, \beta_k, \gamma_k > 0, 0 \leqslant \delta_k < 2\pi$ and γ_k distinct. The γ_k are the eigen frequencies of the system. The β_k determine the rate of decay of the oscillations and are often redefined as

$$\beta_k = \gamma_k/(2Q_k) \tag{20}$$

in terms of Q_k damping factors.

A traditional means of estimating the γ_k of equation (19) in geophysics has been the searching for peaks in the periodograms, or smoothed periodograms, calculated from the geophysical time series. The usual numerical procedure has been to calculate the amplitude Fourier spectrum only, using a fast Fourier transform (FFT) algorithm. A less usual method, called *maximum entropy*, involves the fitting of a long autoregressive scheme to the digital record (see Bolt and Currie, 1975).

14.3.2 Complex demodulation

The recommended procedure depends heavily on the ability not only to locate as precise a value of an eigen-frequency as the data permit, but to allow an assessment of whether difficulties in resolution are arising from such physical causes as multiple energy sources or splitting of peaks due to Earth inhomogeneities and rotation. We suggest the method of complex demodulation conceived by J. Tukey in 1961 and especially the use of the instantaneous phase spectrum for decisions on resolution. This is a sensitive method that provides an informative way of comparison between demodulate estimates of the amplitudes, frequencies and damping factors of the oscillations and estimates obtained by the technique of non-linear regression (see §14.3.3). The latter technique allows the relative uncertainties between individual calculated eigen-frequencies to be estimated. The former gives a way to select the most closely resolved modes.

Given a record $X(T), t = 1,..., T$, the complex demodulate at frequency λ of that record is the time series $W(t, \lambda), t = 1, 2,...,$ that results from low-pass filtering the series $X(t) \exp\{- i\lambda t\}$. The complex demodulate $W(t, \lambda)$ will be much smoother than the original time series. The technique is described in detail in Brillinger (1975), for example.

In the present application, suppose the low-pass filter adopted has impulse response $b(t)$, transfer function $B(\lambda)$ with small bandwidth, and suppose λ is near an eigenfrequency γ_k. The demodulate may be written

$$W(t, \lambda) = \sum b(t - u)X(u)\exp\{- i\lambda t\}. \tag{21}$$

For the signal $s(t, \theta)$ of equation (19), the result of demodulating is

$$Z(t, \lambda) \approx \tfrac{1}{2}B(0)\alpha_k \exp\{- \beta_k t\} \exp\{i(\gamma_k - \lambda)t + i\delta_k\}. \tag{22}$$

Standardise the low-pass filter by making $B(0) = 2$. Then, from the complex demodulate, the following defines the instantaneous phase

$$\arg Z(t, \lambda) \approx (\gamma_k - \lambda)t + \delta_k \tag{23}$$

and the logarithm of the instantaneous amplitude function is

$$\ln |Z(t, \lambda)| \approx -\beta_k t + \ln \alpha_k. \tag{24}$$

It follows that plots of arg $W(t, \lambda)$ and log $|W(t, \lambda)|$ against t can provide evidence of the presence of a damped periodicity in a time series of interest. Indeed, successive variations of the demodulate frequency λ lead to parameter trajectories from which α, β, γ and δ can be estimated in some optimal sense. If the plots (see Fig. 14.6) of arg $W(t, \lambda)$ and $\ln |W(t, \lambda)|$ are made nearly linear over the record duration T, especially where the signal amplitude is large, the damped vibration is close to the adopted model. If the plot of $W(t, \lambda)$ is erratic, there is a suggestion that the record is just noise. If the plots have regular non-linear behaviour, there is some violation of the basic simple model, perhaps beating between signal and noise harmonics with nearly equal frequencies, perhaps the injection of new energy into the system by applied forces (perhaps an aftershock arrived), perhaps there is time dependent dispersion.

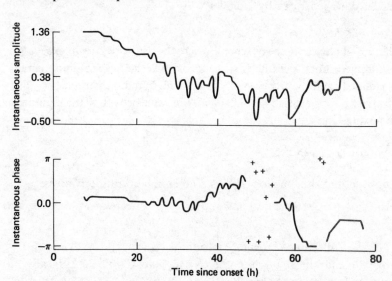

Fig. 14.6. Instantaneous amplitude and phase as a function of time from normal oscillations $_0T_4$ recorded by a Trieste north–south horizontal pendulum after the 1960 Chilean earthquake. The phase remains more or less constant until about 30 h, when it becomes irregular, in agreement with the amplitude.

The slope of the logarithm of the instantaneous amplitude curve gives an estimate of the decay constant β_k; the intercept gives the log instantaneous amplitude of the oscillation at the beginning of the movement. Similarly, the intercept of the instantaneous phase plot yields the relative phase of the oscillation. In addition, it should be noted that some idea of the uncertainty of these estimates is given by the variation of the complex demodulate curves about the fitted straight lines over the selected time interval.

14.3.3 Calculation of eigen-frequency, amplitude, phase and Q

The suggested approach is multi-stage. Assume that one of the traditional Fourier methods has been used to determine frequencies that perhaps correspond to eigen-vibrations. Then complex demodulation is carried out at the determined frequencies. Examination of the results of complex demodulation suggests whether an individual frequency is reasonable and allows initial estimation of a precise value for the frequency, decay, phase and amplitude.

Finally, non-linear regression, based on the Fourier transform values in the neighbourhood of a given frequency, is carried out in order to determine best estimates of the spectral parameters and their standard errors. Consider data generated by a model

$$y_j = f_j(\theta) + e_j,$$

$j = 1,..., J$ where the y_j are observed, where the $f_j(\theta)$ are known except for the K-dimensional parameter θ, and where the e_j are unobserved, uncorrelated random errors with mean 0 and common variance σ^2. The least squares estimate (see § 10.4.2) of θ is the value providing the minimum of the expression

$$\sum_{j=1}^{J} |y_j - f_j(\theta)|^2.$$

Suppose that the function $f_j(\theta)$ is differentiable with derivatives

$$g_{jk}(\theta) = \frac{\partial f_j(\theta)}{\partial \theta_k},$$

$k = 1,..., K$. Collect the y_j together into the J-vector $\mathbf{y}$, the $f_j(\theta)$ into the J-vector $\mathbf{f}(\theta)$ and the $g_{jk}(\theta)$ into the $J \times K$ matrix $\mathbf{g}(\theta)$. One means of determining an extreme value of θ is through the Gauss–Newton iteration procedure

$$\theta^{n+1} = \theta^n + [\mathbf{g}(\theta^n)^{\mathrm{T}}\mathbf{g}(\theta^n)]^{-1}\mathbf{g}(\theta^n)^{\mathrm{T}}[\mathbf{y} - \mathbf{f}(\theta)], \tag{25}$$

$n = 0, 1, 2,...,$ having started with some initial value θ^0.

Under certain conditions this estimate will be approximately normal with mean θ and covariance matrix (see § 7.6.3) that may be estimated by

$$[\mathbf{g}(\theta^n)^\mathrm{T}\mathbf{g}(\theta^n)]^{-1} \sum_j |y_j - f_j(\theta^n)|^2/(J - K). \tag{26}$$

14.3.4 Observations

In Table 14.2, a list of selected observational periods is shown for reference. Standard errors of the estimates vary throughout the spectrum listed since estimation methods were not uniform from observer to observer. While the error can be put at about 0.1 per cent as a rule, certain modes have higher uncertainties; for example, the value for $_0T_2$ depends strongly on the Trieste spectrum of the 1960 Chilean earthquake obtained by Bolt in several studies and may be in error by several seconds.

Highly reliable estimates of Q are still not numerous. We mention here results obtained by Hansen (1982) from time series following two large earthquakes, the Philippine earthquake of 16 August 1976 ($M_S = 7.9$ BRK),

Table 14.2 *Eigen-periods*

Mode	Period (s)	Mode	Period (s)
$_0S_0$	1230.64	$_2S_2$	905.47
$_0S_2$	3232.20	$_3S_0$	305.84
$_0S_4$	1547.30	$_3S_{21}$	209.59
$_0S_6$	964.17	$_4S_0$	243.59
$_0S_{10}$	579.78	$_6S_1$	348.41
$_0S_{15}$	426.15	$_0T_2$	2637.75
$_0S_{20}$	347.66	$_0T_3$	1704.89
$_0S_{29}$	268.44	$_0T_6$	925.82
$_0S_{45}$	193.95	$_0T_{12}$	536.84
$_0S_{65}$	143.14	$_0T_{18}$	390.94
$_0S_{79}$	120.61	$_0T_{31}$	249.85
$_0S_{90}$	107.56	$_0T_{45}$	181.17
$_1S_0$	613.48	$_0T_{66}$	128.63
$_1S_2$	1471.64	$_1T_3$	695.18
$_1S_4$	852.68	$_1T_8$	438.50
$_1S_7$	603.93	$_1T_{33}$	168.82
$_1S_{14}$	338.14	$_2T_7$	363.01
$_1S_{16}$	299.92	$_3T_{19}$	179.31
$_1S_{30}$	184.55	$_4T_{16}$	176.82
$_2S_0$	398.54		

Table 14.3 Q *estimates for* $_0S_1$ *and standard errors*

	Indonesia earthquake		
	JAS	WDC	BKS
$_0S_9$	243 ± 105	244 ± 30	
$_0S_{10}$	232 ± 48		
$_0S_{11}$			254 ± 75
$_0S_{12}$	218 ± 35	288 ± 82	
$_0S_{13}$	264 ± 72		
$_0S_{14}$		357 ± 50	
$_0S_{15}$	263 ± 41		190 ± 44
$_0S_{16}$			
$_0S_{17}$		350 ± 33	
$_0S_{18}$	350 ± 85	208 ± 45	282 ± 51
$_0S_{19}$	275 ± 28	239 ± 28	224 ± 27
$_0S_{21}$	237 ± 19	213 ± 21	204 ± 39
$_0S_{22}$		219 ± 29	
$_0S_{23}$		247 ± 71	215 ± 42
$_0S_{24}$	236 ± 37		243 ± 32
$_0S_{25}$	330 ± 81		369 ± 174
$_0S_{26}$		378 ± 65	
$_0S_{27}$	194 ± 76		283 ± 151
$_0S_{28}$	325 ± 105	267 ± 154	
$_0S_{29}$		362 ± 84	
$_0S_{30}$		283 ± 91	316 ± 113
$_0S_{31}$	239 ± 91		254 ± 97
$_0S_{34}$	176 ± 76		
$_0S_{35}$	266 ± 109		
$_0S_{36}$			
$_0S_{37}$			
$_0S_{40}$			192 ± 101

and the Indonesian earthquake of 19 August 1977 ($M_S = 8.0$ BRK). Both earthquakes were recorded on magnetic tape at three different sites in northern California on broadband displacement seismometers: a three-component system in Berkeley (free pendulum period $T_0 = 100$ s), a vertical component at Jamestown ($T_0 = 40$ s), and a vertical component at Whiskeytown ($T_0 = 40$ s). The distances between these three sites range from 200 to 350 km. From these recordings, estimates were made for Q for 27 fundamental spheroidal modes and 12 fundamental torsional modes.

Fig. 14.1 shows the computed Fourier amplitude spectrum of one of these earthquakes. No winnowing was applied to the above data; however,

Table 14.4 Q estimates for $_0T_l$ and standard errors for the Indonesia earthquake

	BKS
$_0T_8$	295 ± 124
$_0T_{11}$	230 ± 81
$_0T_{12}$	213 ± 47
$_0T_{14}$	270 ± 121
$_0T_{15}$	170 ± 124
$_0T_{16}$	168 ± 67
$_0T_{17}$	340 ± 162
$_0T_{20}$	175 ± 61
$_0T_{23}$	150 ± 37
$_0T_{24}$	321 ± 159
$_0T_{35}$	244 ± 33
$_0T_{45}$	360 ± 203

the four largest Earth tides, called $M2$, $S2$, O, and K, were subtracted from the data prior to the spectral analysis. When a well-excited overtone mode had a period very near that of a fundamental mode of different order, splitting from modes due to terrestrial asphericity or heterogeneities (see §14.2.3) introduced interference, even though they were not resolvable as two peaks in the spectrum. The result was broadening of the Fourier amplitude spectral peaks in addition to that due to wave attenuation and the introduction of a bias in estimates of Q to low values. When scalloping of demodulates (cf. Fig. 14.6) was observed, care was taken in assigning uncertainties to the estimates obtained.

The analysis suggests that estimates of Q from eigen-spectra should not be accepted at full weight unless specified coherency tests are satisfied. By repetition at successive steps in the demodulating frequency, the instantaneous amplitude and phase plots allow a decision to be made on the best available eigen-frequency resolution and quality of the Q values. When these criteria have been met, complex Fourier transform values may be estimated using an FFT algorithm applied to a record length previously determined to have signals above the noise. For the California regional network, values for Q and their standard errors obtained from spectra of the Indonesia earthquake are listed in Tables 14.3 and 14.4. The standard errors are such that there are no significant differences in Q values between stations at the 99 per cent confidence level for the few modes that can be used for comparison.

Recent inversions show the capacity of observations of long-period oscillations of the Earth to discriminate fairly finely between different Earth models. In applying the observations to improve resolution and precision of the internal structure of the Earth, it is in theory desirable to set up a considerable number of Earth models, compute all the periods of their free oscillations for a long run of values of both n and l (cf. § 5.6.4), and check against the observations. Models could then be steadily eliminated until only a small range survives. In practice, the work starts from existing models, seeking to amend them by successive inversion until full compatibility with the observations is achieved within the uncertainties of the observations. Questions of uniqueness of course still remain (cf. § 13.5.3).

15

Earthquake statistics and prediction

In the preceding three chapters, we have shown how seismic data provide inferences on the Earth's interior. Working back further we can derive information on the occurrence of earthquakes and, in particular, conditions near the sources of earthquakes.

15.1 Energy released in earthquakes

An earthquake is generated by the rapid release of energy E_T inside the Earth (see § 16.1), and we need to determine the quantity of this released energy.

If, in an earthquake, we measure the ground motion at a number of points of the Earth's outer surface, it is then possible, using formulae based on (8.10), (8.19) and (8.20) to estimate the total seismic energy E_S released in the earthquake. In practice, there are severe difficulties in the way of making precise estimates of either E_T or E_S. First, let us note simply that $E_T > E_S$ and we will consider in this chapter only E_S. (The question of estimating E_T is discussed in § 16.1.) There is, for instance, the difficulty of estimating wave damping and the appropriate transmission factor for each refraction of a wave through an internal discontinuity (see § 8.4). Hence, approximate methods must be used which yield estimates of the seismic energy released, often uncertain to order about 10.

We give here a simplification which is of fairly wide applicability. In general, it is to be expected that a moderate fraction of the total energy in an earthquake will leave the focus in waves of *SH* type. Now, on the theory of chapter 6, *SH* waves are reflected and refracted only into *SH* waves, and conversely reflected and refracted *SH* waves arise only from incident *SH* waves. Accordingly, it is reasonable to estimate the order of the total energy in many earthquakes from observations only of the horizontal earth movements during the passage of *S* waves.

We now discuss representative formulae that have been used to estimate the energies of various types of earthquakes.

15.1.1 Case of near earthquakes

In the Jersey earthquake of 1926, Jeffreys observed that the records traced at observatories within 500 km showed relatively large Sg waves, and that most of the horizontal movement in these waves was at right angles to the line joining an observatory to the epicentre. The energy calculations were therefore made from these waves alone. Jeffreys assumed that the energy travelled out symmetrically with a cylindrical wave front in a granitic layer, of density ρ g/cm^3 and thickness H cm. Let Δ degrees be the epicentral distance. Assuming the waves to constitute a simple harmonic group, it would then follow from (8.10) that the energy in the Sg waves would be

$$(2\pi H r_0 \sin \Delta)(2\pi^2 \rho L A^2 \tau^{-2}), \tag{1}$$

where r_0 is the radius of the Earth, L is the length of the wave train, A is the amplitude, and τ is the period. This expression is readily reduced to

$$\pi \rho H L v_m^2 r_0 \sin \Delta, \tag{2}$$

where v_m is the maximum velocity of the ground in the train of waves.

The total seismic energy released in the earthquake (including that sent out in P and SV waves) would of course be a few times greater than the value (about 10^{19} ergs) given by the calculations.

It should be noted that the body waves in earthquakes do not normally constitute a simple harmonic group, and a more accurate formula than (1) would be

$$4\pi^3 \rho r_0 \sin \Delta \int H c A^2 \tau^{-2} dt, \tag{3}$$

where c is the wave velocity, and the integration is taken over the interval of time occupied by the passage past the particular observatory of the group of waves.

15.1.2 Assumption of spherical symmetry about the source

Galitzin obtained a formula for estimating the order of the released energy E_S from observations at a single station on the assumption that the disturbance would spread out symmetrically in all directions from the focus. This formula, as later modified by Jeffreys, is equivalent to

$$4\pi^3 \rho (2r_0 \sin \tfrac{1}{2}\Delta)^2 \int c A^2 r^{-2} dt. \tag{4}$$

(Sometimes a factor of the form $\exp k\Delta$ is included in formulae such as (4) to

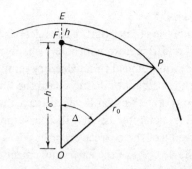

Fig. 15.1. Correction for focal depth.

allow for absorption of energy during transmission – cf. §4.5). In the derivation of (4), focal depth is neglected, and it is assumed that the (assumed spherical) spreading out of the disturbance is essentially confined to the lower side of the focus.

Also in (4) A denotes the amplitude in the incident waves, but it is assumed that this amplitude is of the same order as that of the observed ground motion. It may be noted that in the case of SH waves the amplitude of the incident waves is one-half that of the surface earth movement; with other incident waves the corresponding amplitude ratio varies with the angle of emergence and therefore with Δ (see §8.3). The assumption of spherical symmetry about the focus is of course not realised with actual earthquakes; but (4) yields useful estimates of the energy in P or S body waves as recorded at stations for which Δ does not exceed about 90°.

For an earthquake of appreciable focal depth h, it is necessary to replace (4) by

$$8\pi^3 \rho \{h^2 + 4r_0(r_0 - h)\sin^2 \tfrac{1}{2}\Delta\} \int cA^2\tau^{-2}dt; \qquad (5)$$

the coefficient 4 in (4) is replaced by 8 in (5) in order to allow for energy which would spread upwards as well as downwards from the focus; the factor in curled brackets corresponds to the square of the length FP (where F is the focus and P a point of the Earth's surface) as indicated in Fig. 15.1.

If P is at the epicentre E, (5) degenerates to the formula

$$8\pi^3 \rho h^2 \int cA^2\tau^{-2}dt. \qquad (6)$$

A formula equivalent to (6) was made the basis of a method of Gutenberg and Richter for assessing energy in earthquakes.

15.1.3 Use of surface waves

With earthquakes in which the greater part of the released energy goes into forming surface waves, and therefore with many shallow-focus earthquakes, use of the formula (4) will not lead to satisfactory estimates of the total energy released. For such earthquakes, Jeffreys used the Rayleigh wave formulae (5.18). The conditions of the simpler Rayleigh wave theory of §5.2 are not realised in the Earth (see chapter 12), but their assumption is nevertheless likely to lead to a result of the correct order.

By (5.18) the components of the velocity of the earth movement during the passage of a group of Rayleigh waves are given by

$$\dot{u}_1 = a\kappa c\{-\exp(0.85\kappa x_3) + 0.58\exp(0.39\kappa x_3)\}\cos\{\kappa(x_1 - ct)\},$$

$$\dot{u}_3 = a\kappa c\{-0.85\exp(0.85\kappa x_3) + 1.47\exp(0.39\kappa x_3)\}\sin\{\kappa(x_1 - ct)\},$$

where the suffixes 1, 3 indicate components parallel to the direction of propagation of the waves, and vertically upwards, respectively; a is a constant determining the amplitude, $2\pi/\kappa$ is the wave-length, c is the wave velocity, and $x_3 = 0$ is taken as the equation of the Earth's outer surface. We deduce that the mean value over one wave-length of $(\dot{u}_1^2 + \dot{u}_3^2)$ is

$$a^2\kappa^2 c^2\{0.86\exp(1.70\kappa x_3) - 1.82\exp(1.24\kappa x_3)$$
$$+ 1.24\exp(0.78\kappa x_3)\}. \tag{7}$$

The mean kinetic energy per unit volume is $\frac{1}{2}\rho$ times the expression (7), and the total mean energy per unit volume is (by §3.3.6) double this. Integrating this result over the range $0 \geqslant x_3 \geqslant -\infty$, we find that the total energy per unit earth surface area is $0.63\rho a^2\kappa c^2$, i.e. $1.26\pi\rho a^2\lambda\tau^{-2}$, by (3.40), where τ is the period and $\lambda = 2\pi/\kappa$ is the wave-length. It follows that the order of the energy in the Rayleigh waves will be given from observations at an observatory whose epicentral distance is Δ by

$$2.5\pi^2\rho r_0\sin\Delta\int ca^2\lambda\tau^{-2}dt. \tag{8}$$

Now the amplitude of the horizontal component of the surface motion in Rayleigh waves is seen by (5.19) to be $0.42a$, $= A$ say. The formulae (3) and (8) then become formally identical if we put $2.5 \times (0.42)^{-2}\lambda = 4\pi H$, i.e. $H = 1.1\lambda$. Thus the energy in the Rayleigh waves may be estimated from observations of the horizontal component of the surface movements at any single observatory, by using the formula (3) and interpreting H as corresponding to an 'equivalent layer' of thickness roughly equal to the wave-length of the passing Rayleigh waves. The formula (3) was first used in this way by Jeffreys in estimating the energy of the Pamir earthquake of 1911 February 18.

15.2 Earthquake magnitude

In 1935, Richter set up a 'magnitude scale' of earthquakes, in which the magnitude of an earthquake M is defined as the logarithm (to base 10) of the maximum amplitude A_0 (measured in micrometres; $1 \, \mu m = 10^{-4} \, cm$) traced on a seismogram by a standard torsion horizontal-component seismograph (see §9.9), distant 100 km from the epicentre. Reduction of amplitudes observed at various distances to the expected amplitudes at the standard distance of 100 km is made by empirical tables set up on the assumption that the ratio of the maximum amplitudes at two given distances is the same for all earthquakes considered, and independent of the azimuth. The scale is applied directly only to earthquakes of shallow focal depth.

We have

$$M = \log A(\Delta) - \log A_0(\Delta), \tag{9}$$

where A is the maximum trace amplitude at distance Δ.

Richter's empirical formula for southern California earthquakes is, approximately,

$$\log_{10} A_0 = 5.12 - 2.56 \log_{10}\Delta, \tag{10}$$

with A_0 in μm, Δ in km and $10 < \Delta < 600$ km.

Substitution of (10) in (9) and allowing for the nominal magnification of the Wood–Anderson seismograph yields

$$M = \log_{10} A + 2.56 \log_{10} \Delta - 1.67, \tag{11}$$

where A is the ground motion amplitude in μm.

Richter first applied the magnitude scale to earthquakes recorded in the Californian region within 600 km of the epicentre. Later, Gutenberg and he set up further empirical tables whereby observations made at distant stations and on seismographs of other than the standard type could be used. The empirical tables were extended to cover earthquakes of significant focal depth and to enable independent magnitude estimates to be made from body and surface wave observations.

Thus the *surface wave magnitude* M_s is defined, for shallow-focus earthquakes, as

$$M_s = \log_{10}A + \alpha \log_{10}\Delta + \beta \tag{12}$$

where A is the maximum amplitude of the ground motion for surface waves having 20 s period. Representative values of α and β for the horizontal component of Rayleigh waves from shallow earthquakes are 1.66, 1.82 respectively. For deep earthquakes, (12) is not applicable and body wave

Table 15.1 *Moments and magnitude for some of the largest recorded earthquakes*

Date	Region	M_s	$M_0(\times 10^{27}\,\text{dyne cm})$	M_w
31 Jan 1906	Ecuador	8.6	204	8.8
18 April 1906	California	$8\frac{1}{4}$	10	7.9
1 Feb 1938	Banda Sea	8.2	70	8.5
15 Aug 1950	Assam	8.6	100	8.6
4 Nov 1952	Kamchatka	$8\frac{1}{4}$	350	9.0
9 March 1957	Aleutian Islands	$8\frac{1}{4}$	585	9.1
22 May 1960	Chile	8.3	2000	9.5
28 March 1964	Alaska	8.4	820	9.2
4 Feb 1965	Aleutian Islands	$7\frac{3}{4}$	125	8.7

Source: after Kanamori (1977)

magnitudes must be defined. The form usually used is

$$m_b = \log_{10}(A/T) + Q(h, \Delta), \tag{13}$$

where T is the measured wave period and Q is an empirical function of focal depth, h, and Δ.

An approximate relation between m_b for P waves and M_s is, for shallow-focus earthquakes,

$$m_b = 2.5 + 0.63 M_s. \tag{14}$$

The above magnitude scales are found to be a strong function of wave frequency; in particular, M_s tends to an upper limit for great earthquakes. Of course, the M_s definition can be extended to longer period waves (say, 100 s period). However, recently, the *moment magnitude* M_w has been defined to provide a more uniform scale. Kanamori has proposed

$$M_w = \tfrac{2}{3}\log_{10} M_0 - 10.7, \tag{15}$$

where M_0 is the seismic moment of the earthquake (cf. § 16.3) in dyne cm. The moment magnitude M_w has the advantages (as a measure of size in earthquakes) that it does not saturate at the top of the scale and has a sounder theoretical basis (see § 16.3) than M_s. With the above definitions, the 1964 great Alaskan earthquake has the estimated values: $M_s = 8.4$, $M_0 = 820 \times 10^{27}$ dyne cm, $M_w = 9.2$ (see Table 15.1).

The various extensions introduced some confusion into the use of magnitude and steps are being taken towards standard scales for international adoption. Usually, the symbol M relates to Richter's original scale in which $M \approx 8.9$ for the greatest known earthquakes. For most moderate shallow-focus damaging earthquakes, it is sufficient for engineering purposes to take M_L, M_s, and M_w to be roughly the same.

Many special magnitude scales for various continental regions have been developed. An illustration is the m_{bLg} relation used to classify central US earthquakes by O. Nuttli. In continental areas the surface waves Lg (see § 11.1.4) propagates efficiently and is often clearly recorded on short-period seismographs. Based on (12), and a plausible theoretical model for attenuation (see § 4.5.3), the empirical relation valid for periods near 1.0 s is

$$m_{bLg} = 3.81 + 0.83 \log_{10} \Delta + \gamma (\Delta - 0.09) \log_{10} e + \log_{10} A, \quad (16)$$

where Δ is in degrees, A is in μm, and γ is a regional attenuation constant.

Representative values are $\gamma = 0.07 \text{ deg}^{-1}, 0.11 \text{ deg}^{-1}$, and 0.53 deg^{-1} for the central, northeastern and central western United States.

15.2.1 Magnitude and energy
Gutenberg and Richter sought to connect the magnitude M_S with the energy E_S of an earthquake by the form

$$aM_S = \log_{10} (E_S/E_0),$$

and after several revisions using (6), (13) and (14) arrived in 1956 at the equation

$$\log_{10} E_S = 11.8 + 1.5 M_S. \quad (17)$$

Båth, working on independent material, obtained a very similar result.

The Gutenberg–Richter formula gives $E_S = 6.3 \times 10^{11}$ and 1.4×10^{25} ergs for earthquakes of $M_S = 0$ and 8.9, respectively. A unit increase in M_S thus corresponds to a 32-fold increase in energy. Negative magnitudes correspond to the smallest instrumentally recorded earthquakes, 1.5 to the smallest felt earthquakes, 3 to those felt at distances up to 20 km; those of 4.5 cause slight damage near the epicentre; those of 6 are destructive over a restricted area; those of 7.5 are at the lower limit of *major earthquakes*.

15.2.2 Magnitude – frequency of occurrence relation
Gutenberg and Richter developed empirical relations for the frequencies of earthquakes of various magnitudes. Let N be the average number of shocks per year for which the magnitude lies in the range $M \pm \Delta M$. They found that

$$\log_{10} N = a - bM_S \quad (18)$$

fits the data well both globally and for particular regions. For example, for the whole world, they gave for shallow earthquakes: $a = 6.7$, $b = 0.9$ when $M_S > 6.0$. The frequency for these earthquakes thus increases by a factor ranging from about 10 when the magnitude is diminished by one unit. The increase in frequency with reduction in M_S falls short, however, of matching the decrease in the energy E. Thus larger earthquakes are overwhelmingly responsible for most of the total seismic energy release. The number of earthquakes per year strong enough to be felt is of the order of 10^5, and the total number per year with $m_b > 4.0$ may reach 20000.

Gutenberg and Richter estimated that the total annual energy released in all earthquakes is about 10^{25} ergs, corresponding to a rate of work between 10 and 100 million kilowatts. This is of the order of 0.001 of the annual amount of heat escaping from the Earth's interior. Ninety per cent of the total seismic energy comes from earthquakes of magnitude 7.0 and more, i.e. whose energy is of the order of 10^{23} ergs or more.

In a recent reassessment, based on seismic moment considerations (see § 16.3), Kanamori estimates the average total seismic wave energy release to be 4.5×10^{24} ergs/year from 1920 to 1976. Fluctuations occur, however, of up to an order of magnitude; correlations with various planetary variations have been suggested, such as the polar wobble of the Earth.

For use in prediction and comparative mechanism studies, standard errors of b must be supplied for statistical tests. Because earthquakes are stochastic processes and the b value is a random variable, the probability distribution and the variance of b are essential in studying its temporal and spatial variation. The b value can be calculated by least-squares regression, but the presence of even a few large earthquakes influences the resulting b value significantly. As an alternative, the maximum likelihood method has been used to estimate b because it yields a more robust value when the number of infrequent large earthquakes changes. There will be cases, however, such as estimating the probability of the largest magnitude of earthquakes, where the least-squares method is more suitable.

From (18), the reader can readily show (see § 17.6.1) that the probability density of magnitude M earthquakes is exponential with mean $1/\beta$, i.e.

$$\beta \exp\{-\beta(M - M_0)\} \quad (M > M_0). \tag{19}$$

The maximum likelihood estimate of $\beta(b = \beta \log e)$ is thus

$$\hat{\beta} = \frac{1}{\bar{M} - M_0}, \tag{20}$$

where $\bar{M}$ is the mean magnitude and M_0 is the smallest (or 'threshold') magnitude considered.

The distribution of $\hat{\beta}$ can then be shown to be

$$2n\beta/\chi_{2n}^2 = \beta F_{\infty,2n} \tag{21}$$

where χ_n is the chi-squared distribution for n degrees of freedom and F is Fisher's distribution.

In the general case, a large sample distribution of $\hat{\beta}$ can be derived in the usual way. Let $X_i = M_i - M_0$. Then $\hat{\beta} = 1/\bar{X}$. If X_i has mean μ and variance σ^2, it can be shown that $\hat{\beta}$ is asymptotically normal with mean $1/\mu$ and variance

$$(1/\mu^2)^2 \sigma^2/n = \beta^4 \sigma^2/n. \tag{22}$$

Here σ^2 may be estimated from $\sum_{i=1}^{n} (M_i - \bar{M})^2/(n-1)$. Note that $\log \hat{\beta}$ is also asymptotically normal with mean $\log \beta$ and variance $\beta^2 \sigma^2/n$. It also follows (see Shi and Bolt, 1982) that, for large n,

$$\sigma(\hat{b}) = \frac{b^2}{\log e} \sigma(\bar{M}) = 2.30 b^2 \sigma(\bar{M}). \tag{23}$$

Thus, in practice, starting from an extensive earthquake catalogue, we can estimate the statistics of b by maximum likelihood from the magnitudes alone using (20) and (23).

Because the value of b in a region is observed to change with time and location, its mean and variance might also be expected to change. In other words, b should be regarded, in general, as a nonstationary stochastic process. When sampling with a small time and space window, however, b can usually be taken as stationary and with constant expectation so that the above formulas hold.

15.3 Seismicity

15.3.1 Geography of shallow earthquakes

The epicentres of the Earth's major earthquakes have long been known to concentrate mainly in belts. This pattern is discernible in early catalogues of felt earthquakes such as those of Montessus de Ballore, and by maps of instrumentally determined hypocentres based on ISS and ISC data, such as the thorough study in 1954 by Gutenberg and Richter of the Earth's present seismicity.

Subsequently, the ability to estimate precisely and uniformly the location of earthquakes around the world has greatly improved. The key ingredients were the establishment of the Worldwide Standardised Network of seismographs (see § 1.3.2) and allied observatory improvement and the use of optimum algorithms for computation. A recent map of worldwide seismicity is given in Fig. 15.2.

One earthquake belt passes round the Pacific Ocean and affects countries with coastlines bordering on this ocean, for instance New Zealand, New Guinea, Japan, the Aleutian Islands, Alaska, and the western regions of North and South America; Gutenberg and Richter estimated that 80 per cent of the energy at present released in earthquakes comes from earthquakes whose epicentres are in this belt. The seismic activity is by no means uniform throughout the belt, and in places there are a number of branches.

A second belt passes through the Mediterranean region eastward through Asia and joins the first belt in the East Indies. The energy released in earthquakes from this belt is about 15 per cent of the total.

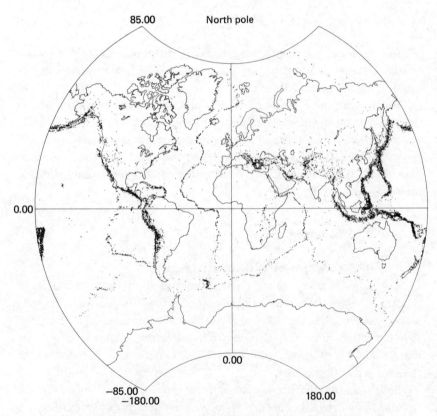

Fig. 15.2. Epicentres of about 16,000 earthquakes $M_s = 4.5–5.5$ from 1965 to 1975. (Compiled by M.K. Hutcheson and P.D. Lowman, Goddard Space Flight Center).

There are also striking connected belts of seismic activity, mainly along mid-oceanic ridges, including ones in the Arctic Ocean, the Atlantic Ocean, the western Indian Ocean, and East Africa (see § 15.3.3). Most other parts of the world experience at least occasional shallow earthquakes.

The geographic distribution of the many lesser earthquakes is less precisely determined partly because of the dependence of the relevant data on the geographical distribution of observatories.

15.3.2 Distribution of deep-focus earthquakes

Reference to the focal depths of earthquakes has already been made in chapter 10 in connection with the setting up of travel-time tables. The great majority of earthquakes originate within 60 km of the Earth's outer surface; these are the shallow earthquakes. We now discuss deeper-seated earthquakes that are of special importance in a number of geophysical problems.

Early attempts to measure focal depth were made by Galitzin and Walker, using observations of the apparent angles of emergence (see § 8.2) of earthquake waves at various epicentral distances. But the results were subject to wide uncertainty because of the dependence of the angles of emergence on the properties of the crustal layers.

In 1922, Turner observed that in the case of some earthquakes, P waves appeared to arrive at stations near an earthquake's anti-centre (see § 10.9.1) significantly earlier than would be the case with a normal earthquake. This led Turner to put forward the hypothesis that the earthquakes in question had foci at depths appreciably greater than the normal.

In 1928, Wadati found that, with Japanese earthquakes, the difference between the arrival-times of P and S waves near an epicentre was sometimes abnormally great. This observation also suggested an abnormal depth of focus. Further, both Turner and Wadati independently found about the same focal depth for some earthquakes.

Jeffreys then pointed out that a crucial test of the hypothesis of deep focus could be made from eigen-theory by examining whether the surface waves from earthquakes suspected to be of deep focus were abnormally small (see § 3.2.5). In 1931, Stoneley discovered that phases that had earlier been recorded as 'L' for earthquakes suspected to be of deep focus mostly arrived much earlier than the L phases in normal earthquakes, the difference reaching as much as 10 min in some cases. Stoneley showed that these 'L' phases were really associated with body waves whose amplitudes exceeded those of the actual surface waves. With this discovery, the existence of deep-focus earthquakes was placed beyond doubt.

Table 15.2 *Distribution of focal depths*

Region	Depth (km)												
	100	150	200	250	300	350	400	450	500	550	600	650	700
Mexico, Central America	20	4	2	—	1	—	—	—	—	—	—	—	—
South America	46	22	18	7	2	—	—	—	—	1	5	11	—
New Zealand, Tonga, Samoa	11	3	3	1	2	1	4	—	6	9	10	5	1
New Hebrides to New Guinea	22	16	9	1	2	3	3	1	—	—	—	—	—
Sunda Islands	21	12	9	1	—	1	3	—	1	—	10	1	5
Celebes to Mindanao	3	6	10	2	4	—	1	—	2	1	3	1	1
Luzon to Kyushu	10	7	4	3	—	—	—	—	—	—	—	—	—
Japanese Islands	33	26	12	5	12	27	25	12	14	14	6	1	—
Hindu Kush	1	—	14	25	—	—	—	—	—	—	—	—	—
Others	11	13	1	1	—	—	—	—	—	—	—	—	—
Total	178	109	82	46	23	32	36	13	23	25	34	19	7

Source: after Gutenberg and Richter (1954)

Another crucial test was provided in the studies of phases such as pP, sP, etc. Scrase and Stechschulte estimated focal depths from observations of the differences between the arrival-times of the phases P and pP, P and sP, and S and sS, the main method still used in computer algorithms today.

Table 15.2 shows the distribution, for a thirty-year period, of earthquakes whose focal depths were 80 km or more. The depths given in the table are in kilometres below the outer surface, a range of depth of ± 25 km being involved in each case.

Gutenberg and Richter distinguish between 'intermediate' focal depths ranging from about 60 to 300 km and greater focal depths. Of the total energy released in earthquakes, 12 per cent comes from 'intermediate' earthquakes and 3 per cent from deeper earthquakes.

The table shows that: (i) the frequency falls off rapidly with increasing focal depth in the intermediate range, while below this the distribution in depth is fairly uniform until the greatest focal depths are approached; (ii) the greatest focal depth is not more than about 700 km; (iii) there are characteristic focal depths in particular geographical regions.

Deep-focus earthquakes commonly occur in patterns, called *Benioff zones*, that dip into the Earth. Dip angles range about 45°, with some shallower dipping zones and others with nearly vertical dips. Benioff zones are found under tectonically active island arcs, such as Japan, the New Hebrides, the Tonga Islands, and Alaska, and they are normally but not always (e.g. Romania, Hindu Kush) associated with deep ocean trenches, such as along the South American Andes. In most Benioff zones, intermediate and deep earthquake foci lie in a narrow layer, although recent precise hypocentral locations in Japan and elsewhere show two distinct parallel bands of foci 20 km apart. Careful estimation gives about 680 km for the deepest depths globally (Stark and Frohlich, 1985).

15.3.3 Tectonic associations

There is a marked correspondence between the geographical distributions of major earthquakes and of volcanic activity, particularly in the circum-Pacific earthquake belts and along mid-oceanic ridges. However, volcanic vents are generally some hundreds of kilometres distant from the majority of the epicentres of major shallow earthquakes and many earthquake sources occur nowhere near active volcanoes. Earthquakes of 'intermediate' focal depth frequently occur directly below structures marked by volcanic vents, but there is probably no immediate causal connection between these earthquakes and the volcanic activity, both being likely due to the same tectonic processes. Reference has been made (§ 15.3.1) to the association of earthquakes with arcuate structures.

Seismicity patterns such as those discussed above and in § 15.3.2 had no strong global theoretical explanation until a dynamical model called *plate tectonics* was developed in the late 1960s. This theory holds that the Earth's upper shell or lithosphere consists of several (about eight) large and quasi-stable slabs called *plates*. The thickness of each plate extends to a depth of about 80 km; the plates move horizontally, relative to neighbouring plates on a layer of softer rock. The rate of movement ranges from a centimetre to ten centimetres a year over a lower strength shell, called the *asthenosphere*. At the plate edges where there is contact with adjoining plates, boundary tectonic forces operate on the rocks causing physical and chemical changes in them. New lithosphere is created at mid-oceanic ridges by the upwelling and cooling of magma from the Earth's mantle. In order to conserve mass, the horizontally moving plates are believed to be absorbed at the ocean trenches where a subduction process carries the lithosphere downwards along the Benioff zones into the Earth's interior.

Seismological evidence is broadly in agreement with this kinematical model (see also § 13.2.2). Earthquake sources are concentrated along mid-oceanic ridges where mechanisms are found to be of normal or dip-slip type (see § 16.1.3), consistent with the horizontal moving apart of lithospheric plates. At the subduction zones, the intermediate and deep-focus earthquakes in the Benioff zone mark the location of the upper part of a dipping plate. The focal mechanisms indicate that stresses are aligned with the dip of the lithosphere underneath the adjacent continent or island arc.

Some earthquakes associated with mid-oceanic ridges are confined to strike-slip faults which offset the ridge crests. These horizontal shear faults were named *transform faults* by J. Tuzo Wilson. Most earthquakes located on these transform faults are characterised by strike-slip motions with the correct sense of horizontal slip for the divergent plate hypothesis. The plate theory is consistent with high seismicity along the edges of the interacting plates (*interplate* earthquakes) and low seismicity within the plate boundaries. Small to large earthquakes do occur, however, well within plates and such *intraplate* earthquakes must be explained by other mechanisms. More detailed discussion of the correlations between seismological evidence and plate tectonics is given by Isacks, Oliver, and Sykes (1968).

15.3.4 Reservoir-induced earthquakes

As well as occurring naturally, earthquakes are sometimes caused by human activities. Well-documented case histories exist of such events associated with the injection of fluids in deep wells, the excavation of mines, the filling of large reservoirs and the detonation of large underground nuclear explosions. In all cases, except deep mining, the induction mechanism is thought to be regional strain release, triggered by small changes in the local strain field leading to fracture or fault slip. The case of underground explosions is discussed by Bolt (1976).

Not only are the largest induced earthquakes triggered by large reservoirs but in many ways this case presents the most important theoretical and practical problems. Over 20 cases have now been documented in which local seismicity has increased after the impounding of large reservoirs behind high dams. There are other cases claimed, but the necessary instrumentation to compare seismicity before and after the filling is lacking. It must be stressed that the induction effect is most marked for large reservoirs defined as exceeding 100 metres in depth and 1 km^3 in volume; the majority of such large reservoirs are, however, aseismic. Of the 26 largest reservoirs, only about 5 undoubtedly have induced earthquakes, including Kariba in Zambia and the high Aswan dam in Egypt.

A simple calculation shows that the stress due to the load of the water in the reservoir is too small to fracture competent rock. The most viable hypothesis is that the rocks in the vicinity of the reservoir are already strained from tectonic forces so that existing faults are almost ready to slip and either the reservoir adds a perturbation which triggers a slip or the increased water pressure resulting from the impoundment lowers the strength of the fault so that it slips under the applied tectonic stress. A simple argument can be given for induced earthquakes with $M_L \geqslant 5$.

In this case (on the calculations of §16.1.2) the minimum volume of the strained region would require a focal depth z greater than 1 km. A fracture at this depth occurs under lithostatic stresses with the greatest principal stress (see §2.1.4) $p_3 = \rho g z$ acting vertically and (see §2.4.4)

$$p_1 = p_2 = \{\sigma/(1 - \sigma)\}p_3, \tag{24}$$

where σ is Poisson's ratio. The shear strength of a vertical fracture with coefficient of friction μ is then $S = \mu g \rho z \sigma/(1 - \sigma)$ with inclined fractures being stronger (see §16.1.3). For normal values of the parameters and $\mu \approx 0.5$, the minimum shear strength for a dry vertical fracture is thus about 50 bars at 1 km depth. If water, density ρ_w, permeates the fracture, the fault strength is reduced by the ratio $(\rho - \rho_w)/\rho$.

Thus the minimum shear strength for a wet vertical fracture at 1 km depth is about 30 bars. By contrast, a reservoir 150 metres deep produces a maximum shear stress of about 5 bars in the rock, which is an order of magnitude too small to induce earthquakes directly. It follows that pre-existing tectonic stresses are needed of well over 10 bars so that faults are almost at the failure point.

The earthquake mechanisms in the case of reservoir induction have been calculated in a few cases. For the main shock at Koyna Dam in India, the preponderance of evidence favours strike-slip motion, and at Hsinfengkian Dam in China, the principal shock was also strike-slip with a stress drop of about 10 bars. At Kariba, and Kremasta Dams in Greece, the mechanism was dip-slip on normal faults. In contrast, thrust mechanisms have been determined for earthquakes at the Nurek Reservoir in Tadjikistan, USSR.

At Nurek, more than 1800 earthquakes ($1.4 < M < 4.6$) occurred during the first nine years of filling of the 300 metre deep reservoir. This rate is four times the average in the region prior to filling. Increased seismicity occurred in a series of swarms; the two most dense accompanied rapid increases in water level during the first two stages of filling to 105 metres in 1972 to 205 in 1976. All periods of high seismicity occurred when the water level was higher than it had been previously or within 10 metres of its previous maximum. All of the largest earthquakes followed decreases in the rate of filling of the reservoir. As the reservoir approached its maximum size extending 40 km upstream from the dam, the area of induced seismicity increased also. The first stage of activity in 1971–2 had low b values and included the largest earthquakes ($M = 4.6$ and $M = 4.3$ in November 1972) that occurred when the water level first exceeded 100 metres.

A full review of induced reservoir seismicity has been given by D.I. Gough (1978).

15.4 Foreshocks and aftershocks

Frequently, series of earthquakes occur within a time interval of the order of a few days or weeks (sometimes longer), all located approximately in the same crustal volume (see Fig. 15.3).

15.4.1 Aftershocks

Usually a major or even moderate shallow earthquake is followed by a host of lesser earthquakes close to the original source region. This is to be expected since the disruption producing a major earthquake will not relieve all of the accumulated strain energy at once (see § 16.1); further, this disruption is liable to cause an increase in the stress and strain at a number of places in the vicinity of the focal region, bringing crustal rocks at some points close to the stress at which fracture occurs. In some cases, the

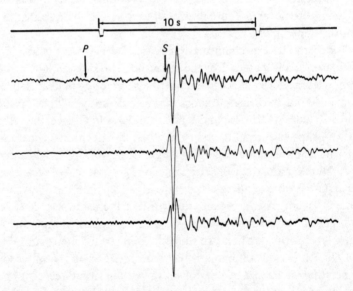

Fig. 15.3. Similarity of wave forms in a foreshock, $M_L = 2.1$ at top, and two aftershocks, $M_L = 2.8$ and 2.4, in a central California sequence with mainshock $M_L = 5.5$.

frequency of aftershocks may be for a time of the order of a hundred or more a day.

It sometimes happens that a large earthquake is succeeded by another at approximately the same focus within an hour or a day. The extreme case is that of multiple earthquakes (see § 4.3). But in the great majority of cases the first principal earthquake of a series is much more intense than all the aftershocks.

In general, the number of aftershocks per day decreases with increasing time. Work of Omori indicated that the aftershock frequency is roughly inversely proportional to the time t since the occurrence of the major earthquake of the series; formulae of the form $a/(t + b)$, where a, b are constants, fit closely particular series of earthquake aftershocks.

15.4.2 Foreshocks

Major earthquakes sometimes occur without detectable warning from less intense foreshocks; this is why there is often great loss of life attending the greater earthquakes.

However, some earthquakes are preceded by foreshocks. For example, in the North-Idu earthquake of 1930, there were foreshocks during the preceding three weeks, increasing steadily in frequency to 70 felt shocks on the day prior to the occurrence of the main earthquake. Imamura found that the tendency for foreshocks to occur is limited to particular seismic zones, and stated that in Japan not more than 20 per cent of major earthquakes are preceded by foreshocks.

15.4.3 Swarms

It sometimes happens that large numbers of lesser earthquakes occur in a region over an interval of time which may extend to some months, without a major earthquake occurring. In the Matsushio region of Japan for instance, there occurred between August 1965 and 1967 a series of hundreds of thousands of earthquakes some sufficiently strong (up to local magnitude 5) to cause property damage but no casualties. The maximum frequency was 6780 small earthquakes on 17 April 1966. Such series of earthquakes are called *earthquake swarms*. Earthquakes associated with volcanic activity often occur in swarms, but swarms also occur in many non-volcanic regions.

The recent concept of a *fractal set* to describe either non-random or random fragmented or irregular patterns has been introduced by B.B. Mandelbrot. Application has been made to such irregular processes as the wiggliness of non-rectifiable curves (e.g. broad-band seismograms) and space-filling patterns of 'trees' and 'islands' (e.g. fracture centres in an earthquake swarm). The concept, while not providing a physical basis, has also been linked to the seismic b value (see §15.2.2).

15.5 Earthquake prediction

15.5.1 Prediction theory

As in the case of other natural phenomena such as weather, tides, and astronomy, a complete understanding of earthquakes requires the ability to give descriptions, explanations, and predictions concerning them. Aspects of earthquake prediction that receive most attention are forecasting the place, the size, and the time of the earthquake. A less well-known aspect is the forecasting of strong ground motion which is likely at a particular site (see §17.6).

In an effort to clarify the theory involved, let us define the types of prediction which occur in science. First, as generally understood, a prediction of an event is a foresaying before its *observation*, but not necessarily before its *occurrence*. Prediction of certain phenomena which precede an earthquake may be checked by going back to records of past earthquakes. Seismograms from seismographic stations, geodetic surveys from last century and historical reports of earthquakes provide many of the basic observations for checking predictions. One difficulty with such *a posteriori* checks is that certain techniques that seem promising in locating earthquake precursors depend on instrumentation which has only lately become available.

Apart from hunches that something may be the case, three main kinds of prediction are generally recognised. The first type is called *descriptive prediction*. If a pattern is suspected by observing events, then, using processes of interpolation and extrapolation, certain other events may be predicted to round out the description. In this type of prediction, gaps in a suspected order lead to forecasts of future events. In earthquake prediction, it has been suggested that in seismically active regions, such as along the San Andreas fault and along the trenches of the circum-Pacific belt, zones of low seismicity ('seismic gaps') are the most likely locations for the next large earthquakes. The argument is based on the assumption of uniformity of crustal displacements. Although uniformity has been found over the short run in certain regions, there is contrary evidence in others. For example, the historical records from China, which go back to about 700 BC, and from the Middle East suggest that earthquake activity often proceeds in apparently uneven cycles, with periods of hundreds of years.

A second type of prediction is *inductive*. In this type of forecasting, a general theory is needed together with a method of recognising that additional premises must be provided to explain the known facts.

Since the 1906 San Francisco earthquake, a great deal of seismological work has been based on the elastic rebound theory of earthquake generation proposed by H.F. Reid (see § 16.1). This theory not only gives a mechanical explanation for earthquake generation but also contains other elements concerning the interval before the rupture (i.e. the build-up of strain) and the level of shaking generated. The theory, however, is not specifically predictive of the time and place of onset of rupture. Additional details or premises must be added to it. Recently, fruitful new premises have been recognised, largely on the basis of laboratory tests of rock samples under strain. Specifically, it is the additional material on *dilatancy* (see § 15.5.5) which provides a fuller predictive technique.

The third type of prediction is *deductive*. In deductive prediction, consequences are worked out as logical, sometimes even mathematical, steps from the general physical theory. Thus, a truly general theory of earthquakes would not only explain the variation of strain in crustal rocks and the mechanism of initiation and completion of source energy radiation, but would also predict as implications, for instance, the occurrence of electromagnetic variations in the focal region and the details of ground shaking.

No one has yet sharply formulated such a general theory in seismology, and at present the implications of a number of competing premises are being tested.

Of course, because of geological complexities in the Earth's crust, even a general theory of earthquakes may not yield practical modes of inductive or deductive prediction in all seismically hazardous regions. However, it would no doubt provide a valuable framework for fitting various clues together so that decisions on likelihood of occurrence could be strengthened.

15.5.2 Periodicities and correlations. Seismicity patterns

The search for earthquake periodicities in space and time goes back many years. After the turn of the century, the availability of statistical tests for hidden periodicities led to attempts to find return periods of large earthquakes in global seismicity catalogues. Generally speaking, cyclical tendencies in time and space of major earthquakes have not been widely accepted. A basic difficulty with long runs of historical earthquakes is that the catalogues are inhomogeneous with lacunae present and non-random selection due to the lack of a well-distributed network of seismographic stations before about 1910.

The longest historical record is that of northern China which begins about 700 BC. This 2700-year period contains between 500 and 1000 destructive earthquakes about which some amount of information is known. The distribution of the intensity of shaking in many of these can be assessed from dynastic and temple records. For example, the Sanho earthquake (the greatest known near Peking) of 2 September 1679, is mentioned in the records of 121 cities with information on damage, ground cracks, and other geological features. From such isoseismal information (see § 17.2.2), an idea of the size or magnitude of the earthquake can be assessed. Already certain periodicities have been estimated, as well as the discovery of irregular long intervals of quiescence. The earthquake occurrence is modelled as a time series and the techniques of statistical analysis of point processes used to locate periodicities. Some tentative predictions for restricted regions in China have been made on the basis of the statistical analysis, but up to the present time the forecasts seem essentially exploratory.

Another approach to the probabilistic occurrence of earthquakes involves the postulation of external trigger forces which initiate the rupture. Such forces have been attributed to severe weather conditions, volcanic activity, and tidal forces. Usually correlations are made between the physical phenomena assumed to provide the trigger and the repetition of earthquakes. Of course, it has long been understood that correlations of this kind, while sometimes useful in bringing to light interactions, require careful use of significance tests. Even when such tests are successful, further enquiry must be made to discover whether a causative link is actually present. About a decade ago a claim was published that the gravitational attraction of the distant planet Uranus had induced periodicities in earthquakes. Although the attractive force of Uranus on the Earth is tiny compared with that of the moon, statistical tests seemed to indicate in this case that periodicities were dependable. The difficulty is that if many intervals from samples of two types of observations are tested for intercorrelations without any fixed hypotheses, sooner or later a correlation with a high level of significance will be found by chance alone! Up to the present time, no trigger mechanism, at least for moderate to large earthquakes, has been found which satisfies the various criteria necessary to establish mechanical connection. (By contrast, tidal triggering of moonquakes (§ 1.3.4) is supported by several lines of evidence.)

Statistical methods have also been tried with populations of regional or local earthquakes. If N is the number of earthquakes, magnitude greater than M, in a given region per unit of time, then the linear regression (18) can be used to indicate broadly the time of return of an earthquake in a given magnitude range for the region (see § 17.6.1). Further, it has been suggested that the slope b of the regression line for a region may change characteristically with time. Specifically, the b value for the population of foreshocks of a major earthquake may be significantly smaller than the mean b value for the region averaged over a long interval of time (see equation (23)). A difficulty in application of the scheme is the estimation of the size of the main earthquake that will follow the foreshocks.

15.5.3 Changes in seismic velocities

For prediction of the time of earthquake occurrence, a proposal is that precursory changes in a region will cause the velocity of seismic waves through the region to change. Thus, if appropriate travel-time residuals are plotted as a function of time, fluctuations will provide a forewarning. The first published work of this kind seems to have been carried out by Kondratenko and Nersesov in 1962 for seismicity in the Tadjikistan Republic of the Soviet Union.

The usual data treatment, originally due to K. Wadati, is to plot the difference in time of arrival of S and P waves $(t_s - t_p)$ on a seismogram against the time of arrival of $P(t_p)$. The slope of the curve (usually nearly linear) is $t_s/t_p - 1$, which gives at once t_s/t_p and hence α/β, the ratio of the apparent mean P to S velocity. Some studies indicate that for earthquakes in the magnitude range 3 to 5, the ratio α/β decreased by about 5 per cent and then increased again to a more normal value just before the main shock occurred. Certain questions arise. Just how closely did the seismic waves penetrate the strained regions involved? What were the precisions of the measurements on the seismograms? Was the selection of test earthquakes exhaustive? What is the effect of such systematic changes as focal depth in location of the test earthquakes?

The first US work along these lines in 1971, using quite small earthquakes in the Adirondacks in New York, detected an α/β anomaly before a magnitude 2.6 earthquake on 3 August 1973. In this case, a successful prediction using the Wadati diagrams was made. Of course, the *a priori* probability of such an earthquake in an ongoing local sequence was not negligible, even without a special study.

Because there are difficulties in estimating the location of earthquakes from which t_s and t_p must be measured, an alternative experiment would be to use travel times from quarry blasts with known position and origin time.

Such a test was carried out in central California along the San Andreas fault by McEvilly and L. Johnson in 1974 using stations of the University of California telemetry network, in which relative timing between stations is precise. Their study indicated that for the years 1961 to 1973 fluctuations in travel times amounted to about ± 0.3 s over paths of 25 to 100 km. All fluctuations could be accounted for simply by reading errors and changes of shot location in the quarry. They concluded that there were no detectable premonitory travel-time changes prior to 17 earthquakes in the region with magnitudes between 4.5 and 5.4.

This negative finding led to the suggestion that t_s/t_p indicators may not work for strike-slip earthquakes or low stress-drop earthquakes (see § 16.1.2).

When all known t_s/t_p anomalies before earthquakes are compared, two properties emerge. First, earthquake magnitude is not correlated with the change in t_s/t_p. Secondly, by contrast, the duration of assumed anomalies is dependent on the earthquake magnitude, at least up to magnitude about 7, beyond which there are no reliable points. The mean linear curve connecting the logarithm of the duration with magnitude indicates that anomaly durations of 3 months, 1 year, and 6 years may be expected before magnitude 5, 6 and 7 earthquakes, respectively. The long precursory duration for damaging earthquakes presents a practical difficulty in using the method for precise prediction.

15.5.4 Changes in strain

The elastic rebound theory for the occurrence of earthquakes (see § 16.1) has the mark of a true scientific theory in that it allows rough prediction. H.F. Reid gave a crude forecast of the next great earthquake near San Francisco. (The theory also forecast, of course, that the *place* would be along the San Andreas or associated fault.) The geodetic data indicated that during a lapse of 50 years relative displacements of 3.2 m had occurred at distant points across the fault. The maximum elastic-rebound offset along the fault in the 1906 earthquake was 6.5 m. Therefore, $(6.5/3.2)$ × 50 or about 100 years would again elapse before sufficient strain accumulated for a repetition of the 1906 earthquake. The premises are that the regional strain will grow uniformly and that various constraints are not altered by the great 1906 rupture itself (e.g. by the onset of slow fault slip or creep).

Prediction research has for many years been influenced by the basic arguments set forward by Reid. Because crustal strain was the determining factor, it was natural to measure, by geodetic means, crustal deformations in the horizontal direction along active faults (trilateration and triangulation) and in the vertical direction by precise levelling and tiltmeters. In particular, such work was carried out in Japan, where the vertical motions (more than 1.5 m) in the 1923 Kwanto earthquake and tilts of the ground before the 1964 Niigata earthquake had been quite remarkable, and in California, where horizontal displacements along the San Andreas fault and changes in elevation have been striking.

With the advent of tidal gauges and, more particularly, with the application of precise levelling methods, many instances of precursory vertical deformation have been recorded, mainly in Japan and in Southern California, near Palmdale.

Changes in ground-water level prior to earthquakes have been reported mainly from China. Water levels in wells respond to a complex array of factors such as rainfall, changes in atmospheric pressure, tides, human use of water, and changes in the state of stress in the crust. Only the last factor is directly related to earthquake activities. In order to study changes of water level relating to earthquake events, effects due to other factors need to be removed.

A change in the state of stress in the crust may affect ground-water level in several ways. It may change the porosity of the rock to cause pore pressure of ground water to vary; it may change the passageways of ground water such that initially isolated channels become connected and initially connected channels become isolated from each other; it may also tilt the ground to cause ground water to flow in seeking new equilibrium under gravity.

15.5.5 Dilatancy model

The theory of dilatancy of rock prior to rupture occupies a central position in recent discussions of the premonitory phenomena of earthquakes. Many solids show an increase in volume during deformation. In 1901, Osbourne Reynolds used the term *dilatancy* to mean the increase of volume of granular masses due to deformation. The term is now generally taken to describe the increase in volume relative to an elastic change. Bridgman in 1949 first observed that some rocks become dilatant when stressed to fracture.

For earthquake prediction, the important aspect of dilatancy is its effects on the various measurable quantities of the Earth's crust, such as the seismic velocities, electric resistivity, ground and water levels, etc. Among these, the best studied is its effect on the seismic velocities.

The influence of internal cracks and pores on the elastic properties of rocks has been recognised since the early part of this century. The effects are most clearly demonstrated in laboratory measurements of the elastic properties of rock as a function of hydrostatic pressure, usually over a range of several kilobars.

For saturated rocks, experiments predict that, at *low pressure*, α will be much greater than that for dry rock. On the other hand, since water is without rigidity, β will not be different from that for dry rocks. As pressure is raised, two cases may be distinguished. In the first case, the pore fluid may flow away to an interconnected reservoir at a low pressure; cracks and pores in the stressed rock will then gradually close and both α and β will rise and approach the values for the dry rock. In the second case, the pore fluid remains in the rock, either because it is trapped in the cracks or because it is in equilibrium with the external pressure. In this latter case, neither α nor β will change much; a gradual rise in both quantities with increasing pressure may be expected as a result of changes in the intrinsic rock properties and the pore fluid.

The above information forms the basis for the dilatancy–diffusion model of A. Nur, developed in order to interpret variations in seismic velocities premonitory to some earthquakes. Briefly, for shallow earthquakes, as a portion of crust is stressed to failure, dilatancy occurs, which brings down the velocities of seismic waves. Recovery of velocity is brought about by subsequent rise of pore pressure of water. The rise of pore pressure also has the effect of weakening the rock and consequently resulting in an earthquake. This sequence of events has a larger effect on the velocity of compressional waves than on that of shear waves, giving the apparent variation in their ratio.

The validity of the model hinges on the assumptions that (1) *in situ* stress near the earthquake source is large enough to cause rocks to dilate prior to the occurrence of the earthquake, (2) water is available in the crust down to several tens of kilometres, and (3) the permeability of rocks under crustal conditions is large enough to allow water to diffuse into the dilatant source region in relatively short lengths of time. The first assumption is reasonable because laboratory data show that rocks often dilate prior to failure, even under high confining pressures. The validity of the second and the third assumptions, however, remains to be established.

15.5.6 Other field parameters. Liquefaction

The effects of strain in the focal region on other observable properties, such as radon concentration, ground and water levels, etc., will not be considered here in detail. Some simple conclusions, however, may be drawn. As cracks open there would be an uplift of ground level, and as water flows into the dilated cracks, the local water table would drop. Furthermore, because electrical conductivity of rocks depends largely on the interconnected water channels in the rocks, resistivity would increase before the cracks became resaturated. On the other hand, as open cracks close there would be a drop in ground level, and as pore fluid is expelled from the closing cracks the local water table would rise and concentration of gases such as the radioactive radon would increase.

If it is assumed that any anomalous changes in seismic travel times and ground level, etc., are results of changes in the mechanical characteristics of the region around an earthquake source, it may be expected that such changes affect other rock properties as well, such as electrical resistivity. Laboratory data for the resistivity of rocks stressed to fracture show that the electrical resistivity of water-saturated crystalline rocks such as granite and diabase does change significantly prior to fracture.

Geological methods of extending the seismicity record back from the present are also being explored. Field studies indicate that the time series of surface ruptures along major active faults associated with large earthquakes can sometimes be constructed. A well-supported example comes from displacement and liquefaction effects (see § 17.4.4) preserved in beds of sand and peat along the San Andreas fault in southern California. K. Sieh found, from mapping and dating by radioactivity methods such beds in trenches excavated across the fault, evidence for a sequence of at least nine *palaeoearthquakes* extending back more than 1400 years to A.D. 545. Each earthquake, produced by fault slip, liquefied water-saturated sand lenses and deposited them as a surficial layer that was subsequently buried by stream-borne alluvial soils. The shallowest buried layer corresponded with the historical Fort Tejon earthquake of 1857 and the average inter-occurrence time in the sequence was about 165 years.

Less well-based precursors to earthquakes also draw attention, particularly earthquake lights and animal behaviour. Many reports of unusual lights in the sky and abnormal animal behaviour prior to earthquakes are known to seismologists, most in ancedotal form. From a theoretical standpoint, it is necessary to start with physical models based on proper experimental evidence and capable of explaining reported observations. The usual suggested explanations of both phenomena are in terms of the pre-earthquake release of gases and electric and acoustic stimuli of various types. The problem with such hypotheses is the very low signal-to-noise ratio of the suggested precursors and the severe attenuation of high-frequency acoustic waves.

Finally, it needs to be emphasised that, in the present state of knowledge of earthquake causes, low probability should be attached to joint predictions of time, place and magnitude that go beyond the above exploratory analysis. In due course, knowledge may include a fuller understanding of conditions below the Earth's surface and of the forces which build up elastic strain energy. The advent of precise strain measurements, detailed fault studies, and the great effort being put into the investigation of the Earth's crust by explosion methods, are steps in this direction. But it will be a long time before predictions can be made, at least in most seismically active regions, with sufficient precision and sufficient reliability to mitigate all disastrous effects of great earthquakes.

A detailed description of theoretical, laboratory and field studies aimed at earthquake prediction can be found in the book by T. Rikitake (1976).

15.6 Exercises

1   For a spherically symmetric body wave, spreading from a surface focus in the Earth (assumed perfectly elastic), derive the expression (4) for the seismic wave energy contained in the wave packet defined.

2   Assume that the difference in the elastic strain energy (§2.4.5) before and after an earthquake is $W = pAD$, where D is the average fault offset, A the ruptured area, and p the average stress during faulting. Prove that for complete stress drop Δp, the (minimum) strain energy drop is

$$W_0 = \Delta p \, M_0/2\mu,$$

where M_0 is the seismic moment.

For incomplete stress drop, if p_1, p_2 are the final and frictional stress during faulting, show that the seismic wave energy E can be written

$$E = W_0 + AD(p_1 - p_2).$$

Hence, from (15), derive an expression for the moment magnitude M_w as a function of moment M_0, assuming $p_1 = p_2$.

3   Define the mean magnitude (see §15.2.2) as

$$\bar{M} - m_0 = \frac{1}{n} \sum_{i=1}^{n} (M_i - m_0)$$

$$= \frac{1}{n} \sum_{i=1}^{n} e_i/\beta,$$

where e_i are independent unit exponentials that are distributed like $\chi_2^2/2$. Then, from the additivity property of χ^2, show that $\bar{M} - m_0$ is distributed like $\chi_{2n}^2/2n\beta$.

Hence, prove that (see Abramowitz and Stegun, 1965, §26.6.3)

$$E(\hat{\beta}) = \beta n/(n - 1) \quad (n > 1)$$

$$\text{var } \hat{\beta} = \beta^2 n^2/[(n - 1)^2 (n - 2)] \quad (n > 2).$$

Show that the simple formula $\beta/n^{\frac{1}{2}}$ for the standard error of b is a biased approximation.

4   There is evidence of slow migration of strain in the Earth's crust with velocities several orders of magnitude less than the velocities of seismic waves in perfectly elastic material. Rather than the wave equation, the diffusion equation applied to a non-elastic material seems more appropriate.

Consider a viscous layer (viscosity η, thickness h_2) sandwiched between an elastic surface plate (rigidity μ and thickness h_1) and a semi-infinite rigid substratum.

The horizontal shear displacement $u(x, t)$ of the plate in the horizontal x-direction due to boundary tractions satisfies

$$\frac{\partial^2 u}{\partial x^2} = \sigma \frac{\partial u}{\partial t}, \tag{1}$$

where inertial forces are neglected, and $\sigma = \eta/h_1 h_2 \mu$. For a sinusoidal pressure $P = P_0 \sin \omega t$ applied uniformly to the elastic plate end at $x = 0$ show that an appropriate solution to (1) is

$$u(x, t) = -(P_0/\sqrt{2k\mu})\exp(-kx)\cos(\omega t - kx + \pi/4)$$

and

$$P(x, t) = P_0 \exp(-kx)\sin(\omega t - kx),$$

where $k = (\omega\sigma/2)^{\frac{1}{2}}$.

The migration of strain can be defined by a speed $v = \omega/k$ and distance X to penetrate to a point where $u(X)/u(0) = e^{-1}$. Show that

$$v = (2\omega/\sigma)^{\frac{1}{2}}$$

and

$$X = (2/\omega\sigma)^{\frac{1}{2}}.$$

Hence calculate X and v for a pressure period of 1 year, $h_1 = 80$ km, $h_2 = 250$ km, $\eta = 2 \times 10^{21}$ poise and $\mu = 10^{12}$ dyne/cm^2.

5   Scale-invariant objects look the 'same' at any magnification. Seismicity swarms share this property with other natural objects, such as coastlines. In order to understand a fractal, define D, the *fractal* dimensionality. Cover the object with a set of discs of radius r. This takes a number of discs, $N(r)$. Now change r and see how N changes. Almost always $N(r) = Cr^{-D}$, where C is a constant. For a straight line, $D = 1$ and for a square, $D = 2$; in these cases D corresponds to the usual dimensionality.

Devise a pattern of epicentres so that dividing the radius by 3 multiplies N by 5. Thus $D = \log(5)/\log(3)$. Such an object whose D is not given by its topology is a fractal. Show that, in general, $1 < D < 2$ for epicentres of aftershock sequences.

16

The earthquake source

16.1 Elastic rebound model

We now give an introduction to the elastodynamic models that have been developed to represent the physical conditions in the rocks that produce seismic waves. In previous chapters (e.g. §§4.1, 5.1 and 5.6) we considered the appropriate solutions of the wave equation without including the seismic source. In certain earlier sections (e.g. §§3.1.3 and 4.2) an indication was given, however, of the way that the seismic source is included in the overall theory.

The basic mechanical representation of a seismic source is a confined region with spatial dimension L and time duration L/c where c is rupture velocity. When the epicentral distance $\Delta \gg L$, and wave-lengths are relatively long ($\lambda/L \gg 1$), or, more rigorously $\lambda\Delta/2 \gg L^2$, the region can be taken as a point at which there is equilibrium of force and moment systems (see §2.1). In 1927, Love made one of the first analyses of systems of forces and couples at a seismic point source and the reader is referred to this basic work.

16.1.1 Causes of earthquakes

We have already pointed out that earthquakes arise through the sudden release of energy within some confined region of the Earth. This energy is gravitational potential energy, kinetic energy, chemical energy, or elastic strain energy. The release of such energy may be regarded as the immediate cause of an earthquake. A fundamental geodynamical question is what are the processes which give rise to the accumulation of this energy.

A variety of evidence indicates that, of the four types of energy listed above, only elastic strain energy could be released in sufficient quantity to cause major earthquakes. Earthquakes caused by the release of elastic strain energy are called *tectonic earthquakes.*

Direct evidence of the slow accumulation and subsequent release of such energy comes from field studies of earthquake phenomena. H.F. Reid's notable study of the geodetic measurements along the San Andreas fault before and after its rupture in the San Francisco earthquake of 1906 April 18 led him to put forward in 1911 his 'elastic rebound theory', which drew attention to the significance of elastic strain energy in connection with earthquakes. A tectonic earthquake occurs when the stresses in some region inside the Earth have accumulated to the point of exceeding the strength of the material, leading rapidly to fracture. There is a series of rock fractures that produce the ensuing seismic waves, all tending in the same direction and sometimes extending over many kilometres; in the case of the 1906 earthquake there was slipping along the San Andreas fault for 430 km with a maximum fault offset of 6 m.

Volcanic earthquakes are observed to be associated with volcanic activity in volcanic regions of the Earth. In this case the released energy may be of hydrodynamical origin, as when motion of a mass of magma below a volcano is suddenly stopped,; in other cases the immediate cause may be an excessive accumulation of gas pressure. Magma emplacement can also be a secondary cause of earthquakes by producing stress and strain in adjacent regions, leading to the occurrence of tectonic earthquakes.

All earthquakes which arise from volcanic activity are relatively small; dense swarms may produce continuous shaking called *harmonic tremor*. Actually major earthquakes also sometimes originate near volcanoes, but field investigations indicate that these earthquakes are not immediately connected with the local volcano. (See also § 15.3.) Other natural earthquakes (all of relatively small energy) include those caused by the release of gravitational potential energy, for example, those due to the collapse of caverns near the Earth's surface, and those due to landslides. In addition, chemical and nuclear explosions produce seismic waves (see § 1.3.1).

The elastic rebound theory of earthquake genesis entails that near the rupturing fault there will be a fling of the ground as the major rupture edge passes by the point, thereby allowing each side of the fault to spring unidirectionally back to a less strained position. Reid stated that

It is possible that the whole movement at any point does not take place at once but proceeds in irregular steps. The more or less sudden stopping of the movement, and the friction, gives rise to the vibrations which are propagated to a distance. The sudden starting of the motion would produce vibrations just as would its sudden stopping, and vibrations are set up by the friction of the moving rock just as the vibrations of a violin string are caused by the friction of the bow.

We see here the recognition of the likelihood of irregular motions associated with the fault rupture caused by intermittent locking, stress variations, and roughness of the faulting surfaces. These properties of fault rupture are now included in modelling earthquake sources. The fault roughnesses are sometimes referred to as *asperities* and, at places where the rupture slows greatly or stops, the fault properties are said to present a *barrier*.

In 1964 and subsequently, N. Haskell developed a model in which the fault motion was represented by a coherent wave only over segments of the fault; radiations from adjacent sections were assumed to be statistically independent. More recently, in 1977, Das and Aki considered types of faults with barriers. Rupture may start at the focus near a barrier and propagate, either unilaterally or bilaterally, over the fault plane until it is brought to rest or slowed at the next barrier. Barriers may be broken or unbroken by the dislocation; sometimes the dislocation re-initiates on the far side of a barrier and continues; sometimes a barrier is not broken initially but, due to repartitioning of stresses and non-linear effects, it eventually breaks, perhaps with the occurrence of aftershocks.

In summary, the physical model for the tectonic earthquake source is now usually conceived as extending over a fault plane in the rocks. The fault rupture proceeds by a series of dislocations which initiates at the focus and spreads out with various rupture velocities. The dislocation front changes speed as it passes through patches of roughness or barriers on the fault. At the dislocation edge, there is a finite time for a slip of adjacent fault faces to take place and the form of the slip is an elastic rebound of each side of the fault leading to a decrease of overall strain. The slip can have vertical components as well as horizontal components and can vary along the fault itself. The waves are produced near the dislocation front due to release of the strain energy during the slippage.

This model resembles radio waves being radiated from a finite antenna. In the far field, the theory of radio propagation gives complete solutions for the reception of radio signals through stratified media. However, when the receiver is very near to the extended antenna, the signal becomes jumbled due to the finiteness of the source and wave interference arising from end effects. Such an end effect, for example, may produce a *stopping* or *breakout* phase associated with the sudden cessation of rupture or its breaking through an individual rock layer.

16.1.2 Strain energy before an earthquake

There is a connection between the energy released in an earthquake and the strength of the materials near the focus.

Suppose that P_{ij} and E_{ij} are the deviatoric stress and strain tensors (as defined in §2.5) at any point P of a region in which there is strain leading to subsequent fracture causing an earthquake. Then by (2.70), the corresponding distortional strain energy, U say, is given by

$$U = \iiint \mu E_{ij}^2 d\tau, \tag{1}$$

where the integral is taken through the volume of the region in question. We shall write

$$P_{ij}^2 = (\alpha S)^2, \tag{2}$$

where S^2 is the value which the function P_{ij}^2 would take if the material at P were on the point of fracturing; thus $0 \leqslant \alpha \leqslant 1$, and S may be taken as an index of the strength (see §2.6.5). Then since $P_{ij} = 2\mu E_{ij}$ (equation (2.68)), we have, neglecting variation in μ and S through the region,

$$4\mu U = S^2 \iiint \alpha^2 d\tau = S^2 T, \quad \text{say}, \tag{3}$$

where T would be the volume of the strained region if α were equal to unity throughout.

In addition to the distortional strain energy U, there will in general be some dilatational strain energy, V say, corresponding to the first term on the right-hand side of (2.70); this may be expected to be of the order of U. In general, because of frictional losses and aftershocks (see §15.4.1), the energy E_S released in seismic waves in a major earthquake will be less than $E_T = U + V$. Thus U/E_S is at least $1/2$, and may be appreciably greater. Nevertheless (3) gives a useful connection between the orders of the released energy, the strength of the rocks along the fault, and the extent of the strained region prior to the earthquake. The rigidity μ of competent rock ranges from about 0.4×10^{12} to 1.5×10^{12} dyne/cm². Hence for magnitude 7 earthquakes, taking the released seismic energy E_s as 10^{22} ergs, $S^2 T$ is between 10^{34} and 10^{35} dyne²/cm. It follows that with these earthquakes the strength of the material in the vicinity of the focus and also the extent of the strained region must be considerable. Laboratory experiments indicate a strength of order 10^9 dyne/cm² for rocks which predominate in the Earth's crustal layers and may be much less in fault zones. Taking $S = 10^8$ dyne/cm² would then give T approaching 10^{18} cm³, i.e. 10^3 km³, in the case of large earthquakes.

On this calculation, the minimum volume of the strained region just prior to the earthquake would be of the order of that of a sphere of diameter 10 km, and if the strain were confined to this sphere the material would have to be at breaking point throughout the whole volume. Hence the actual volume inside which substantial prior strain existed would be appreciably greater than $10^3 \, \mathrm{km}^3$.

Since T could hardly exceed so large a value by much more than a factor of 10, the calculation sets both a severe lower limit to the strength of the Earth's materials where the largest earthquakes originate, and also a severe upper limit to the energy E_T that can be released in an earthquake. The strength in the focal region may be less than $10^9 \, \mathrm{dyne/cm}^2$ but E cannot be much greater than 10^{25} ergs. This calculation when first made in 1953 by Bullen was one factor leading to a large reduction in an estimate, current at the time, of 10^{27} ergs for the energy of the largest earthquakes.

Earthquakes with nearly the greatest energy include ones with focal depths up to 150 km, and the deepest earthquakes can have energies approaching one-tenth of the greatest value. Maximum M does not decrease between 200 and 650 km. It follows that the strength of materials in deep focus zones up to a depth of 600–680 km is comparable with the strength in the crust.

16.1.3 Faults and fracture

In 1776, Coulomb postulated that a brittle material under stress fractures along a plane of greatest tangential stress. Let p_1, p_2, p_3, where $p_1 > p_2 > p_3$, be the principal stresses just before a fracture. On the Coulomb postulate the plane of fracture, which is also the plane of greatest shear stress, passes through the direction of p_2 and bisects the angle between the directions of p_1 and p_2, thus making an angle of $\pm \frac{1}{4}\pi$ with p_1; the magnitude of the greatest shear stress is $\frac{1}{2}(p_1 - p_3)$.

In 1905, Anderson suggested that the fracture starts when the stress-difference $p_1 - p_3$ exceeds $S + \mu p_n$, where S is a measure of the strength, p_n is the normal stress at the plane of fracture, and μ is a friction coefficient. Let θ be the angle between the plane of fracture and the direction of p_1. Then $\mu = 0$ gives the Hopkins result $\theta = \pm \frac{1}{4}\pi$, while $\mu = 1$ gives $\theta = \frac{1}{8}\pi$ or $\frac{3}{8}\pi$. (For proof, see Jeffreys, 1970, Appendix A.)

Laboratory experiments show that for a bar of brittle material under longitudinal thrust (the behaviour is very different under tension) θ is about 45° agreeing with the Hopkins result. In geological conditions, where there is a superimposed lithostatic load on the material under thrust, θ may be reduced to 30° or less, and Anderson's theory becomes relevant. Jeffreys suggests that the Coulomb–Hopkins theory may apply to good approximation at the initiation of fracture, and the Anderson theory after slip has begun.

Tectonic earthquakes are associated with fractures below the Earth's surface, and the above theories are relevant to the production of new faults.

When the accumulated strain energy becomes excessive near an already existing fault, however, slipping is likely to take place along that fault, irrespective of the preceding theory, since the strength near such a fault is likely to be lower than that elsewhere.

The introduction of the crack in an elastic medium leads to a sudden reduction of stress Δp and strain Δe. From elastodynamic fracture theory for an infinitely long strip, $\Delta e = \xi D/h$, where D is the maximum fault slip and h is the fault width (or half-width of the strip). The geometry of the fracture leads to different values of the constant ξ: for example, for a circular ($\lambda = \mu$) crack of diameter h, $\xi = 7\pi/24$, for dip-slip ($\lambda = \mu$) or strike-slip along an infinitely long fault $\xi = 1/3$ or $1/4$, respectively. In terms of the deforming shear stresses, tractions across the fault at a point P increase to a value S at which time slip commences and the stress at P drops to the dynamic friction stress S_f. Usually the effective stress $S - S_f$ equals Δp. Therefore, from (2.8), the average stress drop at the time of a strike-slip earthquake is given by

$$\Delta p = 2\mu\Delta e = \frac{\mu D}{2h}, \tag{4}$$

Substitution of the 1906 earthquake fault rupture values gives a stress drop of order 10 bars when $h = 10$ km. This cannot be considered a large pressure drop; it is equivalent to the hydrostatic pressure at a depth of only about 500 m in the crust. Stress drops for many earthquakes have been estimated; the values range from the order of 100 bars to as low as 1 bar.

Crack dynamics also provides formulae for the elastic strain energy lost by the introduction of the crack. Based on work of Starr in 1928, the total elastic energy released by a crack of length L in an infinite strained body with Poisson's relation is of the form

$$E_T = \gamma\pi\mu D^2 L, \tag{5}$$

where γ is a constant, appropriate to the fault geometry and stress drop. The formula may be derived in an elementary way from Reid's elastic rebound theory by integration around the crack of the work done by the shear stresses. For a thin strike-slip fault Knopoff in 1958 found that $\gamma = 1/16$ when the fault plane is stress-free after slipping. Thermodynamical and other effects at a fault rupture will limit the energy E_S available for propagation as seismic waves. If η is the efficiency factor then

$$E_S = \eta E_T, \quad \text{where } \eta < 1. \tag{6}$$

For the 1906 San Francisco earthquake substitution of measured values in (5) yields $E_T \approx 10^{24}$ ergs, compared with the estimated value $E_S = 1.8 \times 10^{24}$ ergs given by equation (15.17). In this case, the order of the estimates is the same, which is mechanically doubtful.

The geological interpretation of a fault is given in terms of standard geometries (Fig. 16.1). The usual fault model has a *strike* ϕ (direction of surface trace from north) and *dip* δ (angle between direction of steepest slope and horizontal). The *hanging* wall lies over the *foot* wall.

Relative offsets parallel to the strike produce *strike-slip* faulting; those parallel to the dip, *dip-slip* faulting. Strike-slip faults are *right* or *left* lateral, depending on whether the block on the opposite side of the fault from the observer moves to his right or left. Dip-slip faults are *normal* if the hanging-wall block moves downwards relative to the foot-wall block; the opposite motion gives *reverse* or *thrust* faulting. A mixed offset gives *oblique-slip* faulting measured either by the plunge or by the slip or rake angle.

The plunge is the angle between the horizontal plane and the direction of slip, measured in the vertical plane while the rake λ is the angle between the strike and the slip direction measured in the plane of the fault.

Observed faults are assumed to be the seat of one or more past earthquakes, although movements along faults are often slow and aseismic. The actual faulting in an earthquake may be very complex, and it is often not clear whether in a particular earthquake the main energy issues from a single fault plane.

Observed geological faults sometimes show overall relative displacements of the order of hundreds of kilometres, whereas the amplitudes of

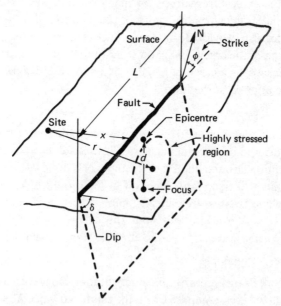

Fig. 16.1. Focal conditions and fault description.

seismic waves reach only several centimetres. In a single earthquake, surface fault slip up to the order of metres is observed. Jeffreys estimated that a fault fling at depth of 100 m at a single stage might be expected to form a layer some 15 cm thick of a rock such as pseudotachylyte made molten by the heat generated. If no pseudotachylyte is formed, for his faulting model the maximum slip is about 4 cm in a single movement. Because most faults show little or no pseudotachylyte, Jeffreys concluded that many small movements are required to produce the larger known faults. The presence of water may greatly affect this inference.

16.1.4 Double couple model

It is now routine to infer the character of faulting in an earthquake from observed distributions of the directions of the first onsets in waves arriving at the Earth's surface. Onsets have been called *anaseismic* (compressional) or *kataseismic* (dilatational) according as the direction is away from or towards the focus, respectively.

A polarity pattern becomes recognisable when the directions of the *P* onsets are plotted on a map: there are broad areas in which the first onsets are predominantly compressions, separated from predominantly dilatational areas by nodal curves near which the *P* wave amplitudes are abnormally small. Udias (1980) has given an interesting account of the early focal mechanism work.

In 1923, Nakano theoretically examined the patterns of first movements that would occur at the surface of a homogeneous Earth model, taking various model representations of the forces at the earthquake focus. The representation that has since been most considered has a set of four equal coplanar forces constituting a pair of couples of equal and opposite moments, the forces of one couple being at right angles to the forces of the other. (See Fig. 16.3).

In 1926, Byerly sought to use patterns of *P* onsets over the whole globe to infer the orientation of the fault-plane in a large earthquake. The polarity method yields two *P* nodal curves at the Earth's surface. For a homogeneous Earth, one curve is in the plane containing the assumed fault, and the other is in the plane (called the *auxiliary plane*) which passes through the focus and is perpendicular to the forces of the couple. For the actual Earth, the nodal curves are displaced from these locations because of curvature of the rays between focus and surface, but knowledge of the paths of seismic rays enables allowance to be made for this.

Given an adequately well determined pattern of first *P* movements, it is possible to locate two planes, one of which is the plane containing the fault.

Given which is which, the direction of motion on the fault is also indicated. Unfortunately the *P* observations do not give a unique physical identification of the fault plane. Also, discrimination cannot be made easily from observations of *S* waves, for which in a double-couple model the amplitudes are a maximum in both the auxiliary and fault planes. In modern practice, computer algorithms are used with either selected *P* and *S* polarities or actual pulse shapes from the digital seismograms to determine the nodal lines, and fault identification comes from field evidence and the distribution of aftershocks.

16.2 Source mechanism estimation

16.2.1 Method of fault-plane solutions

Consider the point-source model for the faulting mechanism corresponding to an infinitesimal shear dislocation that is expressible either as a double-couple source without moment or a system of compressive and tension forces acting at 45° to the dislocation plane (see Fig. 16.2). If there is

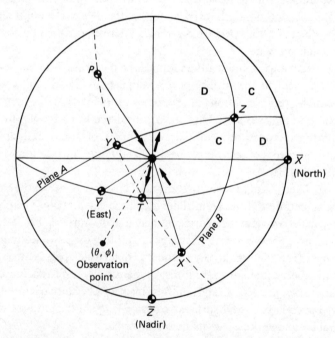

Fig. 16.2. Definition of orientations of fault axes *X*, *Y*, and the principal stress axes *P*, *T* in relation to the focal sphere.

no internal friction these resultant forces can be considered as the greatest and least axes of the principal stresses (see § 2.1.4).

The directions of first motions of *P* waves determine two orthogonal nodal planes which determine uniquely the direction of the principal axes of stress. In terms of the moment tensor (cf. § 16.4.1 below) representation of the point source, first motion data alone determine the eigenvectors of the moment tensor, but not its eigenvalues.

It is statically equivalent to use as variables in the determination of the focal mechanism the orientation of either the force-axes *X* and *Y* or the principal stress axes *T* and *P*. Moreover, there are some advantages in using the latter system apart from the more direct physical meaning. The *T* and *P* axes are unambiguously defined by the quadrants of compression and dilatations of the data; this is not the case for the *X* and *Y* axes. Since the *T* and *P* axes are in the direction of the maxima of the *P*-wave amplitudes, it is sometimes possible to estimate an approximate orientation for these axes, even in the case of scarce data, by picking the average orientation of the bulk of compressions for the *T* axis and that of the dilatations for the *P* axis. It is usually more difficult to try to separate the polarities which pass through sectors, near nodal planes where the signals are weak. In addition, for most regional studies, the orientation of regional stresses is required and it seems better to calculate them directly in the program, rather than derive them from the orientations of the *X* and *Y* axes.

The problem of determining the orientation of the focal mechanism is greatly simplified by the use of a focal sphere. Observing stations are projected upon points on the surface of the *focal sphere* by ray tracing back to the source. For stations at teleseismic distances, this procedure does not involve any serious difficulty, but, for near stations, a detailed knowledge of the velocity–depth distribution in the crust and upper mantle is needed to project the ray appropriately. Sometimes lateral variations of the structure, when not taken into account, distort the orthogonality of the source mechanism and detection of the effect is difficult.

In a case of a symmetrical source mechanism, only half of the focal sphere is needed. Because rays arriving at most stations, except those near to the source, are downgoing, the lower hemisphere is a natural choice. Observations that record upgoing rays may also be projected upon the lower hemisphere.

In the lower hemisphere of the focal sphere (Fig. 16.2), we define the Cartesian coordinate axes $\bar{X}, \bar{Y}, \bar{Z}$ in the directions north, east, and nadir. A point on the surface of the focal sphere has coordinates $(\bar{x}, \bar{y}, \bar{z})$. Because the radius of the focal sphere is taken to be unity, these are related to the spherical coordinates (θ, ϕ) measured from the positive direction of $\bar{X}$ and $\bar{Z}$ by

$$\bar{x} = \sin\theta\cos\phi$$
$$\bar{y} = \sin\theta\sin\phi$$
$$\bar{z} = \cos\theta$$

where

$$0° \leqslant \theta \leqslant 90°; \quad 0° \leqslant \phi \leqslant 360°.$$

The direction of the principal stresses form the set of coordinate axes T, P, Z, and the direction of the forces form the axes X, Y, Z. The transformation of the coordinates of a point from the geographical axes to either of these two is done through the matrix of the direction cosines

$$\mathbf{B} = \begin{pmatrix} \alpha_T & \beta_T & \gamma_T \\ \alpha_P & \beta_P & \gamma_P \\ \alpha_z & \beta_z & \gamma_z \end{pmatrix}; \quad \mathbf{A} = \begin{pmatrix} \alpha_x & \beta_x & \gamma_x \\ \alpha_y & \beta_y & \gamma_y \\ \alpha_z & \beta_z & \gamma_z \end{pmatrix}, \tag{7}$$

where $\alpha_g = \sin\Theta_g\cos\Phi_g$, $\beta_g = \sin\Theta_g\sin\Phi_g$, $\gamma_g = \cos\Theta_g$; $g = X, Y, Z, T, P$. Thus,

$$\begin{pmatrix} x' \\ y' \\ z' \end{pmatrix} = \mathbf{B}\begin{pmatrix} \bar{x} \\ \bar{y} \\ \bar{z} \end{pmatrix} \text{ and } \begin{pmatrix} x \\ y \\ z \end{pmatrix} = \mathbf{A}\begin{pmatrix} \bar{x} \\ \bar{y} \\ \bar{z} \end{pmatrix}. \tag{8}$$

The direction cosines of the axes X and Y are easily derived from those of P, T, and Z by

$$\begin{pmatrix} \alpha_x \\ \beta_x \\ \gamma_x \end{pmatrix} = \mathbf{B}^T\begin{pmatrix} 1/\sqrt{2} \\ 1/\sqrt{2} \\ 0 \end{pmatrix}; \quad \begin{pmatrix} \alpha_y \\ \beta_y \\ \gamma_y \end{pmatrix} = \mathbf{B}^T\begin{pmatrix} 1/\sqrt{2} \\ -1/\sqrt{2} \\ 0 \end{pmatrix}. \tag{9}$$

For polar coordinates referred to the geographical axes the position of each of the axes P, T, X, Y, Z, is defined by the angles Θ and Φ. Because of the orthogonality condition, each of the system of axes is uniquely defined by only three angles Θ_T, Φ_T, Φ_P or Θ_x, Φ_x, Φ_y.

A normalized expression for the amplitude of the P-wave displacement in the far field at a point of coordinates (θ, ϕ) referred to the principal stress axes is

$$u_p \propto x'^2 + y'^2$$

and referred to the axes of the force system

$$u_p \propto 2xy.$$

The orientations and slip of the two fault planes, A (normal to X) and B

(normal to Y), given the strike (ϕ) and dip (δ) and slip (λ) angles, are (see §15.1.3)

$$\phi_A = \Phi_x + 90°$$
$$\delta_A = \Theta_x$$
$$\lambda_A = \sin^{-1}\left\{\frac{\cos \Theta_y}{\sin \Theta_x}\right\}, \tag{10}$$

and

$$\phi_B = \Phi_y + 90°$$
$$\delta_B = \Theta_y$$
$$\lambda_B = \sin^{-1}\left\{\frac{\cos \Theta_x}{\sin \Theta_y}\right\}. \tag{11}$$

A simple illustration of the method is shown in Fig. 16.3 and Fig. 16.4 for vertical strike-slip and reverse-oblique faulting. Arrows show directions of

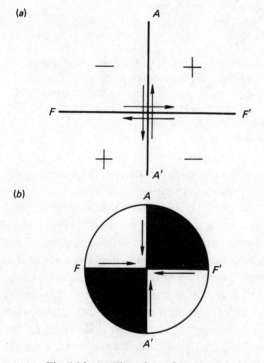

Fig. 16.3. (*a*) Plan view of horizontal displacement on vertical fault $A-A'$ or $F-F'$ and resulting distribution of compressions ($+$) and dilatations ($-$); (*b*) corresponding fault-plane diagram.

(a)

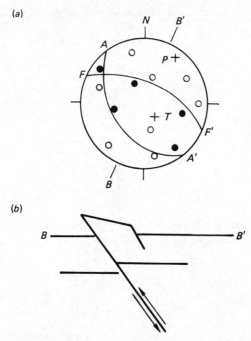

(b)

Fig. 16.4. (a) Resolved radiation pattern, (lower hemisphere plot) indicating reverse-left oblique displacement on fault $F-F'$. The fault strikes N 65° W and dips 50° NE; $A-A'$ auxiliary nodal plane; $B-B'$, line of section; solid circles, compression; open circles, dilatation; T, tension axis; P, compression axis; slip plunges 45° N 50° E. (b) Geologic section along $B-B'$.

relative moment and the double-couple representation; the shaded quadrants and circles denote compressional (positive) P polarity.

The procedure follows four steps: (a) the directions of motion are read from seismograms at each recording station; (b) the observed first motion for each station is projected back to the earthquake focus and plotted in its correct geometric relation on the focal sphere, represented by an equal-area or other suitable stereographic projection of the lower hemisphere; (c) the resulting radiation pattern of first-motion polarities is then separated by use of a stereonet into quadrants of compressional and dilatational onsets so that adjacent quadrants have opposite polarities (Fig. 16.3); (d) the inferred stress axes (P, compression; T, tension) can then be found. Once the fault and auxiliary planes are identified, the slip vector (i.e. the axis of slip in the fault plane) and the proportion of vertical to horizontal displacement can be determined. The reader should note that the P and T axes do not necessarily correspond exactly to the tectonic stresses that caused the earthquake; the latter can be determined, however, approximately once the fault plane and slip vector are known.

Fig. 16.4 shows a characteristic radiation pattern, in which the focal planes are constrained by polarities that ideally are uniform within quadrants. The quality of resolution is dependent on the number and distribution of reliable recorders relative to the hypocentre and the adequacy of the velocity model used to trace the rays, as well as proper polarity of recorders.

16.2.2 Probability model for group fault-plane solutions

Traditionally, fault-plane solutions are determined for individual earthquakes using the polarity of first P motions (supplemented by S polarisation vectors (§4.1.1)) at all stations. In the case of sparse polarity data, readings from a number of earthquakes in the region or from an earthquake sequence are combined to allow a 'composite' solution. Most solutions are made subjectively with little indication of uncertainty. We clearly require not only an efficient numerical algorithm for this important technique, but an objective method of analysis.

Many fault-plane solutions are still made visually by graphical methods despite the availability of computer programs. The first step toward an analytical treatment appears to have been that of Homma who in 1941 proposed fitting recorded amplitudes by least-squares. Knopoff was the first to formulate the problem in terms of a probability function, describing the proper polarity of first motion in the presence of noise, and in 1963 Kasahara wrote a computer program based on Knopoff's probability function that allowed routine machine solutions to be computed. This program became a routine tool at the Dominion Observatory in Canada and many results were published.

Further important developments in the field were published in the 1970s by Keilis-Borok, Dillinger, Pope and others describing computer algorithms using maximum likelihood for both P and S waves, and extending the method to the joint use of P-wave polarities and S-wave polarization angles. A general treatment of the focal-mechanism problem developed by Brillinger, Udias and Bolt is now described.

Several probability distributions for the observed first motion are compared and for simplicity, only P-wave first motions are considered. (Extensions to include S-wave polarisation angles are not difficult). The fundamental extension from the earlier work is that fault-plane solutions for a group of earthquakes in a region are obtained simultaneously. An immediate advantage is the ability to measure the fit by comparing the number of inconsistent polarities for each focal-plane solution in the group. This is particularly important when seismograms are not directly read and measurements of the signal-to-noise ratios are not available. We consider a set of I earthquakes in a particular region with polarities recorded at various stations. These earthquakes can constitute a regional sequence, local seismicity, or aftershocks. The case of only a few stations and many earthquakes, leading to a need for composite solutions, is contained in the present treatment.

A seismogram, observed at a given station, may be viewed as a noise disturbed record of a signal, $s(t)$, of interest. That is

$$Z(t) = s(t) + \varepsilon(t) \tag{12}$$

with $Z(t)$ the seismometer displacement recorded at time t and with $\varepsilon(t)$ an additive noise series. The basic measurement is the sign of first P motion. Specifically, let τ denote the time of the apparent first motion and set

$$Y = \begin{cases} +1 & \text{if the first motion is recorded as positive } (Z(\tau) > 0) \\ -1 & \text{if the first motion is recorded as negative } (Z(\tau) < 0). \end{cases}$$

Then let us consider the model (with $P\{\cdot\}$ denoting a probability function)

$$P\{Y = 1\} = \gamma + (1 - 2\gamma)P\{Z(\tau) > 0\},$$

with γ small and meant to handle reader and recorder errors. The initial amplitude, $s(\tau)$, of the signal may be viewed as proportional to the theoretical amplitude, A, for the event and station, as a function of the fault-plane parameters. Setting $s(\tau) = \alpha A$ and supposing that the noise value $\varepsilon(\tau)$ has distribution function $G(\cdot)$ leads to

$$P\{Z(\tau) > 0\} = P\{\varepsilon(\tau) > -\alpha A\} = 1 - G(-\alpha A). \tag{13}$$

Suppose, specifically, that $\varepsilon(\tau)$ is normally distributed with mean 0 and variance σ^2. Then equation (13) may be written

$$P\{Z(\tau) > 0\} = \Phi(\rho A),$$

where $\Phi(\cdot)$ is the cumulative normal and $\rho = \alpha/\sigma$. In summary, the above arguments lead to the statistical model

$$P\{Y = 1\} = \gamma + (1 - 2\gamma)\Phi(\rho A)$$
$$P\{Y = -1\} = 1 - P\{Y = 1\} \tag{14}$$

for an observed first motion Y, where $\frac{1}{2} > \gamma > \gamma_0$.

Precise data correspond to γ small, σ small (and hence ρ large). Imprecise data correspond to γ near $\frac{1}{2}$ or ρ near 0. If $\rho = \infty$, then equation (14) only involves the sign of A, and not its magnitude. The model has the property that the larger the magnitude of A, the more likely is the sign of first motion to be observed correctly.

For group analysis, data are available for a number of earthquakes with approximately the same focus as recorded at a number of stations. The observatories may vary with the event. Let $i = 1, ..., I$ index the events to be studied, and $j = 1, ..., J_i$ index the stations within events. The basic quantities available are then Y_{ij}, the observed sign of first motion at station j of the ith event, and $A_{ij}(\Theta_T, \Phi_T, \Theta_P)$, the corresponding theoretical signed amplitude as a function of the fault-plane parameters Θ_T, Φ_T, Θ_P.

A succession of probability models involving varying assumptions of generally increasing complexity may be contemplated, where

$$\pi_{ij} = P\{Y_{ij} = 1\}.$$

For the ith earthquake and jth station,

$$\pi_{ij} = \Phi(\rho A_{ij}) \tag{15}$$

$$\pi_{ij} = \gamma + (1 - 2\gamma)\Phi(\rho A_{ij}) \tag{16}$$

$$\pi_{ij} = \gamma + (1 - 2\gamma)\Phi(\rho_i A_{ij}) \tag{17}$$

$$\pi_{ij} = \gamma_j + (1 - 2\gamma_j)\Phi(\rho_i A_{ij}). \tag{18}$$

Models (17) and (18) lead to solutions averaging together the various earthquakes with weights proportional to their number of observations J_i. The remaining models allow the events to enter with weights depending on their appropriateness in some sense.

The parameters of a statistical model are estimated by setting down a numerical criterion involving the data and the parameters and then choosing as estimates the parameter values that optimise the criterion. A classical criterion is the principle of least squares as used elsewhere in this text. A suitable function is

$$\sum_{i,j} H(A_{ij} Y_{ij}) = \tfrac{1}{2}\sum(1 + Y_{ij} \operatorname{sgn} A_{ij}), \tag{19}$$

where $H(x) = 1$ if $x > 0$ and $= 0$ if $x < 0$. The criterion counts the number of agreements between the observed and theoretical polarities of first motion. The fault-plane parameters are chosen to maximise this criterion.

The likelihood function employed is

$$L = \tfrac{1}{2}\sum_{i,j}\{(1 + Y_{ij})\ln \pi_{ij} + (1 - Y_{ij})\ln(1 - \pi_{ij})\}, \tag{20}$$

corresponding to the method of maximum likelihood. This criterion is a differentiable function of the parameters for each of the models (15) to (18), so that quite a variety of efficient optimisation routines are available.

Further, because the procedure is maximum likelihood, expressions are available for the standard errors of the estimates and methods are available for testing hypotheses concerning the parameter values.

The method has been applied to the main shock and 25 aftershocks of the Alaskan earthquake sequence of March 1964. The mechanism of most shocks consists of thrust faulting on a low angle plane or reverse faulting on a nearly vertical one. From the teleseismic data available, only the nearly vertical plane is well defined. The orientations of the P (pressure) and T (tension) axes of preliminary graphical solutions are concentrated about two directions T (45°, 315°) and P (50°, 135°) with the exception of eight events with scattered orientations. The individual graphic solutions indicate that over two-thirds of the events have similar orientations of the principal axes of stress and come from a common regional stress pattern.

As starting values the three independent angles (Θ_T, Φ_T, $\Phi_P = 45°, 315°, 135°$) and $\rho_i = 5$ for all events were adopted. The initial value of the logarithm of the likelihood function, given by (20), is $-L = 1322$. After 30 iterations, the program yielded 39°.0, 318°.3, 138°.9 for the axes angles and $-L = 407$. The final solution is

	Θ	Φ
T	$39° \pm 1°$	$318° \pm 15°$
P	$51° \pm 1°$	$139° \pm 14°$

Plane A	Plane B
$\phi = 232°$	$\phi = 49°$
$\delta = 6°$	$\delta = 84°$
$\lambda = 87°$	$\lambda = 90°$.

The total number of polarities used was 1239 and the final solution predicted 984 correctly, i.e. a proportion of 0.79 of correct readings. A noteworthy point is the difference between the standard errors of the plunge $(90° - \Theta)$ and trend Φ angles. The higher uncertainty in Φ is a consequence of the geometry of this particular mechanism and the distribution of the data points on the focal sphere where only two of the four quadrants are covered.

The program adjusted the values of 26 precision parameters ρ_i in each iteration and the final correlation of the values of ρ with the proportion of consistencies, p, for each event is clearly reflected in Fig. 16.5. Negative values of ρ correspond to those events with a proportion of consistencies less than 0.5. In these events, the group solution requires the orientation of the individual P and T axes to have reversed polarity to the regional

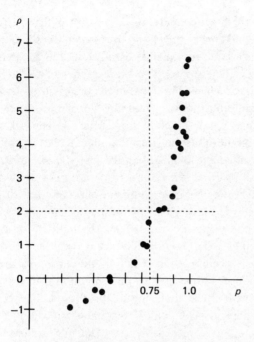

Fig. 16.5. Plot of the event precision parameter ρ, against the consistency ratio p for 26 Alaskan earthquakes.

orientation. Theoretically, for a perfect score $p = 1.0$, the value of ρ would be infinity; for probability 0.5, the weight is zero; for $p < 0.5$ the negative values of ρ reverse the polarities and again give probabilities greater than 0.5. The values of ρ for those events with high scores of correct polarity observations (p near 1.0) depend on the initial values given.

16.3 Moving dislocation source

In §16.2, basic theoretical results were obtained for a seismic source modelled as the release of stress at a fixed point in the strained elastic medium. In this section and in §17.4 more realistic earthquake sources are considered.

16.3.1 Kinematics and dynamics. Near field and far field

There are two main approaches to representing mathematically a realistic fault source. First, the kinematic model assumes that the time-history of the slip on the generating fault is known *a priori*. Several defining parameters may be specified, such as the shape, duration, and amplitude of the source (or source time-function and slip), the velocity of the slip over the fault surface, and the final area of the region over which the slip occurred.

The second approach is to use the differential equations involving the stresses which produce the rupture. The basic model of this dynamic procedure is a shear crack which is initiated in the pre-existing stress field which causes stress concentrations around the tip of the crack. These concentrations, in turn, cause the crack to grow. Much of the work on this subject has been built on the pioneering formulation of Kostrov (1966). For example, in 1969 Burridge and Willis obtained analytic expressions for particle accelerations in given directions from a uniformly growing elliptical crack, although they did not include the effect of crack stopping. Attempts to obtain realistic solutions for the seismological crack problem endeavour to model the physical processes of the fault dislocation to incorporate interaction between the rate of crack growth, the criterion of fracture, and the stress accumulation. Friction properties of the fault surface are also introduced. Most studies on dynamic shear cracks are concerned primarily with the actual rupture process, so the crack is assumed to be embedded in an infinite homogeneous medium. Studies concerned with seismic waves recorded on the Earth's surface need a numerical approach to handle realistic structural conditions. The mathematics required for such complete developments of dynamic source models is complicated and goes beyond the limit set for this book. The reader may pursue the problem further in Aki and Richards (1980).

In the simpler kinematic case, we now give some introductory detail on the mathematical modelling commonly used. The starting point is that information about the source is contained in the seismic pulse shape. The source parameters, slip or stress change on the fault surface are assumed to be similar to a step or Heaviside time function. A more physically plausible time function has a finite initial gradient.

Consider a moving dislocation model, constant rupture velocity v_r, with the geometry similar to that in Fig. 16.1. The x axis is parallel to the strike and the y axis normal to the fault plane. Haskell suggested that the displacement, $u(t)$, on the fault surface be expressed in the form (cf. Fig. 3.2)

$$u(t) = u_\infty G(t - x/v_r), \tag{21}$$

with u_∞ and $G(t)$ denoting the final displacement (half the fault slip D in the simplest case) and the *ramp function*, respectively. This ramp dislocation model is zero at $t < 0$ and increases linearly with time until it reaches unity at $t = T$, which defines the *rise time*. The function in (21) represents, therefore, the model of fault slip.

Next, for simplicity, we ignore the effects of rupture propagation and consider a simultaneous slip along the fault that initiates a shear wave propagating perpendicularly to the fault surface. Let the initial stress time function be of the form

$$p_0(y, t) = pH(t - y/\beta), \tag{22}$$

where $H(t)$ is the Heaviside function, and p is used to denote the effective stress (see equation (4)).

If, from (2.13),

$$\mu\left(\frac{\partial u}{\partial y}\right) = -p_0(y, t)$$

then

$$\dot{u} = (p/\mu)\beta H(t - y/\beta), \tag{23a}$$

and, for $t \geqslant y/\beta$,

$$u = (p/\mu)\beta(t - y/\beta). \tag{23b}$$

Near the fault, therefore, displacement first increases linearly in time as the shear pulse propagates away from the fault before leveling off due to the finite dimension of the fault.

We note that in this model, from (23), the S particle velocity at $y = 0, t = 0$ is, for $p = 10^8$ dyne/cm^2, $\mu = 3 \times 10^{11}$ dyne/cm^2, and $\beta = 3.0$ km/s, $\dot{u}_0 = 100$ cm/s. Strong-motion accelerometers in several earthquakes have recorded particle velocities of this order (see §17.4.1).

At a point near the fault, the S-wave particle velocity given by (23) decreases to zero when the waves from the fault extremities reach the observation point. We follow the important suggestion of Brune (1970) and introduce a time constant, τ, equal to the travel time of this signal, L/β, where L is the equivalent dimension of the fault surface, and replace (23b) by

$$u(y = 0, t) = (p/\mu)\beta\tau(1 - \exp[-t/\tau]), \tag{24}$$

The basic assumption is that the finite seismic source generates a wave pulse shape similar to the ramp model (equation (21)). All such Heaviside time functions have frequency spectra similar to the function sinc $X = (\sin \pi X/\pi X)$ (Kanasewich, 1973). In this context, we define

$$\bar{u}(\omega) = A_0\tau_0(\sin X)/X \tag{25}$$

where $X = (\omega\tau_0/2)$, A_0 is the pulse height, and τ_0 is a measure of pulse width. The spectrum of sinc X is plotted in Fig. 16.6. Its basic structure is a plateau at low frequencies and a downward sloping envelope for high frequencies with slope ω^{-1} represented by a broken line in Fig. 16.6. For a triangular time source function, the Fourier transform is like sinc$^2 X$ and the envelope

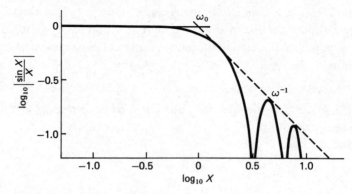

Fig. 16.6. The sinc function.

slope is ω^{-2}. The spectral form (25) arises in descriptions of radio emission from antennas. The seismological analogy of the aperture width of the antenna, given by the central pulse width, is the length L of the fault. The frequency corresponding to the intersection of the horizontal low-frequency asymptote and the sloping high-frequency asymptote is called the *corner frequency* and is generally denoted by ω_0.

The Fourier spectrum of displacement near the fault is, for the model (24),

$$\bar{u}(\omega) = (p\beta/\mu)/\omega(\omega^2 + \tau^{-2})^{1/2}, \tag{26a}$$

with phase

$$\phi(\omega) = \tan^{-1}(\omega\tau)^{-1}. \tag{26b}$$

This spectrum has a similar slope to that in Fig. 16.6. We therefore conclude that if wave attenuation, source radiation pattern and effects of rupture propagation are ignored, the near-field displacement spectrum of the shear wave pulse may be described by three basic parameters: the spectral amplitude at zero frequency, the corner frequency and the spectral slope at frequencies above the corner frequency (on a log–log plot). In the time domain each of these parameters is a measure of a basic property of the seismic source: namely, source power, fault length and source time function, respectively.

It is worth noting that, while for a simple point source the far-field displacement is proportional to a source time function, $G(t)$, say, for a double-couple source the far-field displacement is proportional to the derivative $G(t) \propto \delta(t)$. Thus, in the frequency domain, the far-field spectrum is approximately a constant in this case.

In the above discussion the rupture velocity v_r was not involved directly and the results are most relevant to the seismic motion near to the seismic source. In an elastic medium, comparing seismic Mach number $M = v_r/\beta$ with the shear wave velocity β, crack velocities can be subsonic ($M < 1$) or supersonic ($M > 1$). For seismic sources, measurements show that $v_r \lesssim \beta$; representative estimates from laboratory simulations and seismogram analysis (see e.g. (30)) indicate $v_r \sim 2.5$ km/s for rupture along crustal faults. In the next section we consider the effect of the movement of the dislocation along the fault.

16.3.2 Radiation patterns and directivity

We now consider, as in Fig. 16.7, a source S of seismic waves, velocity c, moving with constant velocity v_r from O to A along the x axis ($OA = L$). For simplicity we consider a receiving site $R(r, \theta)$ in the far field such that the epicentral distance r and characteristic wave-length λ have ratios $L/r, \lambda/r \ll 1$.

The time delay τ_0 of a wave front emitted from S relative to emission at O is, by a binomial expansion for $x/r < 1$,

$$\tau_0 = (x/v_r) + (r - x\cos\theta)/c.$$

Then integration of the wave form along OA gives, as the response at S,

$$u(\omega, t) \propto \frac{1}{L}\int_0^L \exp\left\{i\omega\left[t - \left(\frac{x}{v_r} + \frac{r - x\cos\theta}{c}\right)\right]\right\}dx. \tag{27}$$

$$\propto \frac{\sin X}{X}\exp(-iX)\exp(i\omega(t - r/c)), \tag{28}$$

where

$$X = \frac{\pi L}{\lambda}(M^{-1} - \cos\theta), \tag{29}$$

M is the seismic Mach number, and $\lambda = 2\pi c/\omega$.

The diffraction sinc X factor in (28) represents the finiteness of the source

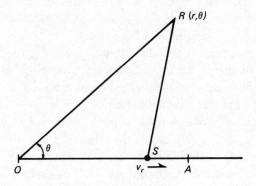

Fig. 16.7. Moving point source.

and produces a deviation from a purely dipole radiation pattern. Its central role in studies of seismic sources was demonstrated by Ben-Menahem. We consider three important aspects.

First, the source finiteness factor can be used to estimate fault length and rupture velocity under appropriate experimental conditions. This result is evident from the diffraction factor which in the wave displacement spectrum at the receiver R will produce spectral nodes (see Fig. 16.6) at

$$L(M^{-1} - \cos \theta) = n\lambda \quad (n = 1, 2, \ldots). \tag{30}$$

This condition corresponds to destructive interference at R and is analogous to Fraunhofer diffraction through a rectangular slit.

Next, consider the ratio $D(\theta)$ of spectral amplitudes of seismic pulses that leave the source OA in opposite azimuth θ and $\theta + \pi$. Then, from (28),

$$D(\theta) = \frac{(M^{-1} + \cos \theta) \sin \dfrac{\pi L}{\lambda}(M^{-1} - \cos \theta)}{(M^{-1} - \cos \theta) \sin \dfrac{\pi L}{\lambda}(M^{-1} + \cos \theta)}. \tag{31}$$

$D(\theta)$, called the *directivity function*, is independent of the source time function and hence provides an experimental measure of L and v_R from seismograms at a station or stations on a great circle path. When multiple-circuit surface waves of the same type (e.g. $G1$, $G2$, $G3$) are recorded on the same seismogram (see Fig. 10.1), location of spectral zeros in spectral ratios (after adjustments for propagation path differences) yields v_R, from (30), for a given fault length. The method can also be used for successive transits of individual modes of acoustic atmospheric waves (cf. Fig. 17.7).

Another application is to use seismograms from stations located at opposite ends of the fault rupture (i.e. $\theta = 0$, π). In this way, Filson and McEvilly in 1967 inferred source parameters of the 1966 Parkfield earthquake from spectral nodes for Love waves ($\Delta = 270$ km) recorded at the Berkeley observatory on an ultra-long-period seismograph. The first zero occurred at $T = 22.5$ s, yielding $v_R \approx 2.2$ km/s.

Thirdly, for a moving source, waves that leave the travelling rupture in opposite directions will have different amplitudes. It follows, either from elementary kinematic considerations for a sound pressure wave at R in Fig. 16.7, or from (31), that the required radiation factor is $(1 - M \cos \theta)^{-1}$. Thus the amplitude in the direction of rupture is larger than that away from the motion by a factor $(1 + M)/(1 - M)$. In addition, there is a frequency shift similar, but not identical, to the Doppler effect in acoustics so that a monochromatic moving point source, frequency ω_0, is observed to emit a frequency $\omega_0/(1 - M\cos\theta)$. Here, the frequency is defined as the time derivative of the phase and the result follows from (28) because wave nodes are phase-shifted by the above factor. The result is interference between waves generated at different parts of the fault plane.

The effect can be seen in the radiation patterns of Fig. 16.8, computed by Hirasawa and Stauder (1965) for P and S waves from double-couple sources with $M = 0.5$ and unilateral rupture (cf. equation (34) below).

Finally, in this section, we turn our attention to the form of the displacement and displacement spectrum due to the *source time function* at a receiver R remote from the rupturing fault. We give only the main results without detailed derivation. The proof of the formulae may be found in several of the advanced seismological texts referenced in the selected bibliography. It is helpful to follow in the first instance the acoustic analogy for the sound pressure (see Exercise 16.5.4) at a point generated by a moving monopole pulsating source (Morse and Ingard, 1968). The central formulae are based on solutions by G.G. Stokes in his 1849 paper on elastic wave radiation and diffraction.

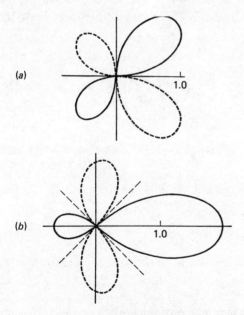

Fig. 16.8. Radiation patterns of the P and S waves from sources with rupture propagation for: (a) $v_r/\beta = 0.5$; (b) $v_r/\beta = 0.9$, where v_r and β denote the rupture and shear wave velocities, respectively. Solid lines indicate positive and dashed lines negative values. (From Hirosawa and Stauder, 1965.)

In 1972, Savage constructed, on the basis of the Haskell model discussed in §16.3.1, the seismic far-field radiation solution in terms of particle velocity as

$$\mathrm{d}\mathbf{u}/\mathrm{d}t = \mathbf{R}_\alpha(\theta, \phi, r)\mu h I_\alpha + \mathbf{R}_\beta(\theta, \phi, r)\mu h I_\beta, \tag{32}$$

with

$$I_c = D \int_0^L \left[\frac{\mathrm{d}^2}{\mathrm{d}t^2} G\left(t - \frac{r}{c} - \frac{\xi}{c} - \left(\frac{c}{v_\mathrm{r}} - \cos\theta \right) \right) \right] \mathrm{d}\xi, \tag{33}$$

Here θ, ϕ and r are the spherical polar coordinates as in Fig. 16.2, ξ is the x coordinate of the point of integration in the fault, $\mathbf{R}_c$ is the radiation field from an appropriate double-couple source. D is the offset, h is the fault width, and c stands for either α or β. The first term in (32) represents the P-wave radiation and the second term the S-wave radiation. The factors $\mathbf{R}_c$ are, for longitudinal shear faulting (slip parallel to x), given by

$$\mathbf{R}_\alpha = \mathbf{e}_r \sin 2\theta \sin \phi/(4\pi\alpha^3\rho r),$$
$$\mathbf{R}_\beta = (\mathbf{e}_\theta \cos 2\theta \sin \phi + \mathbf{e}_\phi \cos \theta \cos \phi)/(4\pi\beta^3\rho r), \tag{34}$$

where $\mathbf{e}_r$, $\mathbf{e}_\theta$ and $\mathbf{e}_\phi$ are unit vectors in the r, θ, and ϕ directions and ρ is the density. Equations (32) and (33) describe the seismic disturbances due to the moving rupture.

Integration of (33) yields

$$I_c = (LD/\tau_0) \left[\frac{\mathrm{d}}{\mathrm{d}t} G\left(t - \frac{r}{c} \right) - \frac{\mathrm{d}}{\mathrm{d}t} G\left(t - \frac{r}{c} - \tau_0 \right) \right], \tag{35}$$

where τ_0 the time width of the pulse is given, from (29), by $L(M^{-1} - \cos\theta)/c$.

Hence, integration of (32) yields the far-field displacement $\mathbf{u}_c$ for each wave as

$$\mathbf{u}_c = \mathbf{R}_c(\theta, \phi, r)(\mu LDh/\tau_0)[G(t - (r/c)) - G(t - (r/c) - \tau_0)]. \tag{36}$$

The amplitude of its Fourier transform is

$$|\mathbf{u}_c| = \mathbf{R}_c(\theta, \phi, r)M_0\omega|\bar{G}||F(\omega, \tau_0)|, \tag{37}$$

where $|\bar{G}|$ denotes the Fourier transform of $G(t)$ and, from (25),

$$F(\omega, \tau_0) = \sin(\omega\tau_0/2)/(\omega\tau_0/2). \tag{38}$$

The $M_0 = \mu h LD$ is called the seismic moment (cf. equation (40)). The sinc function again appears in the frequency amplitude spectrum (see Fig. 16.6) and its low frequency asymptote will give a measure of the seismic moment.

For further details the reader should consult Brune in Lomnitz and Rosenblueth (1976) and Aki and Richards (1980).

In the discussion in §16.3.1 a new parameter, the rise time, was introduced. The exact definition depends on the model used. For example, in the Haskell model (21) it is defined by the transient time of the ramp function or by the time required for 90 per cent of the final slip to occur, but in the Brune model (24), where an exponential form is assumed, it is defined by the time constant, τ, (i.e. the time interval in which the amplitude decreases to $1/e$ of its initial value). The differences are not of critical importance because numerical estimates from observed seismograms are based on the model assumed.

16.3.3 Synthetic seismograms

A number of seismologists have demonstrated that, in important cases, synthetic seismograms for seismic waves can be computed rather realistically (see Fig. 13.3). A number of alternative procedures have been used.

The first is a deterministic method for predicting near-field ground motion. The basic form for a given wave type is defined by equation (36). For a more detailed and complete representation the seismic source is treated as elastic rebound along a fault produced by a moving dislocation and seismograms are calculated from a Green's function representation (see §4.2.2) of the displacements of the medium. Green's functions for the various classifications of faulting have been constructed, and numerous theoretical papers using this approach have been published. (The book by Kasahara and the review edited by Bolt (1986) are references.) The process is a kind of complicated curve fitting whereby the parameters of the source are varied in order to estimate the closeness of fit with the radiated seismic waves.

The second procedure is the numerical solution of the equations of motion by finite element or finite difference representation (see §12.3). Such studies offer the inclusion of more realistic structure around the source, such as a weak zone of gouge material which occurs in many fault zones and a realistic crustal structure near the surface. We should also mention that the moment tensor defined in §16.4.1 has been found effective in computing synthetic seismograms for small earthquakes and for underground nuclear explosions. Numerical calculations are made by using many point sources on the fault, and delaying and summing their contributions at any given point on the surface. Successive approximations yield synthetic records which are close to the observed seismograms.

A drawback of many synthetic seismograms has been their limitation to longer wave-lengths. Because of limitations in theoretical approximations or computers, many constructions have been restricted to seismic waves with periods above about two seconds, and models which fit the data well at such periods have been found not to give acceptable fits to higher frequency accelerations recorded near to the fault. The problem is an important one since for engineering purposes the concern is mainly with acceleration records, rather than with displacements (see §17.3).

Despite the physical complications involved and the limitations on the mathematical models, some remarkable estimates of fault rupture properties have already been obtained by using the above methods and inverse procedures (see §10.4.2). An illustration comes from the use of near-field recordings of ground acceleration in the 1979 Imperial Valley earthquake in California by Olson and Apsel (1982) to map variations in fault rupture velocity and stress drop.

16.4 Seismic moment

Important physical properties of the earthquake source can be found by generalising the basic parameter, moment of a force system, used in rigid body mechanics. There, a moment is defined as the vector product $\varepsilon_{ikm}x_kX_m$ of the force X_m with the position vector x_k of the point of application (see §2.1.3). Two parallel forces produce a couple with definite moment.

Steketee in 1958 proved the *fundamental equivalence theorem* that the displacement field produced by the dislocation $\Delta\mathbf{u}$ on a plane element δS in an elastic body equals that produced by a double couple applied at δS. To illustrate the derivation, consider a shear dislocation at the small element δS, thickness 2ε, length $2c$, in the elastic body analysed in Fig. 2.3. Let δS be perpendicular to the x_2 axis and its centroid be the origin of coordinates. Let Δu_1 be the relative displacement across a cut in the x_1 direction.

Then it can be shown that moments per unit length in the x_3 direction, say, due to forces in the x_2 direction are, from (2.9),

$$\int_{-\varepsilon}^{\varepsilon}\int_{-c}^{c} x_1\rho f_2\,dx_1\,dx_2 = -\int_{-\varepsilon}^{\varepsilon}\int_{-c}^{c} x_1\frac{\partial p_{12}}{\partial x_1}\,dx_1\,dx_2$$

$$= \int_{-\varepsilon}^{\varepsilon}\int_{-c}^{c} \mu\frac{\partial u_1}{\partial x_2}\,dx_1\,dx_2 \quad \text{(by integration by parts)}$$

$$= \mu\int_{-c}^{c} \Delta u_1\,dx_1, \tag{39}$$

where μ is constant over the cut plane and Δu_1 is the relative displacement across δS. Similar consideration of the remaining moments and approach to the limits as $\varepsilon \to 0$, $\delta S \to 0$ proves the required equivalence. In addition, the argument yields the formula for moment for a simple fault source,

$$M_0 = \mu A D, \tag{40}$$

where D is the average slip Δu over the total faulted area A (cf. equation (37)).

16.4.1 Moment tensor

In source theory, the *seismic moment tensor* M_{ij} depends on the seismic source (e.g. fault) orientation and strength. At a point source in Cartesian coordinates (see §2.1.1), there are three force components and three lever arm directions. Therefore, with reference to Fig. 2.2, the equivalent surface traction on a face δS is statically equivalent to a system of nine couples. The reader can easily verify by drawing the force pairs that the three couples which are diagonal elements of M_{ij} are equivalent to extensions along the orthogonal axes (called *vector dipoles*) and the six other couples represent shears. From conservation of angular momentum, M_{ij} is symmetric and there is a correspondence between the moment tensor and the stress tensor (see §2.1.2).

Unlike in §2.1, where body forces and couples were ultimately zero, the presence of a source requires their inclusion in the elastic equation of motion. We therefore consider a solution for displacement u_i produced by such body forces in the form discussed in §3.1.5. We write, for equilibrium of the source volume V, from (3.27),

$$u_i(\xi, \tau) = \int_{-\infty}^{\infty} \int_V G_{ij}(\xi, \tau; \mathbf{x}, t) f_j(\mathbf{x}, t) \, dV \, dt, \tag{41}$$

where G_{ij} is the Green's function for the elastodynamic system (i.e. structural type) and f_j are the equivalent body forces. Let the focus in V be at the origin of coordinates O. We may then expand in a Taylor's expansion for $\mathbf{x}$ about the focus O so that

$$G_{ij}(\xi, \tau; \mathbf{x}, t) = G_{ij}(\xi, \tau; 0, t) + x_k \frac{\partial G_{ij}}{\partial x_k}$$

$$+ \tfrac{1}{2} x_k x_l \frac{\partial^2 G_{ij}}{\partial x_k \partial x_l} + \cdots \tag{42}$$

The series may be interpreted as the superposition of source radiation components. It may be proved (cf. equation (3.28)) that terms in the series decrease rapidly with frequency like $O(l/\lambda)^n$ where l is the dimension of the source ($n > 1$). For most seismological modelling, such as with small fault rupture length, only the first term is retained; additional accuracy arises with larger sources by using, as reference, the centroid of the rupture (see § 11.2.2) rather than the focus.

For convenience, we now define the *moment tensor density* in terms of the equivalent body forces as

$$\frac{\partial m_{ij}}{\partial x_i} = -f_j. \tag{43}$$

Substitute from (43) in (41) and integrate by parts. We obtain

$$u_i(\xi, \tau) = \int_{-\infty}^{\infty} \int_V \frac{\partial G_{ij}}{\partial x_k} m_{kj} \, dV \, dt$$

$$= \frac{\partial G_{ij}}{\partial x_k} * M_{kj}, \tag{44}$$

where the temporal convolution (see e.g. equation (3.27)) is in terms of the moment tensor for the volume and

$$M_{kj}(0, t) = \int_V m_{kj}(\mathbf{x}, t) \, dV. \tag{45}$$

In equation (44), the moment tensor components provide the weights to be applied to the Green's function and the moments of its derivatives in the multipole expansion (42). Further details are given in Kennett (1983). Equation (44) has been successfully applied, for example, by transforming the convolution to the equivalent frequency domain multiplicative form. It then reduces to a set of linear algebraic equations $\mathbf{U} = \mathbf{GM}$, say, which for an appropriate set of Green's functions and seismograph observations give, by inversion of $\mathbf{G}$, the elements of $\mathbf{M}$. The method is based on the algorithm referred to in § 10.4.2.

16.4.2 Estimation of seismic moments

Three estimation procedures are in common use. The first comes simply by substituting field measurements into (40). We may venture a numerical calculation for the 1906 San Francisco earthquake. With $\mu = 3 \times 10^{11}$ dyne/cm^2, rupture length $L = 400$ km, rupture depth $h = 10$ km, fault offset $D = 500$ cm, the overall average moment is about 6×10^{27} dyne cm.

Secondly, a formula first derived for far-field measurements of ground displacement by Keilis-Borok is available. In the case of *SH* waves, for example,

$$M_0 = 4\pi\rho\beta^3 r\Omega_0 / 2R(\theta, \phi) \tag{46}$$

where β is the shear wave velocity, $R(\theta, \phi)$ is the source radiation pattern, r is the source distance to the receiver, and Ω_0 is the spectral amplitude at low frequencies. On a broadband displacement seismogram, the physical meaning of Ω_0 is the product of pulse-width and amplitude. When focal mechanisms are not available, $R(\theta, \phi)$ is taken to be one-half.

The proof of (46) follows from (36) and (34) by assuming a point source with moment M_0 in a spherical polar system. For application to surface recordings on a half-space the factor 2 in the denominator arises from the free surface reflection (see § 6.2.1).

Thirdly, as demonstrated in § 16.3.2, estimates of M_0 come from the long-period asymptote of the spectrum of the wave displacements recorded in the far field. (See equation (38)).

Finally, in observatory work, an empirical method can be used to assess M_0 that parallels the routine observation of local magnitude M_L. The relation used, suggested by (46), is

$$\log M_0 = a + b \log (C \times D \times \Delta^p), \tag{47}$$

where C is the maximum peak-to-peak amplitude read on a Wood–Anderson (or equivalent) seismogram, D is the duration between the S arrival and the onset with amplitude C/d, Δ is epicentral distance, and a, b, p and d are constants. It has been found, for $3.0 \leqslant M_L \leqslant 6.2$, that in Central California a regression form for (47) is

$$\log_{10} M_0 = (16.74 \pm 0.20) + (1.22 \pm 0.14) \log_{10}(C \times D \times \Delta), \tag{48}$$

where M_0 is in dyne cm, C in millimetres, and Δ in km. The corresponding moment–magnitude relation is

$$\log_{10} M_0 = (17.92 \pm 1.02) + (1.11 \pm 0.15)M_L. \tag{49}$$

As an illustration, the seismic moment (M_0) of a 1979 California earthquake ($M_L = 5.8$) was estimated from integration of the SH pulse (Ω_0) from formula (46) to be $M_0 = (6 \pm 1.4) \times 10^{24}$ dyne cm, where $4\pi\mu\beta = 1.2 \times 10^{18}$ g/sec^3. The SH pulse Ω_0 was measured from the ultra-long-period ($T_0 = 100$ s) displacement seismograms (see Fig. 16.9) recorded at Berkeley (BKS).

The direction from the focus to BKS was near a maximum of the radiation pattern for SH (unlike the P phase) so that the SH pulse should provide a reliable measure. The other parameters in (46) are $\Delta = 120$ km, $R_{\theta\phi} = 0.62$. The main sources of uncertainty are the assessment of source–station distance (since the rupture centroid rather than the focus is more appropriate) and the assumed value for $\mu\beta$. The estimated M_0 is probably correct within 20 per cent. It should be noted, however, that equation (46)

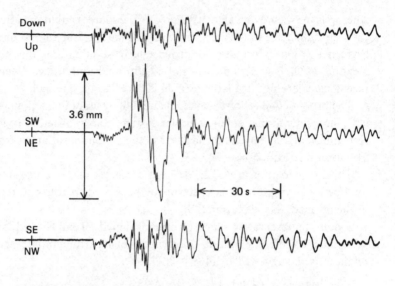

Fig. 16.9. Berkeley broadband seismograms of an earthquake near Coyote Lake, California. Response is flat from 0.1 to 100 s and the magnification in the plot is approximately 9 times the ground motion.

assumes that the *SH* pulse was measured in the far field (distance to BKS is an order of magnitude larger than the rupture length) and that the *SH* pulse does not contain reflections due to crustal structure. The presence of near-field terms or crustal reflections will generally increase the size of the measured *SH* pulse and thus lead to an overestimate of M_0. By comparison, $M_0 = 2.3 \times 10^{24}$ dyne cm, from (49).

16.5 Exercises

1   Draw the following mathematical forms, where $H(t)$ is the Heaviside unit function, and discuss their plausibility as fault rupture models, for rise time T and rupture velocity v_r.

(a) $G\left(t - \dfrac{x_1}{v_r}\right) = \dfrac{t - x_1/v_r}{T} \operatorname{sgn}(t - x_1/v_r).$

(b) $G\left(t - \dfrac{x_1}{v_r}\right) = \dfrac{t - x_1/v_r}{T}\left[H\left(t - \dfrac{x_1}{v_r}\right) - H\left(t - \dfrac{x_1}{v_r} - T\right)\right]$
$+ H\left(t - \dfrac{x_1}{v_r} - T\right).$

(c) $G\left(t - \dfrac{x_1}{v_r}\right) = \exp\left[-\pi(t - x_1/v_r)^2/T^2\right].$

2   Show that the likelihood function

$$\tfrac{1}{2}\sum_{i,j}\{(1 + Y_{ij})\ln \pi_{ij} + (1 - Y_{ij})\ln (1 - \pi_{ij})\},$$

may also be written as

$$\sum_{i,j}\ln \tfrac{1}{2}[1 + (2\pi_{ij} - 1)Y_{ij}] = \sum_{i,j}\ln \tfrac{1}{2}(1 + \psi_{ij}Y_{ij}\mathrm{sgn}\, A_{ij}),$$

making use of the symmetry of π as a function of A, and setting $\psi = 2\pi(|A|) - 1$. (The advantage of this representation is that it shows the dependence on the variate $Y\ \mathrm{sgn}\ A$, which equals 1 if the sign is observed correctly and -1 if not.).

3   Consider principal stresses in the crust of the Earth oriented along vertical, N and E directions. Assume that faulting occurs along planes where the tangential traction is a maximum. From the theory of §2.1, calculate the strike, dip, and type of faulting in the cases:

(a) Maximum principal stress is vertical (gravity). Minimum principal stress is EW.

(b) Maximum principal compressive stress is NS. Intermediate principal stress is vertical.

(c) Maximum principal compressive stress is NS. Minimum principal stress is vertical.

4   A simple acoustic source in an infinite medium can be thought of as a unit sphere at the origin uniformly expanding and contracting. Show that, in spherical coordinates, a general solution for the pressure p (defined by $\rho(\partial\mathbf{u}/\partial t) = -\,\mathrm{grad}\, p$) is

$$p = \frac{1}{r}F(r - ct) + \frac{1}{r}f(r + ct).$$

If the flow outward is $S \exp(-i\omega t)$, show that the fluid velocity $\mathbf{u}$ and power radiated are given by

$$u_r = -\frac{1}{4\pi r^2}(ikr - 1)S \exp ik(r - ct),$$

and

$$P = \rho\omega^2 |S|^2/4\pi c,$$

where

$$k = \omega/c_0 = 2\pi/\lambda.$$

Show that in the far field ($r > \lambda$), the above relations approach those for a plane wave i.e. $u_r = p/\rho c$. In the near field ($r \ll \lambda$) show that u_r has a component out of phase with the pressure.

5   By drawing focal spheres, verify the solutions of the P wave focal mechanism diagrams in Fig. 16.10. The dark areas denote compression and north is to the top.

6   Prove by trigonometric projection that the radiation pattern f for P, SV, and SH waves is given by

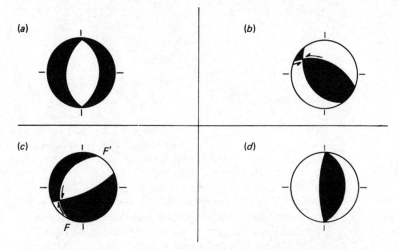

Fig. 16.10. (*a*) Fault-plane diagram showing normal dip-slip on north-striking fault dipping 50° W or 40° E. (*b*) Fault-plane diagram corresponding to thrust left-oblique displacement. (*c*) Fault-plane diagram showing normal right-oblique displacement on fault which strikes N 20° E and dips 30° NW. (d) Fault-plane diagram showing pure reverse slip on north-striking fault that dips 60° W or 30° E.

$$
\begin{aligned}
f_P = \{ &\cos \lambda \sin \delta \sin^2 i \\
&\times \sin 2(\phi - \phi_s) - \cos \lambda \cos \delta \sin 2i \\
&\times \cos(\phi - \phi_s) + \sin \lambda \\
&\times \sin 2\delta [\cos^2 i - \sin^2 i \sin^2 (\phi - \phi_s)] \\
&+ \sin \lambda \cos 2\delta \sin 2i \sin(\phi - \phi_s) \};
\end{aligned}
$$

$$
\begin{aligned}
f_{SV} = &\sin \lambda \cos 2\delta \cos 2i \sin(\phi - \phi_s) \\
&- \cos \lambda \cos \delta \cos 2i \cos(\phi - \phi_s) \\
&+ \tfrac{1}{2} \cos \lambda \sin \delta \sin 2i \sin 2(\phi - \phi_s) \\
&- \tfrac{1}{2} \sin \lambda \sin 2\delta \sin 2i [1 + \sin^2(\phi - \phi_s)];
\end{aligned}
$$

and

$$
\begin{aligned}
f_{SH} = &\cos \lambda \cos \delta \cos i \sin(\phi - \phi_s) \\
&+ \cos \lambda \sin \delta \sin i \cos 2(\phi - \phi_s) \\
&+ \sin \lambda \cos 2\delta \cos i \cos(\phi - \phi_s) \\
&- \tfrac{1}{2} \sin \lambda \sin 2\delta \sin i \sin 2(\phi - \phi_s);
\end{aligned}
$$

with respect to strike ϕ_s, slip λ and slip δ, where ϕ and i are the azimuth and the angle of incidence, respectively, of the ray path at the source. (The coordinate system is with respect to the focal mechanism, where the azimuth ϕ is measured clockwise from the strike plane and the angle of incidence i is measured from the line of intersection between the strike and auxiliary planes.) Show that f_{SH} gives a four-lobe pattern on the surface for a vertical strike-slip fault.

7 Consider the problem of estimating seismic moment (M_0) from the size of the P-, SV- and SH-wave displacement pulses measured in the far field. The appropriate equations of condition are of the form (for a whole space) given in §16.4.

Show from exercise 6 above that the corresponding singular value (see §10.4.2) is given by

$$\Lambda_{M_0}^2 = \sum_{j=1}^{n} [(\beta^6/\alpha^6)\sin^4 i_j \sin^2 2\phi_j$$
$$+ \tfrac{1}{4}\sin^2 2i_j \sin^2 2\phi_j + \sin^2 i_j \cos^2 2\phi_j],$$

where n is the number of stations and the coordinate system is the same as for the focal mechanism problem (here, $\phi_s = 0$ for simplicity). The three terms in the sum correspond to P, SV and SH, respectively. The contributions of the P and SH terms are maximum in the plane $i = 90°$ and the contribution of SV is maximum when $i = 45$ or $135°$.

Hence prove that the variance of the seismic moment is

$$\sigma_{M_0}^2 = \sum_{j=1}^{n} [(\beta^6/\alpha^6)\sigma_P^2 \sin^4 i_j \sin^2 2\phi_j$$
$$+ \tfrac{1}{4}\sigma_{SV}^2 \sin^2 2i_j \sin^2 2\phi_j$$
$$+ \sigma_{SH}^2 \sin^2 i_j \cos^2 2\phi_j]\Lambda_{M_0}^{-4},$$

where from §16.4.2,(46), $\sigma_P = 4\pi\rho\beta^3 r\sigma \{\Omega_0(P)\}$, and two similar expressions for σ_{SV} and σ_{SH}.

Show that the standard error of the seismic moment is minimum when the angle of incidence i is $90°$ and that if $\sigma_{SH} < \alpha^3\sigma_P/\beta^3$, the SH-wave pulse will be the most important type of data for estimating M_0, but if $\sigma_{SH} \approx \alpha^3\sigma_P/\beta^3$, the P- and SH-wave pulse measurements will be equally important in the estimation of M_0.

8 Show that the appropriate moment tensor models are

(a) Point explosion: $M_{ij} = M_0 \delta_{ij}$
(b) Double couple: $M_{13} = M_{31}$, all other $M_{ij} = 0$
(c) Opening crack: $M_{ij} = Au \operatorname{diag}\{\lambda, \lambda, \lambda + 2\mu\}$,

when u is the average crack displacement and λ and μ are the Lamé parameters.

9 From the formulae of §16.1.3 for energy and moment and the moment and magnitude formulae of §16.4.1 and §17.2.3, give the physical basis for the following proportionalities.

(i) $E_T \propto \Delta p \cdot M_0$
(ii) $M \propto \tfrac{1}{3} \log M_0$
(iii) $\log L \propto \tfrac{1}{3} \log M_0$.

17

Strong-motion seismology

17.1 Effects of earthquakes

The effects of earthquakes are many, and include geological effects, damage to buildings, etc., and effects on humans and animals.

The geologist is specially interested in geomorphological and structural changes caused by an earthquake; movements (vertical or horizontal) along fault traces; the production of new fault planes; the raising, lowering and tilting of the ground surface with related effects on the flow of ground water; the liquefaction of sandy ground; landslides; mudflows; etc. The investigation of topographical changes is assisted by geodetical measurements, which are made systematically in a number of countries seriously affected by earthquakes. The geologist relates these strain effects to the geological and tectonic structure of the region.

Engineers and architects investigate the damage caused by an earthquake to buildings, bridges, pipe-lines, railways, embankments and other man-made structures. These effects are related to the instrumentally recorded motions and the behaviour of the foundation soils.

In the most intensely damaged region, called the *meizoseismal area*, the effects of a severe earthquake are usually complicated. The most drastic effects occur chiefly in the neighbourhood of faults along which there is appreciable relative movement. Away from these faults, the effects are dependent on the topography and the nature of the surface materials, and are often more severe on soft alluvium and unconsolidated sediments than on hard rock. At some distance from the source, the main damage is caused by surface waves. In mines, there is often little damage below depths of a few hundred metres although the surface immediately above is considerably affected (see Exercise 8.9.3). Of special importance in some earthquakes are the effects produced on bodies of water such as in tanks, reservoirs, lakes and the sea.

Further effects of interest are the occurrence of earthquake sounds and lights. These sounds are generally low-pitched, and have been likened to the noise of an underground train passing through a station. The occurrence of such sounds implies the existence of significant short periods in the *P* waves in the ground; the intensity of the sound depends on the nature of the local surface materials. Occasionally luminous flashes, streamers and balls are seen in the night sky during earthquakes. Recently, such lights have been photographed in Japan and a theory of electric induction along the source fault rupture developed.

In severe earthquakes, individuals have sometimes reported that they have seen surface ground waves approaching from a distance. In some cases this is probably an illusion, but it is possible that gravity surface waves analogous to waves in water may be set up in soft or muddy areas (see *seiches* in §17.7).

17.2 Macroseismic data

Studies of the various effects referred to in §17.1, i.e. field investigations of earthquakes, yield *macroseismic data*, which supplement the data obtained from seismographs. It is sometimes possible to infer from this data the mechanism of a particular earthquake. We have already referred, for example, to work of Reid (§16.1) in this connection.

17.2.1 Intensity of earthquake effects

The macroseismic data reveal broad features of the variation in the intensity of an earthquake over the affected area. This 'intensity' is not capable of simple quantitative definition, and is estimated by reference to 'intensity scales' that describe the effects in qualitative terms. Efforts have been made to associate the divisions in these scales with accelerations of the local ground shaking; but the intensity depends in a complicated way not only on ground accelerations but also on the periods and other features of the seismic waves.

Many factors determine the intensity at a particular point of the Earth's surface. These include the area and moment of the seismic source, the mechanism of the earthquake, the quantity and frequency spectrum of energy released, the crustal structure in the disturbed region, the distance of the point from the source, the elastic and other properties of the surficial rocks and soils adjacent to the point, and the local geological structure.

A number of different intensity scales have been set up during the past century. For many years, the most widely used was the scale set up by de Rossi and Forel in 1878. The scale now generally used in North America

Table 17.1 *Modified Mercalli scale of intensity (abridged)*

I. Not felt except by a few under especially favourable circumstances. (RF, I.)

II. Felt only by a few persons at rest, especially on upper floors of buildings. Delicately suspended objects may swing. (RF, I–II.)

III. Felt quite noticeably indoors, especially on upper floors of buildings, but many people do not recognise it as an earthquake. Standing motor cars may rock slightly. Vibration like passing of truck. Duration estimated. (RF, III.)

IV. During the day felt indoors by many, outdoors by few. At night some awakened. Dishes, windows, doors disturbed, walls make creaking sound. Sensation like heavy truck striking building. Standing motor cars rocked noticeably. (RF, IV–V.)

V. Felt by nearly everyone, many awakened. Some dishes, windows, etc., broken; a few instances of cracked plaster; unstable objects overturned. Disturbance of trees, poles, and other tall objects sometimes noticed. Pendulum clocks may stop. (RF, V–VI.)

VI. Felt by all; many frightened and run outdoors. Some heavy furniture moved; a few instances of fallen plaster or damaged chimneys. Damage slight. (RF, VI–VII.)

VII. Everybody runs outdoors. Damage negligible in buildings of good design and construction; slight to moderate in well-built ordinary structures; considerable in poorly built or badly designed structures; some chimneys broken. Noticed by persons driving motor cars. (RF, VIII –.)

VIII. Damage slight in specially designed structures; considerable in ordinary substantial buildings, with partial collapse; great in poorly built structures. Panel walls thrown out of frame structures. Fall of chimneys, factory stacks, columns, monuments, walls. Heavy furniture overturned. Sand and mud ejected in small amounts. Changes in well water. Disturbs persons driving motor cars. (RF, VIII + to IX –.)

IX. Damage considerable in specially designed structures; well-designed frame structures thrown out of plumb; great in substantial buildings, with partial collapse. Buildings shifted off foundations. Ground cracked conspicuously. Underground pipes broken. (RF, IX +.)

X. Some well-built wooden structures destroyed; most masonry and frame structures destroyed with foundations; ground badly cracked. Rails bent. Landslides considerable from river banks and steep slopes. Shifted sand and mud. Water splashed (slopped) over banks. (RF, X –.)

XI. Few, if any, (masonry) structures remain standing. Bridges destroyed. Broad fissures in ground. Underground pipe-lines completely out of service. Earth slumps and land slips in soft ground. Rails bent greatly. (RF, X.)

XII. Damage total. Waves seen on ground surfaces. Lines of sight and level distorted. Objects thrown upward into the air. (RF, X +.)

---

and many other countries is the Mercalli scale as modified by Wood and Neumann in 1931, in which the intensity is considered to be more uniformly graded. Alternative scales have been developed in Japan and Europe for local conditions. An abridged form of the modified Mercalli scale is as in Table 17.1, the corresponding Rossi–Forel scale numbers being (approximately) indicated in brackets. The European Medvedev–Sponheuer–Karnik (MSK) scale of twelve grades is similar.

17.2.2 Isoseismal curves and acceleration

With the use of an intensity scale it is possible to summarise the macroseismic data for an earthquake by constructing *isoseismal curves*, which are the loci of points that separate areas of equal intensity (see Fig. 17.1). If there were complete symmetry about the vertical through the earthquake's focus, the isoseismals would be circles with the epicentre as centre. But because of the many unsymmetrical factors influencing the intensity, particularly the local soils and fault rupture length the curves are often far from circular. Moreover, sometimes one or more areas of equal intensity lie entirely inside one of different intensity.

When the isoseismal curves have been constructed for an earthquake, it is possible to estimate the position of the earthquake's source. The most probable position of the epicentre based on the macroseismic data will be at a point inside the area of highest intensity. In some cases, it is verified from instrumental data that the epicentre is well determined in this way. But not infrequently it happens that the true epicentre is outside the area of greatest intensity.

It is evident that, other things being equal, the rate of diminution of intensity with distance from the centre of the disturbed area will be greater the shallower the source of a given earthquake. This suggests the possibility of inferring focal depth from an examination of the spacings between consecutive isoseismal curves.

A number of empirical formulae have been constructed for various regions but should be used with caution. A serviceable form for pre-dominantly circular isoseismals is

$$a(I_0 - I) = \log(1 - r^2/b^2)^{\frac{1}{2}} \tag{1}$$

where I_0 and I are the intensities at the epicentre and distance r from it; a and b are constants to be fit; b may be interpreted as the focal depth.

Two correlations (of similar algebraic form) between empirical seismological parameters have come to play a central role in the forecasting of seismic intensities. We first deal with the correlation of horizontal peak

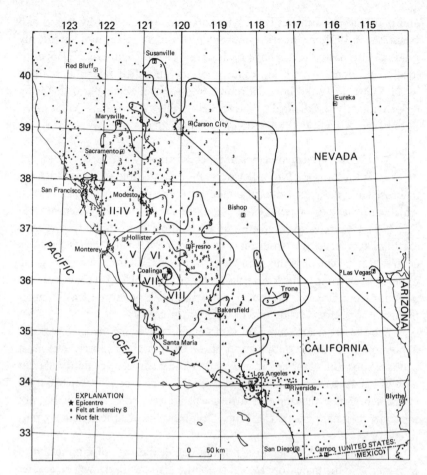

Fig. 17.1. Isoseismals drawn after an earthquake near Coalinga in 1983, magnitude 6.7. (Courtesy NEIS/NOAA.)

acceleration A (cm/s^2) with maximum modified Mercalli intensity I_{MM} of the earthquake, and we take up the correlation with source distance in the next section. Reasons have already been given why it is a fallacy to expect peak horizontal acceleration of the ground to depend on a mixed measure, such as seismic intensity. Clearly, ground accelerations are best estimated from records obtained by accelerometers during earthquakes. In many parts of the world, however, such instrumental measurements are not available, particularly for great historical earthquakes. It is therefore necessary to make use of historical intensity information and to correlate it in a plausible way with a measure of the maximum wave motion. One widely used regression model based on a Western U.S. data set is the linear equation

$$\log_{10}A = 0.01 + 0.30I_{MM}, \quad IV \leqslant I \leqslant X. \tag{2}$$

This fit (Trifunac and Brady, 1975) is close to fits of similar measurements by other seismologists and has been widely used. The data come from earthquakes for which both isoseismal maps and instrumental acceleration records are available.

Theoretical improvements to most available studies are needed. Formal accounts should be taken of the effect of the variance in the statistical distributions of the observed intensity. While the question of estimation of intensity is now under review, what is required is a frequency distribution of intensities constructed from the published felt reports and description of damage. The histograms would give a definitive statistic for I_{MM} as well as an estimate of the variance, whereas in the above formula, I_{MM} is the maximum intensity reported, an erratic measure.

17.2.3 Fault rupture correlations

There are both empirical and theoretical reasons (see equation (16.36)) to expect a direct relation between the dimension of fault rupture and the magnitude (in the form M_s or, better, the form M_w) of the generated earthquake. D. Tocher (1958) was the first to link the observed length of surface fault rupture L with magnitude M using California and Nevada earthquakes. He fitted the form

$$M = a + b\log_{10}L \tag{3}$$

where $a = 5.65$, $b = 0.98$, and L is in kilometres.

Subsequently, similar regressions of M on log L have been made in which the populations of observations of rupture length have been divided into fault type, geological region, and so on. Parallel regressions of log L on M have also been published. In addition, similar magnitude correlations have been worked out with fault slip (log D) and also with products like DL and with fault rupture area. While such correlations have theoretical support, the present paucity of reliable direct observation reduces their value. Nevertheless, such relations have been used to make predictions of the probable size of earthquakes that might shake critical structures. It is therefore unfortunate that such empirical relations have sometimes been used erroneously. For example, it is usually incorrect to use regressions of log L on magnitude M to predict a magnitude given a fault rupture length.

Typically, the regression curves (2) and (3) have been estimated assuming that errors in the independent variable can be neglected and weights have been ignored, i.e. log L is error free. Consider, however, the regression of M on log L (equation (3)) with uncertainty allowed in log L. For simplicity, suppose that the ratio of the standard deviations of M and L is a constant for all i observations. (More plausible constraints can be applied, but the results are much more complicated.) Then, the slope of the regression line can be shown to be

$$b = \frac{\sum w_i x_i \log_{10} y_i}{\sum w_i (\log_{10} y_i)^2 - (n-1)\sigma^2}, \tag{4}$$

where n is the number of observations and σ^2 is the variance of the errors in log y (i.e. in log L). (Values are relative to the means.)

Note that the presence of the measure of scatter in the independent variable decreases the denominator and hence increases the slope. Thus the regression line drawn for M on log L rotates clockwise for $\sigma > 0$. In fact, for large scatter, the line becomes parallel to the abscissa and the regression becomes meaningless.

Field data on 23 strike-slip surface fault ruptures accompanying worldwide earthquakes studied by Bonilla, Mark and Lienkaemper in 1984 yield, when measurement errors are incorporated,

$$M_s = (6.10 \pm 0.25) + (0.70 \pm 0.13)\log_{10} L. \tag{5}$$

17.3 Near-field parameters

The seismological challenge is to explain characteristics of the recorded ground motions and, consequently, to predict ground motions in future specified earthquakes.

The physics of the problem indicates that elastic waves recorded within a few wave-lengths of an extended source (the near field) are likely to be complicated and difficult to interpret uniquely. There are at least four components involved in understanding such near-field motion. The first arises from the generation of the waves by the rupturing fault as the moving dislocation sweeps out an area of slip along the fault plane in a given time. The signature of the faulting mechanism is a function of a finite number of dislocation parameters such as stress drop, fault dimension, and rupture velocity (see § 16.3). Elastic waves of various types radiate from the vicinity of the moving dislocation in all directions. The geometry of the fault affects critically the pattern of radiation from it. Also, it is suspected that the linear elastic constitutive relations usually assumed in elastic wave theory may be defective along the fault zone itself (see § 4.5).

The second component of the problem concerns the passage of the waves through the intervening medium to the site. Although in the near field some wave properties such as dispersion are not as significant as in teleseismic recording, nevertheless attenuation and scattering near the deformed fault zone may significantly affect the wave trains. The third component treats the conditions at the recording site itself, such as topography and highly attenuating soils. As mentioned in §17.4.4, this part of the problem is usually studied separately from the rock sites by soils engineers. The final component of the interpretation of strong ground motion records is the effect of the recording instruments. Fortunately, this can be removed when necessary, using the response equations given in §9.7.

The elastic rebound model defined in §16.1 considers a moving dislocation along a fault plane over which roughnesses of various efficiencies are distributed stochastically. This model is the starting point for the interpretation of near-field wave motion. On this basis studies demonstrate that, in important cases, synthetic seismograms for seismic waves near their source can now be computed rather realistically. For large magnitude earthquakes, the theory must, however, incorporate effects of the moving source (see §16.3) and several defining parameters must be specified: source geometry, slip function, velocity of the rupture, moment, stress drop, and final area of slip.

17.4 Recorded strong ground motion

17.4.1 Peak ground accelerations, velocity, and displacement

Three types of strong-motion record are usually available. The primary seismogram is the accelerogram, but there are also the corresponding records of ground velocity and displacement (see Fig. 17.2). Usually the latter records are computed by integration and filtering of a digital sample of the accelerogram.

Let the accelerogram be represented by

$$a(t) = \ddot{u} + \varepsilon, \tag{6}$$

where $u(t)$ is the real ground displacement and ε is a constant. With two integrations we obtain the seismograms for velocity and displacement, respectively, namely,

$$v(t) = \dot{u} + \varepsilon t + \eta \tag{7}$$

and

$$d(t) = u + \varepsilon t^2/2 + \eta t + \zeta, \tag{8}$$

where η, ζ are constants.

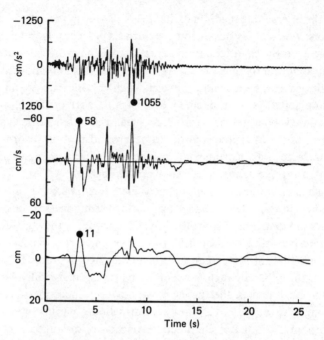

Fig. 17.2. Ground motions recorded on the S 74° W component accelerometer on the abutment of Pacoima Dam in the 1971 San Fernando, California earthquake. From top: acceleration, velocity, displacement.

In the above representation, a non-seismic error term equal to a constant base-line offset was assumed. In the displacement record this corresponds to a quadratic variation, making the seismic displacement curve difficult to obtain in a reliable way. Among the various theoretical and practical procedures suggested is least-squares fitting of a polynomial curve to the original points, followed by point-by-point subtraction. Also various types of filter are commonly used. Caution must be the rule to make sure that aliasing of high frequency wave components or digitisation errors do not occur or these will greatly distort the displacement record (see §9.9.2).

An alternative procedure is to work throughout in the frequency domain. Use of Fourier spectra followed by deconvolution has been found attractive in several studies.

The availability of three time functions is of great assistance, not only for engineering purposes but in the interpretation of strong-motion records. Accelerograms appear more random, with many high-frequency pulses and considerable variability in amplitudes. The integration to ground velocity considerably smooths these records and emphasises frequencies in the middle range of interest. The second integration produces usually smooth displacement records with fewer fluctuations and a simpler pattern of dominant waves, usually with periods beyond one second.

Experience has indicated that accelerograms have a variable pattern in detail, but most have generally regular shapes (except in the case of strongly multiple earthquakes). There is an initial portion of motion made up mainly of *P* waves which are often show strongly on the vertical motion. This is followed by the onset of mainly *S* waves, often associated with a longer period pulse related to the near-site fault slip or *fling*. After the *S* onset, there is enhanced shaking that consists of a mixture of *S* and *P* waves, but the *S* motions become dominant as the duration increases. Later, in the horizontal component, surface waves of both Rayleigh and Love type dominate, mixed with some *S* body waves. Depending on the distance of the site from the fault and the structure of the intervening rocks and soils, surface waves are dispersed into trains. The coda is significantly affected by the focal depth of the faulted surface. Portions of the record contain pulses that can be explained in terms of the special properties of a finite extended source, such as the *breakout phase* and the *stopping phase* (§16.1.2).

These expectations are largely based on theoretical modelling and the observed wave patterns normally are found to contain a residual portion of unexplained wave motion, particularly for wave frequencies above 5 Hz. The unexplained portion is usually dealt with stochastically in modelling studies, as suggested, for example, by Haskell (§16.1.1). Theoretically, this random component of strong ground motion can be thought to arise primarily from the distribution of roughness along the fault and consequently the *roughness density distribution* $\Phi(x)$. If this could be specified, the stochastic problem would become a deterministic one. As yet, no roughness distribution densities have been proposed for different classes of earthquakes and rather *ad hoc* assumptions must be made. Suggested ways to estimate this function involve geological field measurements, fits of the model to both observed near- and far-field seismograms, and the use of recordings from small regional earthquakes to scale upwards to larger fault rupture.

It is often valuable to transform the ground motion recorded by strong-motion accelerometers to the equivalent motions as recorded by more common seismographs. In particular, it is useful to transform strong-motion accelerograms to equivalent seismograms written by Wood–Anderson torsion seismographs (see §9.9.1), thus producing a synthetic seismogram which can be read for magnitude in the standard way (see §15.2). The procedure greatly extends the availability of torsional records

Table 17.2 *Sample maximum recorded ground motions*

Earthquake	Dist. (km)	M_L	Horizontal				Vertical			Station
			Accel. ($\times g$)	Vel. (cm/s)	Disp. (cm)	Dur. (s)	Accel. ($\times g$)	Vel. (cm/s)	Disp. (cm)	
San Fernando, California, 1971	3.2	6.4	1.25	113.2	37.7	13	0.72	58.3	19.3	Pacoima
San Fernando, California, 1971	16.9	6.4	0.20	11.6	5.0	7	0.09	5.9	2.3	Los Angeles
Imperial Valley, California, 1940	12.0	6.4	0.36	36.9	19.8	29	0.28	10.8	5.6	El Centro
Imperial Valley, California, 1979	1	6.6	0.72	110.0	55.0	11.8	1.74	50.6	13.0	El Centro No. 6
Kern County, California, 1952	42.0	7.2	0.20	17.7	9.1	20	0.12	6.7	5.0	Taft
Parkfield, California, 1966	16.1	5.5	0.41	22.5	5.5	4	0.17	4.4	1.4	Temblor
Taiwan, 1981	30.0	6.9	0.24	19.3	2.6	3.5	0.09	5.2	0.9	SMART 1

appropriate for the estimation of magnitude, especially near the rupturing source where regular Wood–Anderson seismographs, if available, would be off-scale and would provide no useable record.

The method is straightforward. Let $a(t)$ be the time function of record acceleration and $\bar{a}(\omega)$ its Fourier transform. By (9.21), the actual ground acceleration is $\bar{a}(\omega)/R(\omega)$, where R is the accelerograph transfer function. Usually R is constant for relevant frequencies. The Fourier transform of ground displacement is thus proportional to $-\bar{a}(\omega)/\omega^2$.

If $\bar{W}(\omega)$ is the known impulse response of the Wood–Anderson instrument (e.g. equation (9.15)), the required Wood–Anderson record $w(t)$ is the Fourier transform of

$$\bar{w}(\omega) = -\bar{a}(\omega)\bar{W}(\omega)/\omega^2. \tag{9}$$

Hundreds of strong-motion accelerograms have now been obtained in earthquakes of varying sizes and magnitudes and at different distances from the source in seismically active countries. Of particular importance are strong-motion records obtained in California, Japan, Italy and China. One important data set of strong-motion records is referred to in §17.4.5. For reference, peak accelerations, velocities and displacements as recorded in six earthquakes are listed in Table 17.2. A discussion of the interpretation of the records in this table from an elementary theoretical viewpoint is given in the report by Bolt (1981).

17.4.2 Duration of shaking

The duration D (in seconds) of strong seismic shaking is one of the most important factors in predicting strong ground motion. Most formulae express the key dependence of D on magnitude M which can be inferred at once from the rupture model of earthquakes (see §16.3.1).

Two definitions have been found useful.

(a) The *bracketed duration*, at a particular frequency, is the elapsed time window between the first and last acceleration excursions greater than a given level (0.05g, say). Particularly for earthquakes with specially complex multiple sources, this definition often leads to a non-physical upper estimate. For instance, spectrograms that show the amplitude spectrum as a function of time often indicate that peak accelerations of 2–3 Hz waves occur separated by reduced motions. For some design and liquefaction analyses, these episodes of weak motion may allow some structural recovery and should be excluded.

(b) The *cumulative duration*, at a particular frequency, is the total time for which acceleration A exceeds a given value. This interval is sometimes measured by cumulatively adding the squared accelerations and adopting the 95 percentile time interval. It may equal the corresponding bracketed duration or be much less and appears to have a greater mechanical significance in some design tests.

For high frequencies, D is limited by attenuation along the propagation path. Suppose, to obtain an upper bound, waves of all frequencies are generated at the moving rupture with an amplitude of $1.0g$. Then, at each frequency, beyond a certain distance on the slipping fault, the source-site distance is too great for the site to continue to receive waves with $A > 0.05g$. For example, after the rupture has propagated to 150 km (corresponding to $M = 7.5$), only a small proportion of 1 Hz (or greater) energy with amplitudes above $0.05g$ will ultimately arrive back at the site.

Measurements of worldwide strong-motion records have been used to fix curves for D versus M that represent nearly the upper bound so as to include 90 per cent of available data. Few values are available for large magnitudes. The 1906 San Francisco earthquake duration is roughly known from timed estimates of 40 s of 'severe shaking' felt by scientists in San Francisco and Berkeley. As well, the EW component of the Ewing seismograph at Lick Observatory wrote an almost continuous record. It suggests that motions with periods less than 2 s had fallen below $A = 0.01g$ after 40 s; smaller fluctuating waves (periods > 3 s) were recorded for 150 s or so.

There is, of course, evidence that longer period waves persist for a minute or more at accelerations $A < 0.05g$, because of sharply lower attenuation and surface wave dispersion (see § 3.3.5). The long-period vibrations, taken with the aftershocks, add to the human propensity to exaggerate the duration of shaking. (Humans can feel $A \geqslant 0.001g$).

Regression of available data indicates that durations of higher frequency shaking do not significantly increase above magnitude 7.5, for $A > 0.05g$, and above magnitude 7, for $A > 0.10g$. Bracketed durations ($f > 1$ Hz) within 25 km of the fault rupture are not likely to exceed the following values for $A > 0.05g$ and $A > 0.10g$, respectively:

$$D = 17.5 \tanh(M - 6.5) + 19.0, \tag{10}$$

and

$$D = 7.5 \tanh(M - 6.0) + 7.5. \tag{11}$$

Bracketed duration values above $0.05g$ for six earthquakes are listed in Table 17.2.

17.4.3 Spectral characteristics

Choice of treatment of ground motion in either the time domain or the frequency domain is a matter of theoretical and practical convenience.

Usually Fourier and response spectra are calculated for all important strong-motion records and published with them (see §9.10 and §17.5).

Often only the amplitude spectrum is considered in strong-motion seismology and earthquake engineering. However, the complementary phase spectrum that defines the pattern of the recorded waves is also of value in interpretation and even in engineering application. For the construction of artificial strong ground motions in the near field, an amplitude spectrum from a magnitude 7.5 earthquake, for example, with adequate maximum amplitudes, can be combined with a phase spectrum from another earthquake (with smaller amplitudes, say, then required) appropriate to the wave pattern for near fault motions. Broadly, the form of the amplitude spectra can be explained according to the theory of §16.3. A permanent offset occurring in a few seconds along a fault produces a seismic pulse rich in longer period components. Spectral curves from near-field records consequently show displacement amplitudes inversely proportional to the frequency of the waves. As the seismic waves travel away from the seismic source, very long period amplitudes (static offsets) are reduced, thereby enhancing relatively the higher frequencies. At larger distances the effect is significantly modified by the attenuation properties of the medium which damp out the higher frequencies more rapidly than the longer ones (see §17.4.5).

As indicated in §16.3, specification of a source model allows the near-field spectrum as well as the far-field spectrum to be interpreted in terms of parameters of the model. For small to moderate earthquakes, some work has been done along these lines, using the long-period spectral level, spectral corner frequency, and the slope of the high-frequency spectral amplitude. It is likely that such identification with physical quantities would be very difficult for moderate to great earthquakes using near-field records.

The high-frequency end of the amplitude spectrum may be scaled, using the high-frequency maximum (peak) acceleration on the accelerogram. For this reason, peak accelerations have come to play a central part in strong-motion seismology. Unfortunately, observation shows that this parameter is not robust. It is easily demonstrated that the high-frequency peak of acceleration can be changed by 10 per cent or more without significantly changing the spectral curve or the overall energy. Also, consideration of the horizontal components of acceleration in many recorded earthquakes demonstrates (i) that they are made up of the superposition of wave types and (ii) the peak accelerations do not necessarily coincide in time on the horizontal components. One suggestion to alleviate the sampling problem is to combine the spectra of the horizontal components in a given earthquake to maximise the resultant spectrum independently of azimuthal orientation.

17.4.4 Local effects. Soil layers and upthrow

From time to time field observations during and after earthquakes and instrumental recordings of strong ground motions bring to light unusual events. Many of these can be explained by local conditions that affect the seismic waves, foundation soils, and shaken structures and systems. In §17.2, we have already dealt with some major local effects that influence the observed intensity. Many local effects are caused by topographic, geological and soil conditions special to the site; in particular, the incoming seismic waves produce landslides, soil lurching and liquefaction which have profound and spectacular consequences. Such effects, so far as they can be treated quantitatively, are the concerns of soil engineers. The reader is referred, for example, to Seed and Idriss (1982).

We now consider several local problems that illustrate how far the theory developed for seismic motion in this book can be applied. First, tests of the analytical predictions of the effects of local soft sediments on seismic waves are dependent largely on seismograms obtained by vertical arrays of strong-motion accelerometers placed in down holes drilled through the soil layers. Recordings from such arrays are still few, but additional controlled experiments are planned.

Theoretically, central aspects of the problem have been considered for elastic waves in layered media in §5.5.1, §8.8, §12.3 and elsewhere. For example, Fig. 5.4 indicates how a 100 m thick alluvial layer will significantly affect surface wave dispersion, duration of shaking, and soil particle motion. In Exercise 8.9.3, an important formula was cited that provides a way to compute amplification factors for vertically propagating *SH* waves, incident on a surficial alluvial or firm soil layer. Further, an indication was given in §12.3.3 of the importance of rapid horizontal changes in rock structure to the local seismic transfer function. Sites on or near ridges or valley margins, for example, may be subject to complex shaking derived from wave mode conversions and interference.

Finally, it is worth demonstrating the rather complicated dynamical effects of the incident seismic waves on mechanical systems (important to explain observed intensities (see §12.2)).

Consider the response during an earthquake of a system of two masses m and M, separated by a perfectly elastic spring; the mass m is initially resting on the ground with the mass M directly above it (Fig. 17.3).

Let L be the length of the unstrained spring and take the x-axis vertically

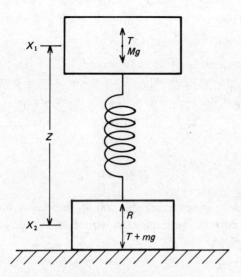

Fig. 17.3. An elastic system of two particles, connected by a spring, and constrained to move freely in the vertical plane.

upward. Let Z be the length of the spring at time t. Therefore, $Z = x_1 - x_2$.

For a spring modulus k, the force T due to strain in the spring is, by Hooke's law,

$$T = -k(Z - L). \tag{12}$$

The equation of motion for the mass m is

$$m\ddot{x}_2 = -mg - T + R, \tag{13}$$

where R is the force exerted on m by the ground. For the mass M, the equation of motion is

$$M\ddot{x}_1 = -Mg + T. \tag{14}$$

If $\ddot{x}$ is the acceleration of the centre of mass, we can write

$$\ddot{x} = -g + \frac{R}{M + m}, \tag{15}$$

which entails that, if $R > (M + m)g$, the centre of mass of the system will rise.

Now consider the motion of M during the first (upward) cycle of ground motion $x = a \sin \omega t$. The amplitude of the quarter cycle is a and its duration is $t = \pi/2\omega$. So, we have from (12)

$$T = -k(x_1 - x_2 - L)$$

$$= -k(x_1 - a \sin \omega t - L) \quad \left(0 < t < \frac{\pi}{2\omega}\right). \tag{16}$$

Hence, from equation (14),

$$M\ddot{x}_1 = -Mg - kx_1 + ak\sin\omega t + kL. \tag{17}$$

Therefore, if the acceleration $\ddot{x}_1$ of M is to exceed $1g$, the right side of equation (17) must be greater than Mg. If we define a new parameter $\Omega^2 = k/M$, the following inequality can be written, necessary for $\ddot{x}_1 > g$,

$$x_1 - a\sin\omega t - L < -\frac{2g}{\Omega^2}. \tag{18}$$

When Ω is large enough, $\ddot{x}_1 > g$ if

$$a\sin\omega t + L > x_1.$$

There appears no difficulty in achieving this condition because a can be arbitrarily large in an earthquake compared with L. Now set $y = x_1 + Mg/k - L$ in (17). Therefore,

$$\ddot{y} + \Omega^2 y = a\Omega^2 \sin\omega t. \tag{19}$$

The initial conditions are that $y = 0$ and $\dot{y} = 0$ at $t = 0$, so the solution is

$$y = x_1 + \frac{g}{\Omega^2} - L = \frac{a\Omega^2}{\Omega^2 - \omega^2}\left(\sin\omega t - \frac{\omega}{\Omega}\sin\Omega t\right). \tag{20}$$

Now let us calculate the displacement x_1, at the end of the first quarter cycle. On rearranging the terms in (20) we find at $t = \pi/2\omega$

$$x_1 = \frac{-g}{\Omega^2} + L + \frac{a\Omega^2}{\Omega^2 - \omega^2}\left(1 - \frac{\omega}{\Omega}\sin\frac{\pi\Omega}{2\omega}\right).$$

Now, we may test if the system will leave the ground by determining the condition for $\ddot{x}_2$ to be positive at $t = \pi/2\omega$ with $R = 0$. From equation (13)

$$m\ddot{x}_2 = -mg + M\Omega^2\left(\frac{-g}{\Omega^2} + \frac{a\Omega^2}{\Omega^2 - \omega^2}\left(1 - \frac{\omega}{\Omega}\sin\frac{\pi\Omega}{2\omega}\right) - a\right). \tag{21}$$

The acceleration of m will be upward if the following inequality holds

$$\frac{\Omega^4}{\Omega^2 - \omega^2}\left(1 - \frac{\omega}{\Omega}\sin\frac{\pi\Omega}{2\omega}\right) > \frac{m + M}{M}\frac{g}{a} + \Omega^2. \tag{22}$$

It is clear that values for the masses and natural frequency of the elastic system can be found such that (22) is satisfied for a given ground motion. Thus, for example, suppose $M \gg m$ and $\Omega = 3\omega$. Then, from (22), the condition for upthrow of m would be

$$\Omega^2 > 3\frac{g}{a}. \tag{23}$$

This is a physically attainable condition.

17.4.5 Attenuation

A key step in specifying earthquake ground motion, in terms of peak dynamic or spectral parameters, is to have a measure of the attenuation with distance between the causative fault and site. Early work on this problem involved curve fitting of available empirical measurements. Usually, fits were made to simple curves with prior assumptions concerning appropriate physical requirements. The fitted observations are measurements from intensity data or strong-motion records of ground acceleration, velocity, or displacement; normally, the maximum (or 'peak') amplitude on such records is selected. Publications of attenuation curves for peak ground motion are many and the reader is referred to reviews by Nuttli (1974), Seed and Idriss (1982), Joyner and Boore (1981) and Brillinger and Preisler (1984).

There is some theoretical and empirical evidence that indicates that the high-frequency peak accelerations are not strong functions of magnitude, particularly if the magnitude is measured from longer period waves. The latter is the case for surface wave magnitude M_s and moment magnitude M_w calculated directly from displacement pulses.

A questionable feature is to use the maximum of the highest accelerations (the 'supremum') as a measure of strong ground motion. Although examples can be cited where the supremum does represent the general level of the highest amplitudes on the records, often suprema recorded on accelerograms in the near field occur as erratic spikes of high-frequency seismic wave motion. Nevertheless, measurements and analysis of suprema, especially for accelerations, have been widespread in seismology and earthquake engineering, mainly because of demands from engineering practice.

Let y be a measure of wave amplitude (e.g. peak acceleration), x a measure of distance from the wave source, and m the measure of earthquake size (e.g. one or another suitable forms of magnitude). We are interested in the form of the relation $y = f(m, x)$.

We have already considered two sources of wave attenuation, geometrical spreading (§ 3.4.2) and damping (§ 8.4.2) with a factor $\exp(-\pi f x / Q c)$ where f is the frequency in Hz. Because attenuation of maximum strong ground motion or intensity involves a mix of wave types and frequencies these theoretical factors can be incorporated only approximately into the empirical fitted forms. A widely used relation was suggested by Milne and Davenport as

$$y = \frac{0.69 \exp 1.64m}{1.1 \exp(1.10m) + x^2}. \tag{24}$$

In this expression, y is in cm/s^2 and x is the epicentral distance in km but the latter measure is not physically appropriate near to an extended fault rupture. The authors noted that with form (24), close to the epicentre acceleration depends exponentially, on magnitude, while at greater distances an inverse-square geometrical spreading law is preserved and magnitude dependence enters through the factor in the numerator. Very few data points existed at that time, however, for $x < 20$ km.

There are arguments that ground accelerations would not be expected to increase without limit as the fault is approached. (We here leave aside questions related to directivity, focusing, and other specific source mechanism effects.) Briefly, high-frequency (> 8 Hz) spikes of acceleration near to the source are predominantly generated by local dislocation or rapid slip on fault surfaces and not by remote dislocation along the rupturing fault. The amplitudes of peak accelerations with high frequencies are governed by these local stress drops and rock properties rather than by the average dynamical fault properties that depend upon overall rupture dimensions. This physical model of wave generation discounts the influence of magnitude (or seismic moment) on the supremum (high frequency) acceleration at a particular site, and suggests a further refinement of the interpretation of the distance parameter x.

Even for a site on the fault rupture itself, the predominant source of the high-frequency wave energy is at a finite depth below the surface. Thus, the interpretation of x in (24) needs extention to the slant distance between the site and the region of energy release, usually by the use of Pythagorean relations of the form $x^2 = x'^2 + h^2$.

Recently, Joyner and Boore obtained, from a two-stage regression analysis of horizontal component acceleration extrema from 23 western North American earthquakes (182 observations), a prediction equation

$$\log_{10} y = -1.02 + 0.249\,M - 0.5\,\log_{10}(x^2 + 7.3^2)$$
$$-\,0.00255\,(x^2 + 7.3^2)^{\frac{1}{2}}, \tag{25}$$

with standard error 0.26, when y is in fractions of g, $5.0 \leqslant M \leqslant 7.7$. The above form was designed to provide less dependence of near source estimates on the more numerous remote observations. However, the commitment in advance to the analytic properties of the assumed functional form is a problem. A general stochastic model and uncertainty estimation have been discussed recently by Brillinger and Preisler. They suggest that a general functional form $\theta(y)$ for seismic wave amplitude y be used such that

$$\theta(y_{ij}) = \phi(M_i) + \psi(x_{ij}) + \varepsilon_i + \varepsilon_{ij}, \tag{26}$$

where indices i and j refer to events and records within events, and ε_i and ε_{ij} are random variation variables between and within earthquakes and θ, ϕ and ψ are functions in the parametric case or graphs in the non-parametric one.

Many other empirical attenuation curves have been published for various regions of the world, for various rock and soil conditions, and for various intensity parameters y_{ij} apart from peak acceleration. For example, for Japanese acceleration recordings on rock sites (in usual units)

$$\log_{10}(0.981\,y) = \left(\frac{x+80}{100}\right)(-4.83 + 0.89M - 0.043\,M^2). \tag{27}$$

The theoretical aspects of the numerous fits do not warrant extensive cataloguing here. A final matter of theoretical import should be mentioned, however.

It has been clearly established that attenuation curves differ significantly from region to region. Seminal studies have been published by O. Nuttli, for example, for the United States stemming from studies of intensity in the great 1811–12 New Madrid earthquakes in the upper Mississippi region. In particular, seismic wave attenuation is significantly less east of the Mississippi River than in California.

He fits observed seismic wave displacements to the form (see §15.1)

$$A(\Delta) = A(\Delta_1)\left(\frac{\sin\Delta_1}{\sin\Delta}\right)^{\frac{1}{2}}\left(\frac{\Delta_1}{\Delta}\right)^{n}\exp\left\{-k(\Delta - \Delta_1)\right\}, \tag{28}$$

where Δ_1 is a reference distance and n and k are attenuation coefficients. For dispersed surface waves, he finds $n = 0.5$ and 0.33 for Airy phases. When $A(\Delta)$ is the spectral amplitude, we have

$$k(f) = \pi f/Q(f)C(f)$$

from (13.36).

For a $Q(f)$ variation with frequency proportional to f^m, where $0 \leqslant m \leqslant 1$, he finds, for the Atlantic coastal plain region, $Q_0 \approx 800$ and $m \approx 0.35$. These values compare with $Q_0 \approx 175$ and $m \approx 0.65$ in coastal California.

17.5 Array analysis

In §9.9.3, the value of arrays of accelerometers to study seismic strong ground motion was explained and the large-scale digital array, called SMART 1, was described. In this section, we outline one approach to

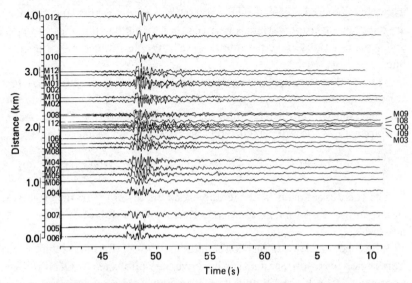

Fig. 17.4. Acceleration wave forms obtained by the SMART 1 array of the north–south component of ground motion in the 29 January 1981 earthquake ($M_L = 6.9$).

the theory needed for detailed spatial and temporal studies of the recorded wave-forms.

Fig. 17.4 is an illustration of recorded wave forms across the SMART 1 array (see Fig. 9.4). In this figure, accelerations in the north–south direction are played back in a form commonly used with exploration geophone spreads. Because absolute time is available, the records can be correlated against epicentral distance in real time. The high P-wave signal strength for this earthquake resulted in both P and S waves being recorded by all 27 operating instruments. The signal amplitudes are different among the three components, with the vertical and north-south components being the weakest and strongest, respectively. As in Fig. 17.4, the wave-forms of individual components vary notably across the array.

The digital format of records makes processing by computer relatively quick. For instance, the Fourier spectra can be calculated *en masse* in a time difficult to match with conventional film accelerograms. Array records can be used also to calculate response spectra and eventually it should be possible by simultaneous inversions to separate the effects of site and source conditions on the spectra. Fig. 9.5 shows the response spectra for the east-west component of the C00 records from the SMART 1 array for the 29 January 1981 earthquake of $M_L = 6.9$. These response spectra are representative of the spectra for most other sites. It was found that the shapes of the response spectra may differ across such an array as SMART 1 in one earthquake for vertical and horizontal components and also between different earthquakes.

Next, consider the phased correlations of a specified portion of such records for a given component at all array elements so that not only the average power in the seismic signal may be estimated but also the speed and direction of any coherently propagating seismic waves. The appropriate mathematical form, from (3.78), with noise δ is

$$y = A \cos \{\kappa(l_i x_i - ct) + \varepsilon\} + \delta; \tag{29}$$

The beam-forming algorithm is to estimate l_i, ε and A by least-squares.

A convenient method is to plot the power spectrum for a particular time window as a function of wave number κ (cycles per kilometre) and azimuth. If the plot is made at a particular frequency ω, the distance from the centre is inversely proportional to the apparent wave speed c (where $\kappa c = \omega$).

A formula for a high-resolution method for frequency–wave–number analysis can be represented in the following way. Let the array contain K sensors with the jth sensor at vector position $\mathbf{r}_j$. Assume the incoming seismic wave has a plane front. Then, for a common component of ground motion and time window Δt, at station j the spectrum is

$$f_j(\omega) = \int_{t_0}^{t_0 + \Delta t} a_j(t) \exp(-i\omega t) dt. \tag{30}$$

Define the cross-spectrum as

$$S_{jl}(\omega) = f_j(\omega) f_l^*(\omega)$$

and compute

$$P(\omega, \kappa) = \left\{ \sum_{j,l} Q_{jl}(\omega) \exp[-i\kappa \cdot (\mathbf{r}_j - \mathbf{r}_l)] \right\}^{-1}, \tag{31}$$

where

$$\{Q_{jl}(\omega)\} = \{S_{jl}(\omega)\}^{-1}.$$

Then normalise the maximum $P(\omega, \kappa)$ to 0 db. Finally, $P(\omega, \kappa)$ is contoured. The contoured diagram can be analysed for average horizontal phase velocity and direction of wave propagation (see Fig. 17.5). For statistical measures of uncertainty, maximum likelihood estimates of A and l_i (given ω) must be made from (29).

As an illustration of the method, in Fig. 17.5 a wave-number spectral plot is shown, computed from 26 SMART 1 north-south component accelero-

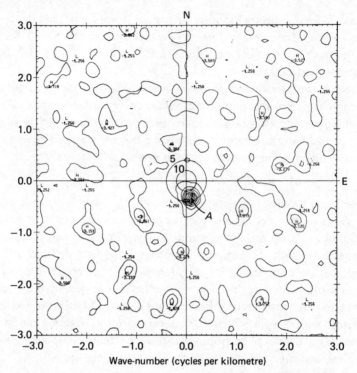

Fig. 17.5. Contours of the summed power at 26 stations of the north–south component of seismic waves at a frequency of 2 Hz plotted against wave-number in the east–west and north–south directions. The circles indicate constant velocities of 5 and 10 km/s. Peak power is at *A*.

grams. The time window chosen spans the first 6 s of the records of Fig. 17.4. The main power (marked A) at 1–2 Hz arrives at an apparent velocity of about 8.3 km/s (appropriate for *P* waves) from an azimuth of E 58°S. (The azimuth of the earthquake focus is E 64°S.)

The above example demonstrates that such arrays have the capability to determine whether, in a given time, the predominant seismic motion is *P* wave, *S* wave or surface wave modes. The method does not make use of the three recorded components of wave motion at each array element but it can be extended to this general case and orbits of the particle motions calculated.

17.6 Seismic risk

17.6.1 Statistical theory. Poisson and hazard distributions

Earthquakes occur as a series of events in space and time, and since temporal earthquake prediction is not now feasible, we must treat them as occurring more or less haphazardly.

It is usual to construct seismic risk models on the assumption that each earthquake can be specified as a *point* in time (the origin-time) and *space* (the focus). On this assumption, the pertinent statistical model is that of *stochastic point processes* about which much has been written. It must be realised, however, that the point process model is hardly satisfactory for extended seismic sources such as the 1906 San Andreas fault rupture and inferences which are based on simple premises may lead to quite false results in such cases.

The basic probability notation we shall use is

$$P(q|H) = \theta \quad (0 \leqslant \theta \leqslant 1); \tag{32}$$

that is, the probability of a proposition q given some background information or data H is equal to a real number θ. The extreme values of θ are zero (*impossibility*) and unity (*certainty*).

We think of the recorded seismicity of a region as a sample of a population of earthquakes – which will never be known. Nevertheless, we would like to estimate properties of the long-term seismicity of the region (e.g. assuming it is stationary in time and space).

Suppose the property of the seismicity that interests us is frequency of earthquake size (α). (This may be magnitude, moment, intensity, peak acceleration, etc.). Then the *probability distribution function* of α is the probability

$$P(\alpha \leqslant a|H) = F(a), \tag{33}$$

and thus,

$$P(\alpha > a|H) = 1 - F(a). \tag{34}$$

These equations allow computation of the probability of the random ground motion parameter *exceeding a particular value* (or not exceeding it). If α has a continuous variation (like acceleration) then $P(\alpha = a|H)$ is zero; only if α is a discrete quantity (like number of earthquakes) is such a point probability useful.

A completely random and time homogeneous series of earthquakes has a Poisson distribution defined per unit time as

$$P(N_t = r|\mu t = \lambda) = \frac{\exp(-\lambda)\lambda^r}{r!}, \tag{35}$$

where N_t earthquakes occur in a time interval of length t at a constant rate μ and numbers of events in disjoint intervals are statistically independent. We may then calculate the probability distribution of the interval between earthquakes.

Let T be the interval from time of origin to the first event. Then no earthquake occurs in $(0, t)$ if and only if $T > t$, so that

$$P(T > t) = P(N_t = 0) = \exp(-\mu t), \tag{36}$$

and, we can write,

$$F(t) = P(T \leqslant t) = 1 - \exp(-\mu t). \tag{37}$$

Equation (37) defines the exponential distribution with parameter μ.

In contrast to the Poisson case, the chance of an earthquake occurring may depend upon the time of the last earthquake in the region, particularly for large magnitude events. Perhaps the simplest conditioning problem is to calculate the probability of the premise q ('an earthquake occurs in $t, t + \mathrm{d}t$') given the premise p ('the last earthquake was at $t = 0$').

Define the *hazard function* $h(t)$ as the frequency function

$$h(t)\,\mathrm{d}t = P(q|p). \tag{38}$$

Let the unconditioned frequency of time T between events be $f(t)$ and the cumulative distribution function be $F(t)$. Then, by the product theorem of probability, the reader can easily show that

$$h(t) = f(t)/(1 - F(t))$$

$$= -\frac{\mathrm{d}}{\mathrm{d}t}\ln(1 - F(t)). \tag{39}$$

For the Poisson process, $h(t) = \mu$; in general, however, the hazard function has a memory and can be estimated from (39) by calculating the tangent to unity minus the cumulative curve $F_n(t)$ when plotted on log paper. After large earthquakes the elastic rebound theory (see § 16.1) would predict that the hazard should increase with T; in aftershock sequences the hazard is high, drops quickly and then levels off.

17.6.2 Probability of exceedence of ground motions

The next basic problem related to risk estimation is the determination of the probability distribution of the maximum motion (e.g. acceleration) from a set of earthquakes.

For a peak acceleration α, we have, from (33),

$$F(a) = P(\alpha \leqslant a|M) \tag{40}$$

for the probability that an acceleration α is less than or equal to a, given that an earthquake of size M (magnitude, moment, intensity, etc.) has occurred. A typical $F(a)$ will be monotonically increasing from 0 to 1 as a increases from 0 to $2g$, say.

Then, assuming accelerations are independent and identically distributed, the cumulative distribution of the maximum observed acceleration for N earthquakes is

$$F_{max}(a) = P(\alpha_{max} \leqslant a | \text{given } N \text{ events})$$
$$= F(a)^N. \tag{41}$$

For Poisson events at rate μ, and a time period of length t,

$$F_{max}(a) = \sum_{i=0}^{\infty} F(a)^i \frac{\lambda^i \exp(-\lambda)}{i!}$$
$$= \exp[-\mu t(1 - F(a))], \tag{42}$$

an exponential distribution.

We can now calculate the return period (or average number of earthquakes) that must occur to obtain an acceleration exceeding a.

We define that return period as

$$R(a) = \frac{1}{1 - F(a)}, \tag{43}$$

and the return period in years is then

$$R_y(a) = \frac{R(a)}{\text{Expected number of earthquakes per year}}. \tag{44}$$

Therefore, from (43),

$$\mu t(1 - F(a)) = t/R_y(a).$$

Substitution in (42) gives

$$F_{max}(a) = \exp(-t/R_y(a)). \tag{45}$$

Usually risk maps depict the maximum acceleration with a 10 per cent probability of being exceeded in 50 years. From (45) we then have

$$\ln(0.90) = -50/R_y(a)$$

and

$$R_y(a) = 475 \text{ years}.$$

We then ask: what are the odds that a maximum acceleration, A, say, will be exceeded during a period of T years? From (42), this probability is given by

$$P(A_{max} > a) = 1 - \exp(-T/R_y(a)) \tag{46}$$

This equation allows contour plots to be made showing the odds that a specified acceleration will be exceeded at least once during the *exposure time T*.

The statistical model above is a simple example from classes of wider models discussed in a number of key papers that have appeared on statistical aspects of the problems of seismic risk assessment and ground motion prediction. We refer the reader to Cornell (1968), Kagan and Knopoff (1977) and Vere-Jones and Smith (1981). The following description follows that of Brillinger (1982).

We define seismic risk assessment as the estimation of the probability that performance variates at a site exceed critical levels, within a specified time period, as a result of earthquakes. The multistage estimation requires critical investigation of four items: (*a*) sources, (*b*) intermediate transmission from the sources, (*c*) the local site of concern and (*d*) the engineered structure of interest. Earthquakes may be thought of as originating at points, on lines or within fault zones (the geometry). They have different sizes and locations. The propagation of the seismic signal involves attenuation with distance depending on the geology. Aspects of the local site include topography and soil type. For any structure the dynamic response must be modelled.

A simple assessment proceeds as follows: assume that (*a*) seismic sources have a frequency of rate μ, and magnitudes exponential $H(M) = P\{m > M\}$, (*b*) intensity falls off with distance and magnitude in accordance with (26). The risk may be evaluated explicitly in this case as (see (42))

$$P[\text{intensity } i \text{ exceeded within time period of } u]$$
$$= 1 - \exp\{-u\mu G(i)\} \tag{47}$$

where

$$G(i) = P\{I(M) + \varepsilon > i\} = E\{H(I^{-1}(i - \varepsilon))\}. \tag{48}$$

Here ε is a normal variate with mean 0 and variance σ^2 and E denotes the expected value. To obtain an estimate of the risk we substitute an estimate of the acceleration (or other intensity parameter) A. To obtain a measure of the sampling variability of this estimated risk we need a measure of the uncertainty of the estimate of A.

Most model assumptions are debatable. The Poisson process arises when events are rare and when many independent earthquake sequences are merged; however, the conditional distribution given the past is the same as the unconditional distribution. Earthquake sequences are often more clustered than sequences from a Poisson process.

As indicated earlier, point processes are relevant to seismic risk prediction and allow a generalisation of the above theory. Formally, a (stochastic) *point process* is a random entity defined by doubly infinite sequences of time points

$$N = \{t_j\}_{j=-\infty}^{\infty}, \tag{49}$$

with $\ldots t_{-1} < t_0 \leqslant 0 < t_1 \ldots$ and t_j is the times of relevant events at the site of interest. A *marked point process* is an extension of the above with a random entity whose occurrences are sequences of pairs (t_j, y_j),

$$\Phi = \{(t_j, y_j)\}_{j=-\infty}^{\infty}, \tag{50}$$

with the t_j as before and with y_j the value or *mark* attached to the jth time point. In risk contexts, y_j denotes the level that the selected performance variate reaches during the jth event: for example, the duration, the damage, or the maximum acceleration, or the amplitude of the response spectrum at a particular frequency and damping of interest. Taking y unity leads to the ordinary point process N. The pairs (t_j, y_j) may result from a number of earthquake faults or recurring events on the same fault.

Suppose that $(t, t + T)$ is the time interval of interest in a risk problem. Let $\Phi = \Phi(t, T, y)$ denote the number of t_j with $t < t_j < t + T$ and $y_j > y$. The function Φ counts the number of times the level y is exceeded in the time interval $(t, t + T)$. Let A denote a conditioning event based on known relevant information. For example, A might be the time and size of the most recent major earthquake in the region.

The (seismic) *risk* may now be defined as

$$P\{\text{level } y \text{ exceeded in time interval } (t, t + T) | A\} = P\{\Phi > 0 | A\}. \tag{51}$$

Below an explicit expression will be given for this probability. In many seismic risk assessments the conditioning event A has been excluded and unconditional probabilities have been evaluated. First, note that a simple useful bound exists for the probability in (51). Because Φ dominates the *indicator function* of the event $\{\Phi > 0\}$ (*that is* $\Phi \geqslant f(\Phi)$, where $f(\Phi) = 1$ if $\Phi > 0$ and $f(\Phi) = 0$ if $\Phi = 0$) we have the inequality

$$P\{\Phi > 0 | A\} \leqslant E\{\Phi | A\}. \tag{52}$$

The evolution of a marked point process may be described by a pair of functions γ and ρ. These functions allow evaluation of all parameters and probabilities of the process. Let $H_t = \{(t_j, y_j) \text{ with } t_j < t\}$ denote the history of the process up to time t. The *conditional intensity function* of the process is defined by

$$P\{\text{there is an event in } (t, t + dt) | H_t\} = \gamma(t | H_t) dt, \tag{53}$$

and the *transition density* by

$$P\{\text{mark at } t \text{ in } (y, y + dy) | \text{point at } t \text{ and } H_t)\} = \rho(t, dy | H_t). \quad (54)$$

Using these functions the risk may be written

$$1 - E\left\{\exp\left\{-\int_t^{t+T} \int_y^\infty \gamma(s | H_t) \rho(s, dy | H_t) \, ds\right\} \Big| A\right\}. \quad (55)$$

Suppose that $\gamma(t | H_t) = \mu$ and that $\rho(t, dy | H_t) = G(dy)$, then Φ is called a (stationary) marked Poisson process. Suppose $A \subseteq H_t$, then from expression (34),

$$P\{\Phi > 0 | A\} = 1 - \exp\left\{-T\mu \int_y^\infty G(dy)\right\}$$
$$= 1 - \exp\{-E\{\Phi\}\}. \quad (56)$$

This has the form of expression (42). An unrealistic feature of this expression is that it does not depend on the conditioning event A.

For distinct earthquake faults, the processes merge at the site to provide the superposed process $\Phi = \Phi_1 + \cdots + \Phi_K$. In the case that the processes are statistically independent, the superposed process is given by $\gamma\rho = \gamma_1\rho_1 + \cdots + \gamma_K\rho_K$. In the Poisson case, the risk becomes (cf. equation (47))

$$1 - \exp\left\{-T \sum_{k=1}^K \mu_k G_k(i)\right\}. \quad (57)$$

Note that, under quite general conditions, when a number of marked point processes are superposed the resultant process is approximately marked Poisson.

Characteristics desired for a seismic process, such as gaps, cycles, trends, and aftershocks, may be built in analytically by the functions γ and ρ defined above. For example, if we assume that hazard increases linearly with the time since the last event, then set $\gamma(t | H_t) = (t - T)/\alpha$.

In many risk studies the mark is defined in terms of seismic variables alone and the performance of the structure is not considered. The studies are applied later by engineers, with the mark used to scale an analysis done for the structure alone. This split approach seems less than optimal because the best final predictor may not be based on seismological proxy variables alone.

17.6.3 Seismic expectancy maps

The form of a contemporary map based on the previous theory is shown in Fig. 17.6. The intensity of ground shaking to be expected, in probabilistic terms, is represented by a parameter called the effective peak acceleration (EPA). Effective peak acceleration can be thought of as the

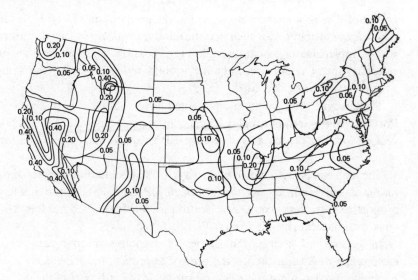

Fig. 17.6. Contour map of effective peak acceleration (EPA) for the United States. Contours represent EPA levels in units of gravity with a non-exceedence probability of between 80 per cent and 95 per cent during a fifty-year period. (Courtesy N.C. Donovan *et al.*, 1978.)

peak acceleration in motion on firm ground after it has been filtered to remove the high frequencies that do not affect large structures.

In order to avoid weaknesses in previous maps, the following general principles were adopted in the preparation of the map: (*a*) the map takes into account not only the size but also the frequency of earthquakes across the country; (*b*) the broad regionalisation pattern uses as a data base historical seismicity, major tectonic trends, acceleration attenuation curves and intensity reports (the basic seismological analysis was done by S.T. Algermissen and colleagues (1976)); (*c*) regionalisation is defined by means of contour lines and design parameters are referred to ordered numbers on neighbouring contour lines; this procedure minimises the sensitivity concerning the exact location of boundary lines between separate zones; such sensitivity is beyond present seismological resolution; (*d*) the map is simple and does not attempt to microzone the country; for this reason, only four highly smoothed contours were selected; (*e*) the mapped contoured surface does not contain discontinuities, so that the level of hazard progresses gradually and in order across any profile drawn on the map.

The next step in the preparation of a ground shaking hazard map requires the assumption of some uniform criteria for the selection of contours. The principal criterion used in the preparation of Fig. 17.6 is to produce contour intervals such that the relative exposure to seismic hazard will be approximately equal throughout the country. This may be done by selecting, in all regions, earthquake ground motions having the same probability of being exceeded.

The effective peak accelerations plotted have a probability (based on a Poisson assumption) of roughly 90 per cent of not being exceeded in a 50-year period. For other levels of risk the EPA values change (see equation (45)).

The map can be used in two main ways. First, in obtaining a scale factor for the elastic response spectrum for design. EPA should be approximately proportional to the spectral acceleration for one-degree-of-freedom systems with natural periods from 0.01 to 0.05 seconds.

In general, of course, the shapes of response spectra vary from earthquake to earthquake. In the present approach, the EPA is a single parameter used as a normalising factor to scale the ordinates of the design response spectra for various locations. For a given seismogram at a site, the EPA might be either larger or smaller than the peak acceleration. High frequency spikes, which have little influence on structural response, will generally have greater acceleration than the EPA.

Because peak velocity is closely correlated in period with the predominant period of response of most structures it is also a desirable scaling factor. A companion seismic regionalisation map based upon ground velocities has also been constructed. The parameter plotted, following the same steps as for acceleration, is effective peak velocity (EPV).

The EPV would be used to scale the response spectra in the period range 0.5 to several seconds. Use of two parameters helps cope with the problem of risk in regions which do not themselves have high local seismicity, but which lie a hundred kilometres or so from active faults. For such locations, the values of EPA might be low but the value of EPV might be relatively high.

The second use of the map is simpler. The contours can be used to define seismic *zones*. These zones might be specified as lists of areas of relative risk (0 to 4) by county, say. Such tables are of value for local zoning and building ordinance purposes.

17.6.4 Design of earthquake-resistant structures

Although earthquakes cause death and destruction from secondary effects such as landslides, tsunamis, fires and fault rupture, their principal hazard is the effect of shaking on structures built on or in the ground. The aim of the best engineering practice is to design and build structures to withstand strong ground motion.

Resistance to forces generated by seismic waves in a structure is developed in design by following building codes or by appropriate methods of analysis. An introduction to the theory is given by Chopra (1980). Many countries require structures in a seismically active region to conform to local building codes while theoretical structural analyses are generally reserved for the larger, more costly or critical buildings. Economic realities usually establish the goal, not of preventing all damage in all earthquakes, but of minimising damage in moderate, more common, earthquakes and ensuring no major collapse at the strongest intensities. An essential part of the evidence that goes into engineering decisions on design and into the development and revision of earthquake-resistant design codes is, therefore, seismological, involving measurement of the strong seismic waves (see § 17.4), field studies of intensity and damage (see § 17.2) and probability of earthquake occurrence (see § 17.5).

From a theoretical point of view there has recently been a closer integration between seismological wave theory in geological structures and seismic wave response theory for engineered structures. Indeed, there is often a significant interaction between the vibrations of the engineered structure (above or below ground) and the incident and reflected fields in the foundation soils and rocks.

In many earthquake countries, a 'seismic constant' is incorporated in building codes, whereby structures are designed to withstand the horizontal forces that would arise from a steady horizontal earth acceleration of $0.1g$. This 'static method' does not fully account for the dynamical behaviour of complicated and tall structures, and ignores many features of the actual motion in an earthquake. But it does lead to much reduction of earthquake damage. Even when the acceleration exceeds that on which the calculations are based, a structure may be undamaged because of quick reversals in the direction of the acceleration.

At present, many theoretical investigations of the dynamical behaviour of particular structures are being carried out, and are assisted by laboratory experiments on shaking tables.

Four major factors are usually considered in determining the capacity of engineering structures to resist earthquakes (apart from foundation failure which involves often difficult questions of interaction between the structure and underlying soil): (*a*) the level of the design spectra; (*b*) the assumed spectral damping; (*c*) the structural stresses and strains permissible; and (*d*) the eigen-periods of vibration of the structure.

It is important to realise that the *design spectrum* is regarded by engineers as an appropriate specification of the amount of seismic design force, or displacement, as a function of modal vibration periods and damping. It is not necessarily the response spectrum (see §9.10) which is a particular description of a particular strong ground motion, or averaged set of such motions. The design spectrum is usually highly smoothed and may take into account shaking from different likely earthquake sources. For example, moderate earthquakes near the site might govern the high-frequency spectral end and larger distant earthquakes the low-frequency portion. Further discussion can be found in the monographs by Housner and Jennings (1982) and by Chopra (1980).

17.7 Tsunamis, seiches, and atmospheric oscillations

Very long water waves in oceans or seas, called *tsunamis*, or sea waves, sometimes of great height, sweep inshore following certain earthquakes.

The immediate cause of a tsunami is a disturbance in the adjacent sea bed, sufficient to cause the sudden raising or lowering of a large body of water. This disturbance may be in the focal region of an earthquake, or may be a submarine landslide arising from an earthquake. Following the initial disturbance to the sea surface, gravity waves travel out in all directions, the order of the speed of travel in deep water being given by $(gh)^{\frac{1}{2}}$, where h is the sea depth; this speed may be considerable, e.g. 100 m/s if h is 1000 m. The amplitude at the surface does not exceed a few metres in deep water, but the principal wave-length may be of the order of some hundreds of kilometres; correspondingly, the principal wave-period may be of the order of some tens of minutes. On account of these features the waves are not noticed by ships well out at sea.

Of course, P waves from an earthquake may pass through the sea following refraction through the sea floor; the speed of these waves is about 1.5 km/s, the speed of sound in water. If these waves meet a ship with sufficient intensity, they give the impression that the ship has struck a submerged object, and the phenomenon is called a *sea-quake*.

When tsunamis approach shallow water the amplitude increases, and in U- and V- shaped inlets and harbours sometimes reaches a height of the order of 20–30 m. It is in lowlying ground around such inlets that great damage is sometimes done. Frequently the wave front in the inlet is nearly vertical as in a tidal bore, and the speed of onrush may be of the order of

10 m/s. In some cases there are several great waves, separated by intervals of some minutes or more, and the first of these waves is not always the greatest. Frequently, the first great wave is preceded by an extraordinary recession of water from the shore which may commence several minutes or even half an hour beforehand.

The initial disturbance causing a tsunami may be some distance from the nearest coast. In the great tsunami of 1896 June 15 which devastated the Sanriku region on the north-east coast of Japan, the place of origin was estimated to be 150–200 km from the shore. Tsunamis are recorded on tide-gauges often in distant countries and sometimes cause damage there.

The transmission of a tsunami across the ocean has been discussed by Jeffreys as wave motion in a dispersive medium. Before the non-linear effects produced by wave run-up on a sloping shore, the tsunami consists of a Jeffreys phase at the beginning, consisting of an exponential rise followed by an oscillatory portion. The whole wave train has the form of the Airy integral (see Fig. 3.7) with only the onset having the $(gh)^{\frac{1}{2}}$ speed.

Organisations, notably in Japan, Siberia, Alaska and Hawaii, have been set up to provide tsunami warning. An internationally supported system designed to reduce loss of life in the Pacific Ocean is called the Seismic Sea Wave Warning System (SSWWS) centered in Honolulu. It issues alerts based on reports of earthquakes from circum-Pacific seismographic stations.

Seiches are rhythmic motions of the water in land-locked bays or lakes, and are sometimes excited by earthquakes and by tsunamis. These oscillations may last for hours or even for one or two days. The name (pronounced saysh) was popularised by F.A. Forel who studied oscillations of the Lake of Geneva whose fundamental period is 72 minutes.

The 1755 Lisbon earthquake caused canals and lakes to go into palpable oscillations as far away as Scotland and Sweden. Seiche surges in Texas in the United States commenced between 30 and 40 minutes after the 1964 Alaska earthquake and were produced by seismic surface waves passing through the area.

The equations of motion for small free oscillations of water in a narrow lake of elliptical shape are, for $-1 < x < 1$,

$$\frac{\partial u}{\partial t} = -g\frac{\partial \zeta}{\partial x}, \quad \frac{\partial}{\partial t}(b\zeta) = -\frac{\partial}{\partial x}(hbu),$$

were u is the velocity, ζ the surface elevation, h is depth, g gravity, and b the breadth.

If $V = hbu$, for periodicity $2\pi/\gamma$, we have

$$b\frac{d}{dx}\left(\frac{1}{b}\frac{dV}{dx}\right) + \kappa^2 V = 0,$$

where $\kappa^2 = \gamma^2/gh$. Using the substitution $V = b^{\frac{1}{2}}U$, we obtain, as with the Love wave equation (see § 5.4.3),

$$\frac{d^2U}{dx^2} + \left(\kappa^2 + \frac{b''}{2b} - \frac{3b'^2}{4b^2}\right)U = 0. \tag{58}$$

The boundary conditions are that $V = 0$ at the ends. We can now proceed to estimate numerically the eigenvalues κ. Jeffreys and Jeffreys (1950) explore the use of Rayleigh's principle in this case (see § 12.3.4). Eigen-periods are clearly dependent on the depth and size of the water body.

Another rather unusual facet of modern seismological research is the use of microbarograph and long-period seismograph recordings of atmospheric acoustic waves caused by major fault movements and volcanic eruptions. Microbarographs and long-period seismographs clearly recorded atmospheric waves and air-coupled Rayleigh waves produced from the sudden vertical fault displacement in the great 1964 Alaskan earthquake. These recordings have been used to compute the rapidity and extent of the large-scale deformation near the earthquake source.

As in illustration, Fig. 17.7 shows air waves corresponding both to direct ($A1$) and antipodean ($A2$) travel paths recorded on a sensitive microbarograph at Berkeley after the violent eruption of Mount St Helens on 18 May 1980. These unusual complementary recordings throw light on the released acoustic energy, the atmospheric oscillations and their attenuation, and the directive properties of the phreatic blast. The principal explosive eruptions followed closely on an earthquake, $M_L = 4.9$, origin time 1532 UT, centred near the volcano. Atmospheric waves and associated magnetic perturbations from these eruptions were recorded by microbarographs, seismographs, and magnetometers around the world.

In Fig. 17.7, the A1 wave train shows an intriguing complexity. The periods of the air waves range from 2 min to almost 20 min. There are two sharp maxima with widths of approximately 3 min, about 6 min apart. Their onsets have group velocities of 308 m/s and 262 m/s, respectively.

Over 33 hours later, a more-or-less monochromatic train of air waves arrives at Berkeley, as shown in the lower part of Fig. 17.7. The train onset corresponds to an acoustic wave with a velocity of 314 m/s. The wave train remains at about 7 min period but with decreasing amplitude until, about 30 minutes later, amplitudes again rise to the maximum in this portion of

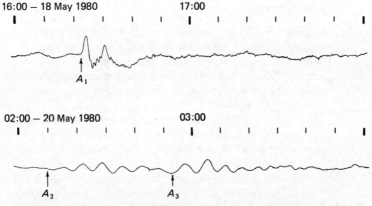

Fig. 17.7. Berkeley barograms of atmospheric waves after the Mount St Helens eruption, 18 May 1980. The top record shows the $A1$ wave train that begins 50 min after 1532 UT. The bottom record shows the antipodean $A2$ train that begins 34 hours and 38 min after 1532 UT. It is followed 40 min later by the $A3$ train.

the barogram. The pulse corresponds to the circumnavigating antipodean train $A3$, with velocity 320 m/s.

The largest pressure variations at Berkeley in the $A1$ and $A2$ pulses were about 0.35 and 0.13 mbar, respectively. The amplitude of the air wave associated with the pulse $A1$ is thus about 3.5 m. Knowledge of the wave amplitudes permits, by integration over the number of cycles and through the volume of atmosphere involved, an estimate to be made of the energy carried by the atmospheric waves. Calculation indicates that this energy in the $A1$ train was at least 10^{22} ergs. From the air pressure pulses recorded after the Krakatoa explosion, 10^{24} ergs was assessed in the atmospheric oscillations, corresponding to an equivalent surface explosion of 100 to 150 megatons.

The relative amplitudes of the A1 and A2 wave trains also provide an estimate of the attenuation (Mikumo and Bolt, 1985). A Q of about 1500 is indicated. This value is consistent with observations of the Mount St Helens air waves by others. The theory of acoustic and seismic waves generated by atmospheric explosions is described by Ben-Menahem and Singh (1981, chapter 9). The atmospheric overpressure depends on a second-order differential equation of type (3.108) with coefficients that are functions of gravity g and the vertical temperature gradient dT/dz.

17.8 Exercises

1   Suppose that the predominant horizontal ground shaking in an earthquake is sinusoidal with frequency 5 Hz. A coin, mass 100 g, on a smooth floor just begins to slip during an earthquake. If the coefficient of friction between the coin and floor is $\mu = 0.5$, find the greatest amplitude of the shaking in the earthquake. What were the greatest acceleration and velocity of the ground?

2   John Milne in 1885 in Japan suggested that the maximum horizontal acceleration of ground motion might be approximately calculated by observing the fall of wooden cylindrical columns (called also Galitzin columns).

Consider a square column side length b and height h and mass m standing on the floor and constrained in such a way that it can overturn but not slip. Show that if a constant acceleration a of the ground perpendicular to one edge is applied the block will overturn if $a = g(b/h)$.

3   Let $f(t)$ be the 'time history' of strong ground acceleration recorded in an earthquake on one horizontal component.

Suppose that $f(t)$ consists of a superposition of $F(t)$, the true ground acceleration from 10 Hz to 0.5 Hz, and a long-term tilting effect $a + bt$, where a and b are constants. Write down integrals for the true ground velocity and true ground displacement and hence show the base line correction for displacement is a cubic curve.

4   Define the Poisson distribution for earthquake occurrence in a region.

Hence, show that the time x between earthquakes that are distributed as the Poisson law is distributed as the exponential distribution $\exp(-\beta x)$. What is the drawback of this distribution in earthquake risk studies?

5   In a seismic region, the number of earthquakes N_M exceeding magnitude M in a year is given by

$$\log_{10} N_M = 4.97 - 0.87\,M.$$

Show that the probability of a magnitude exceeding a given value is also an exponential distribution and hence find the odds of an earthquake of magnitude 6 or greater in one year. What is the expected number of earthquakes with $M > 5$ each year?

6   The equation of motion for long waves in shallow water (tsunamis) is

$$b\frac{\partial^2 z}{\partial t^2} = \frac{\partial}{\partial x}\left(Ag\frac{\partial z}{\partial x}\right)$$

for a channel of breadth b and cross-section area A.

A channel of unit width is of depth h, where $h = kx$, k being a constant. Show that long waves with frequency $p/2\pi$ are possible for which

$$z = BJ_0(ax^{\frac{1}{2}})\cos pt,$$

where $a^2 = 4p^2/kg$, and J_0 is Bessel's function of order zero.

The distance between successive zeros of $J_0(x)$ tends to π when x is large. Show that the wave-length of the corresponding stationary waves increases with increasing values of x.

[Note that this is the problem of tsunamis on a shelving beach or coast.]

7   Show that equation (22) provides, without requiring extreme vertical accelerations, a plausible explanation for the displacements of a fire truck in the 1971 San Fernando earthquake. Suppose that the resonant frequency Ω of the spring system of the truck is near 6 Hz and consider vertical ground accelerations with frequencies in a band around 2 Hz ($\Omega = 3\omega$). Prove that the required amplitude for lift is 21.8 cm and hence calculate the required minimum ground acceleration of the truck.

Appendix

Reference velocities and elastic parameters in two Earth models

In chapter 13, a discussion was given of recent procedures in estimating the elastic properties of the Earth's interior. Here, for reference, we tabulate values for the main parameters for two recent radially symmetric models discussed in §13.2.3. The values refer to some average estimate at the given radius (see §13.5.3) and, properly considered, the formal standard error in each parameter is dependent on the thickness of the shell through which the average is taken. For this reason, it was thought best to compare two independent solutions of the inversion of seismic data, themselves selected independently from the available measurements.

It should be noted that despite the differences in selection and iterative technique, the listed values are within 1 per cent at most depths.

Table A.1 *P-wave velocity (α) and S-wave velocity (β) at various depths. (The first column (I) is for Earth model CAL8 and the second (II) for PREM* (see* § 13.2.3).)

Depth (km)	α (km/s)		β (km/s)	
	I	II	I	II
0–10	4.30	—	2.30	—
10–20	7.50	—	4.30	—
20	7.96	8.11	4.52	4.49
40	7.97	8.10	4.45	4.48
80	8.00	8.08	4.36	4.47
150	8.08	8.03	4.36	4.44
220	8.23	7.99	4.48	4.42
220	8.23	8.56	4.48	4.64
300	8.51	8.69	4.68	4.70
400	9.08	8.91	4.89	4.77
400	9.41	9.13	5.09	4.93
500	9.72	9.65	5.26	5.22
600	9.97	10.16	5.42	5.52
640	10.20	10.22	5.51	5.55
640	10.54	10.22	5.70	5.55
670	10.68	10.27	5.85	5.57
670	10.68	10.75	5.85	5.95
800	11.10	11.11	6.26	6.26
1000	11.48	11.46	6.44	6.40
1200	11.78	11.78	6.54	6.52
1400	12.06	12.06	6.64	6.63
1600	12.32	12.33	6.75	6.74
2000	12.80	12.82	6.96	6.93
2400	13.26	13.28	7.12	7.11
2700	13.61	13.62	7.19	7.24
2780	13.67	13.69	7.17	7.26
2850	13.54	13.71	7.07	7.26
2885	13.37	13.72	6.96	7.26
2885	8.09	8.06	0	0
3000	8.27	8.25	0	0
3200	8.56	8.56	0	0
3400	8.84	8.83	0	0
3800	9.33	9.31	0	0
4200	9.74	9.69	0	0
4550	10.00	9.97	0	0
4800	10.12	10.14	0	0
5000	10.18	10.27	0	0
5155	10.19	10.36	0	0
5155	10.89	11.03	3.49	3.50
5200	10.94	11.05	3.50	3.52
5400	11.13	11.11	3.55	3.56
5600	11.24	11.17	3.58	3.60
6200	11.33	11.26	3.60	3.66
6371	11.33	11.26	3.60	3.67

*From 'Preliminary Reference Earth Model' by A.M. Dziewonski and D.L. Anderson, *Physics of the Earth and Planetary Physics,* **25**, 297–356; 1981.

Table A.2 *Density and gravity at various depths. (The first column (I) is for Earth model CAL8 and the second (II) for PREM (see §13.2.3)).*

Depth (km)	Density (g/cm³) I	Density (g/cm³) II	Gravity (cm/s²) I	Gravity (cm/s²) II
0–10	2.16	2.6*	982	982
10–20	3.26	2.90	983	983
20	3.34	3.38	984	984
40	3.35	3.38	985	984
80	3.36	3.37	986	985
150	3.40	3.37	988	988
220	3.41	3.36	991	990
220	3.42	3.44	991	990
300	3.47	3.48	994	993
400	3.58	3.54	998	997
400	3.64	3.72	998	997
500	3.84	3.85	1000	999
600	4.98	3.98	1002	1000
640	4.02	3.98	1002	1001
640	4.16	3.98	1002	1001
670	4.22	3.99	1002	1001
670	4.22	4.38	1002	1001
800	4.43	4.46	1001	999
1000	4.61	4.58	997	997
1200	4.74	4.69	994	994
1400	4.83	4.81	992	993
1600	4.92	4.91	992	993
2000	5.06	5.12	999	999
2400	5.23	5.32	1020	1018
2700	5.43	5.47	1048	1043
2780	5.52	5.51	1058	1053
2850	5.74	5.56	1066	1065
2885	5.92	5.57	1071	1068
2885	9.82	9.90	1071	1068
3000	10.01	10.07	1046	1044
3200	10.33	10.37	1001	998
3400	10.62	10.54	953	948
3800	11.11	11.11	848	842
4200	11.52	11.51	736	728
4550	11.84	11.79	634	623
4800	12.03	11.97	560	547
5000	12.14	12.09	502	485
5155	12.17	12.17	459	440
5155	13.34	12.76	459	440
5200	13.38	12.79	442	422
5400	13.49	12.88	368	352
5600	13.55	12.96	294	280
6200	13.59	13.08	76	63
6371	13.58	13.09	0	0

*Densities in the crust depend on whether oceans, continents, or a composite crust is being modeled.

Table A.3 *Pressure and elastic moduli at various depths.* (*The first column* (*I*) *is for Earth model CAL8 and the second* (*II*) *for PREM* (*see* §13.2.3).)

Depth (km)	Pressure (kbar)		Rigidity, μ (kbar)		Incompressibility, k (kbars)	
	I	II	I	II	I	II
0	0	0		0		
20	5	6	682	441	1206	1315
80	25	25	639	674	1299	1303
150	48	48	645	665	1356	1287
220	72	71	686	656	1401	1270
220	72	71	686	741	1401	1529
300	99	99	760	769	1500	1618
400	135	134	856	806	1810	1735
400	135	134	921	906	1995	1899
500	172	171	1062	1051	2211	2181
600	211	210	1169	1210	2397	2489
640	227	226	1220	1226	2555	2528
640	227	226	1352	1226	2819	2528
670	239	238	1442	1239	2887	2556
670	239	238	1442	1548	2887	2999
800	296	296	1736	1749	3144	3182
1000	389	386	1912	1874	3526	3519
1200	482	478	2027	1996	3874	3850
1400	577	573	2130	2115	4186	4174
1600	673	669	2242	2232	4479	4494
2000	872	869	2451	2462	5022	5132
2400	1079	1080	2615	2692	5661	5794
2700	1244	1241	2807	2868	6315	6318
2780	1290	1293	2838	2909	6531	6449
2850	1332	1334	2862	2925	6697	6507
2885	1353	1358	2868	2938	6759	6556
2885	1353	1358	0	0	6427	6441
3000	1474	1472	0	0	6846	6581
3200	1683	1681	0	0	7569	7589
3400	1887	1886	0	0	8299	8303
3800	2279	2275	0	0	9671	9633
4200	2638	2631	0	0	10929	10814
4550	2918	2907	0	0	11840	11726
4800	3097	3180	0	0	12320	12306
5000	3225	3204	0	0	12581	12740
5155	3315	3289	0	0	12637	13047
5155	3315	3289	1625	1567	13654	13434
5200	3342	3316	1639	1582	13828	13498
5400	3451	3415	1700	1637	14444	13731
5600	3541	3497	1737	1681	14803	13922
6200	3689	3631	1761	1756	15097	14236
6371	3699	3639	1760	1761	15086	14253

Selected bibliography

Abramovitz, M. and Stegun, I.A. (1965). Handbook of mathematical functions, 8th printing. New York: Dover.

Achenbach, J.D. (1973). *Wave propagation in elastic solids.* Amsterdam: North-Holland.

Aki, K. and Richards, P.G. (1980). *Quantitative seismology–theory and methods.* San Francisco: Freeman.

Båth, M. (1973). *Introduction to seismology.* New York: Wiley.

Ben-Menahem, A. and Singh, S.J. (1981). *Seismic waves and sources.* Heidelberg: Springer.

Birch, F., Schairer, J.F. and Spicer, H.C. (Editors) (1942). *Handbook of physical constants.* Spec. Pap. Geol. Soc. Amer., **36**.

Bland, D.R. (1960). *The theory of linear visco-elasticity.* New York: Pergamon.

Bolt, B.A. (Editor) (1972a, 1972b, 1973). *Methods in computational physics*, Vols. 11, 12, and 13. New York: Academic Press.

Bolt, B.A. (1976). *Nuclear explosions and earthquakes: the parted veil.* San Francisco: Freeman.

Bolt, B.A. (1978). *Earthquakes: a primer.* San Francisco: Freeman.

Bolt, B.A. (1982). *Inside the earth.* San Francisco: Freeman.

Bolt, B.A. (Editor) (1986). *Seismic strong motion synthetics.* New York: Academic Press.

Bracewell, R.B. (1965). *The Fourier transform and its applications.* New York: McGraw-Hill.

Brekhovskikh, L.M. (1960). *Waves in layered media.* New York: Academic Press.

Brillinger, D.R. (1975). *Time series data analysis and theory.* New York: Holt, Rinehart and Winston.

Budden, K.G. (1961). *The wave-guide mode theory of wave propagation.* London: Logos Press.

Bullen, K.E. (1975). *The Earth's density.* London: Chapman and Hall, Halsted Press.

Cagniard, L. (1962). *Reflection and refraction of progressive seismic waves,* translated and revised by E.A. Flinn and C.H. Dix. New York: McGraw-Hill.

Červený, V. and Ravindra, R. (1971). *Theory of seismic head waves.* University of Toronto Press.

Červený, V., Molotkov, I.A. and Psencik, I. (1977). *Ray method in seismology.* Prague: Univerzita Karlova.

Chopra, A.K. (1980). *Dynamics of structures: a primer.* Berkeley: Earthquake Engineering Research Institute.

Claerbout, J.F. (1976). *Fundamentals of geophysical data processing: with application to petroleum prospecting.* New York: McGraw-Hill.

Conrad, V. (1932). Die zeitliche Folge der Erdbeben und bebenauslösende Ursachen. *Handbuch der Geophysik,* **4**, 1007–1190. Berlin.

Cook, A.H. (1980). *Interiors of the planets.* Cambridge University Press.

Dahlman, O. and Israelson, H. (1977). *Monitoring underground nuclear explosions.* Amsterdam: Elsevier.

Davison, C. (1927). *The founders of seismology.* Cambridge University Press.

Davison, C. (1938). *Studies on the periodicity of earthquakes.* (Reviewed by R. Stoneley, *Bull. Seismol. Soc. Amer.,* **29**, 559–62, 1939). London: Murby.

Deans, S.R. (1983). *The Radon transform and some of its applications.* New York: Wiley.

Dobrin, M.B. (1976). *Introduction to geophysical prospecting,* 3rd ed. New York: McGraw-Hill.

Ewing, M., Jardetzky, W.S. and Press, F. (1967). *Elastic waves in layered media.* New York: McGraw-Hill.

Fung, Y.C. (1965). *Foundations of solid mechanics.* New Jersey: Prentice-Hall.

Galitzin, B. (1914). *Vorlesungen über Seismometrie.* Leipzig: Teubner.

Gantmacher, F.R. (1959). *The theory of matrices.* Translated by K.A. Hirsch. New York: Chelsea.

Grant, F.S. and West, G.F. (1965). *Interpretation theory in applied geophysics.* New York: McGraw-Hill.

Gu Gungshu, Lin Tinghuang and Shi Zenliang (editors) (1983). *The seismicity of China, 1931 B.C. to 1969 A.D.* State Seismological Bureau. Beijing: Science Publisher.

Gutenberg, B. and Richter, C.F. (1954). *Seismicity of the Earth and associated phenomena,* 2nd ed. Princeton: University Press.

Gutenberg, B. (1959). *Physics of the Earth's interior.* New York: Academic Press.

Housner, G.W. and Jennings, P.C. (1982). *Earthquake design criteria.* Berkeley: Earthquake Engineering Research Institute.

Hudson, D.E. (1982). *Reading and interpreting strong motion accelerograms.* Berkeley, California: Earthquake Engineering Research Institute.

Hudson, J.A. (1980). *The excitation and propagation of elastic waves.* Cambridge University Press.

Jacobs, J.A. (1975). *The Earth's core.* New York: Academic Press.

Jeffreys, Sir H. (1970). *The Earth,* 5th ed. Cambridge University Press.

Jeffreys, Sir H. and Jeffreys, B.S. (1950). *Methods of mathematical physics,* 2nd ed. Cambridge University Press.

Kanasewich, E.R. (1981). *Time sequence analysis in geophysics.* Edmonton: University of Alberta Press.

Karnik, V. (1969). *Seismicity of the European area,* Vol. 1. Dordrecht, Holland: D. Reidel.

Karnik, V. (1971). *Seismicity of the European area,* Vol. 2. Dordrecht, Holland: D. Reidel.

Kasahara, K. (1981). *Earthquake mechanics.* Cambridge University Press.

Kennett, B.L.N. (1983). *Seismic wave propagation in stratified media.* Cambridge University Press.

Knott, C.G. (1908). *The physics of earthquake phenomena.* Oxford: Clarendon Press.

Kondorskaya, N.V. and Shebalin, N.V. (Editors-in-chief) (1982). *New catalog of strong earthquakes in the U.S.S.R. from ancient times through 1977.* Report SE-31. Translated and published by World Data Center A for Solid Earth Geophysics, EDIS, Boulder, Colorado.

Lamb, H. (1932). *Hydrodynamics,* 6th ed. Cambridge University Press.

Lanczos, C. (1961). *Linear differential operators.* London: Van Nostrand.

Landau, L.D. and Lifschitz, E.M. (1977). *Elastizitätstheorie.* Berlin: Akademie Verlag.

Lapwood, E.R. and Usami, T. (1981). *Free oscillations of the Earth.* Cambridge University Press.

Lee, S.P. (Editor) (1956). *Chronological tabulation of Chinese earthquake records,* 2 vols. Peking: Chinese Academy of Sciences.

Lee, W.H.K. and Stewart, S.W. (1981). *Principles and applications of microearthquake networks.* New York: Academic Press.

Lighthill, M.J. (1960). *Fourier analysis and generalized functions.* Cambridge University Press.

Lighthill, J. (1978). *Waves in fluids.* Cambridge University Press.

Lomnitz, C. and Rosenblueth, E. (Editors) (1976). *Seismic risk and engineering decisions.* Developments in Geotechnical Engineering, Vol. 15. Amsterdam: Elsevier.

Love, A.E.H. (1911). *Some problems of geodynamics.* Cambridge University Press.

Love, A.E.H. (1945). *The mathematical theory of elasticity,* 4th rev. ed. Cambridge University Press.

Macelwane, J.B. and Sohon, F.W. (1932, 1936). *Theoretical seismology,* Parts I and II. New York: Wiley.

Milne, J. (1939). *Earthquakes and other Earth movements,* rev. by A.W. Lee. London: Routledge, Kegan Paul.

Moiseiwitsch, B.L. (1966). *Variational principles.* New York: Interscience Publishers, Wiley.

Montessus de Ballore, F. (1924). *La géologic sismologique.* Paris: Armond Colin.

Morse, P.M. and Feshbach, H. (1953). *Methods of theoretical physics.* New York: McGraw-Hill.

Morse, P.M. and Ingard, K.U. (1968). *Theoretical acoustics.* New York: McGraw-Hill.

Murnaghan, F.D. (1951). *Finite deformation of an elastic solid.* New York: Wiley.

Noble, B. and Daniel, J. (1975). *Applied linear algebra.* Englewood Cliffs, New Jersey: Prentice-Hall.

Pestel, E.C. and Leckie, F.A. (1963). *Matrix methods in elastomechanics.* New York: McGraw-Hill.

Pilant, W.L. (1979). *Elastic waves in the Earth.* Amsterdam: Elsevier.

Richter, C.F. (1958). *Elementary seismology.* San Francisco: Freeman.

Rikitake, T. (1976). *Earthquake prediction.* Amsterdam: Elsevier.

Rothé, J.P. (1969). *The seismicity of the earth 1953–1965.* Paris: UNESCO.

Savarensky, E.F. and Kirnos, D.P. (1955). *Elements of seismology and seismometry.* Moscow.

Seed, H.B. and Idriss, I.M. (1982). *Ground motions and soil liquefaction during earthquakes*. Berkeley: Earthquake Engineering Research Institute.

Sieberg, A. (1930). *Die Erdbeben; Erdbebengeographie*. Handb. Geophys. 4, 527–1005. Berlin.

Sokolnikoff, I.S. (1956). *Mathematical theory of elasticity*, 2nd ed. New York: McGraw-Hill.

Tolstoy, I. (1973). *Wave propagation*. New York: McGraw-Hill.

U.S. Earthquake Observatories: Recommendations for a New National Network (1980). Washington: National Academy Press.

Verhoogen, J. (1980). *Energetics of the Earth*. National Academy of Sciences, Washington.

White, J.E. (1965). *Seismic waves–radiation, transmission and attenuation*. New York: McGraw-Hill.

Wiegel, R.L. (Coordinating editor) (1970). *Earthquake engineering*. Englewood Cliffs, New Jersey: Prentice-Hall.

References

Adams, L.H. and Williamson, E.D. (1923). The composition of the Earth's interior. *Smithson Rep.* 241–60.

Agnew, D., Berger, J., Buland, R., Farrell, W. and Gilbert, F. (1976). International deployment of accelerometers: a network for very long period seismology. *EOS, Trans. Amer. Geophys. Union,* **57**, 180–8.

Aki, K. (1967). Scaling law of seismic spectrums. *J. Geophys. Res.* **72**, 1217–31.

Aki, K. (1982). Strong motion prediction using mathematical modelling techniques. *Bull. Seismol. Soc. Amer.* **72**, 529–42.

Aki, K. and Chouet, B. (1975). Origin of coda waves: source attenuation and scattering effects. *J. Geophys. Res.* **80**, 3322–42.

Algermissen, S.T. and Perkins, D.M. (1976). A probabilistic estimate of maximum acceleration in rock in the contiguous United States. U.S. Geological Survey Open File Report 76–416.

Alterman, Z., Jarosch, H. and Pekeris, C.L. (1959). Oscillations of the Earth. *Proc. Roy. Soc. Lond.* A252, 80–95.

Anderson, D.L. (1964). Universal dispersion tables I. Love waves across oceans and continents on a spherical earth. *Bull. Seismol. Soc. Amer.* **54**, 681–726.

Anderson, D.L. (1984). The Earth as a planet: paradigms and paradoxes. *Science,* 4634, 347–55.

Anderson, D.L. and Harkrider, D.G. (1968). Universal dispersion tables II. Variational parameters for amplitudes, phase velocity and group velocity for first four Love modes for an oceanic and a continental Earth model. *Bull. Seismol. Soc. Amer.* **58**, 1407–99.

Anderson, D.L. and Hart, R.S. (1978). *Q* of the Earth *J. Geophys. Res.* **83**, 5869–82.

Anderson, D.L. and Kovach, R.L. (1969). Universal dispersion tables III. Free oscillation variational parameters. *Bull. Seismol. Soc. Amer.* **59**, 1667–93.

Anderson, J.A. and Wood, H.O. (1925). Description and theory of the torsion seismometer. *Bull. Seismol. Soc. Amer.* **15**, 1–72.

Archambeau, C.B. (1968). General theory of elastodynamic source fields. *Rev. Geophys.* **6**, 241–88.

Backus, G.E. and Gilbert, F. (1970). Uniqueness in the inversion of inaccurate gross Earth data. *Phil. Trans. Roy. Soc. Lond.* A26, 123–92.

Backus, G.E. and Mulcahy, M. (1976). Moment tensors and other phenomenological descriptions of seismic sources–I. Continuous displacements. *Geophys. J.* **46**, 341–61.

478

Båth, M. (1954). The density ratio at the boundary of the Earth's core. *Tellus*, **6**, 408–13.

Benioff, H. (1935). A linear strain seismograph. *Bull. Seismol. Soc. Amer.* **25**, 283–309.

Benioff, H., Press, F. and Smith, S. (1961). Excitation of the free oscillations of the Earth by earthquakes. *J. Geophys. Res.* **66**, 605–19.

Ben-Menahem, A. (1961). Radiation of seismic surface waves from a finite moving source. *Bull. Seismol. Soc. Amer.* **51**, 401–35.

Berry, M.J. and West, G.F. (1966). Reflected and head wave amplitudes in a medium of several layers. In *The Earth Beneath Continents*, Geophys. Monograph 10, American Geophysical Union.

Bessonova, E.N. Fishman, V.M., Shnirman, M.G., Sitnikova, G.A. and Johnson, L.R. (1976). The tau method for the inversion of travel times–II. Earthquake data. *Geophys. J.* **46**, 87–108.

Birch, F. (1938). The effect of pressure upon the elastic parameters of isotropic solids, according to Murnaghan's theory of finite strain. *J. Appl. Phys.* **9**, 279–88.

Birch, F. (1952). Elasticity and constitution of the Earth's interior. *J. Geophys. Res.* **57**, 227–86.

Blandford, R.R. (1982). Seismic event discrimination. *Bull. Seismol. Soc. Amer.* **72**, 569–88.

Bolt, B.A. (1957). Earth models with chemically homogeneous cores. *Mon. Not. R. Astr. Soc.*, Geophys. Suppl., **7**, 360–8.

Bolt, B.A. (1960). The revision of earthquake epicentres, focal depths and origin times using a high-speed computer. *Geophys. J., R. Astr. Soc.* **3**, 433–40.

Bolt, B.A. (1981). The interpretation of strong motion seismograms. Misc. Report 17, U.S. Army Engineer Waterways Experiment Station S-73-1, 1–215.

Bolt, B.A. (1982). The constitution of the core: seismological evidence. *Phil. Trans. Roy. Soc.* **A306**, 11–20.

Bolt, B.A. and Brillinger, D. (1979). Estimation of uncertainties in eigenspectral estimates from decaying geophysical time series. *Geophys. J., Roy. Astr. Soc.* **59**, 593–603.

Bolt, B.A. and Currie, R.G. (1975). Maximum entropy estimates of Earth torsional eigenperiods from 1960 Trieste data. *Geophys. J.* **40**, 107–14.

Bolt, B.A. and Dorman, J. (1961). Phase and group velocities in a spherical, gravitating Earth. *J. Geophys. Res.* **66**, 2965–81.

Bolt, B.A., Doyle, H.A. and Sutton, D.J. (1958). Seismic observations from the 1946 atomic explosions in Australia. *Geophys. J., R. Astr. Soc.* **1**, 135–45.

Bolt, B.A. and Marussi, A. (1962). Eigenvibrations of the Earth observed at Trieste. *Geophys. J., R. Astr. Soc.* **6**, 299–311.

Bolt, B.A. and O'Neill, M.E. (1964). Times and amplitudes of the phases *PKiKP* and *PKIIKP*. *Geophys. J., R. Astr. Soc.* **9**, 223–31.

Bolt, B.A. and Qamar, A. (1970). An upper bound to the density jump at the boundary of the Earth's inner core. *Nature* **228**, 148–50.

Bolt, B.A. and Uhrhammer, R.A. (1975). Resolution techniques for density and heterogeneity in the Earth. *Geophys. J.* **42**, 419–35.

Bolt, B.A. and Uhrhammer, R.A. (1981). The structure, density and homogeneity of the Earth's core. In *Evolution of the Earth*, Geodynamics Series, Amer. Geophys. Union 5, 28–37.

Boore, D.M. (1972). Finite difference methods for seismic wave propagation in heterogeneous materials. In Bolt (1972a), 1–37.

Brennan, B.J. and Smylie, D.E. (1981). Linear viscoelasticity and dispersion in seismic wave propagation. *Rev. Geophys. and Space Phys.* **19**, 233–46.

Brillinger, D.R. (1982). Seismic risk assessment: some statistical aspects. *Earthq. Predict. Res.* **1**, 183–95.

Brillinger, D., Udias, A. and Bolt, B.A. (1980). A probability model for regional focal mechanism solutions. *Bull. Seismol. Soc. Amer.* **70**, 149–70.

Brillinger, D. and Preisler, H.K. (1984). An exploratory analysis of Joyner–Boore attenuation data. *Bull. Seismol. Soc. Amer.* **74**, 1441–50.

Brune, J.N. (1970). Tectonic stress and the spectra of seismic shear waves from earthquakes. *J. Geophys. Res.* **75**, 4997–5009.

Brune, J.N. and Oliver, J. (1959). The seismic noise of the Earth's surface. *Bull. Seismol. Soc. Amer.* **49**, 349–53.

Buchwald, V.T. (1961). Rayleigh waves in transversely isotropic media. *Quart. J. Mech. Appl. Math.* **14**, 293–317.

Buland, R. (1976). The mechanics of locating earthquakes. *Bull. Seismol. Soc. Amer.* **66**, 173–87.

Bullen, K.E. (1937, 1938, 1939). The ellipticity correction to travel-times of P and S earthquake waves. *Mon. Not. R. Astr. Soc.*, Geophys. Suppl., **4**, 143–57, 317–31, 332–5, 469-71.

Bullen, K.E. (1940). The problem of the Earth's density variation. *Bull. Seismol. Soc. Amer.* **30**, 235–50.

Bullen, K.E. (1942). The density variation of the Earth's central core. *Bull. Seismol. Soc. Amer.* **32**, 19–29.

Bullen, K.E. (1950). An Earth model based on a compressibility–pressure hypothesis. *Mon. Not. R. Astr. Soc.*, Geophys. Suppl., **6**, 50–9.

Bullen, K.E. (1955). Features of seismic pP and PP rays. *Mon. Not. R. Astr. Soc.* Geophys. Suppl., **7**, 49–59.

Bullen, K.E. (1961). Seismic ray theory. *Geophys. J., R. Astr. Soc.* **4**, 93–105.

Bullen, K.E. (1963). An index of degree of chemical inhomogeneity in the Earth. *Geophys. J., R. Astr. Soc.* **7**, 584–92.

Burridge, R. and Willis, J. (1969). The self-similar problem of the expanding elliptical crack in an anisotropic solid. *Proc. Camb. Phil. Soc.* **66**, 443–68.

Byerly, P. (1926). The Montana earthquake of June 28, 1925. *Bull. Seismol. Soc. Amer.* **16**, 209–65.

Capon. J. (1973). Signal processing and frequency–wavenumber analysis for a large aperture seismic array. In Bolt (1973).

Chapman, C.H. (1974). Generalized ray theory for an inhomogeneous medium. *Geophys. J.* **36**, 673–704.

Chapman, C.H. and Phinney, R.A. (1972). Diffracted seismic signals and their numerical solution. In Bolt (1972b), 165–230.

Choy, G.L. (1977). Theoretical seismograms of core phases calculated by frequency-dependent full wave theory and their interpretation. *Geophys. J.* **51**, 275–312.

Choy, G.L. and Cormier, V.F. (1983). The structure of the inner core inferred from short-period and broadband GDSN data. *Geophys. J., R. Astr. Soc.*, **72**, 1–21.

Cleary, J. (1974). The D″ region. *Phys. Earth and Planet. Inter.* **9**, 13–27.

Cormier, V.F. (1982). The effect of attenuation on seismic body waves. *Bull. Seismol. Soc. Amer.* **72**, 5169–200.

Cornell, C.A. (1968). Engineering seismic risk analysis. *Bull. Seismol. Soc. Amer.* **58**, 1583–1606.

Crampin, S. (1981). A review of wave motion in anisotropic and cracked elastic media. *Wave Motion* **3**, 343–91.

Crosson, R.S. (1976). Crustal structure modelling of earthquake data I. Simultaneous least squares estimation of hypocenter and velocity parameters. *J. Geophys. Res.* **81**, 3036–46.

Dahlen, F.A. and Smith, M.L. (1975). The influence of rotation on the free oscillations of the Earth. *Phil. Trans. R. Soc. Lond.* **A279**, 583–624.

Das, S. and Aki, K. (1977). Fault plane with barriers: a versatile earthquake model. *J. Geophys. Res.* **82**, 5658–70.

Derr, J.S. (1969). Free oscillation observations through 1968. *Bull. Seismol. Soc. Amer.* **59**, 2079–99.

Dewey, J.W. (1972). Seismicity and tectonics of Western Venezuela. *Bull. Seismol. Soc. Amer.* **62**, 1711–51.

Dewey, J.W. and Byerly, P. (1969). The early history of seismometry (to 1900). *Bull. Seismol. Soc. Amer.* **59**, 183–277.

Donovan, N.C., Bolt, B.A. and Whitman, R.V. (1978). Development of expectancy maps and risk analysis. *J. Structural Div.*, ASCE 104, 1179–92.

Dorman, J. and Prentiss, D. (1960). Particle amplitude profiles for Rayleigh waves on a heterogeneous Earth. *J. Geophys. Res.* **65**, 3805–16.

Douglas, A. (1967). Joint epicenter determination. *Nature.* **215**, 47–8.

Drake, L.A. (1972). Love and Rayleigh waves in nonhorizontally layered media. *Bull. Seismol. Soc. Amer.* **62**, 1241–58.

Drake, L.A. and Bolt, B.A. (1980). Love waves normally incident on a continental boundary. *Bull. Seismol. Soc. Amer.* **70**, 1103–23.

Dziewonski, A.M. and Anderson, D.L. (1981). Preliminary reference Earth model. *Phys. Earth Planet. Inter.* **25**, 297–356.

Dziewonski, A.M. and Gilbert, F. (1976). The effect of small aspherical perturbations or travel-times and a re-examination of the corrections for ellipticity. *Geophys. J.R. Astr. Soc.* **44**, 7–18.

Engdahl, E.R., Peterson, J. and Orsini, N.A. (1982). Global digital networks current status and future directions. *Bull. Seismol. Soc. Amer.* **72**, 5243–60.

Evernden, J.F. (1982). Earthquake prediction: what we have learned and what we should do now. *Bull. Seismol. Soc. Amer.* **72**, 5343–50.

Ewing, M. and Press, F. (1959). Determination of crustal structure from phase velocity of Rayleigh waves. Part III. *Bull. Geol. Soc. Amer.* **70**, 229–44.

Filson, J. and McEvilly, T.V. (1967). Love wave spectra and the mechanism of the 1966 Parkfield sequence. *Bull. Seismol. Soc. Amer.* **57**, 1245–59.

Fuchs, K. (1968). The reflection of spherical waves from transition zones with arbitrary depth-dependent elastic moduli and density. *J. of Physics of the Earth*, Special Issue **16**, 27–41.

Fuchs, K. and Müller, G. (1971). Computation of synthetic seismograms with the reflectivity method and comparison with observations. *Geophys. J.* **23**, 417–33.

Geller, R.J. (1976). Body force equivalents for stress-drop seismic sources. *Bull. Seismol. Soc. Amer.* **66**, 1801–4.

Gerver, M. and Markushevich, V. (1967). On the characteristic properties of travel-time curves. *Geophys. J.* **13**, 241–6.

Gilbert, F. and Backus, G. (1966). Propagator matrices in elastic wave and vibration problems. *Geophysics* **31**, 326–32.

Gilbert, F. and Dziewonski, A.M. (1975). An application of normal mode theory to the retrieval of structural parameters and source mechanisms from seismic spectra. *Phil. Trans. R. Soc. Lond.* **A278**, 187–269.

Gough, D.I. (1978). Induced seismicity. In *Assessment and Mitigation of Earthquake Risk*, UNESCO, Paris.

Gutenberg, B. (1943). Seismological evidence for roots of mountains. *Bull. Geol. Soc. Amer.* **54**, 473–98.

Gutenberg, B. (1948). On the layer of relatively low wave velocity at a depth of about 80 kilometers. *Bull. Seismol. Soc. Amer.* **38**, 121–48.

Gutenberg, B. and Richter, C.F. (1934, 1935, 1936, 1939). On seismic waves. *Beitr. Geophys.* **43**, 56–133; **45**, 280–360; **47**, 73–131; **54**, 94–136.

Gutenberg, B. and Richter, C.F. (1952). Earthquake magnitude, intensity, energy and acceleration. *Bull. Seismol. Soc. Amer.* **32**, 163–91.

Haddon, R.A.W. and Bullen, K.E. (1969). An earth model incorporating free Earth oscillation data. *Phys. Earth and Planet. Inter.* **2**, 37–49.

Haddon, R.A.W. and Cleary, J.R. (1974). Evidence for scattering of PKP waves near the core-mantle boundary. *Phys. Earth Plan. Inter.* **8**, 211–34.

Hald, O.H. (1980). Inverse eigen-value problems for the mantle. *Geophys. J. Roy. Astr. Soc.* **62**, 41–8.

Hales, A.L. and Roberts, J.L. (1970). The travel-times of S and SKS. *Bull. Seismol. Soc. Amer.* **60**, 461–89.

Hanks, T.C. and Kanamori, H. (1979). A moment-magnitude scale. *J. Geophys. Res.* **84**, 2348–50.

Hannon, W.J. (1964). An application of the Haskell–Thompson matrix method to the synthesis of the surface motion due to dilational waves. *Bull. Seismol. Soc. Amer.* **54**, 2067–79.

Hansen, R.A. (1982). Simultaneous estimation of terrestrial eigenvibrations, *Geophys. J. Roy. Astr. Soc.* **70**, 155–72.

Hansen, R.A. and Bolt, B.A. (1980). Variations between Q values estimated from damped terrestrial eigenvibrations. *J. Geophys. Res.* **85**, 5237–43.

Harkrider, D.G. (1964). Surface waves in multilayered elastic media. I Rayleigh and Love waves from buried sources in a multilayered elastic half-space. *Bull. Seismol. Soc. Amer.* **54**, 627–79.

Haskell, N.A. (1953). The dispersion of surface waves on multilayered media. *Bull. Seismol. Soc. Amer.* **43**, 17–34.

Haskell, N.A. (1969). Elastic displacements in the near-field of a propagating fault. *Bull. Seismol. Soc. Amer.* **59**, 865–908. (Corrections, *Bull. Seismol. Soc. Amer.* **67**, 1215, 1977.)

Helmberger, D.V. and Harkrider, D.G. (1978). Modelling earthquakes with generalized ray theory. In Miklowitz and Achenbach (1978).

Herrin, E. (1968). 1968 seismological tables for *P* phases. *Bull. Seismol. Soc. Amer.* **58**, 1193–1241.

Hirasawa, T. and Stauder. W. (1965). On the seismic waves from a finite moving source. *Bull. Seismol. Soc. Amer.* **55**, 1811–42.

Honda, H. (1959). The elastic waves generated from a spherical source. *Sci Rep. Tohoku University*, 5th ser. **11**, 178–83.

Hron, F. (1972). Numerical methods of ray generation in multilayered media, in Bolt (1972*b*), 1–34.

Hudson, J.A. (1977). Scattered waves in the coda of *P. J. Geophys.* **43**, 359–74.

Inglada, V.O. (1928). Calculo de las coordenadas del foco y del instante inicial de un sismo por medio de las horas de las ondas *S* registradas en las estaciones proximas. *Revista de la Real Academia de Ciencias exactas, fisicas y naturales*, de Madrid, **24**, 175–201.

Isacks, B., Oliver, J. and Sykes, L.R. (1968). Seismology and the new global tectonics. *J. Geophys. Res.* **73**, 5855–900.

Jacob, K.H. (1970). Three-dimensional seismic ray tracing in a laterally hetero-geneous spherical Earth. *J. Geophys. Res.* **75**, 6675–89.

Jeffreys, H. (1926). On near earthquakes. *Mon. Not. R. Astr. Soc.*, Geophys. Suppl. **1**, 385–402.

Jeffreys, H. (1930). The thermodynamics of an elastic solid. *Proc. Camb. Phil. Soc.* **26**, 101–6.

Jeffreys, H. (1939). The times of *P*, *S* and *SKS* and the velocities of *P* and *S. Mon. Not. R. Astr. Soc.*, Geophys. Supp. **4**, 498–533.

Jeffreys, H. (1939). The times of the core waves. *Mon. Not. R. Astr. Soc.*, Geophys. Suppl. **4**, 548–61, 594–615.

Jeffreys, H. (1954). The times of *P* in Japanese and European earthquakes. *Mon. Not. R. Astr. Soc.*, Geophys. Suppl. **6**, 557–65.

Jeffreys, H. (1961). Small corrections in the theory of surface waves. *Geophys. J., R. Astr. Soc.* **6**, 115–17.

Jeffreys, H. (1977). *P* and *S* beyond 95°. *Geophys. J.* **51**, 387–92.

Jeffreys, H. and Bullen, K.E. (1935). Times of transmission of earthquake waves. *Bur. Centr. Seism. Internat.* A, Fasc. **11**, 202 pp.

Johnson, L.E. and Gilbert, F. (1972). Inversion and inference for teleseismic ray data. In Bolt (1972*b*), 231–66.

Johnson, L.R. (1969). Array measurements of *P* velocities in the lower mantle. *Bull. Seismol. Soc. Amer.* **59**, 973–1008.

Johnson, L.R. (1974). Green's function for Lamb's problem. *Geophys. J.* **37**, 99–131.

Jones, G.M. (1977). Thermal interaction of the core and the mantle and long term behavior of the geomagnetic field. *J. Geophys. Res.* **82**, 1703–9.

Jordan, J.H. (1979). Structural geology of the Earth's interior. *Proc. Natl. Acad. Sci. U.S.A.* **76**, 4192–200.

Joyner, W.B. and Boore, D.M. (1981). Peak horizontal acceleration and velocity from strong motion records including records from 1979 Imperial Valley, California earthquake. *Bull. Seismol. Soc. Amer.* **71**, 2011–38.

Julian, B.R. and Gubbins, D. (1977). Three-dimensional seismic ray tracing. *J. Geophys.* **43**, 95–113.

Kagan, Y. and Knopoff, L. (1977). Earthquake risk prediction as a stochastic process. *Phys. Earth and Planet. Inter.* **14**, 97–108.

Kanamori, H. (1977). The energy release in great earthquakes. *J. Geophys. Res.* **82**, 2981–7.

Kanamori, H. and Anderson, D.L. (1975). Theoretical basic of some empirical relations in seismology. *Bull. Seismol. Soc. Amer.* **65**, 1073–95.

Karal, F.C. and Keller, J.B. (1959). Elastic wave propagation in homogeneous and inhomogeneous media. *J. Acoust. Soc. Am.* **31**, 694–705.

Keilis-Borok, V.I. (1959). An estimation of the displacement in an earthquake source and of source dimensions. *Ann. Geofis.* **12**, 205–14.

Kelley, K.R., Ward, R.W., Trietel, S. and Alford, R.M. (1976). Synthetic seismograms: a finite-difference approach. *Geophysics*, **41**, 2–27.

Kennett, B.L.N. (1976). A comparison of travel-time inversions. *Geophys. J.* **44**, 517–36.

Kirons, D.P. (1955). Some problems of instrumental seismology. *Trans. Geophys. Inst. Acad. Sci. U.S.S.R.* **27**, 154.

Knopoff, L. (1958) Energy release in earthquakes. *Geophys. J., R. Astr. Soc.* **1**, 44–52.

Knopoff, L., Schwab, F. and Kausel, E. (1973). Interpretation of *Lg*. *Geophys. J.* **33**, 389–404.

Kostrov, B.V. (1966). Unsteady propagation of longitudinal shear cracks. *J. Appl. Math. Mech.* **30**, 1241–8.

Kostrov, B.V. (1974). Crack propagation at variable velocity. *J. Appl. Math. Mech.* **38**, 511–19.

Lahr, J.C. (1979). HYPOELLIPSE: a computer program for determining local earthquakes hypocentral parameters, magnitude and first motion pattern. U.S.G.S. Open File Report 79–431, 1–239.

Lamb, H. (1904). On the propagation of tremors over the surface of an elastic solid. *Phil. Trans. A.* **203**, 1–42.

Langston, C.A. (1978). Moments, corner frequencies, and the free surface. *J. Geophys. Res.* **83**, 3422–6.

Lapwood, E.R., Hudson, J.A. and Kembhavi, V.K. (1973). The passage of elastic waves through an anomalous region–I. Transmission of body waves through a soft layer. *Geophys. J.* **31**, 457–67.

Ledoux, P. (1958). Stellar stability. In *Handbuch der Physik* (ed. S. Flúgge) **51**, 605–15. Berlin: Springer.

Lehmann, I. (1936). *P'. Bur. Centr. Seism. Internat. A*, **14**, 3–31.

Lehmann, I. (1961). *S* and the structure of the upper mantle. *Geophys. J., R. Astr. Soc.* **4**, 124–38.

Levenberg, K. (1944). A method for the solution of certain non-linear problems in least squares: *Quart. Appl. Math* **2**, 164–8.

Longuet-Higgins, M.S. (1950). A theory of the origin of microseisms. *Phil. Trans. A.* **243**, 1–35.

Louderback, G.D. (1942). Faults and earthquakes. *Bull. Seismol. Soc. Amer.* **32**, 305–30.

Lysmer, J. and Drake, L.A. (1972). A finite element method for seismology. In Bolt (1972*a*), 181–216.

Madariaga, R. (1978). The dynamic field of Haskell's rectangular dislocation fault model. *Bull. Seismol. Soc. Amer.* **68**, 869–87.

Marquardt, D.W. (1963). An algorithm for least squares estimation of non-linear parameters. *J. Soc. Ind. Appl. Math.* **11**, 431–41.

Masters, G. and Gilbert, F. (1981). Structure of the inner core inferred from observations of its spheroidal shear modes. *Geophys. Res. Lett.* **8**, 569–71.

McCowan, D.W. (1976). Moment tensor representation of seismic surface wave studies. *Geophys. J.* **44**, 595–9.

McEvilly, T.V. and Johnson, L.R. (1974). Stability of P and S velocities from central California quarry blasts. *Bull. Seismol. Soc. Amer.* **64**, 343.

Miklowitz, J. and Achenbach, J.D. (Editors) (1978). *Modern problems in elastic wave propagation.* New York: Wiley.

Mikumo, T. (1981). A possible rupture process of slow earthquakes on a frictional fault. *Geophys. J.R. Astr. Soc.* **65**, 129–53.

Mikumo, T. and Miyatake, T. (1979). Earthquake sequence on a functional fault model with non-uniform strengths and relaxation times. *Geophys. J., R. Astr. Soc.* **59**, 497–522.

Mikumo, T. and Bolt, B.A. (1985). Excitation mechanism of atmospheric pressure waves from the 1980 Mount St. Helens eruption. *Geophys. J.R. Astr. Soc.* **81**, 445–61.

Mitchell, B.J. (1976). Anelasticity of the crustal upper mantle beneath the Pacific Ocean from the inversion of observed surface wave attenuation. *Geophys. J.* **46**, 521–33.

Mohorovičić, A. (1909). Das Beben vom 8.x.1909. *Jb. met. Obs. Zagreb*, **9**, 1–63.

Molodenski, M.S. (1955). Density and elasticity within the Earth. *Trans. Geophys. Inst. Acad. Sci. U.S.S.R.* **26**.

Mooney, H.M. and Bolt, B.A. (1966). Dispersive characteristics of the first three Rayleigh modes for a single surface layer. *Bull. Seismol. Soc. Amer.* **56**, 43–67.

Mueller, S. (1977). A new model of the Earth's crust. In *The Earth's Crust*, Geophysical Monograph 20, American Geophysical Union, Washington.

Müller, G. (1973). Amplitude studies of core phases. *J. Geophys. Res.* **78**, 3469–90.

Nakano, H. (1923). Notes on the nature of forces which give rise to earthquake motions. *Seismol. Bull., Centr. Meteor. Obs. of Japan*, **1**, 92–120.

Ness, H.F., Harrison, J.C. and Slichter, L.B. (1961). Observations of the free oscillations of the Earth. *J. Geophys. Res.* **66**, 621–9.

Northwood, T.D. and Anderson, D.V. (1953). Model seismology. *Bull. Seismol. Soc. Amer.* **43**, 239–46.

Nur, A., Dilatancy, pore fluids, and premonitory variations of t_s/t_p travel-times. *Bull. Seism. Soc. Am.*, **62**, 1217, 1972.

Nuttli, O.W. (1974). Seismic wave attenuation and magnitude relations for east and north America. *J. Geophys. Res.* **78**, 876–85.

Oldham, R.D. (1906). Constitution of the interior of the Earth as revealed by earthquakes. *Quart. J. Geol. Soc.* **62**, 456–75.

Oliver, J. (1962). A summary of observed seismic wave dispersion. *Bull. Seismol. Soc. Amer.* **52**, 81–6.

Olson, A.H. and Apsel, R.J. (1982). Finite faults and inverse theory with applications to the 1979 Imperial Valley earthquake. *Bull. Seism. Soc. Amer.* **72**, 1969–2002.

Peterson, J., Butler, H.M. Holcomb, L.G., and Hutt, C.R. (1976). The seismic research observatory. *Bull. Seismol. Soc. Amer.* **66**, 2049–68.

Phinney, R.A. and Alexander, S.S. (1969). The effect of a velocity gradient at the base of the mantle on diffracted *P* waves in the shadow. *J. Geophys. Res.* **74**, 4967–71.

Press, F. (1956). Determination of crustal structure from phase velocity of Rayleigh waves. Part I. Southern California. *Bull. Geol. Soc. Amer.* **67**, 1647–58.

Press, F. (1966). Seismic velocities. In: S.P. Clark, Jr. (Editor), Handbook of physical constants. *Geol. Soc. Amer. Memoire No.* **97**, New York.

Press, F. and Harkrider, D. (1962). Propagation of acoustic-gravity waves in the atmosphere. *J. Geophys. Res.* **67**, 3889–902.

Qamar, A. (1973). Revised velocities in the Earth's core. *Bull. Seis. Soc. Amer.* **63**, 1073–1105.

Radau, R.R. (1885). Sur la loi des densités à l'intérieur de la Terre. *C.R. Acad. Sci. Paris*, **100**, 972.

Randall, M.J. (1976). Attenuative dispersion and frequency shifts of the Earth's free oscillations. *Phys. Earth and Planet. Inter.* **12** P1–P4.

Rayleigh, Lord (Strutt, J.W.) (1885). On waves propagated along the plane surface of an elastic solid. *Proc. Lond. Math. Soc.* **17**, 4–11.

Reid, H.F. (1911). The elastic-rebound theory of earthquakes. *Bull. Dep. Geol. Univ. Calif.* **6**, 412–44.

Richards, P. G. (1973). Calculation of body waves for caustics and tunneling in core phases. *Geophys. J.* **35**, 243–64.

Richards, P.G. (1976). On the adequacy of plane-wave reflection/transmission coefficients in the analysis of seismic body waves. *Bull. Seis. Soc. Amer.* **66**, 701–17.

Richter. C.F. (1935). An instrument earthquake magnitude scale. *Bull. Seismol. Soc. Amer.* **25**, 1–32.

Ringdal, F. and Husebye, E.S. (1982). Application of arrays in the detection, location, and identification of seismic events. *Bull. Seismol. Soc. Amer.* **72**, S201–24.

Rodgers, P.W. (1968). The response of the horizontal pendulum seismometer to Rayleigh and Love waves, tilt, and the free oscillation of the Earth. *Bull. Seismol. Soc. Amer.* **58**, 1384–1406.

Sabatier, P.C. (1978). Spectral and scattering inverse problems. *J. Math. Phys.* **19**, 2410–25.

Saito, M. (1967). Excitation of free oscillations and surface waves by a point source in a vertically heterogeneous Earth. *J. Geophys. Res.* **72**, 3689–99.

Savage, J.C. (1972). Relation of corner frequency to fault dimensions. *J. Geophys. Res.* **2**, 3788–95.

Savage, J.C. (1974). Relation between *P*- and *S*-wave corner frequencies in the seismic spectrum. *Bull. Seismol. Soc. Amer.* **64**, 1621–7.

Scholte, J.G.J. (1956). On seismic waves in a spherical Earth. *Kon. Ned. Meteorol. Inst.* **65**, 1–551.

Scrase, F.J. (1932). The characteristics of a deep focus earthquake; a study of the disturbance of February 20, 1931. *Phil. Trans.* A **231**, 207–34.

Seed, H.B. (1982). The selection of design earthquakes for critical structures. *Bull. Seismol. Soc. Amer.* **72**, 57–512.

Sieh, K.E. (1978). Prehistoric large earthquakes produced by slip on the San Andreas Fault at Pallet Creek, California. *J. Geophys. Res.*

Shi, Y.L. and Bolt, B.A. (1982). The standard error of the magnitude–frequency *b* value. *Bull. Seismol. Soc. Amer.* **72**, 1677–87.

Shimamura; H., Tomoda, Y. and Asada, T. (1975). Seismographic observation at the bottom of the Central Basin fault of the the Philippine sea. *Nature (London)* **253**, 177–9.

Sipkin, S.A. (1982). Estimation of earthquake source parameters by the inversion of wave form data. *Phys. Earth and Planetary Int.* **30**, 242–59.

Sipkin, S.A. and Jordan, T.H. (1976). Lateral heterogeneity of the upper mantle determined from the travel times of multiple *ScS*. *J. Geophys. Res.* **81**, 6307–20.

Smith, W.D. (1974). A non-reflecting plane boundary for wave propagation problems. *J. of Computational Physics* **15**, 492–503.

Smith, W.D. (1975). The application of finite element analysis to body wave propagation problems. *Geophys. J.* **42**, 747–68.

Smith, W.D. and Bolt, B.A. (1976). Rayleigh's principle in finite element calculations of seismic wave response. *Geophys. J. Roy. Astro. Soc.* **45**, 647–55.

Stark, P.B. and Frohlich, C. (1985). The depths of the deepest deep earthquakes. *J. Geophys. Res.* **90**, B2, 1859–69.

Stauder, W. (1960). *S* waves and focal mechanisms: the state of the question. *Bull. Seismol. Soc. Amer.* **50**, 333–46.

Steketee, J.A. (1958). Some geophysical applications of the elasticity theory of dislocations. *Can. J. Phys.* **36**, 1168–98.

Stokes, G.G. (1880). Propagation of an arbitrary disturbance in an elastic medium. *Cambridge Mathematical Papers*, **2**, 257–80.

Stoneley, R. (1924). Elastic waves at the surface of separation of two solids. *Proc. Roy. Soc. A.* **106**, 416–28.

Stoneley, R. (1934). The transmission of Rayleigh waves in a heterogeneous medium. *Mon. Not. R. Astr. Soc.*, Geophys. Suppl., **3**, 222–32.

Stoneley, R. (1949). The seismological implications of aeolotropy in continental structure. *Mon. Not. R. Astr. Soc.*, Geophys. Suppl. **5**, 343–53.

Stump, B.W. and Johnson, L.R. (1977). The determination of source properties by the linear inversion of seismograms. *Bull. Seismol. Soc. Amer.* **67**, 1489–1502.

Takeuchi, H. (1950). On the earth tide in the compressible Earth of varying density and elasticity. *Trans. Amer. Geophys. Un.* **31**, 651–89.

Takeuchi, H. and Saito, M. (1972). Seismic surface waves. In Bolt (1972a), 217–95.

Tanimoto, T. and Bolt, B.A. (1983). Coupling of torsional modes in the Earth. *Geophys. J. Roy. Astro. Soc.* **74**, 83–95.

Tocher, D. (1958). Earthquake energy and ground breakage. *Bull. Seismol. Soc. Amer.* **48**, 147–52.

Trifunac, M.D. and Brady, A.G. (1975). On the correlation of seismic intensity scales with the peaks of recorded strong ground motion. *Bull. Seismol. Soc. Amer.* **65**, 139–62.

Turner, H.H. (1922). On the arrival of earthquake waves at the antipodes and on the measurement of the focal depth of an earthquake. *Mon. Not. R. Astr. Soc.*, Geophys. Suppl. **1**, 1–13.

Udias, A. (1980). Development of earthquake source mechanism studies. *Publ. Inst. Geophys. Pl. Acad. Sci.*, **A-10**, 7–32.

Uhrhammer, R.A. (1979). Shear-wave velocity structure for a spherically averaged Earth, *Geophys. J.R. Astr. Soc.* **58**, 749–67.

Uhrhammer, R.A. (1980). Analysis of small seismographic station networks. *Bull. Seismol. Soc. Amer.* **70**, 1369–80.

Vere-Jones, D. and Smith, S. (1981). Stochastic models for earthquake risk. *Geophys. J. Roy. Astr. Soc.* **42**, 811–26.

Wadati, K. (1928, 1929, 1931). On shallow and deep earthquakes. *Geophys. Mag., Tokyo*, **1**, 162–202; **2**, 1–36; **4**, 231–83.

Walton, K. (1974). The seismological effects of elastic pre-straining within the earth. *Geophys. J. Roy. Astr. Soc.* **36**, 651–77.

Whitcomb, J.H. and Anderson, D.L. (1970). Reflection of *P'P'* seismic waves from discontinuities in the mantle. *J. Geophys. Res.* **75**, 5713–28.

Wiechert, E. (1903). Theorie der automatischen Seismographen. *Abh. Ges. Wiss. Göttingen*, Math-Phys. Klasse, **2**, 1–128.

Wiechert, E. (1907). Über Erdbebenwellen. I. Theoretisches über die Ausbreitung der Erdbebenwellen. *Nachr. Ges. Wiss. Göttingen*, Math.-Phys. Klasse, 415–529.

Williamson, E.D. and Adams, L.H. (1923). Density distribution in the Earth, *J. Wash. Acad. Sci.* **13**, 413–28.

Willmore, P.L. (1961). Some properties of heavily damped electromagnetic seismographs. *Geophys. J. Roy. Astr. Soc.* **4**, 389–404.

Wood, H.O. and Neumann, F. (1931). Modified Marcalli intensity scale of 1931. *Bull. Seismol. Soc. Amer.* **21**, 277–83.

Woodhouse, J.H. and Dahlen, F.A. (1978). The effect of a general aspherical perturbation on the free oscillations of the Earth. *Geophys. J. Roy. Astr. Soc.* **53**, 263–80.

Unit conversion table

Physical quantity	cgs unit	Equivalent in SI
Mass	g	10^{-3} kg
Length	cm	10^{-2} m
Time	s	1 s
Force	dyne	10^{-5} N (newton)
Pressure and	dyne/cm^2	10^{-1} Pa (pascal)
elastic moduli	bar	10^5 Pa
Energy	erg	10^{-7} J (joule)
Density	g/cm^3	10^3 kg/m^3
Viscosity	poise	10^{-1} Ns/m^2

Index